Mechanical Engineering Series

Series Editor

Francis A. Kulacki, Department of Mechanical Engineering,
University of Minnesota, Minneapolis, MN, USA

The Mechanical Engineering Series presents advanced level treatment of topics on the cutting edge of mechanical engineering. Designed for use by students, researchers and practicing engineers, the series presents modern developments in mechanical engineering and its innovative applications in applied mechanics, bio-engineering, dynamic systems and control, energy, energy conversion and energy systems, fluid mechanics and fluid machinery, heat and mass transfer, manufacturing science and technology, mechanical design, mechanics of materials, micro- and nano-science technology, thermal physics, tribology, and vibration and acoustics. The series features graduate-level texts, professional books, and research monographs in key engineering science concentrations.

More information about this series at http://www.springer.com/series/1161

Maurizio Bottoni

Physical Modeling and Computational Techniques for Thermal and Fluid-dynamics

Practical Numerical Mathematics

Springer

Maurizio Bottoni
Bruchsal, Germany

ISSN 0941-5122 ISSN 2192-063X (electronic)
Mechanical Engineering Series
ISBN 978-3-030-79716-4 ISBN 978-3-030-79717-1 (eBook)
https://doi.org/10.1007/978-3-030-79717-1

This Springer imprint is published by the registered company Springer Nature Switzerland AG
The registered company address is: Gewerbestrasse 11, 6330 Cham, Switzerland

This book is dedicated to the memory of Prof. Giuseppe Zwirner, Ordinary of Mathematics at the Ferrara University, unsurpassed in the art of explaining difficult concepts in simple terms.

Preface

The text of computational techniques for thermal and fluid-dynamics problems arose from the author's experience gained over the years (1974–1990) at KfK (Kernforschungszentrum Karlsruhe, actually Karlsruhe Institute of Technology) and at the Argonne National Laboratory (ANL,1990–1996) as researcher of fast breeder reactors nuclear safety. Teaching courses of about fifty hours were organized during one sabbatical year at the Research Laboratory for Nuclear Reactors (RLNR) of the Tokyo Institute of Technology (TITech) and in subsequent years, between 1999 and 2002, at the Tsinghua University in Beijing. Excerpts of these courses were adopted between 2003 and 2006 for teaching numerical methods in atmospheric physics at the Physic Department of the University of Ferrara, Italy.

The present lecture notes are divided into eight chapters. A scholastic approach is followed in Chaps. 1 and 2, devoted to the numerical solution of ordinary differential equations (ODE) of first order, and of partial differential equations of first and second order, respectively, and in Chap. 4, where the basic concepts of consistency, stability and convergence are revised.

Other parts follow a less conventional approach. Among them is Chap. 3, where the residual method of orthogonal collocations is presented in several variants, ranging from the classical Galerkin method to point and domain collocations, applied to the numerical solution of partial differential equations of first order.

Solutions of fluid dynamic problems often are led, through the discretization process, to the numerical solutions of large linear systems. Problems of large dimensions require iterative algorithms. For these cases, the classical theory of successive over relaxation (SOR) by Young and Frankel is revised in Chap. 5. Variational methods, based on minimization of functionals, are given with numerical examples.

Numerical solution of elliptic problems plays an ever-increasing role in fluid dynamics. Several CFD (computational fluid dynamics) codes use the pressure method, consisting of combining momentum and continuity equations to derive a Poisson-like equation for pressure. Several variants of solutions of a Poisson equation are dealt with in Chap. 6. Besides discussing classical solution methods, among them the Ritz and Galerkin methods, a variant of the residual method of orthogonal collocations (RMOC) is proposed as a valuable alternative.

The last two chapters lead the student to more advanced problems, at the edge of current scientific research. The case of momentum equations describing a two-phase flow is treated in all details which would enable implementation of the algorithm into a computer program.

Chapter 8 introduces turbulence modelling, with a brief overview of the most important variants used in the past decades. It explains the numerical solution of Poisson-like equations arising from the discretization of the transport equations for the components of the Reynolds stresses. The case of homogeneous turbulence is treated in short terms only for didactical purposes.

Bruchsal, Germany Maurizio Bottoni

Acknowledgements

The author wishes to thank Prof. Andrea Corli of the Department of Mathematics and Informatics of the University of Ferrara, Italy, for the revision of Chap. 1; the former colleagues Dr. Christoph Homann and Mr. Burkhardt Dorr for their contributions to the development and testing of the computer program BACCHUS-3D/TP; Mrs. Chen Lixing, General Manager of the Beijing Tian Di Xing Communication Software Company, for editing figures of the text; Dr. Simone Mantovani and Dr. Mario Cavicchi of the MEEO (Meteorological and Environmental Earth Observation) in Ferrara, Italy, for running and checking the test cases of the Appendix; Dr. Wolfgang Sengpiel, former colleague at KfK (Kernforschungszentrum Karlsruhe, actually Karlsruhe Institute of Technology) for the final revision of several parts of the text.

Contents

List of Figures

List of Tables

List of Acronyms

AAAS	American Association for the Advancement of Science
ADI	Alternating direction implicit
ANL	Argonne National Laboratory (USA)
ASM	Algebraic stress model
ASWR	Asymmetric separated region weighted residual
ATM	Anisotropic turbulence model
AVA	Aerodynamische Versuchsanstalt (Germany)
CAAM	Computational and Applied Mathematics
CENG	Centre d'Etudes Nucleaires de Grenoble (France)
CFD	Computational Fluid Dynamics
CFL	Courant-Friedrichs-Lewy condition
CR	Convergence rate
COMMIX	COMponent MIXing
CRFBR	Clinch River Fast Breeder Reactor
DEM	Department of Mechanical Engineering (Tsinghua University, Beijing)
DNS	Direct numerical simulation
DOE	Department of Energy (USA)
DS	Direct simulation
ENS	Ecole Normale Supérieure (France)
FDM	Finite difference method
FRAM	Filtering Remedy and Methodology
FTCS	Forward Time Central Space
GALCIT	Guggenheim Aeronautical Laboratory at the California Institute of Technology (USA)
GI	Gaussian integration
GRS	Gesellshaft für Reaktorsicherheit (Germany)
HEM	Homogeneous Equilibrium Model
ICE	Implicit Continuous Eulerian
IFR	Integral fast reactor

INET	Institute of Nuclear Energy Technology (Tsinghua University, Beijing)
JPL	Jet Propulsion Laboratory (USA)
KfK	Kernforschungszentrum Karlsruhe (Germany)
KIT	Karsruhe Institute of Technology (Germany)
LANL	Los Alamos National Laboratory (USA)
LASL	Los Alamos Scientific Laboratory (USA)
LES	Large eddy simulation
LOF	Loss of flow
MEEO	Meteorological and Environmental Earth Observation
MIT	Massachusetts Institute of Technology (USA)
ML-ADI	Messina-Londrillo variant of the ADI algorithm
NASA	National Aeronautics and Space Administration (USA)
OC	Orthogonal collocations
ODE	Ordinary differential equation
PDE	Partial differential equation
PNC	Power Reactor and Nuclear Fuel Company (Japan)
QUICK	Quadratic Upstream Interpolation for Convective Kinematics
RMOC	Residual Method of Orthogonal Collocations
SGS	Subgrid-scale modeling
SM	Slip model
SOR	Successive over-relaxation
SPM	Separated phases model
TDMA	Tridiagonal matrix algorithm
UVET	Unequal phase velocity, equal phase temperature
VOF	Volume of fluid

About the Author

Maurizio Bottoni was born in 1941 in Ferrara, Italy. After humanistic studies in Ferrara and basic courses in mathematics and physics at the Ferrara University, he studied nuclear engineering at the University of Bologna where he obtained a PhD in December 1965, with a thesis in neutron physics, which was awarded the RIGHI Prize of the Bologna University. For about 25 years, he worked at the former Nuclear Research Centre of Karlsruhe (actually Karlsruhe Institute of Technology), Germany, and at the Argonne National Laboratory of Chicago, Illinois, on fast nuclear reactor safety analysis.

After the demise of the fast reactor projects, both in Germany and in the USA, he was a visiting professor at the Tokyo Institute of Technology (1996–1997) and the Tsinghua University in Beijing (1999–2002).

The experience gained with the research work in fast breeder reactor theory was collected in a series of monographs presented for the first time in 1994 at the Chinese Institute of Atomic Energy (CIAE) in Beijing. These preliminary teaching courses were extended and reorganized as a full course of about fifty teaching hours and formed the bulk of the teaching material for the lectures given over years at the Tsinghua University.

Excerpts of these courses were adopted between 2003 and 2006 for lectures held on numerical methods in atmospheric physics and molecular dynamics at the Physics Department of the University of Ferrara, Italy.

At present, the author is consultant at the Meteorological and Environmental Earth Observation (MEEO) Company in Ferrara and is involved with numerical simulations of space physics in the upper atmosphere with the computational methods of molecular dynamics.

Chapter 1
Ordinary Differential Equations

In this chapter we consider ordinary differential equations (ODEs), namely, differential equations which include one or more functions of only one independent variable. In Sect. 1.1 we consider ODEs of first order, in which the maximum order of the derivatives is one, in Sect. 1.2 we will consider ODEs of second and higher order.

1.1 Ordinary Differential Equations of First Order

1.1.1 Introduction

In this section we introduce the basic methods for numerical integration of an ordinary differential equation (ODE) of first order for an unknown function $y = y(x)$, namely:

$$F(x, y, y') = 0. \tag{1.1}$$

We assume that F is a continuous function of the three variables x, y and y' defined in a three-dimensional space. We further assume at first that Eq. (1.1) can be written in the normal form:

$$y' = f(x, y), \tag{1.2}$$

solved for the first-order derivative. This case is treated in Sects. 1.2.1 through 1.2.9. ODEs which cannot be written in the normal form (1.2) are treated in Sects. 1.2.10 through 1.2.13.

In general we will be interested in finding a particular integral of Eq. (1.2) satisfying the initial condition:

M. Bottoni, *Physical Modeling and Computational Techniques for Thermal and Fluid-dynamics*, Mechanical Engineering Series,
https://doi.org/10.1007/978-3-030-79717-1_1

$$y_0 = y(x_0), \tag{1.3}$$

therefore in finding an integral curve passing through the given point $P(x_0, y_0) = P(x_0, y(x_0))$. Assuming that the function $f(x, y)$ is defined in a rectangle $R = (a \leq x \leq b, c \leq y \leq d)$ of the two-dimensional space (x,y) and that it is continuous in R, the point $P(x_0, y_0)$ is an arbitrary, but fixed, internal point in R.

Before proceeding to the explanation of the basic numerical integration methods we give in the following sections an overview of the analytical techniques applicable to obtain exact numerical solution of the Eq. (1.2) with a given initial condition. This overview will avoid application of numerical methods to cases in which analytical solutions can be obtained and also allow to construct analytical solutions useful to test numerical methods.

1.1.2 Overview of Analytical Solution Methods

The main types of ordinary differential equations of first order which can be integrated analytically are reviewed in this section.

The classification proposed in the following sections, from Section (1.1.2.1) through Section (1.1.2.13), follows the organization of the classical book by Zwirner (1977), with further contributions from Davis (1962) and Spataro and Tribulato (1980).

Ordinary Differential Equations of first order which can be written in the normal form (1.2) – Sects. 1.2.1 through 1.2.9.

1.1.2.1 Exact Differential Equations

The differential equation (1.2), which can be written:

$$dy - f(x, y)dx = 0, \tag{1.4}$$

is a particular case of the more general differential equation:

$$M(x, y)dx + N(x, y)dy = 0. \tag{1.5}$$

We assume that the functions $M(x,y)$ and $N(x,y)$ are continuous with their first derivatives in a rectangular domain around an arbitrary point $P(x_0, y_0)$ and that $M(x_0, y_0) \neq 0$ and $N(x_0, y_0) \neq 0$.

The differential equation (1.5) is exact when its first member is an exact differential. This occurs if it exists a continuous function $F(x,y)$ such that:

$$dF = \frac{\partial F}{\partial x}dx + \frac{\partial F}{\partial y}dy = M(x,y)dx + N(x,y)dy, \tag{1.6}$$

hence if:

$$\frac{\partial F(x,y)}{\partial x} = M(x,y), \tag{1.7}$$

$$\frac{\partial F(x,y)}{\partial y} = N(x,y). \tag{1.8}$$

The necessary and sufficient condition for the existence of such a function $F(x,y)$ is that:

$$\frac{\partial M(x,y)}{\partial y} = \frac{\partial N(x,y)}{\partial x}. \tag{1.9}$$

If this condition is satisfied, the general integral of Eq. (1.5) is given by either one of the following formulas:

$$F(x,y) = \int_{x_0}^{x} M(x,y)dx + \int_{y_0}^{y} N(x_0,y)dy = c, \tag{1.10}$$

$$F(x,y) = \int_{x_0}^{x} M(x,y_0)dx + \int_{y_0}^{y} N(x,y)dy = c. \tag{1.11}$$

where c is an arbitrary constant. The constant c can be determined imposing the condition that the function $F(x,y)$ assumes a given value $F(x_1,y_1)$ in the given point P (x_1,y_1). As remarked above, the point $P(x_0,y_0)$ is arbitrary, but it can be chosen so that the calculation of the second integral in Eq. (1.10) [or of the first integral in Eq. (1.11)] is simplified.

If Eq. (1.5) is defined in a domain of arbitrary shape, but simply connected, around point $P(x_0,y_0)$, the general integral is given by

$$F(x,y) = \int_{\gamma} [M(x,y)dx + N(x,y)dy] = c, \tag{1.12}$$

where γ is any regular curve connecting the point $P(x_0,y_0)$ to $P(x,y)$.

Let a, b be the abscissas of the extreme points of the curve γ and $y = \varphi(x)$ be the equation of the curve γ. The curvilinear integral is computed by

$$\int_\gamma \left[M(x,y)dx + N(x,y)dy\right] = \int_a^b \{M[x,\varphi(x)] + N[x,\varphi(x)]\varphi'(x)\}dx. \quad (1.13)$$

If the curve γ is given by the parametric equations $x = x(t)$, $y = y(t)$, $(t_0 \leq t \leq t_1)$, the curvilinear integral is computed by

$$\int_\gamma \left[M(x,y)dx + N(x,y)dy\right] = \int_{t_0}^{t_1} \{M[x(t),y(t)]x'(t) + N[x(t),y(t)]y'(t)\}dt. \quad (1.14)$$

If the integrand $M(x,y)dx + N(x,y)dy$ is an exact differential the integral (1.12) depends only on the coordinates of the extreme points $P(x_0,y_0)$ and $P(x_1,y_1)$, but it is independent of the curve γ which connects them. One immediate result arising from this property is that the curvilinear integral (1.12) computed over a closed curve is always equal to zero.

An alternative to the formulas (1.10) or (1.11) is given by the following algorithm (Zwillinger, 1996): Let $m(x,y) = \int M(x,y)dx$, and then the general integral is given by:

$$F(x,y) = m(x,y) + \int \left(N(x,y) - \frac{\partial m(x,y)}{\partial y}\right)dy + c. \quad (1.15)$$

Physical interpretation of an exact differential equation. Exact differential equations have several physical applications and interpretations. We present here one interpretation related to the description of a two-dimensional incompressible and inviscid flow with the so-called "$\Psi - \omega$" method ($\Psi$is the stream function and ω the vorticity of the flow).

Let us consider a two-dimensional domain $D(x,y)$ with an incompressible [with constant density ($\rho = c$) and zero divergence ($\nabla \cdot \overline{u} = 0$), $\overline{u} = (u,v)$)] and inviscid (with viscosity $\mu = 0$) flow. We further assume that the flow is irrotational:

$$\omega = \varsigma\overline{k} = \left(\frac{\partial v}{\partial x} - \frac{\partial u}{\partial y}\right)\overline{k} = 0, \quad (1.16)$$

where ς is the component of the vorticity $\overline{\omega} = \nabla \times \overline{u}$ along the z coordinate axis.

Under these hypotheses it is possible to define a stream function $\Psi(x,y)$ that satisfies identically the continuity equation:

$$\frac{\partial u}{\partial x} + \frac{\partial v}{\partial y} = 0, \quad (1.17)$$

by letting:

$$u = \frac{\partial \Psi}{\partial y}, \tag{1.18a}$$

$$v = -\frac{\partial \Psi}{\partial x}. \tag{1.18b}$$

Because the flow is irrotational, from Eq. (1.16) one obtains:

$$\nabla^2 \Psi = \frac{\partial^2 \Psi}{\partial x^2} + \frac{\partial^2 \Psi}{\partial y^2} = 0. \tag{1.19}$$

Writing the continuity equation in the form $\frac{\partial(-v)}{\partial y} = \frac{\partial u}{\partial x}$ and comparing with Eq. (1.9) one can set $M(x, y) = -v$ and $N(x, y) = u$, hence define the exact differential

$$d\Psi = -v dx + u dy. \tag{1.20}$$

A streamline in the flow is characterized by a constant value of the stream function $\Psi(x, y) = c$ and its equation is given by

$$\left(\frac{dy}{dx}\right)_{\Psi=c} = \frac{v}{u}. \tag{1.21}$$

According to Eq. (1.12) the distribution of the stream function in the domain is given by

$$\Psi(x, y) = \int_{\gamma} [(-v)dx + u dy] + c. \tag{1.22}$$

Example 1.1
Let us consider the differential $(2xy + y^2)dx + (x^2 + 2xy)dy$ *defined in the (x,y) plane. We want to:*

(i) *Verify that it is an exact differential.*
(ii) *Choosing* $P(x_0, y_0) = P(0, 0)$*, calculate the general integral F(x,y).*
(iii) *Determine the integration constant so that a particular integral passes through the point P(1,1).*

(i) $\frac{\partial M(x,y)}{\partial y} = \frac{\partial N(x,y)}{\partial x} = 2x + 2y$, and therefore the given differential is exact.
(ii) Using Eq. (1.10) we obtain the general integral

$$
\begin{aligned}
F(x,y) &= \int_{x_0}^{x} M(x,y)dx + \int_{y_0}^{y} N(x_0,y)dy = c,\\
F(x,y) &= \int_{0}^{x} (2xy + y^2)dx + \int_{0}^{y} (0)dy = c, \qquad (1.23)\\
F(x,y) &= x^2y + xy^2 = c.
\end{aligned}
$$

As a control, from Eq. (1.6) one obtains the given differential equation:

$$
dF = \frac{\partial F}{\partial x}dx + \frac{\partial F}{\partial y}dy = \left(2xy + y^2\right)dx + \left(x^2 + 2xy\right)dy.
$$

(iii) From Eq. (1.23) one has $F(1,1) = 2 = c$. The particular integral sought is therefore $F_1(x,y) = x^2y + xy^2 = 2$.

Example 1.2

Let us consider the differential.

[sin(*x* + *y*) + *x* cos (*x* + *y*)]*dx* + [*xcos*(*x* + *y*)]*dy, defined in the square domain* $(0 \le x \le 1, 0 \le y \le 1)$. *We want to:*

(i) *Verify that it is an exact differential.*
(ii) *Calculate the general integral F(x,y).*
(iii) *Find the particular integral though the point* $P_1\left(\sqrt{2}/2, \sqrt{2}/2\right)$.
(iv) *Prove that the curvilinear integral of F(x,y) depends only on the coordinates of the extreme points of the integration curve and not on the path itself, by integrating F(x,y) along (see* Fig. 1.1*):*

Path γ_1, *from point O(0,0) to* $P_1\left(\sqrt{2}/2, \sqrt{2}/2\right)$, *along the line* $y = x$;

Path γ_2, *from O(0,0) to* $P_1\left(\sqrt{2}/2, \sqrt{2}/2\right)$, *along the parabola* $y = \sqrt{2}x^2$;

Path $\gamma_3 = \gamma_a + \gamma_b$, *where* γ_a *is the segment O(0,0) – A(1,0) along the x axis;* γ_b *is the arc of the circle of equation* $x^2 + y^2 = 1$ *from point A(1,0) to* $P_1\left(\sqrt{2}/2, \sqrt{2}/2\right)$, *oriented in counterclockwise direction.*

(i) Being $M(x,y) =$ sin $(x + y) + x$ cos $(x + y)$ and $N(x,y) = x$ cos $(x + y)$ one has:

$$
\frac{\partial M(x,y)}{\partial y} = \frac{\partial N(x,y)}{\partial x} = \cos(x+y) - x\sin(x+y), \qquad (1.24)
$$

and therefore the given differential is exact.

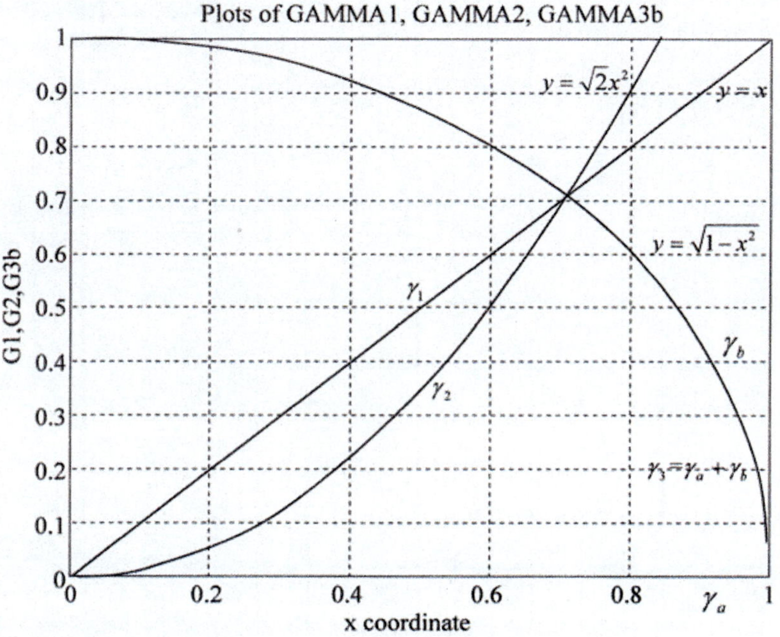

Fig. 1.1 (Example 1.2) Sketch showing the domain in which the given exact differential must be integrated along the paths γ_1, γ_2, $\gamma_3 = \gamma_a + \gamma_b$

(ii) Using Eq. (1.10) with, $x_o = 0$, $y_o = 0$, one obtains the general integral:

$$
\begin{aligned}
F(x,y) &= \int_{x_0}^{x} M(x,y)dx + \int_{y_0}^{y} N(x_0,y)dy = c, \\
F(x,y) &= \int_{0}^{x} [\sin(x+y) + x\cos(x+y)]dx + \int_{0}^{y} (0)dy = c, \qquad (1.25)\\
F(x,y) &= -\cos(x+y) + \int_0^x x\cos(x+y)dx = c.
\end{aligned}
$$

The integral $\int_0^x x\cos(x+y)dx$ can be computed by parts assuming <x> as finite factor and < cos (x + y)> as differential factor:

$$
\begin{aligned}
\int x\cos(x+y)dx &= x\sin(x+y) - \int \sin(x+y)dx \\
&= x\sin(x+y) + \cos(x+y). \qquad (1.26)
\end{aligned}
$$

[Cfr. Table 1.2, Nr. 2].

Table 1.2 Summary of the indefinite integrals used in the examples of Sect. 1.1.2

1. $\int xe^x dx = e^x(x-1)$	(Example 1.18)
2. $\int x \cos(x+y) dx = x \sin(x+y) + \cos(x+y)$	(Example 1.2)
3. $\int x \cos(2x) = \frac{1}{2} x \sin(2x) + \frac{1}{4}\cos(2x)$	(Example 1.2)
4. $\int x \cos x dx = x \sin x + \cos x$	(Example 1.2)
	(Example 1.14)
5. $\int \frac{\sin x}{\cos^2(x)} dx = \frac{1}{\cos x}$	(Example 1.4)
6. $\int \frac{\cos x}{\sin^2(x)} dx = -\frac{1}{\sin x}$,	(Example 1.4)
7. $\int \frac{1}{ax+bx^3} dx = \frac{1}{2a} \log\left(\frac{x^2}{a+bx^2}\right)$, $a \neq 0$,	(ZW, Pag. 364, Nr. 70)
7b. Hence for a = 1, b = 1: $\int \frac{1}{x+x^3} dx = \frac{1}{2} \log \frac{x^2}{1+x^2}$	(Example 1.6)
8. $\int \frac{1}{a+bx} dx = \frac{1}{b} \log(a+bx)$, $b \neq 0$, $(a+bx) > 0$,	(ZW, Pag. 362, Nr. 27)
8b. Hence for a = 5, b = 4: $\int \frac{1}{5+4x} dx = \frac{1}{4} \log(5+4x)$, $(5+x) > 0$	(Example 1.9)
9. $\int \frac{x}{a+bx} dx = \frac{x}{b} - \frac{a}{b^2} \log(a+bx)$ $b \neq 0$	(ZW, Pag. 362, Nr. 30)
9b. Hence for a = - 3, b = 1: $\int \frac{x}{-3+x} dx = x + 3\log(-3+x)$	(Example 1.8)
9c. Hence for a = 5, b = 4: $\int \frac{x}{5+4x} dx = \frac{x}{4} - \frac{5}{16} \log(5+4x)$	(Example 1.9)
10. $\int \frac{1}{a+bx+cx^2} dx = \frac{1}{\sqrt{-q}} \log\left(\frac{2cx+b-\sqrt{-q}}{2cx+b+\sqrt{-q}}\right)$, $(q = 4ac - b^2)$	(ZW, Pag. 367, Nr. 108)
10b. Hence for a = 1, b = - 1, c = - 2, q = - 9: $\int \frac{1}{1-x-2x^2} dx = \frac{1}{3} \log\left(\frac{2+2x}{-1+2x}\right)$	(Example 1.7)
11. $\int \frac{x}{a+bx+cx^2} dx = \frac{1}{2c} \log(a+bx+cx^2) - \frac{b}{2c} \int \frac{1}{a+bx+cx^2} dx = \frac{1}{2c} \log(a+bx+cx^2) - \frac{b}{2c} \frac{1}{\sqrt{-q}} \log\left(\frac{2cx+b-\sqrt{-q}}{2cx+b+\sqrt{-q}}\right)$ $(q = 4ac - b^2 < 0)$	(ZW, Pag. 367, Nr. 112)
11b. Hence for a = 1, b = - 1, c = - 2. q = - 9: $\int \frac{x}{1-x-2x^2} dx = -\frac{1}{4} \log(1-x-2x^2) - \frac{1}{12} \log\left(\frac{2+2x}{-1+2x}\right)$	(Example 1.7)
12. $\int (e^x \cos x) dx = \frac{e^x}{2} (\sin x + \cos x)$	(Example 1.10)
13. $\int \frac{1}{\sqrt{1-x^2}} dx = \arcsin x$	(Example 1.15)

ZW refers to Zwillinger (1996)

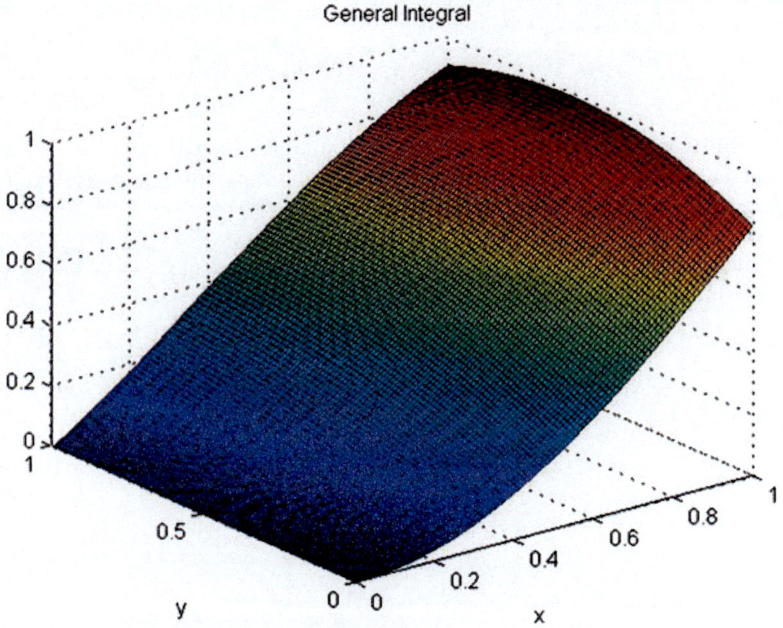

Fig. 1.2 (Example 1.2) perspective view of general integral $F(x, y) = x \sin(x + y)$, with $c = 0$

Inserting (1.26) in (1.25), one obtains:

$$\begin{aligned} F(x,y) &= -\cos(x+y) + x\sin(x+y) + \cos(x+y) = c, \\ F(x,y) &= x\sin(x+y) = c. \end{aligned} \tag{1.27}$$

As a control, from Eq. (1.6) one obtains the given differential equation:

$$dF = \frac{\partial F}{\partial x}dx + \frac{\partial F}{\partial y}dy = [\sin(x+y) + x\cos(x+y)]dx + [x\cos(x+y)]dy.$$

z = $F(x, y)$ is the equation of the integral surface. A section of this surface with a plane z = constant gives a particular integral as if it were projected from the plane z = c upon the (x,y) plane.

Figure 1.2 gives a perspective view of the integral surface. Along the y-axis $F(0, y) = 0$. Some values are given: $F(0.5, 0) = 0.239$; $F(1, 0) = 0.841$; $F(1, 0.5) = 0.997$; $F(1, 1) = 909$; $F(0.5, 1) = 0.498$.

(iii) From Eq. (1.27) one has:

$$F\left(\frac{\sqrt{2}}{2}, \frac{\sqrt{2}}{2}\right) = \frac{\sqrt{2}}{2}\sin\left(\frac{\sqrt{2}}{2} + \frac{\sqrt{2}}{2}\right) = \frac{\sqrt{2}}{2}\sin\left(\sqrt{2}\right) = c = 0.698455\,999.$$

The particular integral sought is therefore

$$F_1(x, y) = x \sin(x + y) = 0.698\ 455\ 999. \tag{1.28}$$

(iv) The following curvilinear integrals are computed in part analytically; all of them were also computed numerically, using Gaussian integration (GI) of order 40 (see, for instance, Krylov, 1962).

1. Integration over the path γ_1, from point O(0,0) to $P_1\left(\sqrt{2}/2, \sqrt{2}/2\right)$, along the line $y = x$;

Using Eq. (1.13) with $y = \varphi(x) = x$, $y^{'} = \varphi^{'} = 1$, one has

$$I1 = \int_{\gamma} \left[M(x, y)dx + N(x, y)dy \right] = \int_0^{\sqrt{2}/2} \{M[x, \varphi(x)] + N[x, \varphi(x)]\varphi'(x)\}dx =$$

$$= \int_0^{\sqrt{2}/2} [\sin(x + x) + x\cos(x + x) + x\cos(x + x)]dx == \int_0^{\sqrt{2}/2} [\sin(2x) + 2x\cos(2x)]dx =$$

$$= \left[-\frac{1}{2}\cos(2x) + x\sin(2x) + \frac{1}{2}\cos(2x) \right]_0^{\sqrt{2}/2} =$$

$$= [x\sin(2x))]_0^{\sqrt{2}/2} = \frac{\sqrt{2}}{2}\sin\left(\sqrt{2}\right) = 0.698\ 455\ 998 \tag{1.29}$$

In (1.29) the integral $\int_0^{\sqrt{2}/2}[2x\cos(2x)]dx$ has been computed by parts taking <x> as finite factor and < cos (2x)> as differential factor [Cfr. Table 1.2, Nr. 3].

2. Integration along path γ_2, from O(0,0) to $P_1\left(\sqrt{2}/2, \sqrt{2}/2\right)$, along the parabola $y = \sqrt{2}x^2$. Using Eq. (1.13) with $y = \varphi(x) = \sqrt{2}x^2$, $y' = \varphi' = 2\sqrt{2}x$, letting $H(x) = x + \sqrt{2}x^2$, one has:

$$I2 = \int_{\gamma} \left[M(x, y)dx + N(x, y)dy \right] = \int_0^{\sqrt{2}/2} \{M[x, \varphi(x)] + N[x, \varphi(x)]\varphi'(x)\}dx =$$

$$= \int_0^{\sqrt{2}/2} \left\{ \sin H(x) + x\cos H(x) + x\cos H(x) \cdot 2\sqrt{2}x \right\} dx =$$

$$= \int_0^{\sqrt{2}/2} \left\{ \sin H(x) + \left(x + 2\sqrt{2}x^2\right) \cos H(x) \right\} dx$$

$$= 0.698\ 455\ 999, \tag{1.30}$$

which is almost identical with I1. I2 was computed with Gaussian integration (GI).

3. Integration along path $\gamma_3 = \gamma_a + \gamma_b$

   (a) Integration along path γ_a, the segment O(0,0) – A(1,0) along the x axis. Using again Eq. (1.13) with $y = \varphi(x) = 0$, $y' = \varphi' = 0$, one has:

$$I3a = \int_\gamma \Big[M(x,y)dx + N(x,y)dy\Big] = \int_0^1 \{M[x,\varphi(x)] + N[x,\varphi(x)]\varphi'(x)\}dx =$$

$$= \int_0^1 [\sin x + x\cos x]dx = [-\cos x + x\sin x + \cos x]_0^1 =$$

$$= [x\sin x]_0^1 = \sin(1) = 0.841\,470\,984. \tag{1.31}$$

In (1.31) the integral $\int_0^1 [x\cos x]dx$ has been computed by parts taking <x> as finite factor and < cos x> as differential factor. [Cfr. Table 1.2, Nr. 4].

(b) Integration along path γ_b, the arc of the circle of equation $x^2 + y^2 = 1$ from point $A(1,0)$ to $P_1\left(\sqrt{2}/2, \sqrt{2}/2\right)$. Using the parametric equations of the circle $x = \cos(t), y = \sin(t)\ (0 \leq t \leq \pi/4)$, from Eq. (1.14), letting $H(t) = \cos(t) + \sin(t)$, one obtains.

$$\begin{aligned} I3b &= \int_0^{\pi/4} \{M[x(t),y(t)]x'(t) + N[x(t),y(t)]y'(t)\}dt = \\ &= \int_0^{\pi/4} \{[\sin H(t) + \cos(t)\cos H(t)](-\sin(t) + \cos(t)\cos H(t)\cos(t)\}dt. \end{aligned} \tag{1.32}$$

Gaussian integration yields

$$\text{I3b} = -0.143\,014\,986. \tag{1.33}$$

The curvilinear integral along path $\gamma_3 = \gamma_a + \gamma_b$ is then obtained as the sum of integrals I3a and I3b:

$$\text{I3} = \text{I3a} + \text{I3b} = 0.841\,470\,984\text{-}0.143\,014\,986 = 0.698\,455\,998, \tag{1.34}$$

which is identical with integral I1.

It is therefore proved that the integrals I1, I2, and I3 are identical and that their common value is equal to the difference of the values of the general integral (1.27) in the points $P_1\left(\sqrt{2}/2, \sqrt{2}/2\right)$ and O(0,0):

$$\text{I1} = \text{I2} = \text{I3} = F\left(\sqrt{2}/2, \sqrt{2}/2\right)\text{-}F(0,0) = 0.698\,455\,999. \tag{1.35}$$

1.1.2.2 Differential Equations with Separate Variables

Differential equations with separate variables are equations of the type:

$$M(x)dx + N(y)dy = 0, \tag{1.36}$$

in which the coefficients multiplying the differential dx and dy are functions of the only variables x and y, respectively. The differential (1.36) is also exact and the general integral is given by.

$$F(x,y) = \int_{x_0}^{x} M(x)dx + \int_{y_0}^{y} N(y)dy = c, \tag{1.37}$$

or alternatively by

$$F(x,y) = \int M(x)dx + \int N(y)dy = c, \tag{1.38}$$

which differs from (1.37) only by a constant.

Example 1.3
Let us consider the differential $xdx + ydy = 0$ defined in the whole (x,y) plane. This differential is exact, with separate variables, and has general integral $x^2 + y^2 = c$ which represents the family of circles of equation $x^2 + y^2 = R^2$ with center in the origin of the coordinate system. The particular integral passing through the point (1,0) is determined by the constant $c = 1$ and has equation $x^2 + y^2 = 1$.

The integral surface $z = x^2 + y^2$ is a paraboloid of revolution.

Figure 1.3 gives a prospective view of this surface in the square domain $(-1 \le x \le 1; -1 \le y \le 1)$..

1.1.2.3 Differential Equations with Separable Variables

An ordinary differential equation of first order is of the separable variables type when the coefficients of the differentials dx and dy are products of one function of the only variable x and another function of the only variable y. The differential is therefore of the type

$$A(x)B(y)dx + C(x)D(y)dy = 0. \tag{1.39}$$

Assuming $C(x)$ and $B(y)$ are different from zero in the definition domain, one obtains an equation with separate variables:

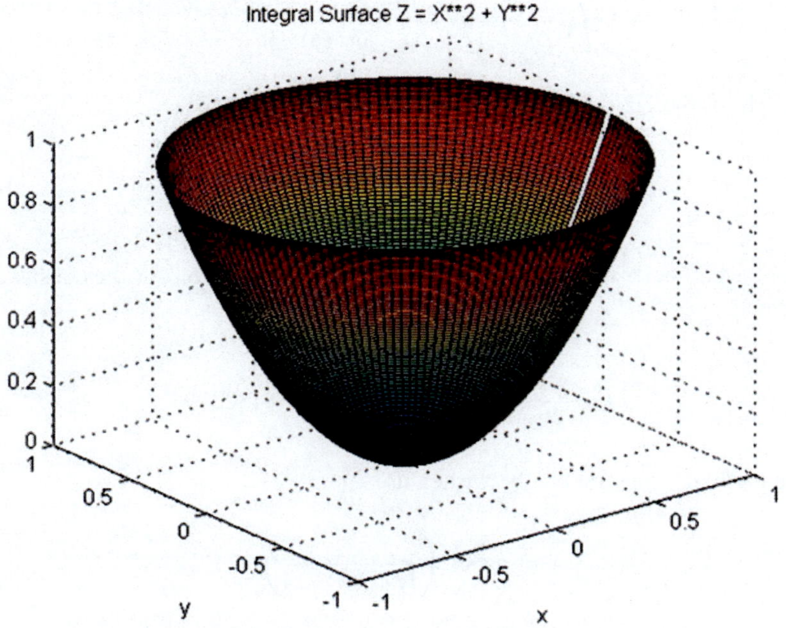

Fig. 1.3 Example 1.3. Prospective view of the integral surface $z = x^2 + y^2$

$$\frac{A(x)}{C(x)}dx + \frac{D(y)}{B(y)}dy = 0, \tag{1.40}$$

which can be integrated as in case of Sect. 1.1.2.2.

Example 1.4
Let us consider the differential

$$\sin x \sin^2(y)dx + \cos^2(x)\cos y dy = 0, \tag{1.41}$$

defined in the (*x*,*y*) plane. Determine:

(i) *The general integral.*
(ii) *The particular integral through the point* $P_1(0.6, 0.8)$.

(i) The differential (1.41) is of the type (1.39) with:

$A(x) = \sin(x)$, $B(y) = \sin^2(y)$, $C(x) = \cos^2(x)$, $D(y) = \cos y$. Assuming $C(x)$ and $B(y)$ are different from zero, the following differential equation with separable variables is obtained:

$$\frac{\sin x}{\cos^2(x)}dx + \frac{\cos y}{\sin^2(y)}dy = 0.$$

Using (1.38) the general integral is given by:

$$\int \frac{\sin x}{\cos^2(x)} dx + \int \frac{\cos y}{\sin^2(y)} dy = 0,$$

hence [Cfr. Table 1.2, Nr. 5,6]

$$\frac{1}{\cos x} - \frac{1}{\sin y} = c. \tag{1.42}$$

(ii) For the particular integral passing through the point P1(0.6,0.8), the constant c is given by

$$c1 = \frac{1}{\cos(0.6)} - \frac{1}{\sin(0.8)} = -0.182379$$

and the particular integral can be written as

$$y = \arcsin\left(\frac{1}{1/\cos x - c1}\right). \tag{1.43}$$

Figure 1.4 shows this particular integral and the line $y = x\,0.8/0.6$, which cross in $P_1(0.6, 0.8)$.

Figure 1.5 shows the particular integral over a larger interval ($0 < x < 50$) of the x-axis. The particular integral (1.43) is not defined in the intervals of the x-axis where $1/\cos x - c1 < -1$.

1.1.2.4 Equations of the Type $y' = f(ax + by)$

An equation of the type:

$$y' = f(ax + by), \tag{1.44}$$

with a and b constants, both different from zero, can be reduced to an equation with separable variables letting $z = ax + by$, hence:

$$y = \frac{z - ax}{b}, \tag{1.45}$$

$$y' = \frac{z' - a}{b}. \tag{1.46}$$

With this change of variables the original Eq. (1.44) becomes:

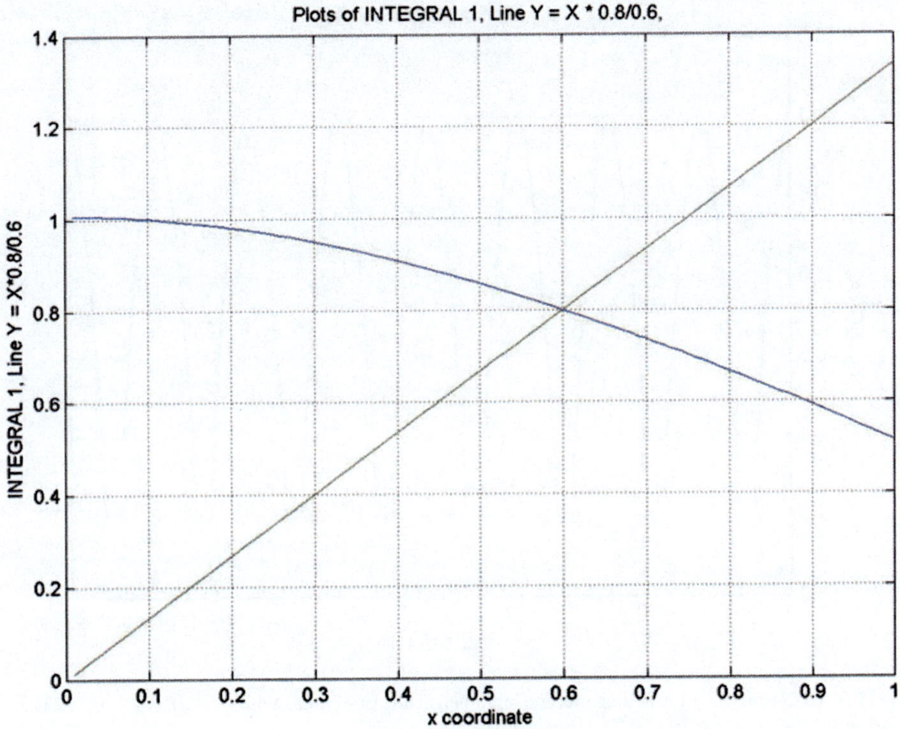

Fig. 1.4 (Example 1.4) Particular integral obtained from Eq. (1.42) with $c = -0.182379$ intersecting the line $y = x$ 08/0.6 in point $P_1(0.6, 0.8)$

$$\frac{z' - a}{b} = f(z), \tag{1.47}$$

hence:

$$z' = a + bf(z), \tag{1.48}$$

$$\frac{dz}{a + bf(z)} = dx, \tag{1.49}$$

which is of the separate variables type. To every integral $z(x)$ of Eq. (1.49) corresponds an integral of the original equation given by Eq. (1.45).

Example 1.5
(i) *Find the general integral of the equation*

$$y' = 3x + y, \tag{1.50}$$

defined in the domain $(0 < x, 0 < y)$;

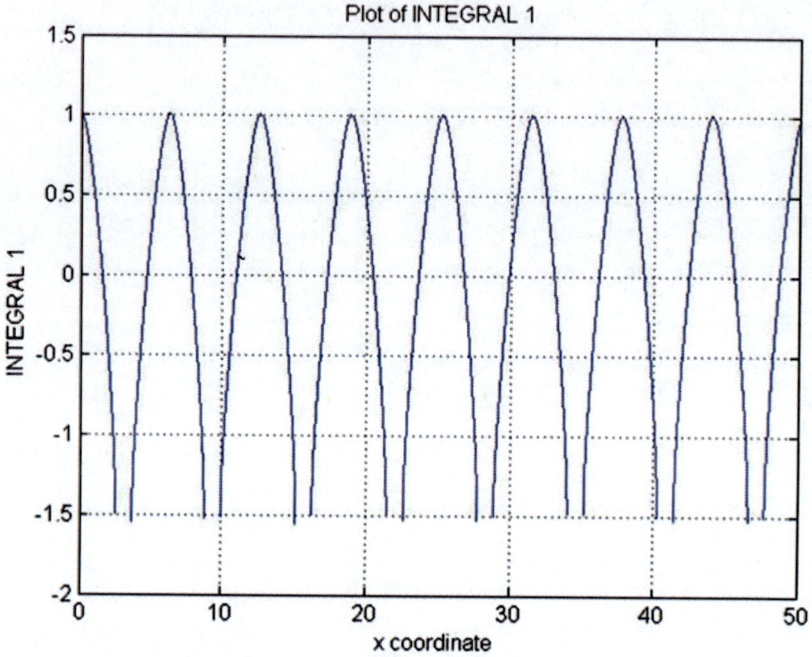

Fig. 1.5 (Example 1.4) Particular integral obtained from Eq. (1.42) with $c = c1 = -0.182379$ passing through point $P_1(0.6, 0.8)$. [Larger interval of x axis]

(ii) *Find the particular integral passing through the point* $P_1(1, 1)$.

(i) Let $z = 3x + y$, hence $y = z - 3x$, $y' = z' - 3$. Inserting in (1.50) one has. $z' - 3 = z$, $\frac{dz}{dx} = 3 + z$, $\frac{dz}{3+z} = dx$, which is an equation with separate variables.

Integrating one obtains

$$\begin{aligned} \int dx - \int \frac{dz}{3+z} &= c, \\ x - \log(3 + z) &= c, \\ x - \log(3 + 3x + y) &= c, \end{aligned}$$

which is the general integral of the given differential equation. It can also be written

$$\begin{aligned} 3 + 3x + y &= ce^x, \\ y &= -3 - 3x + ce^x, \end{aligned} \tag{1.51}$$

(ii) From (1.51) the constant which determines the particular integral through the point $P_1(1, 1)$ is given by $7 = c_1 e$; hence $c_1 = 7/e = 2.575\,156$. The particular integral has equation

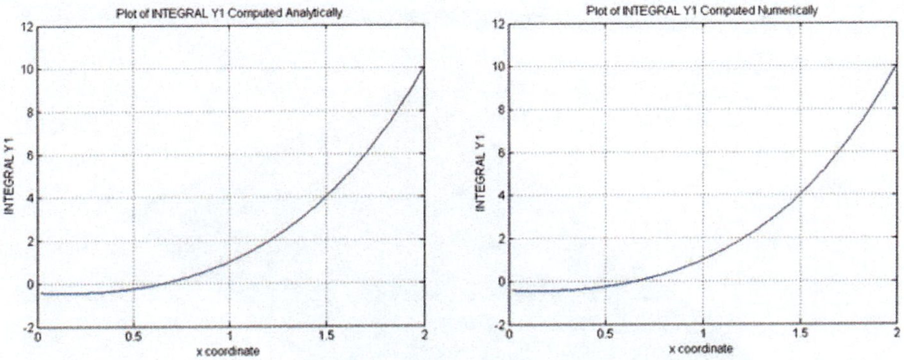

Fig. 1.6 Example 1.5. Left plot: particular integral through the point $P_1(1, 1)$ computed analytically; Right plot: particular integral computed numerically

$$y = -3 - 3x + c_1 e^x. \tag{1.52}$$

The particular integral (1.52) is plotted at the left side of Fig. 1.6 for the domain $(0 < \underline{x} < 2)$ divided in 100 intervals. At the right side of the same figure is the plot of the particular integral computed numerically with the method of bisection explained in Sect. 1.1.2.6. In short, the method consists in finding the root of the equation $f(y) = y + 3 + 3x - c_1 e^x = 0$ for a given value of the abscissa x. The method requires an estimation ("bracketing") of the range in which the root will be found. In this example the numerical method would not be of interest because the particular integral can be computed analytically. The numerical method has been applied only to show that the numerical results match well the analytical ones. The plots of the left and right figures are practically superimposed. With an accuracy of 10^{-9} required for the bisection method, the maximum relative deviation between the results plotted in the left and right figures was $3.038x10^{-7}$.

The integral surface $Z = F(x, y) = x - \log(3 + 3x + y)$ is plotted in Fig. 1.7. The prospective view of the integral surface is important to ascertain whether or not it has ripples. If the integral surface has no ripples, the numerical method of bisection can be implemented easily. Otherwise, if there are ripples, the application of the bisection method, which requires "bracketing", becomes much more laborious.

1.1.2.5 Homogeneous Equations

A function $F(x,y)$ is homogeneous of degree n if $F(kx, ky) = k^n F(x, y)$. An equation of type (1.5) is homogeneous when the functions $M(x,y)$ and $N(x,y)$ are homogeneous functions of the same degree, because then in the equation written in the normal form $y' = -M(x, y)/N(x, y)$ the second member is a homogeneous function of zero degree.

An ordinary differential equation of first order is said to be homogeneous when it is of the type:

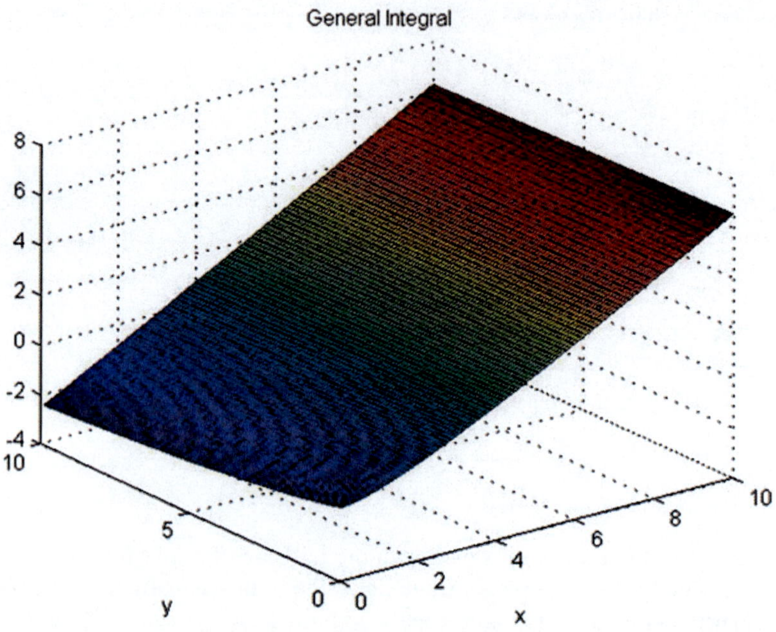

Fig. 1.7 Example 1.5. Integral surface $Z = F(x, y) = x - \ln(3 + 3x + y)$

$$y' = f\left(\frac{y}{x}\right), \tag{1.53}$$

where f is a homogeneous function of zero degree.

We let $z = y/x$, where $z = z(x)$ is considered to be a function of x. Hence:

$$y = xz, \tag{1.54}$$

$$y' = z + xz'. \tag{1.55}$$

With this change of variable, Eq. (1.53) becomes:

$$z + xz' = f(z), \tag{1.56}$$

hence:

$$z' = \frac{f(z) - z}{x}, \tag{1.57}$$

which is of the separable variables type and yields:

$$\frac{dz}{f(z) - z} - \frac{dx}{x} = 0. \tag{1.58}$$

The general integral is:

$$\int \frac{dz}{f(z) - z} - \int \frac{dx}{x} = c. \tag{1.59}$$

To every integral $z(x)$ of the Eq. (1.59) corresponds an integral of the original equation given by Eq. (1.54).

Example 1.6
Given the differential equation:

$$y' = -\left(\frac{y}{x}\right)^3. \tag{1.60}$$

find:

(i) *The general integral.*
(ii) *The particular integral passing through the point* $P_1(1, 1)$.

(i) Letting $z = y/x$, $y = xz$, $y' = z + xz'$ and replacing in (1.60), one obtains:

$z + xz' = -z^3$; hence:

$$z' = \frac{dz}{dx} = -\frac{z + z^3}{x},$$

$\frac{dz}{z+z^3} = -\frac{dx}{x}$, which has general integral.

$$\int \frac{dz}{z + z^3} + \int \frac{dx}{x} = c. \tag{1.61}$$

The first integral can be found in Zwillinger (1996), (Pag. 364, Integral Nr. 70) [Cfr. Table 1.2, Nr. 7]. One gets the following general integral:

$$\frac{1}{2} \log \frac{z^2}{1 + z^2} + \log x = c,$$

or, using again $z = y/x$

$$\frac{1}{2} \log \frac{(y/x)^2}{1 + (y/x)^2} + \log x = c. \tag{1.62}$$

Figure 1.8 plots the function $z = F(x, y) = \frac{1}{2} \log \frac{(y/x)^2}{1+(y/x)^2} + \log x$

(ii) We want now to determine the particular integral passing through the point $P_1(1, 1)$. From (1.62) one obtains for the constant the value $c_1 = (1/2) \ln (1/2) = -0.346573$. Introducing this constant in (1.62) and solving for y one obtains the explicit equation of the particular integral considered, plotted in Fig. 1.9:

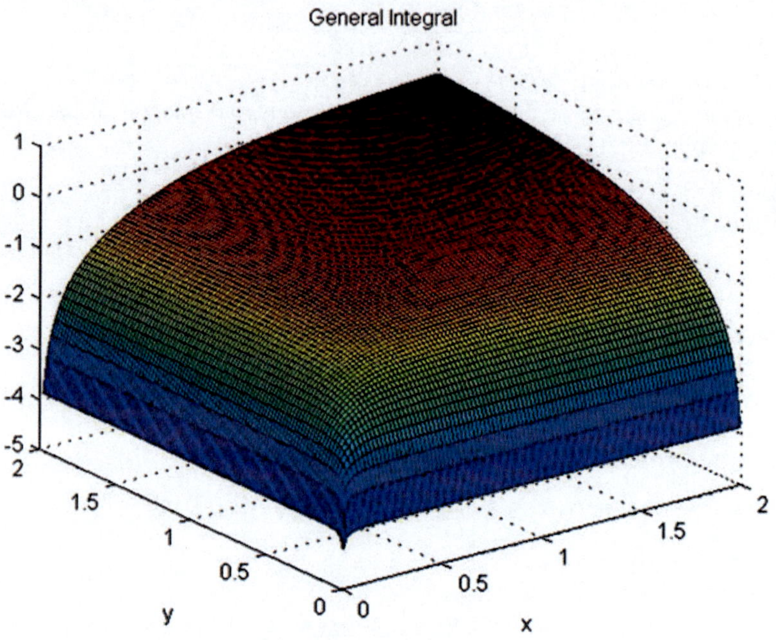

Fig. 1.8 (Example 1.6). Plot of integral surface (1.62)

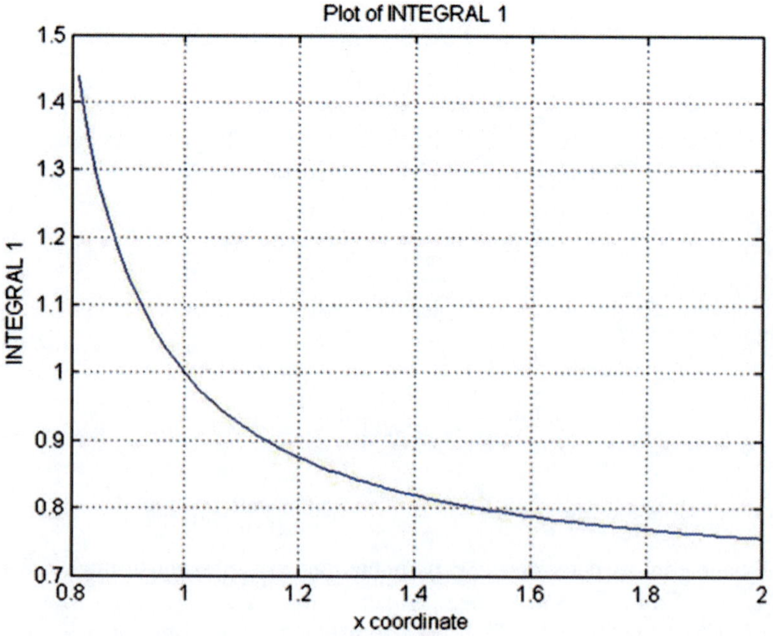

Fig. 1.9 (Example 1.6) Particular integral (1.63) with $c = -0.346573$ passing through the point $P_1(1, 1)$

$$y = \sqrt{\frac{e^{2c1}x^2}{x^2 - e^{2c1}}}. \tag{1.63}$$

The domain of definition of this integral is imposed by the condition that $x^2 > e^{2c1}$ or $x \geq 0.707$..

1.1.2.6 Equations of the Type $y' = f\left(\frac{ax+by+c}{\overline{a}x+\overline{b}y+\overline{c}}\right)$

In the equation of the type:

$$y' = f\left(\frac{ax + by + c}{\overline{a}x + \overline{b}y + \overline{c}}\right) \tag{1.64}$$

we assume that f is a continuous function. We distinguish two cases: In the first case the lines of equations $ax + by + c = 0$ and $\overline{a}x + \overline{b}y + \overline{c} = 0$ intersect in a point (α, β), which occurs if $a\overline{b} - \overline{a}b \neq 0$. In the second case $a\overline{b} - \overline{a}b = 0$ and the lines are parallel.

First Case: The Lines Intersect in a Point (α, β)

We replace the variables x and y with new variables u and v, given by:

$$x = u + \alpha, \tag{1.65}$$

$$y = v + \beta. \tag{1.66}$$

Hence we obtain:

$$y' = \frac{dy}{dx} = \frac{d(v + \beta)}{d(u + \alpha)} = \frac{dv}{du}, \tag{1.67}$$

$$ax + by + c = au + bv, \tag{1.68}$$

$$\overline{a}x + \overline{b}y + \overline{c} = \overline{a}u + \overline{b}v. \tag{1.69}$$

In Eqs. (1.68) and (1.69), the fact has been used that the point (α, β) belongs to both lines $ax + by + c = 0$ and $\overline{a}x + \overline{b}y + \overline{c} = 0$.

With these changes of variables the original differential Eq. (1.64) becomes:

$$\frac{dv}{du} = f\left(\frac{au + bv}{\overline{a}u + \overline{b}v}\right) = f\left(\frac{a + b\frac{v}{u}}{\overline{a} + \overline{b}\frac{v}{u}}\right), \tag{1.70}$$

which is a homogeneous equation of zero degree and can be integrated as explained in the previous section. To every integral $v = v(u)$ of Eq. (1.70) corresponds an integral $y = v + \beta$ of Eq. (1.64).

In this case therefore the integration procedure goes through two steps: (i) In the first step with a change of variables, a homogeneous equation is obtained, and (ii) in the second step, the homogeneous equation is integrated.

Example 1.7
Given the differential equation:

$$y' = \frac{x + 2y + 3}{3x + 2y + 1}. \tag{1.71}$$

find:

(i) *The general integral.*
(ii) *The particular integral passing through the point $P_1(1, 1)$.*

(i) The lines of equations $x + 2y + 3 = 0$ and $3x + 2y + 1 = 0$ intersect in the point $(\alpha, \beta) = (1, -2)$. We set:

$$x = u + 1, \tag{1.72a}$$

$$y = v - 2, \tag{1.72b}$$

from which one derives $dx = du$, $dy = dv$, $y' = \frac{dy}{dx} = \frac{dv}{du} = v'$ and

$$x + 2y + 3 = u + 2v, \tag{1.73a}$$

$$3x + 2y + 10 = 3u + 2v, \tag{1.73b}$$

$$y' = \frac{x + 2y + 3}{3x + 2y + 1} = v' = \frac{dv}{du} = \frac{u + 2v}{3u + 2v} = \frac{1 + 2\frac{v}{u}}{3 + 2\frac{v}{u}}. \tag{1.74}$$

Equation (1.74) can now be treated as a homogeneous equation. We let, recalling also Eq. (1.72a and 1.72b):

$$z = \frac{v}{u} = \frac{y + 2}{x - 1}, \tag{1.75}$$

considering $z = z(u)$ as function of u. Hence:

$$\begin{aligned} v &= uz \\ v' &= z + uz'. \end{aligned}$$

Recalling Eq. (1.74) one obtains:

$$\begin{aligned} z + uz' &= \frac{1+2z}{3+2z} = f(z), \\ z' &= \frac{dz}{du} = \frac{f(z)-z}{u} = \frac{\frac{1+2z}{3+2z} - z}{u} = \frac{\frac{1-z-2z^2}{3+2z}}{u}, \\ &\frac{3+2z}{1-z-2z^2}dz = \frac{du}{u}, \\ &\int \frac{3+2z}{1-z-2z^2}dz - \int \frac{du}{u} = c. \end{aligned} \tag{1.76}$$

The first integral in (1.76) can be separated in the fundamental integrals listed in Zwillinger (1996) (Pag. 367, Integrals Nr. 108 and 112, respectively) [Cfr. Table 1.2 Nr. 10, 11] and is given by:

$$\begin{aligned} \mathrm{I1} &= \int \frac{3}{1-z-2z^2}dz = \log \frac{2+2z}{-1+2z}, \\ \mathrm{I2} &= \int \frac{2z}{1-z-2z^2}dz = -\frac{1}{2}\log\left(1-z-2z^2\right) - \frac{1}{6}\log \frac{2+2z}{-1+2z}. \end{aligned}$$

The first integral in (1.76) becomes:

$$\begin{aligned} \int \frac{3+2z}{1-z-2z^2}dz = \mathrm{I1} + \mathrm{I2} &= \log\frac{2+2z}{-1+2z} - \frac{1}{2}\log\left(1-z-2z^2\right) - \frac{1}{6}\log\frac{2+2z}{-1+2z} = \\ &= \frac{5}{6}\log\frac{2+2z}{-1+2z} - \frac{1}{2}\log\left(1-z-2z^2\right), \end{aligned}$$

and the general integral (1.76) can be written, in terms of the auxiliary variable z, as:

$$\frac{5}{6}\log\frac{2+2z}{-1+2z} - \frac{1}{2}\log\left(1-z-2z^2\right) - \log(u) = c. \tag{1.77}$$

Equation (1.77) can be written in the alternative, more compact forms:

$$\begin{aligned}
\frac{5}{6}\log\frac{2+2z}{-1+2z}+\frac{6}{6}\log\left(\frac{1}{1-z-2z^2}\right)^{1/2}-\frac{6}{6}\log(u) &= c,\\
\frac{1}{6}\left[\log\left(\frac{2+2z}{-1+2z}\right)^5+\log\left(\frac{1}{1-z-2z^2}\right)^3-\log(u)^6\right] &= c,\\
\log\left[\left(\frac{2+2z}{-1+2z}\right)^5\left(\frac{1}{1-z-2z^2}\right)^3(u)^{-6}\right] &= c,\\
\left(\frac{2+2z}{-1+2z}\right)^5\frac{1}{(1-z-2z^2)^3}(u)^{-6} &= c,
\end{aligned}
\qquad (1.78)$$

Recalling (1.75), Eq. (1.78) becomes, in terms of the primitive x, y variables:

$$\left(\frac{2+2\frac{y+2}{x-1}}{-1+2\frac{y+2}{x-1}}\right)^5\frac{1}{\left(1-\frac{y+2}{x-1}-2\left(\frac{y+2}{x-1}\right)^2\right)^3}(x-1)^{-6}=c.$$

With a few calculations one obtains the form:

$$\left(\frac{x+y+1}{-x+2y+5}\right)^5\frac{1}{\left[(x-1)(x-y-3)-(y+2)^2\right]^3}=c. \qquad (1.79)$$

Figure 1.10 plots the general surface of equation:

$$Z=\left(\frac{x+y+1}{-x+2y+5}\right)^5\frac{1}{\left[(x-1)(x-y-3)-(y+2)^2\right]^3}. \qquad (1.80)$$

(ii) Let us determine the integration constant which identifies the integral curve through the point $P_1(1,1)$. One obtains $c_1 = -4.286x10^{-5}$. This integral curve has the equation:

$$\left(\frac{x+y+1}{-x+2y+5}\right)^5\cdot\frac{1}{\left[(x-1)(x-y-3)-(y+2)^2\right]^3}+4.286x10^{-5}=0. \qquad (1.81)$$

It is not easy to reduce (1.81) to an explicit form $y=f(x)$. A plot of the particular integral can however be obtained with a numerical method which illustrates the transition from analytical to numerical mathematics (or from continuous to discrete mathematics).

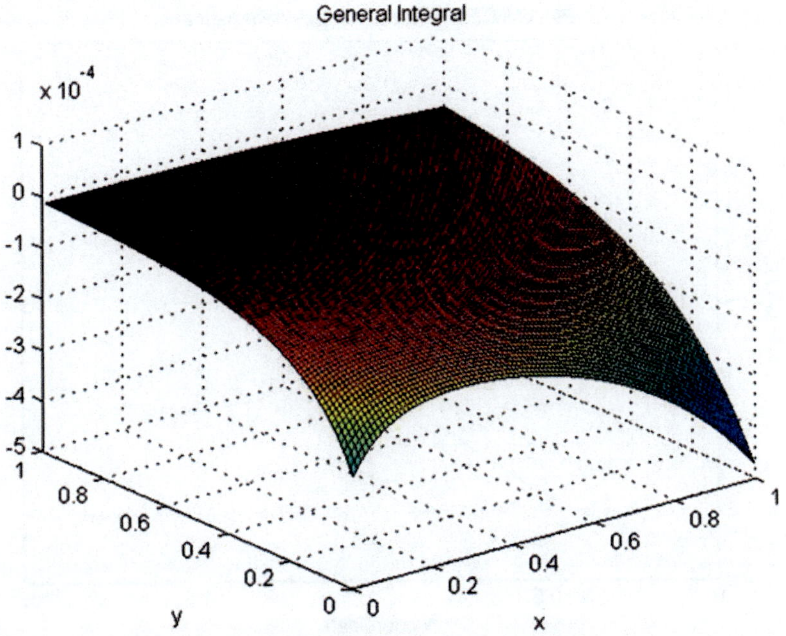

Fig. 1.10 (Example 1.7) Plot of the general surface of equation (1.80)

Because the general surface is smooth and has no ripples, it is appropriate to plot the particular integral computing the coordinates of a series of points with the method of bisections. We used the subroutine RTBIS (RooT with the BI-Section method) given in Press et al. (1986). Given the abscissa $\overline{x}$ of a point of the particular integral, and inserting this value in (1.81), one obtains implicitly an equation of type $f(y, \overline{x}) = 0$, which must be solved by finding its root $\overline{y}$. The bisection method requires that the root be "bracketed," i.e., one estimates the domain ($y_{\mathbf{min}}$, $y_{\mathbf{max}}$) in which the root is expected to be found. This might be the difficulty in the application of the method but the difficulty is alleviated by the knowledge of the general surface. In fact the particular integral can be considered as the projection upon the (x,y) coordinate plane of the cross section of the general surface with a plane $z = c$. With this knowledge bracketing can be made without great difficulties. Using (1.81) in this way, we obtained the plot of the particular integral shown in Fig. 1.11 for the interval $(0 < x < 2)$.

Second Case: The Lines Are Parallel

In this case the integration procedure is simpler because, with a change of variables, the given equation is transformed into an equation with separable variables, without passing through the step of a homogeneous equation. We recall Eq. (1.64):

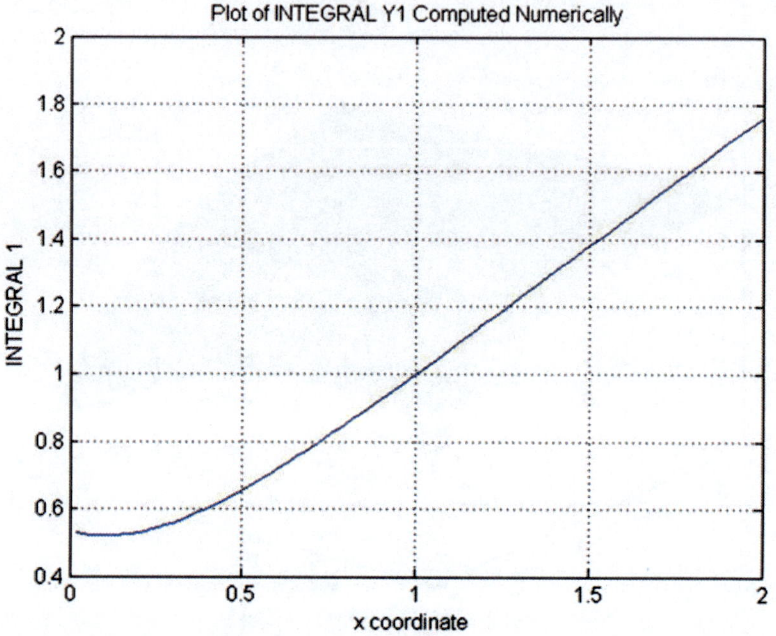

Fig. 1.11 (Example 1.7). Particular integral through the point $P_1(1, 1)$ obtained numerically from Eq. (1.81) with the bisection method

$$y' = f\left(\frac{ax + by + c}{\overline{a}x + \overline{b}y + \overline{c}}\right). \tag{1.82}$$

where now $a\overline{b} - \overline{a}b = 0$ and the coefficients $\overline{a}, \overline{b}$ are proportional to the cefficients a, b:

$$\frac{\overline{a}}{a} = \frac{\overline{b}}{b} = h. \tag{1.83}$$

Introducing a new variable:

$$X = ax + by, \tag{1.84}$$

one derives:

$$\overline{a}x + \overline{b}y = h(ax + by) = hX. \tag{1.85}$$

Differentiating Eq. (1.84) one obtains:

$$dX = adx + bdy, \tag{1.86}$$

$$dx = \frac{dX - bdy}{a}. \tag{1.87}$$

Inserting Eqs. (1.86) and (1.87) into Eq. (1.82) yields:

$$dy = f\left(\frac{X + c}{hX + \overline{c}}\right)dx = f\left(\frac{X + c}{hX + \overline{c}}\right)\frac{dX - bdy}{a}. \tag{1.88}$$

Equation (1.88) is an equation with separable variables which can be integrated straightly.

Example 1.8
Find the general integral of the differential equation:

$$y' = \frac{-3x - 3y + 1}{x + y + 1}, \tag{1.89a}$$

$$(3x + 3y - 1)dx + (x + y + 1)dy = 0. \tag{1.89b}$$

In this case $a\overline{b} - \overline{a}b = (-3)1 - 1(-3) = 0$ and the lines of equations $3x + 3y - 1 = 0$ and $x + y + 1 = 0$ are parallel.

Letting

$$X = 3x + 3y, \tag{1.90}$$

differentiating Eq. (1.90), one obtains:

$$dX = 3dx + 3dy,$$

$$dx = \frac{dX - 3dy}{3}.$$

Inserting Eq. (1.90) in Eq. (1.89b), one obtains:

$$(X - 1)\frac{dX - 3dy}{3} + \left(\frac{X}{3} + 1\right)dy = 0. \tag{1.91}$$

With a few calculations Eq. (1.91) simplifies to the equation with separate variables:

$$\frac{X - 1}{X - 3}dX - 2dy = 0. \tag{1.92}$$

Integrating one obtains:

$$\begin{aligned}
\int \frac{X-1}{X-3} dX - 2\int dy &= c, \\
-\int \frac{1}{X-3} dX + \int \frac{X}{X-3} dX - 2\int dy &= c \qquad \text{[Cfr.Table 1.2, Nr.9]} \\
X + 2\log(X-3) - 2y &= c.
\end{aligned} \tag{1.93}$$

Using (1.90) one returns to the pristine variables:

$$3x + 3y + 2\log(3x + 3y - 3) - 2y = c,$$

and, with a few calculations, one obtains the general integral in the form:

$$3x + y + 2\log(x + y - 1) = c. \tag{1.94}$$

As a proof of the general integral (1.94), we let $f(x, y) = 3x + y + 2\ln(x + y - 1)$ and compute the total differential:

$$df(x, y) = \frac{\partial f}{\partial x} dx + \frac{\partial f}{\partial y} dy = 0, \tag{1.95}$$

with:

$$\frac{\partial f}{\partial x} = 3 + \frac{2}{x + y - 1} = \frac{3x + 3y - 1}{x + y - 1}, \tag{1.96}$$

$$\frac{\partial f}{\partial y} = 1 + \frac{2}{x + y - 1} = \frac{x + y + 1}{x + y - 1}. \tag{1.97}$$

Inserting (1.96) and (1.97) in (1.95) one recovers (1.89b).

Example 1.9
Find the general integral of the differential equation:

$$y' = \frac{x + 2y + 1}{2x + 4y + 3}, \tag{1.98a}$$

$$(x + 2y + 1)dx - (2x + 4y + 3)dy = 0. \tag{1.98b}$$

In this case $ab' - a'b = 1 \cdot 4 - 2 \cdot 2 = 0$ and the lines of equations $x + 2y + 1 = 0$ and $2x + 4y + 3 = 0$ are parallel.

Letting:

$$\begin{aligned} X &= x + 2y, \\ dX &= dx + 2dy, \\ dx &= dX - 2dy, \end{aligned} \tag{1.99}$$

and inserting in Eq. (1.98b), one obtains:

$$(X + 1)(dX - 2dy) - (2X + 3)dy = 0$$

hence, with a few calculations:

$$\begin{aligned} \frac{X+1}{4X+5}dX - dy &= 0, \\ \int \frac{X+1}{4X+5}dX - y &= c \end{aligned} \tag{1.100}$$

The first integral in (1.100) can be split in the two following integrals:

$$\int \frac{X}{4X+5}dX = \frac{X}{4} - \frac{5}{16}\log(4X+5), \quad \text{[Cfr.Table I.2, Nr.9]} \tag{1.101}$$

$$\int \frac{1}{4X+5}dX = \frac{1}{4}\log(4X+5). \quad \text{[Cfr.Table I.2, Nr.8]} \tag{1.102}$$

Inserting (1.101) and (1.102) into (1.100), one obtains, with a few calculations:

$$X - \frac{1}{4}\log(4X+5) - 4y = c.$$

Hence, recalling (1.99), one obtains the general integral:

$$\begin{aligned} x + 2y - \frac{1}{4}\log(4x+8y+5) - 4y &= c, \\ \log(4x+8y+5) - 4x + 8y &= c. \end{aligned} \tag{1.103}$$

As a proof of the general integral (1.103), we let $f(x, y) = \log(4x + 8y + 5) - 4x + 8y$ and compute the total differential:

$$df(x,y) = \frac{\partial f}{\partial x}dx + \frac{\partial f}{\partial y}dy = 0, \tag{1.104}$$

with:

$$\frac{\partial f}{\partial x} = \frac{4}{4x + 8y + 5} - 4 = \frac{-16x - 32y - 16}{4x + 8y + 5}, \tag{1.105}$$

$$\frac{\partial f}{\partial y} = \frac{8}{4x + 8y + 5} + 8 = \frac{32x + 64y + 48}{4x + 8y + 5}. \tag{1.106}$$

Inserting (1.105) and (1.106) in (1.104), one recovers (1.98b).

1.1.2.7 Linear Differential Equations

An ordinary differential equation of first order is linear when it is of first degree with respect to the unknown function $y(x)$ and its first derivative. It can be written in the general form:

$$y' + p(x)y = q(x), \tag{1.107}$$

where we assume that the functions $p(x)$ and $q(x)$ are continuous in the domain of definition. If $q(x) = 0$ the equation is homogeneous; otherwise it is non-homogeneous.

Letting $\int p(x)dx$ be a primitive of the function $p(x)$, Eq. (1.107) is equivalent to:

$$e^{\int p(x)dx}[y' + p(x)y] = q(x)e^{\int p(x)dx}, \tag{1.108}$$

hence equivalent to the equation:

$$\frac{d}{dx}\left[ye^{\int p(x)dx}\right] = q(x)e^{\int p(x)dx}. \tag{1.109}$$

Integrating Eq. (1.109) one obtains:

$$ye^{\int p(x)dx} = \int q(x)e^{\int p(x)dx}dx + c, \tag{1.110}$$

hence the general solution:

$$y = e^{-\int p(x)dx}\left[\int q(x)e^{\int p(x)dx}dx + c\right]. \tag{1.111}$$

Example 1.10
Let us consider the differential equation:

$$y' + y = \cos x. \tag{1.112}$$

Find:

(i) *The general integral.*
(ii) *The particular integral through the point* $P_1(\pi/4, 1)$.

(i) Equation (1.112) corresponds to the general type (1.107) with $p(x) = 1$ and $q(x) = \cos x$. According to Eq. (1.111), its general integral is:

$$y = e^{-\int dx}\left[\int \cos x e^{\int dx} dx + c\right] = e^{-x}\left[\int \cos x \; e^x dx + c\right]. \tag{1.113}$$

The integral in the square brackets is computed easily integrating twice by parts taking e^x as finite factor. The result is $\int \cos x \; e^x dx = \frac{e^x}{2}(\sin x + \cos x)$, which inserted in Eq. (1.113) gives:

$$y = \frac{1}{2}(\sin x + \cos x) + ce^{-x}. \text{ [Cfr.Table 1.2, Nr.12]} \tag{1.114}$$

(ii) The constant which determines the sought particular integral is given by:

$$1 = \frac{1}{2}\Big(\sin(\pi/4) + \cos(\pi/4) + c_1 e^{-\pi/4},$$
$$c_1 = \left(1 - \sqrt{2/2}\right)e^{\pi/4} = 0.6424.$$

Hence the particular integral is:

$$y = \frac{1}{2}(\sin x + \cos x) + \left(1 - \sqrt{2}/2\right)e^{\pi/4}e^{-x}. \tag{1.115}$$

The particular integral (1.115) is shown in Fig. 1.12. The exponential factor, hence the constant c_1, becomes irrelevant for large values of x.

1.1.2.8 Bernoulli Equation

A Bernoulli equation is of the general type:

$$y' + p(x)y = q(x)y^n, \tag{1.116}$$

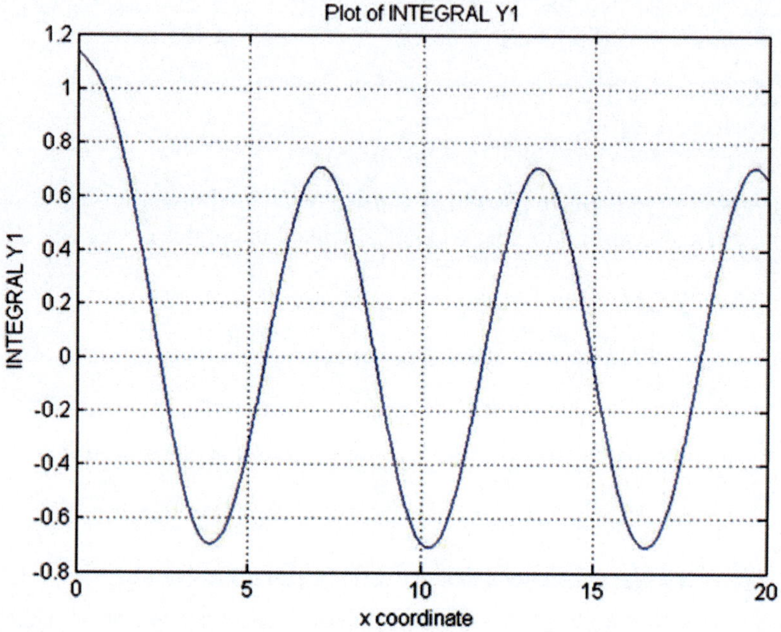

Fig. 1.12 (Example 1.10) Particular integral (1.115) with $c = 0.6424$ passing through the point P1 ($\pi/4$, 1)

where we assume that the functions $p(x)$ and $q(x)$ are continuous in the definition domain and that n is a constant different from zero and one. (If $n = 0$, Eq. (1.116) is linear; if $n = 1$ it is of separable variables type).

Equation (1.116) can be transformed into a linear equation with the change of variable $y^{1-n} = z$, from which one derives:

$$y = z^{\frac{1}{1-n}}, \tag{1.117}$$

$$y' = \frac{1}{1-n} z^{\frac{n}{1-n}} z'. \tag{1.118}$$

With these changes of variables Eq. (1.116) becomes:

$$\frac{1}{1-n} z^{\frac{n}{1-n}} z' + p(x) z^{\frac{1}{1-n}} = q(x) z^{\frac{n}{1-n}}. \tag{1.119}$$

Dividing by $\frac{1}{1-n} z^{\frac{n}{1-n}}$ one obtains:

$$z' + (1-n)p(x)z = (1-n)q(x), \tag{1.120}$$

which is a linear equation. It can be solved using formula (1.111) which yields:

$$z = e^{-(1-n)\int p(x)dx}\left[\int (1-n)q(x)e^{\int (1-n)p(x)dx}dx + c\right] \tag{1.121}$$

hence:

$$y = z^{\frac{1}{1-n}} = \left\{e^{-(1-n)\int p(x)dx}\left[\int (1-n)q(x)e^{\int (1-n)p(x)dx}dx + c\right]\right\}^{\frac{1}{1-n}}. \tag{1.122}$$

Example 1.11

Find the general integral of the equation $y' + \frac{y}{x} = -xy^2$.

Because $n = 2$, we set:

$$y = \frac{1}{z}, \tag{1.123}$$

$$y' = -\frac{1}{z^2}z'. \tag{1.124}$$

Inserting (1.123) and (1.124) in the given equation one obtains:

$$\left(-\frac{1}{z^2}\right)z' + \frac{1}{x}\frac{1}{z} = -x\frac{1}{z^2},$$

and dividing every term by $(-1/z^2)$

$$z' - \frac{1}{x}z = x. \tag{1.125}$$

(1.125) is a linear differential equation of the type (1.107) with $p(x) = -1/x$, $q(x) = x$. Using formula (1.111) one obtains:

$z = e^{-\int p(x)dx}[\int q(x)e^{\int p(x)dx}dx + c] = e^{\ln x}[\int xe^{-\ln x}dx + c] = x[\int xx^{-1}dx + c] = x^2 + cx$

Hence, from (1.123), one obtains the general integral $y = \frac{1}{x^2+cx}$.

1.1.2.9 Integrating Factor

Let us assume that the differential equation:

$$M(x,y)dx + N(x,y)dy = 0 \tag{1.126}$$

is not exact, which is equivalent to say that $\partial M/\partial y \neq \partial N/\partial x$.

A theorem of analysis states that if the expression:

$$\frac{\frac{\partial M(x,y)}{\partial y} - \frac{\partial N(x,y)}{\partial x}}{N(x,y)} \tag{1.127}$$

is a function, say $g(x)$, of the only variable x (or a function of the only variable y), there exists a function $\mu(x)$ of the only variable x (or of the only variable y), such that the differential:

$$\mu(x)M(x,y)dx + \mu(x)N(x,y)dy \tag{1.128}$$

is exact. Then the function $\mu(x)$ is said to be an integrating factor.

According to Eq. (1.9), for (1.128) to be an exact differential, it must hold:

$$\frac{\partial[\mu(x)M(x,y)]}{\partial y} = \frac{\partial[\mu(x)N(x,y)]}{\partial x}, \tag{1.129}$$

$$\mu(x)\frac{\partial M(x,y)}{\partial y} = N(x,y)\frac{d\mu(x)}{dx} + \mu(x)\frac{\partial N(x,y)}{\partial x}, \tag{1.130}$$

$$\frac{1}{\mu(x)}\frac{d\mu(x)}{dx} = \frac{\frac{\partial M(x,y)}{\partial y} - \frac{\partial N(x,y)}{\partial x}}{N(x,y)} = g(x) \text{ (by assumption)}, \tag{1.131}$$

$$\frac{1}{\mu(x)}d\mu = g(x)dx. \tag{1.132}$$

Integrating Eq. (1.132) one obtains the integrating factor:

$$\mu(x) = e^{\int g(x)dx}. \tag{1.133}$$

Conversely, if the expression:

$$\frac{\frac{\partial N(x,y)}{\partial x} - \frac{\partial M(x,y)}{\partial y}}{M(x,y)} \tag{1.134}$$

is a function, say $h(y)$, of the only variable y, Eq. (1.126) has the integrating factor

$$\mu(y) = e^{\int h(y)dy}. \tag{1.135}$$

Example 1.12
Given the differential equation:

$$(3xy + y^2)dx + (x^2 + xy)dy = 0, \tag{1.136}$$

which is not exact, find an integrating factor and then the general integral.
Applying formula (1.127), one obtains a function of the only variable x:

$$\frac{\frac{\partial M(x,y)}{\partial y} - \frac{\partial N(x,y)}{\partial x}}{N(x,y)} = \frac{3x + 2y - (2x + y)}{x^2 + xy} = \frac{x + y}{x(x + y)} = \frac{1}{x} = g(x).$$

Therefore an integrating factor is:

$$\mu(x) = e^{\int g(x)dx} = e^{\int \frac{1}{x}dx} = e^{\ln(x)} = x.$$

The equation

$$x(3xy + y^2)dx + x(x^2 + xy)dy = 0$$

must be exact. In fact, setting $M_b(x, y) = x(3xy + y^2)$ and $N_b(x, y) = x(x^2 + xy)$, one obtains:

$$\frac{\partial M_b(x,y)}{\partial y} = \frac{\partial N_b(x,y)}{\partial x} = 3x^2 + 2xy.$$

Choosing a point $P(x_0, y_0) = P(0, 1)$ and using Eq. (1.10) the general integral is:

$$\begin{aligned}
F(x,y) &= \int_{x_0}^{x} M_b(x,y)dx + \int_{y_0}^{y} N_b(x_0,y)dy + c, \\
F(x,y) &= \int_{0}^{x} (3x^2y + xy^2)dx + \int_{1}^{y} 0dy + c, \\
F(x,y) &= x^3y + \frac{1}{2}x^2y^2 + c.
\end{aligned}$$

It is easy to verify that computing the total differential: $dF(x,y) = \frac{\partial F}{\partial x}dx + \frac{\partial F}{\partial y}dy = 0$, one obtains the given differential equation (1.136).

Ordinary Differential Equations of first order which cannot be written in the normal form (1.2): Sections 1.2.10 through 1.2.13.

1.1.2.10 Riccati Equation

The Riccati equation has the form:

$$y' + A(x)y = B(x)y^2 + C(x). \tag{1.137}$$

We assume that $B(x) \neq 0$ and $C(x) \neq 0$; otherwise one would fall in the following cases:

- If $B(x) = 0$ one has a linear non homogeneous equation; [Cfr. Sect. 1.1.2.7].
- If $C(x) = 0$ one has a Bernoulli equation with $n = 2$; [Cfr. Sect. 1.1.2.8].
- If both $B(x) = 0$ and $C(x) = 0$ one has a linear homogeneous equation. [Cfr. Sect. 1.1.2.7].

The integration of the Riccati equation is possible if a particular integral is known. In such a case, the integration goes through the procedure of transforming the Riccati equation in a Bernoulli equation which in turn transforms in a linear equation. Let us assume that y_1 is a particular integral of the given equation. It therefore satisfies Eq. (1.137):

$$y_1' + A(x)y_1 = B(x)y_1^2 + C(x). \tag{1.138}$$

Subtracting Eq. (1.138) from (1.137), one obtains:

$$y' - y_1' + A(x)(y - y_1) = B(x)\left(y^2 - y_1^2\right). \tag{1.139}$$

Notice that the function $C(x)$ is at first eliminated, but it reenters into the solution via the particular integral y_1 and its derivative.

We introduce an auxiliary variable:

$$z = y - y_1 \tag{1.140}$$

and derive:

$$z' = y' - {y_1}', \tag{1.141}$$

$$\begin{aligned} y^2 - {y_1}^2 &= (y - y_1)(y + y_1) = (y - y_1)(y - y_1 + 2y_1) = z(z + 2y_1) \\ &= z^2 + 2zy_1. \end{aligned} \tag{1.142}$$

Introducing Eqs. (1.140)–(1.142) into Eq. (1.139), one obtains:

$$z' + A(x)z = B(x)\left(z^2 + 2zy_1\right),$$

$$z' + [A(x) - 2y_1B(x)]z = B(x)z^2. \tag{1.143}$$

Equation (1.143) is a Bernoulli equation [Cfr. Eq. 1.116] with $n = 2$. It can be transformed into a linear equation [Cfr. Eq. (1.120)] with the procedure explained in Sect. 1.1.2.8, which leads to the solution given by Eqs. (1.121) and (1.122). To illustrate once more the procedure, we define:

$$P(x) = A(x) - 2y_1B(x), \tag{1.144}$$

$$Q(x) = B(x). \tag{1.145}$$

With these definitions Eq. (1.143) becomes:

$$z' + P(x)z = Q(x)z^2. \tag{1.146}$$

We now introduce the auxiliary variable $t = z^{1-n}$, with $n = 2$, hence:

$$z = \frac{1}{t}, \tag{1.147}$$

$$z' = -\frac{1}{t^2}t'. \tag{1.148}$$

Introducing Eqs. (1.147) and (1.148) into Eq. (1.146) one obtains:

$$-\frac{1}{t^2}t' + P(x)\frac{1}{t} = Q(x)\frac{1}{t^2}, \tag{1.149}$$

and dividing every member by $-1/t^2$ yields:

$$t' - P(x)t = -Q(x), \tag{1.150}$$

which is a linear equation. Its solution is:

$$t = e^{-\int P(x)dx}\left[-\int Q(x)e^{\int P(x)dx}dx + c\right], \tag{1.151}$$

$$t = e^{-\int [A(x)-2y_1B(x)]dx}\left[-\int B(x)e^{\int [A(x)-2y_1B(x)])dx}dx + c\right]. \tag{1.152}$$

Passing from the auxiliary variable t to the variable $z = 1/t$ one has:

$$z = \left\{e^{-\int [A(x)-2y_1B(x)]dx}\left[-\int B(x)e^{\int [A(x)-2y_1B(x)])dx}dx + c\right]\right\}^{-1}. \tag{1.153}$$

Hence we obtain:

$$\begin{aligned} y &= z + y_1 \\ &= \left\{ e^{-\int [A(x) - 2y_1 B(x)]dx} \left[-\int B(x) e^{\int [A(x) - 2y_1 B(x)])dx} dx + c \right] \right\}^{-1} + y_1, \end{aligned} \tag{1.154}$$

which is the general integral of the Riccati equation. The function $C(x)$ enters into the solution (1.154) through the particular integral y_1.

Example 1.13
Given the differential equation:

$$y' - \frac{2}{x} y = -y^2 - \frac{2}{x^2}, \tag{1.155}$$

and its particular integral:

$$y_1 = 1/x, \tag{1.156}$$

find the general integral.

Equation (1.155) is a Riccati equation with $A(x) = -2/x$, $B(x) = -1$, $C(x) = -2/x^2$. The particular integral satisfies Eq. (1.155):

$$y_1' - \frac{2}{x} y_1 = -y_1^2 - \frac{2}{x^2}, \tag{1.157}$$

hence subtracting (1.157) from (1.155), one obtains:

$$y' - y_1' - \frac{2}{x}(y - y_1) = -(y^2 - y_1^2). \tag{1.158}$$

Introducing the variable $z = y - y_1$, and $z' = y' - y_1'$, $y^2 - y_1^2 = z^2 + 2zy_1$ in (1.158), one has:

$$\begin{aligned} z' - \frac{2}{x} z &= -(z^2 + 2zy_1), \\ z' + z\left[-\frac{2}{x} + 2y_1\right] &= -z^2. \end{aligned} \tag{1.159}$$

Equation (1.159) is a Bernoulli equation with $P(x) = -2/x + 2y_1$, $Q(x) = -1$ and $n = -2$. Because in this particular case $P(x) = 0$, (1.159) can be integrated immediately:

$$z' = \frac{dz}{dx} = -z^2; \; -\frac{dz}{z^2} = dx; \; -\int \frac{dz}{z^2} = \int dx + c; \; \frac{1}{z} = x + c; z = \frac{1}{x + c}.$$

However, ignoring that $P(x) = 0$, we want to apply the procedure to obtain from the Bernoulli equation (1.159) a linear equation which can then be integrated by applying formula (1.111).

Let us introduce the variable t defined by $z^{1-n} = z^{-1} = t$; $z = \frac{1}{t}$; $z' = -\frac{1}{t^2}t'$. With this new variable t, Eq. (1.159) becomes:

$$-\frac{1}{t^2}t' + \frac{1}{t}P(x) = \frac{1}{t^2}Q(x).$$

Dividing every term by $-1/t^2$, one obtains:

$$t' - tP(x) = -Q(x). \tag{1.160}$$

Equation (1.160) is a linear equation and its general integral is (recalling that $P(x) = 0$, $Q(x) = -1$):

$$t = e^{-\int P(x)dx}\left[-\int Q(x)e^{\int P(x)dx}dx + c\right] = \int dx + c = x + c.$$

Hence, in terms of the variable z:

$$z = \frac{1}{t} = \frac{1}{x+c},$$

which is the result that had been obtained above by direct integration. Recalling that $y = z + y_1$ one obtains the general integral of the given Riccati equation (1.155):

$$y = \frac{1}{x+c} + \frac{1}{x}. \tag{1.161}$$

An overview of the integration methods explained in the Sects. 1.2.1 through 1.2.10 is given in Table 1.1.

1.1.2.11 Equations of the Form $x = f(y')$ or $y = f(y')$

The general integral of equations of these types can be found in parametric form, with the independent variable x and the dependent variable y being functions of a parameter $p : x = x(p)$, $y = y(p)$. In some cases it is possible to eliminate the parameter p and obtain the general integral in the usual form $y = f(x) + c$.

Let us refer to the first form of the differential equation:

$$x = f(y'). \tag{1.162}$$

Let:

Table 1.1 Overview of the integration methods for ordinary differential equations of first order treated in Sects. (1.2.1) through (1.2.10)

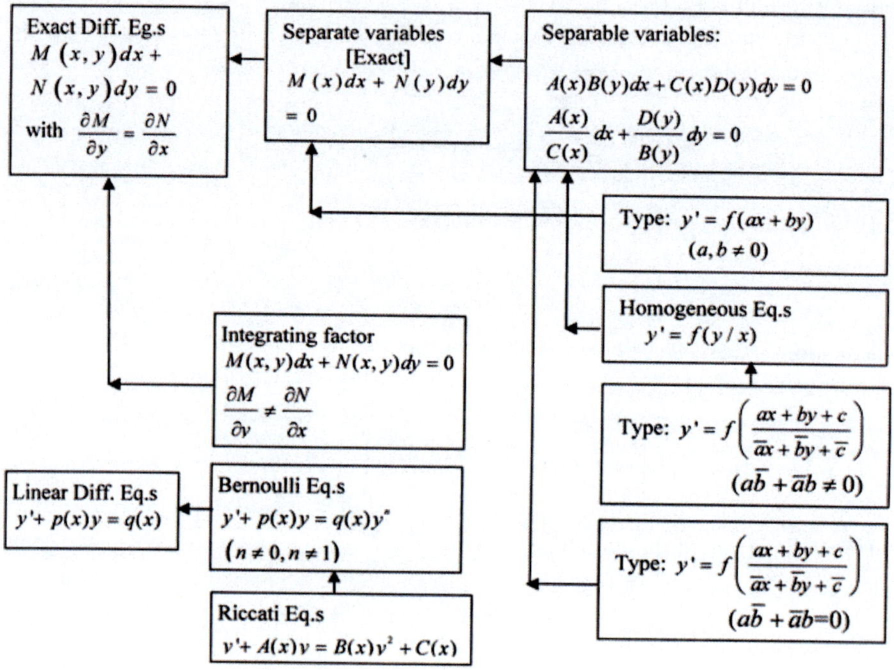

$$y' = p, \quad dy = pdx, \quad dx = dy/p. \tag{1.163}$$

From Eq. (1.162) we have:

$$x = f(p), \tag{1.164}$$

$$dx = f'(p)dp. \tag{1.165}$$

From the third of Eq. (1.163), one has:

$$dy = pdx = pf'(p)dp, \tag{1.166}$$

$$y = \int pf'(p)dp + c. \tag{1.167}$$

Equations (1.164) and (1.167) are the general integral of Eq. (1.137) in parametric form. In some cases, as it will be shown in the following example, it is possible to eliminate the parameter p to obtain the general integral in the form $y = f(x) + c$.

Example 1.14
Find the general integral of the equation:

$$x = \sin(y'), \tag{1.168}$$

(i) *Find the particular integral passing through the point* $P_1\left(\sqrt{2}/2, \sqrt{2}/2\right)$..

(i) Let:

$$y' = p, \quad dy = pdx, \quad dx = dy/p. \tag{1.169}$$

From (1.168) one has:

$$\begin{aligned} x &= \sin(p), \\ dx &= \cos(p)dp, \end{aligned} \tag{1.170}$$

and from the third of (1.169)

$$dy = pdx = p\cos(p)dp,$$

$$y = \int p\cos(p)dp + c = p\sin(p) + \cos(p) + c. \text{ [Cfr.Table 1.2, Nr.4]} \tag{1.171}$$

Equations (1.170) and (1.171) are the general integral in parametric form. In this case it is easy to eliminate the parameter $p = \arcsin(x)$ substituting it in (1.158):

$$\begin{aligned} y = (\arcsin x)\sin(\arcsin x) + \cos(\arcsin x) + c &= x(\arcsin x) + \cos(\arcsin x) + c \\ &= x(\arcsin x) + \sqrt{1 - x^2} + c. \end{aligned} \tag{1.172}$$

Deriving (1.172) one obtains $y' = \arcsin x$ which matches the given differential equation (1.168).

(ii) Setting $x = y = \sqrt{2}/2$ in (1.172) one obtains the value of the constant which determines the particular integral through the point $P_1\left(\sqrt{2}/2, \sqrt{2}/2\right)$. One obtains $c_1 = -0.555$. The particular integral is therefore:

$$y = x(\arcsin x) + \sqrt{1 - x^2} - 0.555. \tag{1.173}$$

Figure 1.13 shows a plot of the particular integral in the domain $(0 < x < 1)$.

Example 1.15
(i) *Find the general integral of the equation:*

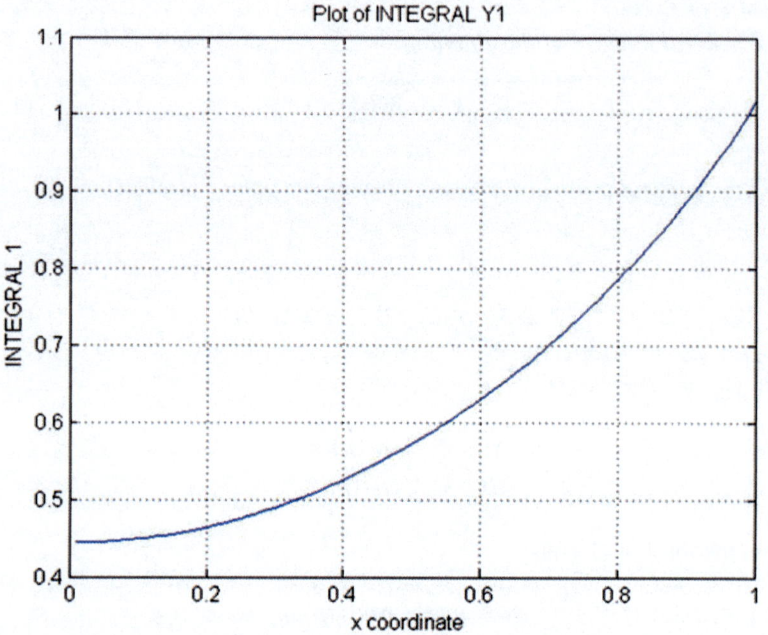

Fig. 1.13 Example (1.14). Plot of the particular integral through the point $P_1\left(\sqrt{2}/2, \sqrt{2}/2\right)$

$$y^2 = 1 - y'^2, \tag{1.174}$$

(ii) *Find the particular integral passing through the point* $P_1(0, 1/2)$.

(i) Let

$$y' = p, \quad dy = pdx, \quad dx = dy/p. \tag{1.175}$$

From (1.174) one has:

$$\begin{aligned} y &= \sqrt{1-p^2}, \\ dy &= -\frac{pdp}{\sqrt{1-p^2}}, \end{aligned} \tag{1.176}$$

and from the third of (1.175)

$$
\begin{aligned}
dx &= \frac{dy}{p} = -\frac{1}{\sqrt{1-p^2}}dp, \quad \text{[Cfr.Table I.2, Nr.13]} \\
x &= -\int \frac{1}{\sqrt{1-p^2}}dp + c = -\arcsin p + c.
\end{aligned}
\tag{1.177}
$$

Equations (1.176) and (1.177) are the general integral in parametric form. The parameter p can be easily eliminated: from (1.177) one has $p = \sin(-x + c)$ and introducing in (1.176) one obtains the general integral in explicit form:

$$
y = \sqrt{1 - \sin^2(-x + c)}. \tag{1.178}
$$

Computing the derivative of (1.178):

$$
y' = \frac{1}{\sqrt{1 - \sin^2(-x + c)}} \sin(-x + c) \cos(-x + c) \tag{1.179}
$$

and introducing the square of both (1.178) and (1.79) in (1.174), one proves that the given differential equation is satisfied.

(ii) Setting $x = 0$, $y = 1/2$ in (1.178), one obtains $c_1 = \pi/3$. The particular integral sought is then

$$
y = \sqrt{1 - \sin^2(-x + \pi/3)}. \tag{1.180}
$$

Figure 1.14 shows a plot of the particular integral in the domain $(0 < x < 1)$.

1.1.2.12 Clairaut Equation

The Clairaut Equation has the form:

$$
y = xy' + f(y'). \tag{1.181}
$$

Its general integral can be found with the same parametric procedure applied in the previous section. Letting $y' = p$ one obtains:

$$
y = xp + f(p), \tag{1.182}
$$

$$
y' = p + xp' + f'(p)p'. \tag{1.183}
$$

Because $y' = p$ one derives from Eq. (1.183):

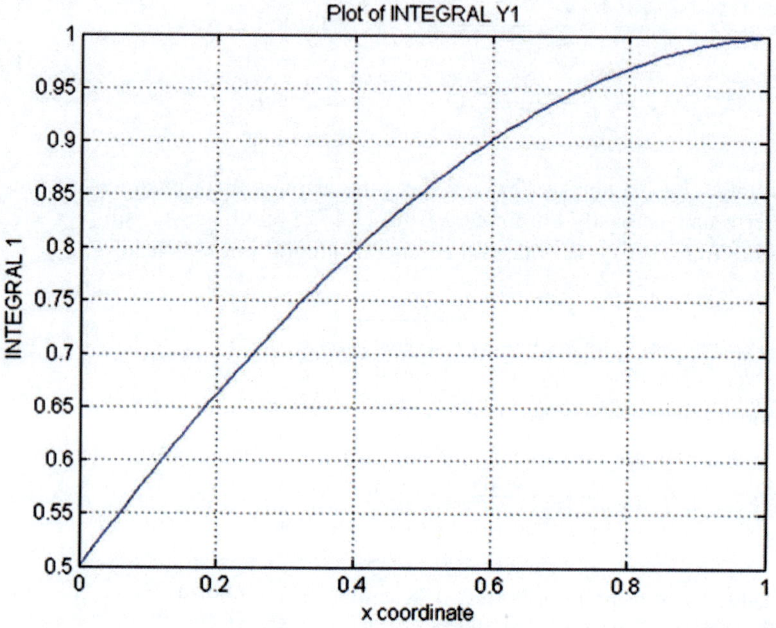

Fig. 1.14 Example (1.15). Plot of the particular integral through the point $P_1(0, 1/2)$

$$0 = p'[x + f'(p)]. \tag{1.184}$$

Equation (1.184) has two solutions: the first is $p' = 0$; hence $p = c$, which, inserted in Eq. (1.181), gives the general integral:

$$y = xc + f(c). \tag{1.185}$$

Remark that the general integral can be obtained immediately from the given equation replacing y' with the constant c.

The second solution of Eq. (1.184) is:

$$x = -f'(p), \tag{1.186a}$$

which leads to a singular integral. This Eq. (1.186a) can also be obtained by deriving Eq. (1.181) with respect to y', which yields:

$$0 = x + \frac{d}{dy'}[f(y')] = x + f'(p) = x + f'(y'). \tag{1.186b}$$

The singular integral is then obtained by eliminating y' between the given Eqs. (1.181) and (1.186b).

As will be shown in the following example, the general integral consists of a family of straight lines tangent to the curve of the singular integral which is an "enveloping curve".

Example 1.16
(i) *Out of the family of straight lines which form the general integral of the equation:*

$$y = xy' + \frac{1}{y'}, \tag{1.187}$$

find those which pass through the points $P_1(4,4)$, $P_2(9,6)$, $P_3(16,8)$.

(ii) *Find the singular integral and show that the three straight lines are tangent to it in the given points.*

(i) The general integral has equation:

$$y = xc + \frac{1}{c}. \tag{1.188}$$

The straight line through the point $P_1(4,4)$ is found inserting $x = 4$, $y = 4$ in (1.187) and solving for c. One finds $c_1 = 1/2$. The equation of the straight line is therefore
$y_1 = \frac{x}{2} + 2$. Similarly one finds $c_2 = 1/3$, $c_3 = 1/4$ and correspondingly $y_2 = \frac{x}{3} + 3$, $y_3 = \frac{x}{4} + 4$.

The singular integral is obtained (letting $p = y'$) from:

$$x = -f'(p) = \frac{1}{p^2}, \tag{1.189}$$

hence $p = y' = 1/\sqrt{x}$ and, integrating:

$$y = \int \frac{1}{\sqrt{x}} dx = 2\sqrt{x}. \tag{1.190}$$

The three straight lines which belong to the family of the general integral and the singular integral (1.190) are plotted in Fig. 1.15.

Example 1.17
Given the equation:

$$y = xy' + \cos(y') \tag{1.191}$$

find:

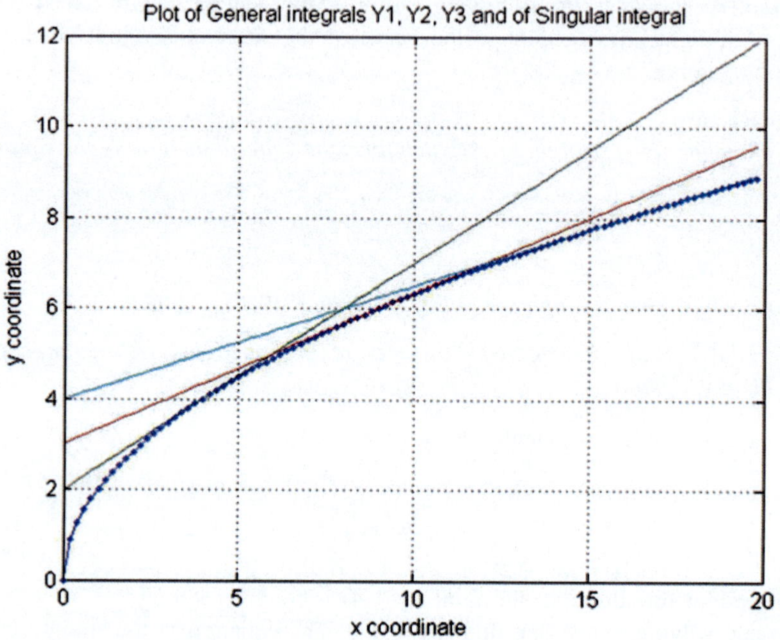

Fig. 1.15 Example 1.16. Plot of the singular integral $y = 2\sqrt{x}$ (point-dotted plot) and of three straight lines belonging to the family of the general integral. The lines with ordinates for $x = 0$ equal to 2, 3, 4, respectively, are tangent to the singular integral in the points $P_1(4, 4)$, $P_2(9, 6)$, $P_3(16, 8)$

(i) *The general integral;*
(ii) *The singular integral;*
(iii) *Consider 3 points of the singular integral with abscissas* $x = 0.4$, $x = 0.6$, $x = 0.8$ *and find the lines, belonging to the general integral, passing through these points.*

(i) The general integral is:

$$y = xc + \cos(c) \tag{1.192}$$

and forms a family of straight lines tangent to the singular integral.

(ii) The singular integral is obtaining eliminating y' between (1.191) and its derivative with respect to y':

$$0 = x - \sin(y'),$$

hence:

$$y' = \arcsin x. \tag{1.193}$$

Introducing (1.193) into (1.191), one obtains the singular integral:

$$y = x \cdot \arcsin x + \cos(\arcsin x) = x \cdot \arcsin x + \sqrt{1 - x^2}. \tag{1.194}$$

(iii) Using (1.194) the ordinates of the 3 points with abscissas $x = 0.4$, $x = 0.6$, $x = 0.8$ are obtained. The coordinates are then $P_1(0.4, 1.081)$, $P_2(0.6, 1.186)$, $P_3(0.8, 1.342)$, which we denote $P_i(x_i, y_i)$, $i = 1,2,3$.

The equations of the general integral lines tangent to the singular integral in these points $P_i(x_i, y_i)$ are:

$$Y_i = X_i c_i + \cos(c_i), \tag{1.195}$$

where $c_i\,(i = 1, 2, 3)$ are the roots of the equations:

$$y_i = x_i c + \cos(c). \tag{1.196}$$

The roots of (1.196) can be found numerically, but there is also an analytical way to compute them. In fact the lines tangent to the singular integral in the points $P_i(x_i, y_i)$ can be written:

$$Y_i = y_i + (X - x_i)\arcsin(x_i) \quad (i = 1, 2, 3) \tag{1.197}$$

The ordinates at the origin ($X = 0$) of (1.197) are $Y0_1 = 0.916$, $Y0_2 = 0.800$, $Y0_3 = 0.600$. Hence from (1.192), for $x = 0$, one derives $c_1 = \arccos(Y0_1) = 0.413$, $c_2 = \arccos(Y0_2) = 0.643$, $c_3 = \arccos(Y0_3) = 0.927$.

Figure 1.16 plots the singular integral (1.194) and the three lines (1.195) of the general integral, tangent to the singular integral in the points $P_i(x_i, y_i)$.

1.1.2.13 D'Alembert – Lagrange Equation

The D'Alembert – Lagrange Equation has the form:

$$y = xf(y') + g(y'), \tag{1.198}$$

where one assumes $f(y') \neq y'$; otherwise one would have the Clairaut equation.

Its general integral can be found with the same parametric procedure applied previously. Letting $y' = p$ one obtains:

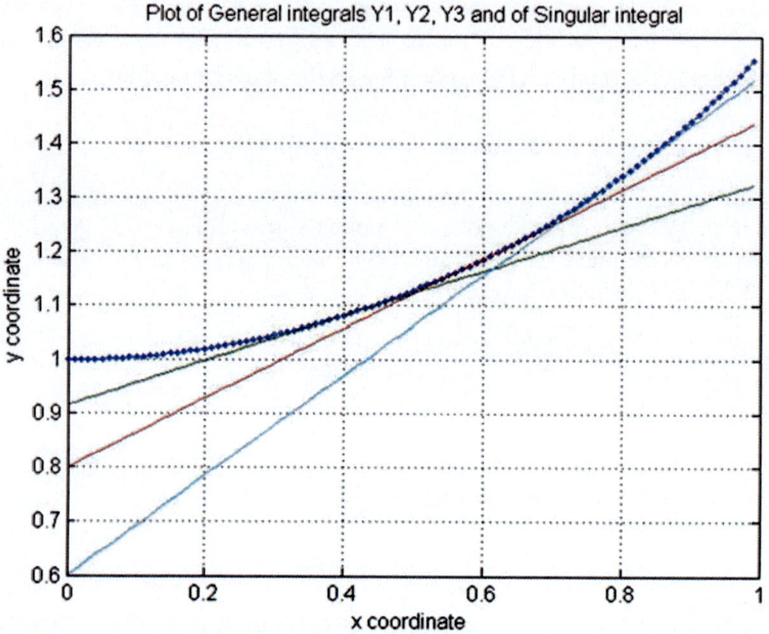

Fig. 1.16 (Example 1.17). Plots of the singular integral $y = x \arcsin x + \sqrt{1 - x^2}$ (point-dotted plot) and of the three lines (1.195) of the general integral, tangent to the singular integral in the points $P_1(0.4, 1.081)$, $P_2(0.6, 1.186)$, $P_3(0.8, 1.342)$

$$y = xf(p) + g(p), \tag{1.199}$$

$$y' = f(p) + xf'(p)p' + g'(p)p'. \tag{1.200}$$

Because $y' = p$ one derives:

$$p - f(p) = [xf'(p) + g'(p)]\frac{dp}{dx}, \tag{1.201}$$

$$x' = \frac{dx}{dp} = x\frac{f'(p)}{p - f(p)} + \frac{g'(p)}{p - f(p)}. \tag{1.202}$$

Equation (1.202) is a linear non-homogeneous differential equation with p as independent variable and x as dependent variable [Cfr. Eq. (1.107)]. Letting $P(p) = \frac{f'(p)}{p - f(p)}$, $Q(p) = \frac{g'(p)}{p - f(p)}$, the general integral of Eq. (1.202) is:

$$x = e^{-\int P(p)dp}\left[\int Q(p)e^{\int P(p)dp}dp + c\right], \tag{1.203}$$

hence from (1.199):

$$y = e^{-\int P(p)dp}\left[\int Q(p)e^{\int P(p)dp}dp + c\right]f(p) + g(p). \tag{1.204}$$

Equations (1.203) and (1.204) are the general integral of Eq. (1.198) in parametric form.

Example 1.18
Find the general integral of the equation:

$$y = x(y' + 1) + (y')^2. \tag{1.205}$$

Letting $y' = p$, $dy = pdx$, one has:

$$y = x(p + 1) + p^2, \tag{1.206}$$

$$\begin{aligned} y' &= (p + 1) + xp' + 2pp', \\ p &= (p + 1) + xp' + 2pp', \\ 0 &= 1 + (x + 2p)\frac{dp}{dx}, \\ \frac{dx}{dp} + x &= -2p. \end{aligned} \tag{1.207}$$

Defining $P(p) = 1, Q(p) = -2p$, the integral of Eq. (1.189) is [Cfr. Eq. (1.203)]:

$$x = e^{-p}\left[-2\int pe^p dp + c\right] = -2p + 2 + ce^{-p}. \ \text{[Cfr.Table I.2, Nr.1]} \tag{1.208}$$

Inserting (1.208) in (1.206) one has, after a few calculations:

$$y = -p^2 + 2 + ce^{-p}(p + 1). \tag{1.209}$$

Equations (1.208) and (1.209) are the general integral of Eq. (1.205) in parametric form.

Table 1.2 summarizes the indefinite integrals which have been used in the examples of Sect. 1.1.2.

Acknowledgments The results of Examples 1.7, 1.8, and 1.11 are given, without proof, in Demidovic (1989).

1.1.3 Peano-Picard Solution Method

Up to this point we have considered ordinary differential equations of the first order which can be integrated analytically, thus obtaining exact solutions. In practice this is possible only in particular cases because usually the coefficients of the differential equations are not constant. When analytical solutions cannot be obtained, it is necessary to look for numerical solutions to approximate as closely as possible the exact, unknown analytical solutions. In this section and in the following ones, we consider numerical methods to obtain approximate solutions.

The Peano-Picard solution method consists in replacing the differential Eq. (1.2) $y' = f(x, y)$ with the initial condition (1.3) $y_0 = y(x_0)$, with the integral equation:

$$y = y_0 + \int_{x_0}^{x} f(x, y)dx, \tag{1.210}$$

which can be solved iteratively with the following algorithm:

$$y_1 = y_0 + \int_{x_0}^{x} f(x, y_0)dx, \tag{1.211a}$$

$$y_2 = y_0 + \int_{x_0}^{x} f(x, y_1)dx, \tag{1.211b}$$

$$y_{r+1} = y_0 + \int_{x_0}^{x} f(x, y_r)dx, \tag{1.211c}$$

where r represents an iteration index.

The numerical scheme represented by Eq. (1.211a, 1.211b, and 1.211c) is known to converge to the exact solution of Eq. (1.2) if the function $f(x, y)$ is continuous in a domain around the point $P(x_0, y_0)$ and in this domain it is a Lipschitz function, satisfying the conditions:

$$|f(x, y_1) - f(x, y_2)| < k|y_1 - y_2| \tag{1.212}$$

for every x and every couple y_1 and y_2 for which it holds:

$$|x - x_0| < \text{h}, \tag{1.213a}$$

$$|y_1 - y_0| < \text{h}, \tag{1.213b}$$

$$|y_2 - y_0| < \mathrm{h}, \tag{1.213c}$$

where h and k are positive constants.

As an example of application of the algorithm (1.211a, 1.211b, and 1.211c), let us consider the differential equation (Cugiani, 1967):

$$y' = 1 + xy \tag{1.214}$$

with initial conditions $x_0 = 0$, $y_0 = y(x_0) = 0$. Successive applications of Eq. (1.211a, 1.211b, and 1.211c) yield:

$$y_1 = y_0 + \int_{x_0}^{x} f(x, y_0)dx = \int_0^x (1 + xy_0)dx = \int_0^x dx = x, \tag{1.215a}$$

$$y_2 = \int_{x_0}^{x} f(x, y_1)dx = \int_0^x (1 + xy_1)dx = \int_0^x (1 + x^2)dx = x + \frac{x^3}{3}, \tag{1.215b}$$

$$\begin{aligned} y_3 &= \int_{x_0}^{x} f(x, y_2)dx = \int_0^x (1 + xy_2)dx = \int_0^x \left(1 + x^2 + \frac{x^4}{3}\right)dx \\ &= x + \frac{x^3}{3} + \frac{x^5}{3 \cdot 5} \end{aligned} \tag{1.215c}$$

$$\begin{aligned} y_4 &= \int_{x_0}^{x} f(x, y_3)dx = \int_0^x (1 + xy_3)dx = \int_0^x \left(1 + x^2 + \frac{x^4}{3} + \frac{x^6}{3 \cdot 5}\right)dx \\ &= x + \frac{x^3}{3} + \frac{x^5}{3 \cdot 5} + \frac{x^7}{3 \cdot 5 \cdot 7} \end{aligned} \tag{1.215d}$$

The general term of the integrand in Eq. (1.215d) is $x^{(2n-2)}/(2n-2)!!$.

Continuing in this way, one finds that the solution can be expressed as the sum of the series:

$$\begin{aligned} \mathrm{y} &= x + \frac{x^3}{3} + \frac{x^5}{3 \cdot 5} + \frac{x^7}{3 \cdot 5 \cdot 7} + \ldots. + \frac{x^{(2n+1)}}{(2n+1)!!} + \ldots \\ &= e^{x^2/2} \int_0^x e^{-x^2/2} dx. \end{aligned} \tag{1.216}$$

We recall that the double factorial of an integer n, denoted $n!!$, is the product of all odd integers smaller than or equal to n. Hence for the odd integer $(2n + 1)$, one has $(2n + 1)!! = 1 \times 3 \times 5 \times \ldots (2n - 1)(2n + 1)$.

Table 1.3 List of double factorials

n	$n!!$
1, 2	1.0000000000000000
3, 4	3.0000000000000000
5, 6	15.000000000000000
7, 8	105.00000000000000
9, 10	945.00000000000000
11, 12	10395.000000000000
13, 14	135135.00000000000
15, 16	2027025.0000000000
17, 18	34459425.000000000
19, 20	654729075.00000000
21, 22	13749310575.000000
23, 24	316234143225.00000
25, 26	7905853580625.0000
27, 28	213458046676875.00
29, 30	6190283353629375.0

A list of the first 30 values of the double factorials is given in Table 1.3. The solution (1.216) can be expressed in terms of the error function:

$$erf(x) = \frac{2}{\sqrt{\pi}} \int_0^x e^{-t^2} dt. \tag{1.217}$$

Hence:

$$erf\left(x/\sqrt{2}\right) = \frac{2}{\sqrt{\pi}} \int_0^{x/\sqrt{2}} e^{-t^2} dt. \tag{1.218}$$

Using Eq. (1.218) and setting $t = x/\sqrt{2}$, we have:

$$\int_0^x e^{-x^2/2} dx = \sqrt{2} \int_0^x e^{-x^2/2} d\left(x/\sqrt{2}\right) = \sqrt{2} \int_0^{x/\sqrt{2}} e^{-t^2} dt$$
$$= \sqrt{\frac{\pi}{2}} erf\left(x/\sqrt{2}\right), \tag{1.219}$$

and Eq. (1.216) becomes:

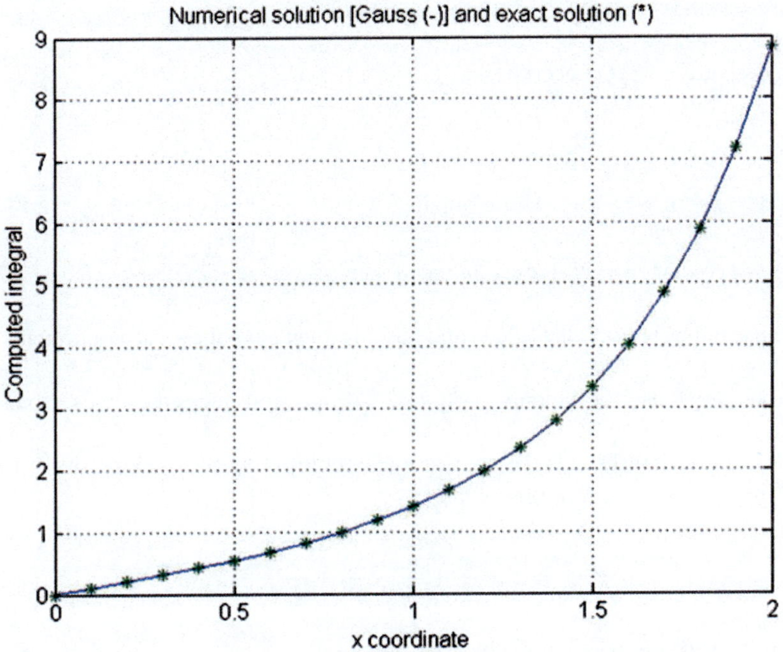

Fig. 1.17 Comparison between numerical (Gaussian integration) and exact solution of the differential equation (1.214)

$$
\begin{aligned}
y &= x + \frac{x^3}{3} + \frac{x^5}{3\cdot 5} + \frac{x^7}{3\cdot 5\cdot 7} + \ldots + \frac{x^{(2n-1)}}{(2n-1)!!} + \ldots \\
&= e^{x^2/2}\int_0^x e^{-x^2/2}dx = \sqrt{\frac{\pi}{2}}e^{x^2/2}erf\left(x/\sqrt{2}\right).
\end{aligned} \tag{1.220}
$$

Numerical solution of Eq. (1.214) with the initial condition $x_0 = 0$, $y(x_0) = 0$ is given in the computer program **PROG_I_1 [PEANO-PICARD]**.

Numerical results obtained with this program with Gaussian integration of the integrand of Eq. (1.215d), with subsequent terms, are compared with the analytical solution, Eq. (1.220), in Fig. 1.17.

The series expansion (1.216) can also be obtained with the method of undetermined coefficients. The unknown function $y(x)$ can be expanded in a series of powers of $(x - x_0)$ in the neighboring of the initial point x_0, y_0:

$$
\begin{aligned}
y =& y_0 + c_1(x-x_0) + c_2(x-x_0)^2 + c_3(x-x_0)^3 + c_4(x-x_0)^4 + c_5(x-x_0)^5 + \ldots \\
&+ c_6(x-x_0)^6 + c_7(x-x_0)^7 + \ldots
\end{aligned} \tag{1.221}
$$

Because $x_0 = 0$, $y_0 = 0$, Eq. (1.221) simplifies to:

$$y = c_1x + c_2x^2 + c_3x^3 + c_4x^4 + c_5x^5 + c_6x^6 + c_7x^7 + \ldots \tag{1.222}$$

The derivative of $y(x)$ becomes:

$$y' = c_1 + 2c_2x + 3c_3x^2 + 4c_4x^3 + 5c_5x^4 + 6c_6x^5 + 7c_7x^7 + \ldots \tag{1.223}$$

Besides, being $y' = 1 + xy$, one obtains:

$$y' = 1 + c_1x^2 + c_2x^3 + c_3x^4 + c_4x^5 + c_5x^6 + c_6x^7 + +c_7x^8 + \ldots \tag{1.224}$$

Because the polynomials (1.223) and (1.224) must have the same coefficients one obtains $c_1 = 1$, $c_2 = 0$, $c_3 = 1/3$, $c_4 = 0$, $c_5 = 1/(3.5)$, $c_6 = 0$, $c_7 = 1/(3.5.7)$ and so on. Introducing these coefficients in Eq. (1.222), one reproduces the series expansion (1.216).

A further check of Eq. (1.216) can be made multiplying series expansions. Using $e^x = x^n/n\,!\ (n = 0, 1, 2, \ldots)$, one obtains:

$$e^{x^2/2} = 1 + \frac{x^2}{2} + \frac{x^4}{8} + \frac{x^6}{48} + \frac{x^8}{384} + \ldots \tag{1.225}$$

From the series expansion of the error function:

$$erf(x) = \frac{2}{\sqrt{\pi}}\left(x - \frac{x^3}{3} + \frac{1}{2!}\frac{x^5}{5} - \frac{1}{3!}\frac{x^7}{7} + \frac{1}{4!}\frac{x^9}{9} - \ldots\right), \tag{1.226}$$

one obtains:

$$\sqrt{\frac{\pi}{2}}erf\left(\frac{x}{\sqrt{2}}\right) = x - \frac{x^3}{6} + \frac{x^5}{40} - \frac{x^7}{336} + \frac{x^9}{3456} - \ldots \tag{1.227}$$

Multiplying term by term the series expansions (1.225) and (1.227), one obtains again the series (1.216).

1.1.4 Euler, Euler-Cauchy and Heun Methods

The method of Peano-Picard, explained in Sect. 1.1.3, although didactically important, is in general rather cumbersome and therefore seldom applied. Much easier to be coded in a computer program are the methods explained in this section, while an increase in precision of the numerical solution is obtained by the Runge-Kutta and predictor-corrector methods explained in the subsequent sections.

(i) The Euler method.

In the Euler method, the simplest for the applications, the solution of Eq. (1.2), with the initial condition (1.3), is approximated in a point x_1 by:

$$y_1 = y_0 + (x_1 - x_0)f(x_0, y_0). \tag{1.228}$$

The approximate solution y_1, in the point of coordinate x_1, can be taken as new initial condition to construct the solution in the point of coordinate x_2 and so on in subsequent points. One therefore applies the algorithm:

$$y_2 = y_1 + (x_2 - x_1)f(x_1, y_1), \tag{1.229a}$$

$$y_3 = y_2 + (x_3 - x_2)f(x_2, y_2), \tag{1.229b}$$

$$y_{k+1} = y_k + (x_{k+1} - x_k)f(x_k, y_k), \tag{1.229c}$$

which allows to construct point by point the integral curve through $P_0(x_0, y_0)$.

The Taylor expansion of the unknown function $y(x)$ in the neighboring of the coordinate x_0 is:

$$y(x) = y_0 + (x - x_0)y'(x_0) + \frac{(x - x_0)^2}{2}y''(x_0) + \ldots \tag{1.230}$$

Equation (1.230) applied for $x = x_1$ and letting $h = x_1 - x_0$ gives:

$$\begin{aligned} y(x_1) &= y_0 + (x_1 - x_0)y'(x_0) + \frac{(x_1 - x_0)^2}{2}y''(x_0) + \ldots = \\ &= y_0 + (x_1 - x_0)f(x_0 y_0) + O(h^2). \end{aligned} \tag{1.231}$$

The difference between Eqs. (1.234) and (1.228), namely:

$$y(x_1) - y_1 = O(h^2) \tag{1.232}$$

shows that the error inherent in the numerical solution (1.228) is of the order of magnitude h^2. In the textbook Cugiani (1967), it is justified the assertion that the accuracy of the Euler scheme, represented by Eq. (1.228), is of order h, or of first order.

A geometrical interpretation of the Euler formula is given in Fig. 1.18. The integral curve through the point $P_0(x_0, y_0)$ has, for $x = x_1$, the ordinate $y(x_1)$ (point B). The Euler formula, considering the tangent t to the integral curve in the point $P_0(x_0, y_0)$, leads to assume as ordinate in x = x_1 the approximate value y_1 (point C).

(ii) The Euler-Cauchy method.

The Euler-Cauchy method is obtained replacing Eq. (1.228) by:

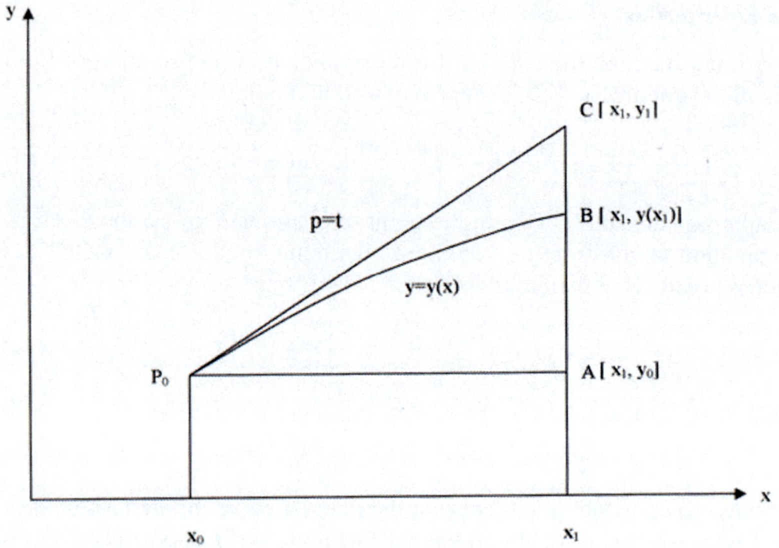

Fig. 1.18 Geometrical interpretation of the Euler formula

$$y_1 = y_0 + hf\left[x_0 + \frac{h}{2}, y_0 + \frac{h}{2}f(x_0, y_0)\right]. \tag{1.233}$$

The function f can be expanded by:

$$\begin{aligned} f\left[x_0 + \frac{h}{2}, y_0 + \frac{h}{2}f(x_0, y_0)\right] = & \\ & = f(x_0, y_0) + \frac{h}{2}f'_x(x_0, y_0) \\ & + \frac{h}{2}f'_y(x_0, y_0)f(x_0, y_0) + O(h^2) \end{aligned} \tag{1.234}$$

Introducing Eq. (1.234) into Eq. (1.233) one obtains:

$$y_1 = y_0 + hf(x_0, y_0) + \frac{h^2}{2}f'_x(x_0, y_0) + \frac{h^2}{2}f'_y(x_0, y_0)f(x_0, y_0) + O(h^3). \tag{1.235}$$

A Taylor series expansion of $y(x_1)$ around point $P(x_0, y_0)$ yields:

$$\begin{aligned} y(x_1) &= y_0 + hy'(x_0) + \frac{h^2}{2}y''(x_0) + O(h^3) \\ &= y_0 + hf(x_0, y_0) + \frac{h^2}{2}\left[f'_x(x_0, y_0) + f'_y(x_0, y_0)y'(x_0)\right] + O(h^3) \\ &= y_0 + hf(x_0, y_0) + \frac{h^2}{2}\left[f'_x(x_0, y_0) + f'_y(x_0, y_0)f(x_0, y_0)\right] + O(h^3) \end{aligned} \quad (1.236)$$

The difference between Eqs. (1.236) and (1.235) yields:

$$y(x_1) - y_1 = O(h^3) \quad (1.237)$$

and shows that the error inherent in the numerical solution (1.233) is of the order of magnitude h^3. The accuracy of the Euler-Cauchy scheme, represented by Eq. (1.233), is of order h^2, or of second order.

The geometrical interpretation of the Euler-Cauchy formula is given in Fig. 1.19. The integral curve through the point $P(x_0, y_0)$ has, for x $= x_1$, the ordinate $y(x_1)$ (point B). The line p is the tangent to the integral curve $y(x)$ in $P(x_0, y_0)$ and determines the point $P_{1/2}\left[x_0 + \frac{h}{2}, y_0 + \frac{h}{2}f(x_0 y_0)\right]$. The dashed curve is the integral curve shifted to pass through $P_{1/2}$. The Euler-Cauchy formula considers the tangent t to the integral curve in the point $P_{1/2}$ and leads to assume as ordinate in $x = x_1$ the approximate value y_1 (point C). The segment $P_0 - C$ is parallel (//t) to the tangent t.

(iii) The Heun method.

The Heun method is obtained replacing Eq. (1.228) by:

$$y_1 = y_0 + h\frac{1}{2}\left[f(x_0, y_0) + f(x_0 + h, y_0 + hf(x_0, y_0))\right]. \quad (1.238)$$

The geometrical interpretation of the Heun formula is given in Fig. 1.20. The tangent p to the integral curve $y(x)$ in $P(x_0, y_0)$ determines the point P with coordinates:

$$P(x_P, y_P) = P(x_0 + h, y_0 + hf(x_0, y_0)). \quad (1.239a)$$

The tangent to the shifted integral curve (dashed curve) in P is t. The parallel line to t, through the point $P(x_0, y_0)$, determines the point R with coordinates:

$$R(x_R, y_R) = R[x_0 + h, y_0 + hf(x_P, y_P)]. \quad (1.239b)$$

The point C is then the midpoint of the segment RP and its coordinates are given by:

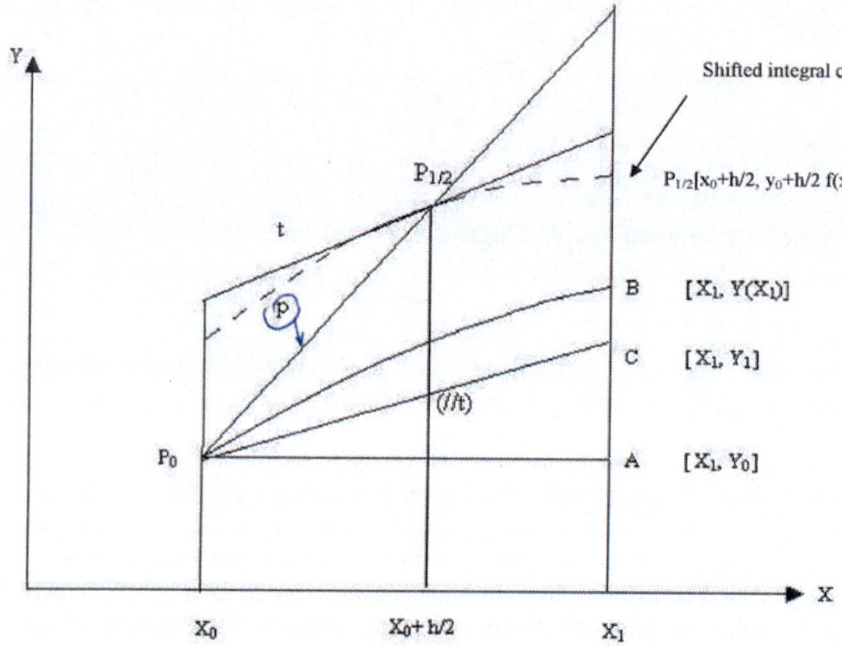

Fig. 1.19 Geometrical interpretation of the Euler-Cauchy formula

$$C(x_1, y_1) = C\left[x_1, y_0 + \frac{h}{2}(f(x_0, y_0) + f(x_P, y_P))\right]. \tag{1.239c}$$

In formula (1.238) one evaluates the tangent to the integral curve as mean value of the slopes in points $P_0(x_0, y_0)$ and $P(x_P, y_P)$.

The tangent to the integral curve in point $P(x_P, y_P)$ can be expanded by:

$$f(x_0 + h, y_0 + hf(x_0, y_0)) = f(x_0 y_0) + hf'_x(x_0, y_0) + hf(x_0, y_0)f'_y(x_0, y_0) \\ + O(h^2). \tag{1.240}$$

Introducing Eq. (1.240) into Eq. (1.238) yields:

$$y_1 = y_0 + hf(x_0, y_0) + \frac{h^2}{2}f'_x(x_0, y_0) + \frac{h^2}{2}f(x_0, y_0)f'_y(x_0, y_0) + O(h^3). \tag{1.241}$$

Subtracting Eq. (1.240) from Eq. (1.236) one obtains again Eq. (1.237). The error of the numerical solution (1.238) is again of the order of magnitude h^3. The accuracy of the Heun method, like that of the Euler-Cauchy method, is of order h^2, or of second order.

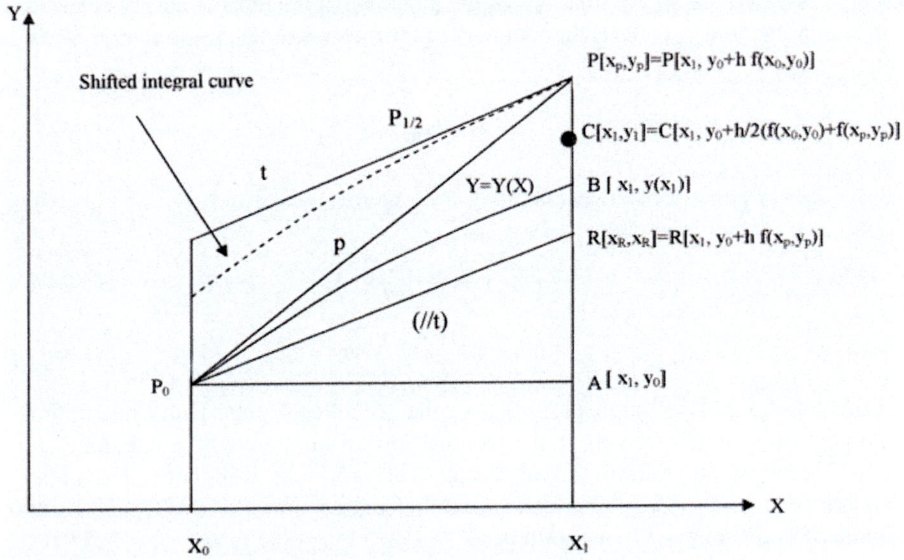

Fig. 1.20 Geometrical interpretation of the Heun formula

1.1.5 *Runge Kutta Methods*

1.1.5.1 Theoretical Background

Methods of the Runge-Kutta family are based on the idea of applying Eq. (1.210) for $x = x_1$ by evaluating numerically the integral:

$$y_1 = y_0 + \int_{x_0}^{x_1} f(x, y(x))dx. \tag{1.242}$$

The integral can be approximated with the Simpson formula:

$$\int_a^b f(x)dx \approx \frac{b-a}{6}\left[f(a) + 4f\left(\frac{a+b}{2}\right) + f(b)\right]. \tag{1.243}$$

Thus, from Eq. (1.242) one obtains:

$$y_1 = y_0 + \frac{h}{6}(k_1 + 2k_2 + 2k_3 + k_4), \tag{1.244}$$

where k_1 is a suitable approximations of the integrand at the left boundary, k_2 and k_3 in the midpoint, and k_4 at the right boundary of the interval of integration $h = x_1 - x_0$, namely:

$$k_1 = f(x_0, y_0), \tag{1.245}$$

$$k_2 = f\left(x_0 + \frac{h}{2}, y_0 + \frac{h}{2} f(x_0, y_0)\right) = f\left(x_0 + \frac{h}{2}, y_0 + \frac{h}{2} k_1\right), \tag{1.246}$$

$$k_3 = f\left(x_0 + \frac{h}{2}, y_0 + \frac{h}{2} k_2\right), \tag{1.247}$$

$$k_4 = f(x_0 + h, y_0 + hk_3). \tag{1.248}$$

Equations (1.244) through (1.248) give the so called Runge-Kutta integration formula of order four, referring to the fact that four evaluations of the integrand are needed for every integration interval. Runge-Kutta formula of higher order is also available (Fehlberg, 1968). One can prove that the accuracy of the Runge-Kutta formula of order four is h^4, or fourth order.

An example of application of the Runge-Kutta method of order four is given for the differential equation $y' = \cos(x)$, with initial condition $y(0) = 0$, with the obvious solution $y(x) = \sin(x)$, in the sample program **PROG_I_2 [TEST_RK4]**. In this small program, the integration is made forward in the interval $(0 - 2\pi)$ and then backward from the point $P(2\pi, 0)$ to $x = 0$. The numerical and analytical solutions are practically superimposed.

In the numerical applications of the Runge-Kutta algorithm, Eq. (1.2) can be a vector equation of the type:

$$\{y'\} = \{f(x, y)\}, \tag{1.249}$$

with:

$$\{y'\} = \{y'_1, y'_2, \ldots y'_N)\}^T, \tag{1.250}$$

$$\{f\} = \{f_1 f_2, \ldots f_N)\}^T, \tag{1.251}$$

where N is the length of the vector and the superscript T denotes transpose.

As an example one may consider the system of differential equations:

$$y' = z(x), \tag{1.252a}$$

$$z' = -y(x), \tag{1.252b}$$

with initial conditions $y(0) = 0$, $z(0) = 1$. The numerical solution can be checked against the analytical solutions $y = \sin(x)$, $z = \cos(x)$.

An application of the Runge-Kutta algorithm of order 4 to a system of 6 equations is given in Chap. 8, Sect. 8.7, for the numerical simulation of homogeneous isotropic turbulence with the application of the Reynolds stress model (RSM).

1.1.5.2 Example of Application to a Fluid-Dynamic Problem

We consider an example of the application of Eq. (1.242) to a fluid dynamic problem in which the distribution of enthalpy in a heated fluid flow is described in a domain D by the following conservation equation:

$$\frac{\partial(\rho h)}{\partial t} + \nabla \cdot \left(\rho h \vec{V}\right) = \frac{\partial p}{\partial t} + \vec{V} \cdot \nabla p - \nabla \cdot \vec{q} + Q. \tag{1.253}$$

In Eq. (1.253) ρ, h, and p are fluid density, specific enthalpy, and pressure, respectively; $\vec{V}$ is the fluid velocity,

$$\vec{q} = -\lambda \nabla T = -\left(\lambda / c_p\right) \nabla h \tag{1.254}$$

is heat flux defined in terms of temperature gradient ∇T or of enthalpy gradient ∇h, with λ thermal conductivity and c_p specific heat at constant pressure; Q is a power source to the fluid and t is time.

Using the identity:

$$\nabla \cdot \left(p\vec{V}\right) = p\left(\nabla \cdot \vec{V}\right) + V \cdot \nabla p, \tag{1.255}$$

Equation (1.253) can be written:

$$\frac{\partial(\rho h)}{\partial t} + \nabla \cdot \left(\rho h \vec{V}\right) = \frac{\partial p}{\partial t} + \nabla \cdot \left(p\vec{V}\right) - p\left(\nabla \cdot \vec{V}\right) - \nabla \cdot \vec{q} + Q. \tag{1.256}$$

With the definition of the vector:

$$\vec{J} = \rho h \vec{V} - p\vec{V} + \vec{q}, \tag{1.257}$$

Equation (1.256) can be written in the more compact form:

$$\frac{\partial(\rho h)}{\partial t} + \nabla \cdot \vec{J} = \frac{\partial p}{\partial t} - p\left(\nabla \cdot \vec{V}\right) + Q. \tag{1.258}$$

The domain D of definition of the fluid dynamic problem can be partitioned in control cells of volume V_f. Integration of Eq. (1.258) over the volume of a cell yields:

$$\int_{V_f} \frac{\partial(\rho h)}{\partial t} dV_f + \int_{V_f} \left(\nabla \cdot \vec{J}\right) dV_f = \int_{V_f} \frac{\partial p}{\partial t} dV_f - \int_{V_f} p\left(\nabla \cdot \vec{V}\right) dV_f + \int_{V_f} Q dV_f. \tag{1.259}$$

Using the Gauss theorem, the volume integral of the divergence of a vector can be transformed into fluxes through the surfaces bounding the control volume. Thus Eq. (1.259) becomes:

$$\int_{V_f} \frac{\partial(\rho h)}{\partial t} dV_f + \int_A \vec{J} \cdot \vec{n} dA = \int_{V_f} \frac{\partial p}{\partial t} dV_f - \int_A p\vec{V} \cdot \vec{n} dA + \int_{V_f} Q dV_f, \tag{1.260}$$

where $\vec{n}$ is an outward-oriented normal unit vector.

The integrand of the first term can formally be split as follows:

$$\frac{\partial(\rho h)}{\partial t} = \rho \frac{\partial h}{\partial t} + h \frac{\partial \rho}{\partial t}, \tag{1.261}$$

and Eq. (1.260) approximated by:

$$\frac{\partial h}{\partial t} = \frac{1}{\rho V_f} \left[-\int_A \vec{J} \cdot \vec{n} dA + \int_{V_f} \frac{\partial p}{\partial t} dV_f - \int_A p\vec{V} \cdot \vec{n} dA + \int_{V_f} Q dV_f \right] - \frac{h}{\rho} \frac{\partial \rho}{\partial t}. \tag{1.262}$$

As it will be shown in detail in Sect. 7.3.6 of this monograph, discretization of the right-hand side of Eq. (1.262) yields an algebraic sum of seven terms. These terms consist of one coefficient, multiplied by the unknown enthalpy value in a central cell, and of six coefficients, multiplied by the respective enthalpy values of the six neighboring cells in the three coordinate directions. Discretization of the right-hand side eliminates the dependence of the enthalpy distribution from the spatial coordinates. The partial derivative of the enthalpy in Eq. (1.262) can be replaced by an ordinary derivative with respect to time. The time axis can be partitioned in interval of length $\Delta t = t^{n+1} - t^n$. We assume that the right-hand side of Eq. (1.262) is known at time point t^n and the resulting ordinary differential equation must be solved over the interval $t^{n+1} - t^n$. Equation (1.262) therefore reduces to the form:

$$\frac{dh}{dt} = \left(\sum_{\alpha=0}^{6} c_\alpha h_\alpha \right)^n + R^n, \tag{1.263}$$

where R^n collects all terms which do not depend on the enthalpy. Subscript $\alpha = 0$ refers to the cell considered, while $\alpha = 1, 2, ...6$ refers to the six neighboring cells.

Writing Eq. (1.263) for all cells in the discretized domain D, one obtains the vector equation:

$$\left\{\frac{dh}{dt}\right\} = \left\{\left(\sum_{\alpha=0}^{6} c_\alpha h_\alpha\right)^n + R^n\right\}, \tag{1.264}$$

or, in compact form:

$$\left\{\frac{dh}{dt}\right\} = \{\, f(h)\}, \tag{1.265}$$

which corresponds to Eq. (1.249) and can be solved, with the Runge-Kutta method, in the following way:

(i) Compute (for every cell):

$$k_1 = f(h^n); \tag{1.266}$$

(ii) Compute:

$$h_a^{n+1/2} = h^n + \frac{\Delta t}{2} k_1, \tag{1.267a}$$

$$k_2 = f\left(h_a^{n+1/2}\right); \tag{1.267b}$$

(iii) Compute:

$$h_b^{n+1/2} = h^n + \frac{\Delta t}{2} k_2, \tag{1.268a}$$

$$k_3 = f\left(h_b^{n+1/2}\right); \tag{1.268b}$$

(iv) Compute:

$$\widetilde{h}^{n+1} = h^n + \Delta t k_3, \tag{1.269a}$$

$$k_4 = f\left(\widetilde{h}^{n+1}\right); \tag{1.269b}$$

(v) Compute:

$$h^{n+1} = h^n + \frac{\Delta t}{6}(k_1 + 2k_2 + 2k_3 + k_4). \tag{1.270}$$

Equation (1.270) is applied subsequently for all time intervals from $t = 0$, starting from a known initial distribution of enthalpy $[h(t = 0)]$, to the end of the time domain. This application of the Runge-Kutta method in a three-dimensional thermal and fluid dynamic computer program was made in the computer code BACCHUS-3D/TP (Bottoni et al., 1992) for single-phase flow calculations. This method allows a gain in computational time by a factor of 2–3 with respect to Gaussian elimination methods or ADI (alternating direction implicit) methods.

1.1.6 Predictor-Corrector Methods

Numerical methods of Runge-Kutta type are widely used because of their simplicity and robustness. In cases in which the ordinary differential equations to be integrated are not stiff and high precision is required, methods of the predictor-corrector type are more appropriate.

A method of the predictor corrector type is based on the following principles.

(i) We assume to know in advance the function to be integrated in a set of points $y_0 = y(x_0)$, $y_1 = y(x_1)$, $y_2 = y(x_2)$, $\ldots y_n = y(x_n)$. At the limit, this set may consist of the first point only. If the points are more than two, they are assumed to be equally spaced. This initial set of points, if not known a priori, may be obtained with another integration method, for instance, with the Runge-Kutta method.
(ii) We assume to know two formulas:

$$y_{n+1} = \varphi(y_0, y_1, y_2, \ldots y_n), \tag{1.271}$$

$$y_{n+1} = \psi(y_0, y_1, y_2, \ldots y_n, y_{n+1}). \tag{1.272}$$

The first formula, referred to as predictor step, relates the known values $y_i (i = 0, 1, 2, \ldots n)$, of the integral function to the first yet unknown value y_{n+1}. The second formula, the corrector step, establishes a relation between the known values $y_i (i = 0, 1, 2, \ldots n)$, the first approximation of y_{n+1} and a refined value of y_{n+1}. In principle, formula (1.272) can be used in an iterative way to further refine the solution y_{n+1}. Letting r be an iteration index, we thus have:

$$y_{n+1}^{r+1} = \psi(y_0, y_1, y_2, \ldots y_n, y_{n+1}^r). \tag{1.273}$$

In most cases, however, iterations are not performed ($r = 1$).

According to the choice of the number of the initial points and of the formulas (1.271) and (1.272), one has several variants of the predictor-corrector method.

(i) Euler-Heun method.

As predictor and corrector steps, one applies the Euler formula (1.228) and the Heun formula (1.238), respectively. Thus one has:

$$\text{Predictor step}: \quad y_1 = y_0 + hf(x_0, y_0), \tag{1.274}$$

$$\begin{aligned}\text{Corrector step}: \quad y_1 &= y_0 + h\frac{1}{2}[f(x_0, y_0) + f(x_0 + h, y_0 + hf(x_0, y_0))] \\ &= y_0 + h\frac{1}{2}[f(x_0, y_0) + f(x_1, y_1)]\end{aligned} \tag{1.275}$$

In this case only one initial point $y_0 = y(x_0)$ needs to be known.

(ii) Milne method.

One assumes to know the integral function $y = y(x)$ in the four nodes $y_0 = y(x_0)$, $y_1 = y(x_1)$, $y_2 = y(x_2)$ and $y_3 = y(x_3)$. The first one is the initial value; the three subsequent points may have been obtained with the Runge-Kutta method. The predictor and corrector steps are as follows:

$$\text{Predictor step}: \quad y_4 = y_0 + \frac{4h}{3}[2f(x_1, y_1) - f(x_2, y_2) + 2f(x_3, y_3)], \tag{1.276}$$

$$\text{Corrector step}: \quad y_4 = y_2 + \frac{2h}{6}[f(x_2, y_2) + 4f(x_3, y_3) + f(x_4, y_4)]. \tag{1.277}$$

The predictor step is the Newton-Cotes formula of open type, applied to the interval $(x_4 - x_0)$ of length 4 h. The corrector step is the Simpson formula applied to the interval $(x_4 - x_2)$ of length 2 h.

(iii) The Adams-Bashforth-Moulton method.

Very wide application has the Adams-Bashforth-Moulton method (Press et al., 1986), based on the following formulas:
Predictor step:

$$y_4 = y_3 + \frac{h}{24}[-9f(x_0, y_0) + 37f(x_1, y_1) + -59f(x_2, y_2) + 55f(x_3, y_3)], \tag{1.278}$$

Corrector step:

$$y_4 = y_3 + \frac{h}{24}[f(x_1, y_1) - 5f(x_2, y_2) + 19f(x_3, y_3) + 9f(x_4, y_4)]. \tag{1.279}$$

1.1.7 *Systems of Ordinary Differential equations of First Order*

In this section we consider linear systems of ordinary differential equations (ODEs) of first order. In the first subsection, we present the general theoretical background which enables to construct the integral solution of the system. In a subsequent subsection we treat the case of stiff systems, namely, systems characterized by very different order of magnitudes of the coefficients of the equations. Stiff systems arise in practical applications of chemical kinetics, transport problems and phase changes. For example, in case of phase changes, with fluid dynamic interactions of a liquid with its vapor phase, very different coefficients of the equations arise from the different density of the liquid and vapor phases. The stream of liquid sodium and sodium vapour in the core of fast reactors under hypothetical accident conditions is a well-known example of problems leading to stiff systems of differential equations.

1.1.7.1 Theoretical Background

Let us consider the following initial value problem:

$$u'(t) = f[t, u(t)], \tag{1.280a}$$

$$u(t_0) = u_0, \tag{1.280b}$$

where $u(t)$ and $f[t, u(t)]$ are vectors of N unknown functions and of given functions, respectively:

$$u(t) = |u_1(t), u_2(t). \dots u_N(t)|^T, \tag{1.281}$$

$$f = |f_1, f_2, \dots f_N|^T. \tag{1.282}$$

We further assume that the system (1.280a and 1.280b) is linear and can therefore be written in the form

$$u'(t) = [A]u(t), \tag{1.283a}$$

$$u(t_0) = u_0, \tag{1.283b}$$

where A is a square $(N \times N)$ matrix with eigenvalues $\lambda_1, \lambda_2, \dots \lambda_N$. We assume that these eigenvalues are real and negative and are labeled so that it holds $|\lambda_1| \leq |\lambda_2| \leq \dots |\lambda_N|$.

The general integral of the system (1.283a and 1.283b) can be built on the basis of the following theorem:

Theorem The general integral of the system (1.283a and 1.283b) is a linear combination of the solutions of the N differential equations $v' = \lambda_k v\,(k = 1, 2, \ldots N)$ and is therefore given by a vector with components:

$$u(t) = c_1 \exp(\lambda_1 t) + c_2 \exp(\lambda_2 t) + \ldots c_N \exp(\lambda_N t). \tag{1.284}$$

The coefficients $c_k(k = 1, 2, \ldots N)$ are determined with the initial conditions (1.283b).

We want to compare in the following the analytical solution (1.284) with a numerical solution obtained with the Euler method explained in Sect. 1.1.4. Let the time interval of interest $[t = (t_0 = 0, t_{\max})]$ be divided in intervals Δt and let denote $u(t_j) = u(j\Delta t)$, $(j = 0, 1, 2, \ldots j_{\max} = t_{\max}/\Delta t)$. The Euler method consists in replacing the derivative at the time point t_j with the tangent to the solution function and leads to the formula:

$$v_{j+1} = v_j + \lambda \Delta t v_j = v_j(1 + \lambda \Delta t). \tag{1.285}$$

Repetitive application of (1.285), starting from the initial condition (1.283b) yields:

$$\begin{aligned} v_1 &= v_0(1 + \lambda\Delta t) \\ v_2 &= v_1(1 + \lambda\Delta t) = v_0(1 + \lambda\Delta t)^2 \\ v_j &= v_0(1 + \lambda\Delta t)^j \end{aligned} \tag{1.286}$$

Corresponding to the analytical solution of Eq. (1.284), the general integral of the system (1.283a and 1.283b), obtained with the numerical Euler method (denoted by subscript "E") is:

$$\begin{aligned} u_E(t_j) = {} & c_1(1 + \lambda_1\Delta t)^j + c_2(1 + \lambda_2\Delta t)^j \\ & + \ldots c_N(1 + \lambda_N\Delta t)^j \ \ (j = 1, 2, \ldots . j_{\max}). \end{aligned} \tag{1.287}$$

We direct the reader to Chap. 4 for a rigorous definition of the concepts of convergence, consistency, and stability of numerical algorithms. We limit ourselves here to a simple qualitative definition. Let u_N be the numerical solution of a differential equation and U be the correct solution. Then the following concepts are introduced:

Convergence Let an equation with spatial and temporal differentials dx, dt be replaced by a difference equation with finite differences Δx, Δt. The difference equation is convergent to the differential equation if the numerical solution u_N tends to the correct solution U when the finite differences Δx, Δt tend to zero.

Consistency Let us define the "truncation error" as the maximum difference between the correct solution and the numerical solution of a differential equation in every point of the definition domain. A finite difference equation is consistent with the differential equation if the truncation error tends to zero when the finite differences Δx, Δt tend to zero.

Stability A numerical scheme for the solution of a difference equation is stable if the cumulative effect of the round-off numerical errors is negligible.

The concept of convergence and consistency concerns the behavior of the numerical solution as the differences Δx, Δt tend to zero. The concept of stability is different because it concerns the behavior of a numerical scheme by given value of the differences Δx, Δt.

The concept of stability has been recalled here because it must be used to examine whether a given value of the time step Δt is appropriate to ensure a stable numerical scheme. In other terms, the concept of stability can be used to set an upper limit to the time step which can be used. In case of Eq. (1.287) to avoid numerical instabilities, namely incorrect increasing of the numerical solution, one has to ensure that $|1 + \lambda_N \Delta t| < 1$, from which one deduces:

$$\Delta t < -2/\lambda_N. \tag{1.288}$$

For instance, in case it were $\lambda_N = -10^3$, the suitable time step should be smaller than 2 milliseconds. Thus the eigenvalue of the matrix A in (1.283a) with the largest absolute value requires the application of a short time step in the numerical solution.

1.1.7.2 Stiff Systems

Systems of ordinary differential equations of the first order are said to be "stiff" when the order of magnitude of the eigenvalues $\lambda_1, \lambda_2, \ldots \lambda_N$ of the matrix A in Eq. (1.283a) is very different so that $|\lambda_N|/|\lambda_1| \gg 1$. In this case the single terms in Eq. (1.287) decrease with time with very different speed. Terms corresponding to large eigenvalues decrease and vanish rapidly, but a small time step is imposed by these terms. One possible solution to cope with the time step constraint is to split, whenever possible, the given system of differential equations in two (or more) systems characterized by more uniform values of the eigenvalues. Thus the two systems can be solved with different time steps, with gain in computational time. More details about these techniques are given in Hofer (1976).

Example 1.19
In the following we consider a numerical example with two equations taken from Grigorieff (1972), but using different values of the initial conditions and of the numerical coefficients of the equations. The system considered is:

$$u' = \frac{1}{2}(q_1 + q_2)u + \frac{1}{2}(q_1 - q_2)v, \tag{1.289a}$$

$$v' = \frac{1}{2}(q_1 - q_2)u + \frac{1}{2}(q_1 + q_2)v. \tag{1.289b}$$

Letting $Q_1 = (q_1 + q_2)/2$, $Q_2 = (q_1 - q_2)/2$, the coefficients of the matrix A are: $a_{11} = Q_1$, $a_{12} = Q_2$, $a_{21} = Q_2$, $a_{22} = Q_1$. Hence one has:

$$\det(A) = Q_1^2 - Q_2^2, \tag{1.290}$$

$$\text{Trace}(A) = 2Q_1 = q_1 + q_2. \tag{1.291}$$

The eigenvalues of matrix A are $\lambda_1 = q_1$, $\lambda_2 = q_2$.

We consider the case $q_1 = -1$, $q_2 = -10$, with initial conditions $u_0 = 5$, $v_0 = 2$. This case with $|\lambda_2|/|\lambda_1| = 10$ is not very stiff, but the results, with components which do not decay to zero too rapidly, are more apt to be visualized. With this choice of parameters one has $Q_1 = -5.5$, $Q_2 = 4.5$, $\det(A) = 10 = \lambda_1\lambda_2$, Trace (A) $= -11 = \lambda_1 + \lambda_2$.

The general integral of the system (1.289a and 1.289b) is:

$$u(t) = c_1 \exp(q_1 t) + c_2 \exp(q_2 t), \tag{1.292a}$$

$$v(t) = c_1 \exp(q_1 t) - c_2 \exp(q_2 t). \tag{1.292b}$$

With the Euler method, the numerical general integral is obtained as:

$$u_E(t_j) = c_1(1 + \lambda_1 \Delta t)^j + c_2(1 + \lambda_2 \Delta t)^j, \tag{1.293a}$$

$$v_E(t_j) = c_1(1 + \lambda_1 \Delta t)^j - c_2(1 + \lambda_2 \Delta t)^j. \tag{1.293b}$$

With the given initial conditions the constant c_1 and c_2 are determined:

$$c_1 = (u_0 + v_0)/2 = 3.5, \tag{1.294a}$$

$$c_2 = (u_0 - v_0)/2 = 1.5. \tag{1.294b}$$

The analytical solution $u(t)$, given by Eq. (1.292a), is shown in Fig. 1.21 as the sum UAN of its two components UAN1 and UAN2. Likewise, the analytical solution $v(t)$, given by q (1.292b), is shown in Fig. 1.22 as the sum VAN of its two components VAN1 and VAN2. The two solutions $u(t)$ and $v(t)$ are plotted together in Fig. 1.23.

Numerical solutions obtained with the Euler method are practically identical with the analytical solutions. According to Eq. (1.288) the time step must not exceed 0.2 s. Due to the time-consuming calculation of the exponentials, the analytical

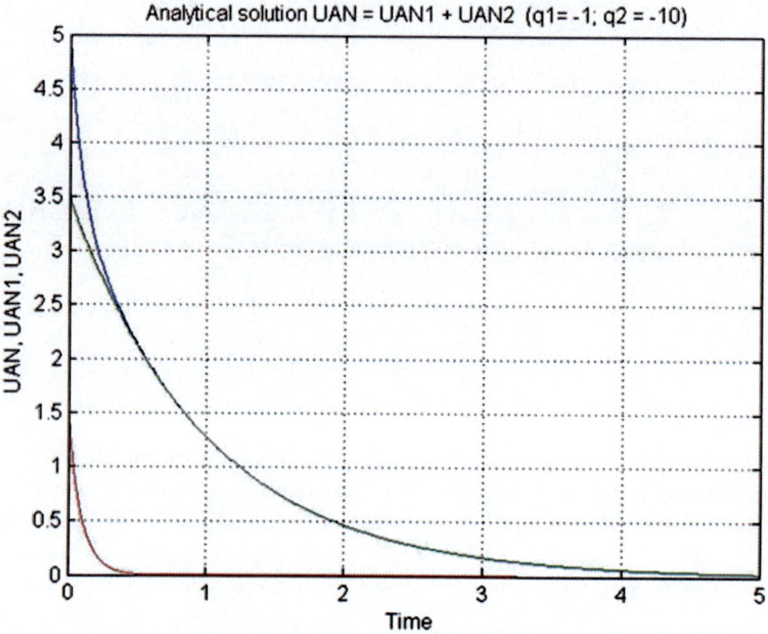

Fig. 1.21 (Example 1.19) Analytical solution u(t) (uppermost plot) as sum of its two components given at the right-hand side of Eq. (1.292a)

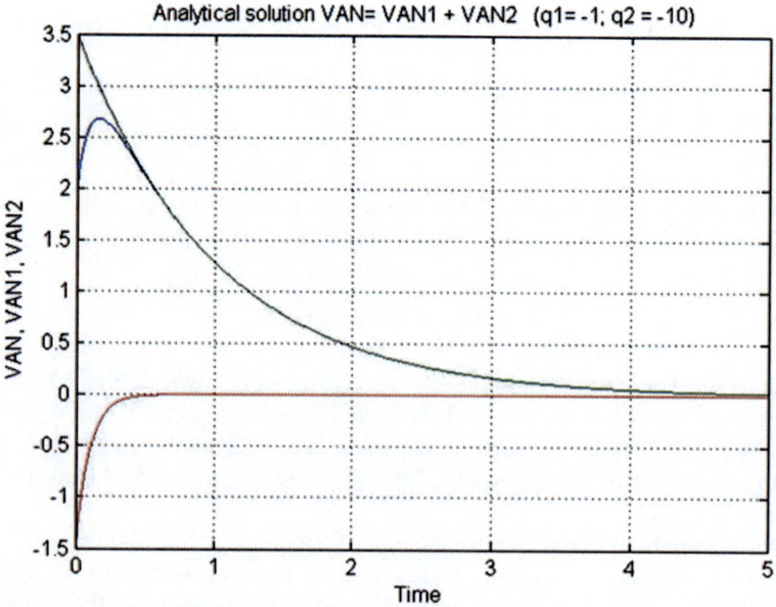

Fig. 1.22 (Example 1.19) Time dependence of the analytical solution v(t) (mid plot) as sum of its two components given at the right hand side of Eq. (1.292b)

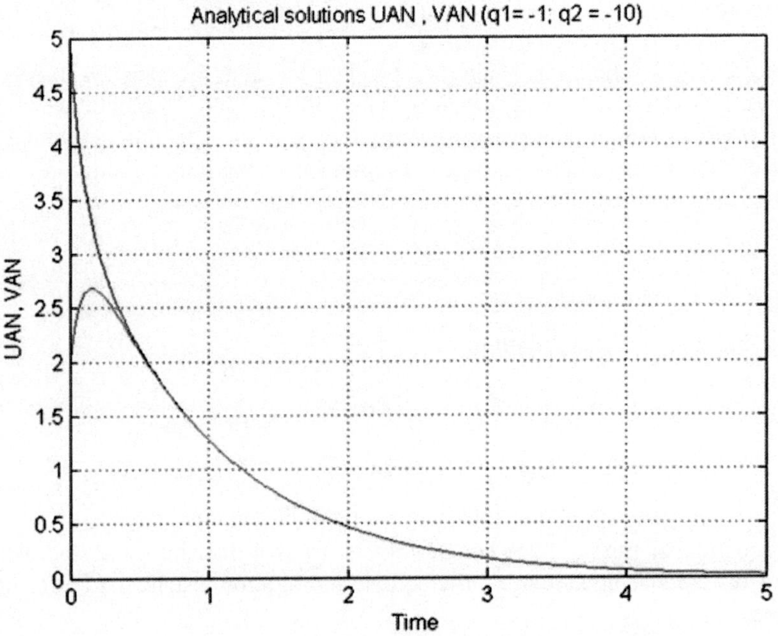

Fig. 1.23 (Example 1.19) Time dependence of the analytical solutions $u(t)$ (uppermost plot) and $v(t)$

solution requires more CPU time (about 30–40% more) than the numerical solution with the Euler method. The numerical method might therefore be preferable for large problems.

1.2 Ordinary Differential equations of Second and Higher Order

1.2.1 Ordinary Differential Equations of Second Order

1.2.1.1 Reduction to a System of Ordinary Differential Equations of First Order

Let us consider an ordinary differential equation of second order and assume it can be solved with respect to the higher-order derivative, hence written in the normal form:

$$y'' = f(x, y, y'), \tag{1.295}$$

where $y = y(x)$ is a function of the independent variable x and is assumed to be continuous with its first derivative in the domain of definition. In general we will not be interested in finding the general integral of Eq. (1.295) but in obtaining a numerical approximation of a particular integral satisfying initial conditions:

$$y(x_0) = c_0, \tag{1.296a}$$

$$y'(x_0) = c_1, \tag{1.296b}$$

or satisfying the boundary conditions:

$$y(a) = \alpha, \tag{1.297a}$$

$$y(b) = \beta, \tag{1.297b}$$

where c_0, c_1, α, β are given constants.

An equation of type (1.295) can always be transformed into a system of two ordinary differential equations of first order, letting, for instance, $y' = z$; hence obtaining:

$$z' = f(x, y, z), \tag{1.298a}$$

$$y' = z, \tag{1.298b}$$

with initial conditions $z(x_0) = c_1$ and $y(x_0) = c_0$ or boundary conditions (1.297a and 1.297b).

Equations (1.298a and 1.98b), with the given initial conditions, are a particular case of the general system:

$$y' = f(x, y, z), \tag{1.299a}$$

$$z' = g(x, y, z), \tag{1.299b}$$

with initial conditions $y(x_0) = y_0$ and $z(x_0) = z_0$.

Numerical solution of the system of Eq. (1.299a and 1.299b) can be obtained, for instance with the Runge-Kutta algorithm of order four, which in this case can be written as follows:

$$y_1 = y_0 + \frac{h}{6}(k_{11} + 2k_{12} + 2k_{13} + k_{14}), \tag{1.300}$$

$$z_1 = z_0 + \frac{h}{6}(k_{21} + 2k_{22} + 2k_{23} + k_{24}), \tag{1.301}$$

with:

$$k_{11} = f(x_0, y_0, z_0), \tag{1.302a}$$

$$k_{21} = g(x_0, y_0, z_0), \tag{1.302b}$$

$$k_{12} = f\left(x_0 + \frac{h}{2}, y_0 + \frac{h}{2}k_{11}, z_0 + \frac{h}{2}k_{21}\right), \tag{1.302c}$$

$$k_{22} = g\left(x_0 + \frac{h}{2}, y_0 + \frac{h}{2}k_{11}, z_0 + \frac{h}{2}k_{21}\right), \tag{1.302d}$$

$$k_{13} = f\left(x_0 + \frac{h}{2}, y_0 + \frac{h}{2}k_{12}, z_0 + \frac{h}{2}k_{22}\right), \tag{1.302e}$$

$$k_{23} = g\left(x_0 + \frac{h}{2}, y_0 + \frac{h}{2}k_{12}, z_0 + \frac{h}{2}k_{22}\right), \tag{1.302f}$$

$$k_{14} = f(x_0 + h, y_0 + hk_{13}, z_0 + hk_{23}), \tag{1.302g}$$

$$k_{24} = g(x_0 + h, y_0 + hk_{13}, z_0 + hk_{23}). \tag{1.302h}$$

1.2.1.2 Finite Difference Method

Let us consider an ordinary differential equation of second order of the general form:

$$y'' + A(x)y' + B(x)y = C(x), \tag{1.303}$$

with initial conditions $y(x_0) = c_0$ and $y'(x_0) = c_1$. The functions $A(x)$, $B(x)$ and $C(x)$ are assumed to be continuous in the domain of definition (a,b) of the problem.

Let us assume that the definition domain is partitioned in a number N of intervals of length $h = (b - a)/N$ and index these intervals with a subscript i, referring to the central node of the interval. Thus for instance $x_i = (i - 1)h + h/2$ $(i = 1, 2, \ldots N)$. The first- and second-order derivatives in the Eq. (1.303) can be replaced by the central finite differences:

$$y' = \frac{1}{2h}(y_{i+1} - y_{i-1}), \tag{1.304}$$

$$y'' = \frac{1}{h^2}(y_{i+1} - 2y_i + y_{i-1}). \tag{1.305}$$

We introduce Eqs. (1.304) and (1.305) into Eq. (1.303) and evaluate the functions $A(x)$, $B(x)$ and $C(x)$ in the midpoints of the respective intervals. Thus, for every interval the discrete form of Eq. (1.303) becomes, upon multiplication of every term by h^2:

$$y_{i+1} - 2y_i + y_{i-1} + \frac{hA(x_i)}{2}(y_{i+1} - y_{i-1}) + h^2 B(x_i) y_i = h^2 C(x_i). \tag{1.306}$$

Rearranging one has:

$$y_{i+1}\left[1 + \frac{h}{2}A(x_i)\right] + y_i\left[-2 + h^2 B(x_i)\right] + y_{i-1}\left[1 - \frac{h}{2}A(x_i)\right] = h^2 C(x_i). \tag{1.307}$$

Equation (1.307) can be written in the compact form:

$$cy_{i+1} + ay_i + by_{i-1} = d, \tag{1.308}$$

with the definition of the coefficients:

$$a = -2 + h^2 B(x_i), \tag{1.309}$$

$$b = 1 - \frac{h}{2}A(x_i), \tag{1.310}$$

$$c = 1 + \frac{h}{2}A(x_i), \tag{1.311}$$

$$d = h^2 C(x_i). \tag{1.312}$$

In terms of finite differences the second initial condition can be written $(y_1 - y_0)/h = c_1$; hence $y_1 = y_0 + hc_1$. From the known values of y_0 and y_1 Eq. (1.308) yields the subsequent value y_2 and so on for all consecutive values. Alternatively Eq. (1.308), written for all N intervals in the definition domain, yields a system of equations with tridiagonal matrix of coefficients. The system can be solved with the Thomas algorithm, given in Sect. 1.3.

1.2.2 *Ordinary Differential Equations of Higher Order*

Let us consider an ordinary differential equation of order n, solved with respect to the higher-order derivative, hence written in normal form:

$$y^{(n)} = f\left(\left[x, y, y', y'', \ldots y^{(n-2)}, y^{(n-1)}\right],\right. \tag{1.313}$$

with initial conditions $y(x_0) = c_0$, $y'(x_0) = c_1$, $y''(x_0) = c_2$, $\ldots y^{(n-1)}(x_0) = c_{n-1}$.

Letting:

$$y = y_1, y' = y_2, y'' = y_3, \ldots\ldots y^{(n-1)} = y_n, \tag{1.314}$$

Equation (1.313) is transformed to a system of n ordinary differential equations of first order:

$$\begin{aligned} y_1' &= y_2, \\ y_2' &= y_3. \\ y_{n-1}' &= y_n, \\ y_n' &= f(x, y_1, y_2, \ldots . y_n). \end{aligned} \tag{1.315}$$

The system of Eq. (1.315) is a particular case of a general system of ordinary differential equations of first order of the type, written in normal form:

$$\begin{aligned} y_1' &= f_1(x, y_1, y_2, \ldots . y_n), \\ y_2' &= f_2(x, y_1, y_2, \ldots . y_n), \\ y_n' &= f_n(x, y_1, y_2, \ldots . y_n), \end{aligned} \tag{1.316}$$

with initial conditions $y_1(x_0) = y_1^0$, $y_2(x_0) = y_2^0, \ldots\ldots y_n(x_0) = y_n^0$, where the y_i^0 $(i = 1, 2, \ldots n)$ are arbitrary constants. The system (1.316) has an unique solution, corresponding to the given initial conditions, if the functions $f_i(x, y_1, y_2, \ldots . y_n)$ $(i = 1, 2, \ldots n)$ are continuous and satisfy the Lipschitz condition with respect to every variable $y_1, y_2, \ldots . y_n$.

For a more complete treatment of these topics we refer to the classical text of Tricomi (1961).

1.3 The Thomas Algorithm

The Thomas algorithm (Thomas, 1949) is a simplified form of Gaussian elimination. It yields a direct solution for a system of equations with tridiagonal matrix of coefficients, like:

$$\begin{aligned} b_1x_1 + c_1x_2 &= d_1 \\ a_2x_1 + b_2x_2 + c_2x_3 &= d_2 \\ a_3x_2 + b_3x_3 + c_3x_4 &= d_3 \\ a_{n-1}x_{n-2} + b_{n-1}x_{n-1} + c_{n-1}x_n &= d_{n-1} \\ a_nx_{n-1} + b_nx_n &= d_n \end{aligned} \tag{1.317}$$

Dividing the first equation of the system by b_1, one obtains:

$$x_1 + c_1^* x_2 = d_1^* \tag{1.318}$$

with $c_1^* = c_1/b_1$ and $d_1^* = d_1/b_1$..

Next we subtract from the second equation of the system Eq. (1.318) multiplied by a_2, thus obtaining:

$$x_2 + c_2^* x_3 = d_2^*, \tag{1.319}$$

with $c_2^* = \frac{c_2}{b_2 - a_2 c_1^*}$ and $d_2^* = \frac{d_2 - a_2 d_1^*}{b_2 - a_2 c_1^*}$..

Proceeding likewise for the following equations we apply the recursive formulas:

$$a_r^* = 0, \quad (r = 2, \dots n) \tag{1.320}$$

$$b_r^* = 1, \quad (r = 1, \dots n) \tag{1.321}$$

$$c_{r+1}^* = \frac{c_{r+1}}{b_{r+1} - a_{r+1} c_r^*}, \quad (r = 1, \dots n - 1) \tag{1.322}$$

$$d_{r+1}^* = \frac{d_{r+1} - a_{r+1} d_r^*}{b_{r+1} - a_{r+1} c_r^*}, \quad (r = 1, \dots n - 1) \tag{1.323}$$

obtaining the following system of equations:

$$\begin{aligned}
x_1 + c_1^* x_2 &= d_1^* \\
x_2 + c_2^* x_3 &= d_2^* \\
x_3 + c_3^* x_4 &= d_3^* \\
x_{n-2} + c_{n-2}^* x_{n-1} &= d_{n-2}^* \\
x_{n-1} + c_{n-1}^* x_n &= d_{n-1}^* \\
x_n &= d_n^*
\end{aligned} \tag{1.324}$$

The solution of the system from Eq. (1.324) is now immediate: from the last equation one obtains $x_n = d_n^*$; from the last but one, one obtains $x_{n-1} = d_{n-1}^* - c_{n-1}^* x_n$ and so on up to x_1. Therefore the solution of the system (1.317) is obtained with just two sweeps of the n equations.

The Thomas algorithm is referred to in the literature as TDMA (Tri-Diagonal Matrix Algorithm). It has been coded in the Subroutine TRIDI, which is listed in the program **PROG_III.1 [FRONT1]**, in Appendix A.III.

References

Bottoni, M., Dorr, B., & Homann, C. (1992, April). The three-dimensional transient two-phase flow computer programme BACCHUS-3D/TP. KfK 4760.

Cugiani, M. (1967). *Metodi dell' Analisi Numerica*. UTET.

Davis, H. T. (1962). *Introduction to nonlinear differential and integral equations*. Courier Corporation.

Demidovic, B. P. (1989). *Esercizi e problemi di analisi matematica (translation from Russian)*. Editori Riuniti.

Fehlberg, E. (1968, October). Classical Fifth-, Sixth-, Seventh- and Eigth-Order Runge-Kutta Formulas with Stepwise Control, NASA-TR-R-287.

Grigorieff, R. D. (1972). *Numerik Gewönlicher Differentialgleichungen, Band 1 Einschrittverfahren*. Teubner Studienbücher.

Hofer, E. (1976, October). A partially implicit method for large stiff systems of ODEs withy only a few equations introducing small time-constants. *SIAM Journal Numerical Analysis, 13*(5).

Krylov, V. I. (1962). *Approximate calculation of integrals*. The Macmillan Company.

Press, W. H., Flannery, B. P., Teukolsky, S. A., & Vetterling, W. T. (1986). *Numerical recipes*. Cambridge University Press.

Spataro, S., & Tribulato, S. (1980). Equazioni Differenziali, Edizioni Tecnos, Milano

Thomas, L. H. (1949). *Elliptic problems in linear differential equations over a network*. New York.

Tricomi, F. G. (1961). *Equazioni Differenziali*. Boringhieri.

Zwillinger, D. (Ed.). (1996). *Standard mathematical tables and formulae*. CRC Press.

Zwirner, G. (1977). Esercizi e Complementi di Analisi Matematica, Parte Seconda, CEDAM, Padova

Chapter 2
Partial Differential Equations and the Method of Characteristics

2.1 Partial Differential Equations of First Order

2.1.1 General Definitions

Let us consider the model partial differential equation of first order:

$$a(x,t)\frac{\partial u}{\partial x} + b(x,t)\frac{\partial u}{\partial t} = c(x,t)u + d(x,t), \tag{2.1}$$

where $u = u(x, t)$ is a differentiable function in the space (x, t). If the coefficients a, b, c, d are constant or functions of x and t alone, Eq. (2.1) is linear; if the coefficients are functions of x, t, u (but not of the derivatives $\partial u/\partial x$, $\partial u/\partial t$), the equation is quasi-linear; if the coefficients are functions of the derivatives $\partial u/\partial x$, $\partial u/\partial t$ or of derivatives of higher order, the equation is nonlinear.

In the following sections, we consider an initial value problem for Eq. (2.1), with initial condition for $t = 0$:

$$u(x,0) = F(x), \tag{2.2}$$

where $F(x)$ is a given function.

2.1.2 Equations of Characteristics and Formal Solution

The general solution of Eq. (2.1), though unknown, can be visualized in principle as consisting of an integral surface in the space (x, t, u) (see Fig. 2.1) which originates from the curve $F(x)$ traced on the $t = 0$ plane. The integral surface consists of integral

M. Bottoni, *Physical Modeling and Computational Techniques for Thermal and Fluid-dynamics*, Mechanical Engineering Series,
https://doi.org/10.1007/978-3-030-79717-1_2

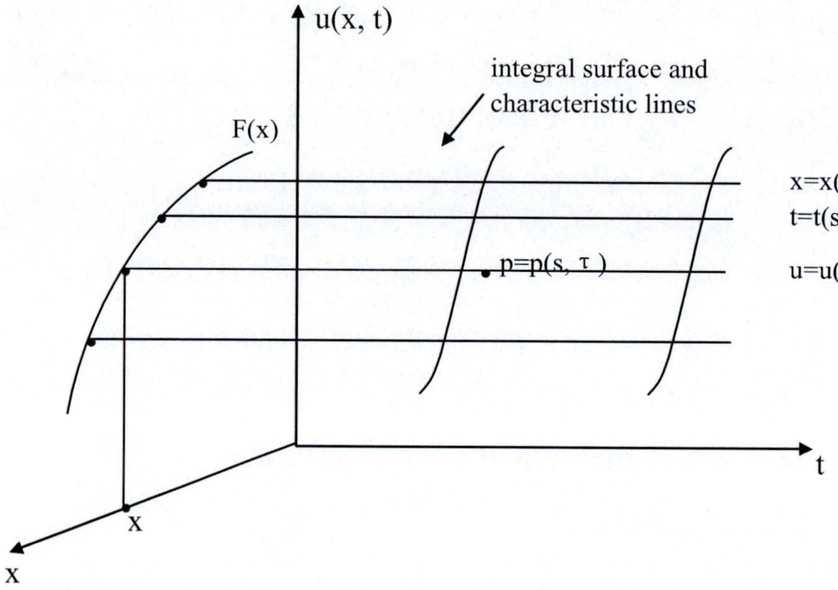

Fig. 2.1 Sketch showing integral surface and characteristic lines of the model equation (2.1)

lines, or characteristics, originating from points of the curve $F(x)$. The characteristic curves can be represented by means of two parameters, s and $\boldsymbol{\tau}$ (specified later) as:

$$x = x(s, \tau), \tag{2.3a}$$

$$t = t(s, \tau), \tag{2.3b}$$

$$u = u(s, \tau). \tag{2.3c}$$

The equations:

$$\frac{dx}{ds} = a(x, t), \tag{2.4}$$

$$\frac{dt}{ds} = b(x, t), \tag{2.5}$$

are the differential forms of Eqs. (2.3a) and (2.3b) and satisfy the given differential equation (2.1). In fact, differentiating the function $u(x, t) = u(x(s), t(s))$, one obtains:

$$\frac{du}{ds} = \frac{\partial u}{\partial x}\frac{dx}{ds} + \frac{\partial u}{\partial t}\frac{dt}{ds} = a\frac{\partial u}{\partial x} + b\frac{\partial u}{\partial t} = cu + d. \tag{2.6}$$

Equation (2.6) is the differential form of Eq. (2.3c).

The method of characteristics reduces the problem of solving a partial differential equation to that of solving a system of ordinary differential equations.

The parameter s can be interpreted as a generalized coordinate along the characteristic curves, and we set the value $s = 0$ on the points of the curve $F(x)$. Therefore the parametric equations of the curve $F(x)$ on the plane $t = 0$, as functions of the parameter τ only, can be written as:

$$t = t(\tau) = 0, \ (s = 0) \tag{2.7a}$$

$$x = x(\tau) = \tau, \quad (s = 0) \tag{2.7b}$$

$$u = F(x) = F(\tau). \quad (s = 0) \tag{2.7c}$$

Equation (2.7b) is interpreted just as a choice of the parameter τ.

Integrating Eqs. (2.4), (2.5), and (2.6) yields:

$$x = as + x(s = 0) = as + \tau, \tag{2.8}$$

$$t = bs + t(s = 0) = bs, \tag{2.9}$$

$$u = \int (cu + d)ds + u(s = 0) = \int (cu + d)ds + F(\tau). \tag{2.10}$$

The integral in Eq. (2.10) is a function of s, say $f(s)$, and therefore $u = u(x, t)$ can be written as:

$$u = f(s) + F(\tau) = \widetilde{F}(s, \tau), \tag{2.11}$$

where $\widetilde{F}$ is some function of s and τ. These functions $f(s)$ and $\widetilde{F}(s, \tau)$ can be determined when the coefficients c and d of Eq. (2.1) are given. Equations (2.8), (2.9), and (2.10) are the explicit forms of the characteristic equations (2.3a, 2.3b, and 2.3c). Equation (2.11) is the sought solution of Eq. (2.1). It can be written in terms of the independent variables x and t by eliminating the parameters s and τ. Using Eq. (2.9) one has $s = t/b$; hence from (2.8) $\tau = x - as = x - at/b$. Therefore the solution becomes:

$$u(x, y) = \widetilde{F}\left(\frac{t}{b}, x - a\frac{t}{b}\right). \tag{2.12}$$

Further details on the method of characteristics are given in Zauderer (1983).

2.1.3 *Solutions of Linear Damped Wave Equation*

Let us consider the linear damped wave equation:

$$\frac{\partial u}{\partial t} + c\frac{\partial u}{\partial x} + \lambda u = 0. \;\; [u = u(x,t), \lambda \geq 0] \tag{2.13}$$

with initial condition $u(x,0) = F(x)$. Comparing Eq. (2.13) with the model Eq. (2.1), one sees that $a = c$, $b = 1$, $c = -\lambda$, $d = 0$.

The characteristic equations in differential form (corresponding to Eqs. (2.4)–(2.6) of the previous section) are:

$$\frac{dx}{ds} = c, \tag{2.14}$$

$$\frac{dt}{ds} = 1, \tag{2.15}$$

$$\frac{du}{ds} = -\lambda u. \tag{2.16}$$

Again for $s = 0$ one obtains the parametric equations of the curve $F(x)$ on the plane $t = 0$, as functions of the parameter τ only:

$$t = 0, \;\; (s = 0) \tag{2.17a}$$

$$x = \tau, \;\; (s = 0) \tag{2.17b}$$

$$u = F(\tau). \;\; (s = 0) \tag{2.17c}$$

Integrating Eqs. (2.14) through (2.16) yields:

$$x = cs + x(s = 0) = cs + \tau, \tag{2.18}$$

$$t = s + t(s = 0) = s, \tag{2.19}$$

$$\log u = -\lambda s + k, \tag{2.20}$$

where k is an integration constant; hence:

$$u = e^{-\lambda s}k. \tag{2.21}$$

For $s = 0$, Eq. (2.21) yields $u = k$ and, from Eq. (2.17c), $u = F(\tau)$; hence: $k = F(\tau)$ and Eq. (2.21) becomes:

$$u = e^{-\lambda s}F(\tau). \tag{2.22}$$

Eliminating $s = t$ and$\tau = x - cs = x - ct$, the solution (2.22) becomes:

$$u(x,t) = e^{-\lambda t}F(x - ct). \tag{2.23}$$

Equation (2.23) represents a wave of amplitude $F(x)$ traveling to the right (for $c > 0$), in the direction of the positive x coordinate. The speed c is given, from Eqs. (2.14) and (2.15), by:

$$\frac{dx}{dt} = c. \tag{2.24}$$

The factor $e^{-\lambda t}$ in Eq. (2.23) represents a damping with time of the wave with time constant λ. If $\lambda = 0$ the wave is not damped.

2.1.4 Quasi-Linear Undamped Wave

Let us consider the quasi-linear equation:

$$\frac{\partial u}{\partial t} + u\frac{\partial u}{\partial x} = 0, \ [u = u(x, t)] \tag{2.25}$$

with initial condition $u(x, 0) = F(x)$. In this case the speed of the wave is not constant but equal to u; hence it is time and space dependent. Comparing Eq. (2.25) with the model equation (2.1), one sees that $a = u$, $b = 1$, $c = d = 0$.

The characteristic equations in differential form are:

$$\frac{dx}{ds} = u, \tag{2.26}$$

$$\frac{dt}{ds} = 1, \tag{2.27}$$

$$\frac{du}{ds} = 0. \tag{2.28}$$

Equation (2.28) implies that u is constant along the characteristic and equal to the value $u = F(\tau)$ on the plane $t = 0$, which corresponds to $s = 0$; hence:

$$u(s, \tau) = u(0, \tau) = F(\tau). \tag{2.29}$$

Integrating Eqs. (2.26) and (2.27) yields:

$$x = us + x(s = 0) = us + \tau, \tag{2.30}$$

$$t = s + t(s = 0) = s. \tag{2.31}$$

From Eq. (2.30), $\tau = x - us = x - ut$. The solution (2.29) becomes:

$$u(x,t) = F(x - ut). \tag{2.32}$$

Equation (2.32) represents a wave of initial shape $F(x)$ which travels with velocity $u = dx/dt$ different from point to point. Points at higher amplitude travel faster than points of lower amplitude. Therefore at some time higher located points of the wave overtake (bypass) lower located points. The function $u(x,t)$ becomes multivalued and loses physical meaning. The wave represented by the function $u(x,t)$ is said to "break" at this point in time, and the time point is called t_{Br} or "breaking point." The time point t_{Br} can be computed because for $t = t_{Br}$,the tangent $\partial u/\partial x$ to the wave $u(x, t)$ becomes infinite. Differentiating u with respect to x, one obtains:

$$\frac{\partial u}{\partial x} = \frac{\partial F}{\partial x} = \frac{\partial F}{\partial \tau}\frac{\partial \tau}{\partial x} = \frac{\partial F}{\partial \tau}\left(1 - \frac{\partial u}{\partial x}t\right), \tag{2.33}$$

having used $\tau = x - ut$, $\partial\tau/\partial x = 1 - (\partial u/\partial x)t$.

Hence, from Eq. (2.33), solving for $\partial u/\partial x$, one derives:

$$\frac{\partial u}{\partial x} = \frac{\partial F/\partial \tau}{1 + t\frac{\partial F}{\partial \tau}}. \tag{2.34}$$

The wave breaks when $1 + t\frac{\partial F}{\partial \tau}$ becomes zero and $\partial u/\partial x$ becomes infinite. This occurs at time:

$$t_{Br} = -\frac{1}{\partial F/\partial \tau}. \tag{2.35}$$

For instance, if the initial condition is $F(x) = \sin(\pi x)$, the solution of Eq. (2.25) is $u = \sin[\pi(x - ut)] = F(x - ut) = F(\tau)$. Hence $\partial F/\partial \tau = \pi \cos(\pi\tau)$ and

$$t_{Br} = -\frac{1}{\pi\cos(\pi\tau)}. \tag{2.36}$$

Equation (2.36) yields the minimum value of t_{Br} for $\cos(\pi\tau) = -1$ and this value is $t_{Br} = 1/\pi \approx 0.318$.

Notice that, from Eq. (2.34), a wave can break only if, at some point of the wave, the derivative $\partial F/\partial \tau$ is negative. If $\partial F/\partial \tau$ is always positive, the wave never breaks.

2.2 Partial Differential Equations of Second Order

2.2.1 General Definitions

Considering a two-dimensional space (x, y) and a differentiable function $u(x, y)$ defined in this space, the most general partial differential equation of second order can be written as:

$$a\frac{\partial^2 u}{\partial x^2}+b\frac{\partial^2 u}{\partial x\partial y}+c\frac{\partial^2 u}{\partial y^2}+d\frac{\partial u}{\partial x}+e\frac{\partial u}{\partial y}+fu+g=0. \tag{2.37}$$

If the coefficients a, b, c…… are constants or functions of x and y only, Eq. (2.37) is linear; if the coefficients are functions of x, y, u, $\partial u/\partial x$, $\partial u/\partial y$ (but not of derivatives of higher order), the equation is quasi-linear; otherwise, when the coefficients are functions of second- or higher-order derivatives, the equation is nonlinear.

Equation (2.37) is classified as hyperbolic if the discriminant $\Delta = b^2 - 4ac > 0$ is positive, as parabolic if $\Delta = 0$, and as elliptic if $\Delta < 0$. In Sect. 2.2.2 we will analyze the reason why the value of Δ plays a crucial role in determining the type of the given differential equation of second order. Notice that only the coefficients of the second-order derivatives (a, b, c) determine the type of the equation.

The Poisson equation $\frac{\partial^2 u}{\partial x^2}+\frac{\partial^2 u}{\partial y^2}+g=0$ and the Laplace equation $\frac{\partial^2 u}{\partial x^2}+\frac{\partial^2 u}{\partial y^2}=0$ are prototypes of the equations of elliptic type, because $\Delta = b^2 - 4ac = -4 < 0$.

The heat diffusion equation, given for a function $u(x,t)$, $\alpha\frac{\partial^2 u}{\partial x^2}-\frac{\partial u}{\partial t}=0$, with $\Delta = 0$, is the prototype of equations of parabolic type.

The vibration equation $\frac{\partial^2 u}{\partial x^2}-\frac{\partial^2 u}{\partial y^2}=0$, as well as the wave equation for a function $u(x,t)$, $\frac{\partial^2 u}{\partial x^2}-\frac{\partial^2 u}{\partial t^2}=0$, both with $\Delta = 4 > 0$, are prototypes of the equations of hyperbolic type.

An equation can be of different types in different regions of its domain of definition. For example, in the equation:

$$y\frac{\partial^2 u}{\partial x^2}+x\frac{\partial^2 u}{\partial x\partial y}+y\frac{\partial^2 u}{\partial y^2}+g=0 \tag{2.38}$$

we have $\Delta = x^2 - 4y^2$. Therefore the equation is hyperbolic for $|x| > 2|y|$, parabolic on the lines $|x| = 2|y|$ (or $y = x/2$ and $y = -x/2$), and elliptic for $|x| < 2|y|$.

Technical or scientific problems described by partial differential equations rarely require the knowledge of the "general integral" of the totality of possible solutions. In most cases particular integrals are sought satisfying additional conditions, and the problem of the solution consists in extracting from the generality of possible solutions the one or those satisfying the conditions imposed.

The character of the partial differential equations is related to both the geometrical configuration of the definition domain and the physical type of problems involved.

Generally speaking boundary value problems are described by elliptical partial differential equations. Initial value problems and wave problems fall in the category of hyperbolic or parabolic partial differential equations.

By problems of mathematical physics, i.e., by mathematical modelling of physical problems, in which, from given boundary or initial conditions, particular integrals are sought, three main requirements are compulsory:

(i) A solution must exist;
(ii) The solution must be uniquely determined;
(iii) The solution must be a continuous function of the data (initial or boundary values).

The first of these conditions is self-evident and does not need to be further discussed. The second condition implies that the problem must be so well defined that its solution is uniquely determined; no ambiguity must be left about the proper solution of the problem. The third condition is related to the fact that in technical or scientific applications the data (initial or boundary values) may not be known with absolute precision but may be affected by uncertainties in some allowed range. To small variations of the data must correspond small and continuous variations of the solution. Only in this case a mathematical problem can be considered a proper model of a physical phenomenon.

When the three mentioned conditions are satisfied, a problem described by partial differential equations is said to be "*properly posed*" or "*well posed*". The concept of "well posedness" is due to the eminent French mathematician Hadamard (1865–1963) who formulated it in 1923. A more detailed treatment of these topics is given in Courant and Hilbert (1968).

At the time of Hadamard, it was accepted as self evident that an improperly posed problem could not have physical relevance, i.e., could not describe any real problem. A typical example of improperly posed problem is the so-called initial value Cauchy problem for elliptical equations.

Given a function $u(x,y)$ defined in a domain D $(0 < x < 1, 0 < y < 1)$, this problem is defined by Forsythe and Wasow (1960):

$$\frac{\partial^2 u}{\partial x^2} + \frac{\partial^2 u}{\partial y^2} = 0 \tag{2.39}$$

$$u(x,0) = f(x) \tag{2.40}$$

$$\frac{\partial u}{\partial y}(x,0) = g(x) \tag{2.41}$$

where $f(x)$ and $g(x)$ are given functions. This problem has no solution.

Further insight into the concept of properly posed problems can be obtained by considering a problem of two-phase flow modelling, defined as follows. Let us consider a stream of two phases, liquid and vapor of water (or liquid sodium and sodium vapor in fast nuclear reactors under hypothetical accident conditions) in a three-dimensional domain D with boundary Γ. The state of the flow is defined by a time-dependent vector Y of unknown independent variables: three components of liquid velocity, three components of vapor velocity, pressure, volume fraction of vapor, and time. Using superscripts "l" and "v" for liquid and vapor, the unknown vector Y can therefore be defined by:

$$Y = Y\left(p, a, u_x^l, u_y^l, u_z^l, u_x^v, u_y^v, u_z^v, t\right) = Y(\overline{x}, t) \tag{2.42}$$

with:

$$\overline{x} = \overline{x}\left(p, a, u_x^l, u_y^l, u_z^l, u_x^v, u_y^v, u_z^v\right) \tag{2.43}$$

The two-phase flow is described by the conservation equations of mass (continuity equation), momentum and energy. The domain of definition D is partitioned in a number N of subdomains, referred to as control volumes or cells. The conservation equations are integrated over the cells to obtain a system of difference equations (linear algebraic equations). Scalar quantities, like density, pressure, and physical properties (viscosity, thermal conductivity), of both phases are usually defined at the center of the cells, while vector quantities, like velocity and mass fluxes, are defined at the center of the surfaces of the cells. Through the partition of the domain in cells and the integration process of the conservation equations over the cell volumes, a system of algebraic equations is obtained with as many equations as the number of cells in the domain.

Without going further into the details of two-phase flow modelling, we state that the two-phase flow can be eventually described by a system of equations which in matrix form can be written as:

$$A\frac{\partial Y}{\partial t} + \sum_{j=1}^{3} \widetilde{B}_j \frac{\partial Y}{\partial x_j} = \widetilde{F} \ \ (j = x, y, z) \tag{2.44}$$

hence, assuming that the matrix A is invertible:

$$\frac{\partial Y}{\partial t} + \sum_{j=1}^{3} B_j \frac{\partial Y}{\partial x_j} = F, \tag{2.45}$$

having defined $B_j = A^{-1}\widetilde{B}_j$ and $F = A^{-1}\widetilde{F}$. In these equations A, $\widetilde{B}_j$ and B_j are square matrices of order N with time-dependent coefficients. $\widetilde{F}$ and F are vectors of length N.

The problem of determining the time-dependent state of the two-phase flow in the given three-dimensional domain is treated as initial value problem, with reference to the time variable, and as boundary problem, with reference to the spatial coordinates:

$$Y(\overline{x}, 0) = G(\overline{x}), \tag{2.46}$$

$$Y(\overline{x}, t) = H(t), \quad (\text{on } \Gamma) \tag{2.47}$$

$$\frac{\partial Y(\overline{x}, t)}{\partial x_j} = T_j(t). \quad (\text{on } \Gamma) \ (j = x, y, z) \tag{2.48}$$

The eigenvalues of the system (2.45) are the roots of the algebraic equation (Forsythe & Wasow, 1960):

$$det\left(\lambda I + \sum_{j=1}^{3} \mu_j B_j\right) = 0. \tag{2.49}$$

Not all the root of this equation are granted to be real; some root may be imaginary, so the problem cannot be classified as hyperbolic but may be of elliptic character.

The question now arises, whether a system should definitely be hyperbolic and have real characteristics in order to have physical significance, or not. A theorem by Lax (1954) states that "*A system of linear partial differential equations with complex characteristics is unproperly posed as initial value problem.*" Such a system has in fact time-dependent pseudo-solutions which increase steadily with time and have not physical significance. There is not however a similar theorem valid for quasi-linear differential equations, which may have real characteristics and physically relevant solutions.

Generally speaking, consideration and modelling of more detailed physical phenomena, like shear stress between the phases, can change the character of the basic equations and lead to systems with real characteristics and physically sound solutions (Trapp, 1986; Travis et al., 1976). Valuable introductions to these topics, which are at research level, are found in Stuhmiller (1976, 1977).

2.2.2 *Equation of Characteristics*

Because only the second-order derivatives are important for determining the type (and the characteristics) of a partial differential equation of second order, we consider the simplified linear equation:

$$a\frac{\partial^2 u}{\partial x^2} + b\frac{\partial^2 u}{\partial x \partial y} + c\frac{\partial^2 u}{\partial y^2} + e = 0, \tag{2.50}$$

where $u = u(x, y)$, $a = a(x, y)$, $b = b(x, y)$, $c = c(x, y)$ and $e = e(x, y)$.

We define:

$$p = p(x, y) = \frac{\partial u}{\partial x}, \tag{2.51}$$

$$q = q(x, y) = \frac{\partial u}{\partial y}, \tag{2.52}$$

$$r = r(x, y) = \frac{\partial^2 u}{\partial x^2}, \tag{2.53}$$

$$s = s(x, y) = \frac{\partial^2 u}{\partial x \partial y}, \tag{2.54}$$

$$t = t(x, y) = \frac{\partial^2 u}{\partial y^2}. \tag{2.55}$$

With these definitions, Eq. (2.50) can be written as:

$$ar + bs + ct + e = 0. \tag{2.56}$$

Differentiating Eqs. (2.51) and (2.52), one obtains:

$$dp = \frac{\partial p}{\partial x} dx + \frac{\partial p}{\partial y} dy = rdx + sdy, \tag{2.57}$$

$$dq = \frac{\partial q}{\partial x} dx + \frac{\partial q}{\partial y} dy = sdx + tdy. \tag{2.58}$$

From Eqs. (2.57) and (2.58), one obtains:

$$r = \frac{dp}{dx} - s\frac{dy}{dx}, \tag{2.59}$$

$$t = \frac{dq}{dy} - s\frac{dx}{dy}. \tag{2.60}$$

Inserting Eqs. (2.59) and (2.60) in Eq. (2.56) and collecting the terms which contain s, one obtains:

$$s\left(-a\frac{dy}{dx} + b - c\frac{dx}{dy}\right) + a\frac{dp}{dx} + c\frac{dq}{dy} + e = 0. \tag{2.61}$$

Multiplying all terms of Eq. (2.61) by $-dy/dx$ yields:

$$s\left[a\left(\frac{dy}{dx}\right)^2 - b\frac{dy}{dx} + c\right] - \left(a\frac{dp}{dx}\frac{dy}{dx} + c\frac{dq}{dx} + e\frac{dy}{dx}\right) = 0. \tag{2.62}$$

We define:

$$A = \left[a\left(\frac{dy}{dx}\right)^2 - b\frac{dy}{dx} + c \right], \tag{2.63}$$

$$B = a\frac{dy}{dx}dp + cdq + edy. \tag{2.64}$$

A is the sum of the terms in the square bracket of Eq. (2.62), while B is the sum of the terms in the round bracket, multiplied by dx. We choose two curves (more precisely, two families of curves) in the real or imaginary space such that their tangent dy/dx in every point satisfies the equation $A = 0$; hence:

$$a\left(\frac{dy}{dx}\right)^2 - b\frac{dy}{dx} + c = 0. \tag{2.65}$$

Equation (2.65) is called equation of characteristics. It has two real distinct roots if $\Delta = b^2 - 4ac > 0$, two real coincident roots if $\Delta = 0$, and two imaginary roots if $\Delta < 0$. The classification of the partial differential equations of second order is given in Sect. 2.2.1.

In the following we assume that Eq. (2.65) has two real and distinct roots so that Eq. (2.50) is of hyperbolic type.

If, for all points along the characteristic curves, equation $A = 0$ is satisfied, from Eq. (2.62) follows $B = 0$, or:

$$a\frac{dy}{dx}dp + cdq + edy = 0. \tag{2.66}$$

Apart from the derivative dy/dx, which we assume to be one of the two solutions of Eq. (2.65), Eq. (2.66) contains only total differentials (dp, dq, and dy). This fact allows one to integrate Eq. (2.66) along the characteristic lines (as shown in Sect. 2.2.3) to determine the solution $u = u(x, y)$, without having to deal with partial differentials.

2.2.3 *The Method of Characteristics (Numerical Application)*

We assume (see Fig. 2.2) that the values of the functions $u = u(x, y)$, $p = p(x, y) = \partial u/\partial x$, $q = q(x, y) = \partial u/\partial y$ are given on a curve Γ of the space (x,y) (boundary value problem). Our goal is to determine the values of the functions p, q and then of u in the domain of definition, assumed, as an example, to be the two-dimensional space above the curve Γ. The two basic equations used for this purpose are recalled here:

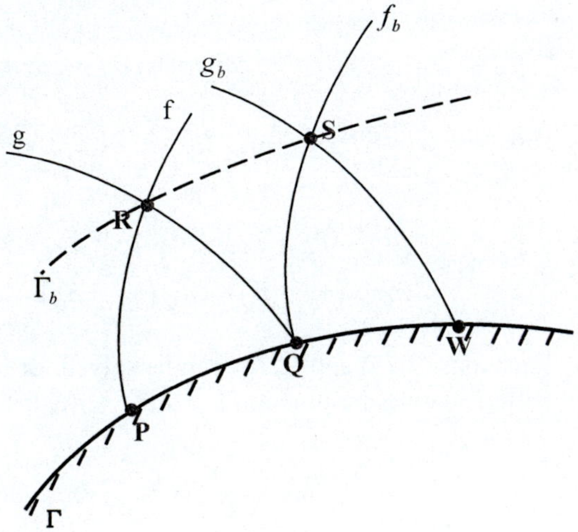

Fig. 2.2 Sketch showing characteristic curves f and g and their use to solve Eq. (2.50)

$$a\left(\frac{dy}{dx}\right)^2 - b\frac{dy}{dx} + c = 0, \tag{2.67}$$

$$a\frac{dy}{dx}dp + cdq + edy = 0. \tag{2.68}$$

We assume that Eq. (2.67) has two real and distinct roots:

$$\frac{dy}{dx} = f, \tag{2.69}$$

$$\frac{dy}{dx} = g. \tag{2.70}$$

These roots, f and g, represent two families of curves (characteristics) such that their tangent in every point satisfies Eq. (2.67). From these families of curves, we consider two, as shown in the sketch, originating from the points P and Q of the curve Γ and crossing each other in point R. First of all, one has to find the coordinates (x_R, y_R) of point R. We apply, and discretize, Eq. (2.69) to the segment $\overline{RP}$ and Eq. (2.70) to the segment $\overline{RQ}$, obtaining:

$$y_R - y_P = (x_R - x_P)f_P, \tag{2.71}$$

$$y_R - y_Q = (x_R - x_Q)g_Q. \tag{2.72}$$

Because f_P,f_Q, x_P,y_P, x_Q,y_Q are known, Eqs. (2.71) and (2.72) can be solved for (x_R, y_R).

We write Eq. (2.66) for the two roots $dy/dx = f$, $dy/dx = g$, obtaining:

$$afdp + cdq + edy = 0, \tag{2.73}$$

$$agdp + cdq + edy = 0. \tag{2.74}$$

Applying and discretizing Eqs. (2.73) and (2.74) to the segments $\overline{RP}$ and $\overline{RQ}$, respectively, yields:

$$a_P f_P(p_R - p_P) + c_P(q_R - q_P) + e_P(y_R - y_P) = 0, \tag{2.75}$$

$$a_Q g_Q(p_R - p_Q) + c_Q(q_R - q_Q) + e_Q(y_R - y_Q) = 0. \tag{2.76}$$

Equations (2.75) and (2.76) can be solved for p_R, q_R.
Differentiating the function $u = u(x, y)$ yields the equation:

$$du = \frac{\partial u}{\partial x}dx + \frac{\partial u}{\partial y}dy = pdx + qdy, \tag{2.77}$$

which could be applied either to the segment $\overline{RP}$ or $\overline{RQ}$. Using $\overline{RP}$ yields:

$$u_R - u_P = \frac{1}{2}(p_R + p_P)(x_R - x_P) + \frac{1}{2}(q_R + q_P)(y_R - y_P). \tag{2.78}$$

from which u_R is computed.

This procedure, represented by Eqs. (2.71) through (2.78), can be repeated for other characteristics, say f_b, g_b, originating from the points Q and W on Γ and crossing in S and so on, till the unknown function $u = u(x, y)$ is determined on a series of points (nodes) of a curve Γ_b. Starting again from Γ_b, the values of the unknown function can be determined on a series of points of a third curve Γ_c and so on for all definition domain.

Several numerical examples of application of the method of characteristics for partial differential equations of second order are given in Smith (1969).

Exercise 2.1
The equation:

$$\frac{\partial^2 u}{\partial x^2} + \frac{\partial^2 u}{\partial x \partial y} - 2\frac{\partial^2 u}{\partial y^2} + 1 = 0 \tag{2.79}$$

is defined in the domain $D = [0 \le x \le 1, 0 \le y]$ and for $y = 0$ hold the boundary conditions:

$$u = x, \tag{2.80}$$

$$q = \frac{\partial u}{\partial y} = x. \tag{2.81}$$

Because $b^2 - 4ac = 9$ (>0), Eq. (2.79) is hyperbolic.

Find the equations of the characteristics through the points $P(x_p, y_p) = P(0.4, 0)$ and $Q(x_q, y_q) = Q(0.5, 0)$ and the solution u_R in the point R in which the characteristics cross each other.

The solution is obtained as follows.

Equation (2.80) gives:

$$p = \frac{\partial u}{\partial x} = 1. \tag{2.82}$$

Comparing with the model equation (2.50), one has for Eq. (2.79): $a = 1, b = 1$, $c = -2$, $e = 1$. Hence Eqs. (2.65) and (2.66) are, in this case:

$$\left(\frac{dy}{dx}\right)^2 - \frac{dy}{dx} - 2 = 0, \tag{2.83}$$

$$\frac{dy}{dx}dp - 2dq + dy = 0. \tag{2.84}$$

Equation (2.83) has the solutions (equations of the characteristics):

$$f = \frac{dy}{dx} = 2, \tag{2.85}$$

$$g = \frac{dy}{dx} = -1. \tag{2.86}$$

In point P one has:
$u_P = 0.4$, [from Eq. (2.80)].
$f_P = 2$, [from Eq. (2.85)].
$p_P = 1$, [from Eq. (2.82)].
$q_P = 0.4$. [from Eq. (2.81)].
In point Q one has:
$u_Q = 0.5$, [from Eq. (2.80)].
$g_Q = -1$, [from Eq. (2.86)].
$p_Q = 1$, [from Eq. (2.82)].
$q_Q = 0.5$. [from Eq. (2.81)].

First we compute the coordinates of point R. Applying Eq. (2.85) to the segment RP of the characteristic f, one has:

$$y_R - y_P = 2(x_R - x_P),$$

$$y_R = 2(x_R - 0.4). \tag{2.87}$$

Applying Eq. (2.86) to the segment RQ of the characteristic g, one has:

$$\begin{aligned} y_R - y_Q &= (-1)(x_R - x_Q), \\ y_R &= (-x_R + 0.5). \end{aligned} \tag{2.88}$$

The system of Eqs. (2.87) and (2.88) has the solutions:

$$x_R = 1.3/3, \tag{2.89a}$$

$$y_R = 0.2/3. \tag{2.89b}$$

Second, we compute the values of p and q in point R. Equation (2.84), written for the characteristic f and applied to the segment RP, yields:

$$\begin{aligned} fdp - 2dq + dy &= 0, \\ 2(p_R - p_P) - 2(q_R - q_P) + (y_R - y_P) &= 0, \\ 2p_R - 2 - 2q_R + 0.8 + 0.2/3 &= 0, \\ p_R &= q_R + 1.7/3. \end{aligned} \tag{2.90}$$

Equation (2.84), written for the characteristic g and applied to the segment RQ, yields:

$$\begin{aligned} gdp - 2dq + dy &= 0, \\ (-1)(p_R - p_Q) - 2(q_R - q_Q) + y_R - y_Q &= 0, \\ -p_R + 1 - 2q_R + 1 + 0.2/3 &= 0, \\ p_R + 2q_R &= 6.2/3. \end{aligned} \tag{2.91}$$

Equations (2.90) and (2.91) yield the solutions:

$$p_R = 3.2/3, \tag{2.92a}$$

$$q_R = 0.5. \tag{2.92b}$$

Finally, Eq. (2.78) yields the solution u_R:

$$
\begin{aligned}
u_R - u_P &= \frac{1}{2}(p_R + p_P)(x_R - x_P) + \frac{1}{2}(q_R + q_P)(y_R - y_P), \\
u_R &= 0.4 + \frac{1}{2}\left(\frac{3.2}{3} + 1\right)\left(\frac{1.3}{3} - 0.4\right) + \frac{1}{2}(0.5 + 0.4)\frac{0.2}{3}, \\
u_R &= \frac{8.34}{18}.
\end{aligned}
\tag{2.93}
$$

Exercise 2.2
The equation:

$$\frac{\partial^2 u}{\partial x^2} - u^2 \frac{\partial^2 u}{\partial y^2} = 0 \tag{2.94}$$

is defined in the domain $D = [0 \leq x \leq 1,\ 0 \leq y]$ and for $y = 0$ hold the boundary conditions:

$$u = 0.2 + 5x^2 \tag{2.95}$$

$$q = \frac{\partial u}{\partial y} = 3x. \tag{2.96}$$

Because $b^2 - 4ac = 4u^2$ (>0), Eq. (2.94) is hyperbolic.

Find the equations of the characteristics through the points $P(x_p, y_p) = P(0.2, 0)$ and $Q(x_q, y_q) = Q(0.3, 0)$ and the solution u_R in the point R in which the characteristics cross each other.

(Solution: $u_R \approx 0.567$).

2.2.4 *Numerical Solution of the Laplace Equation*

This last section of this chapter illustrates some methods for the numerical solution of the Laplace equations $\nabla^2 u = 0$. We direct the reader to Chap. 6 for the numerical treatment, made with several variants, of the Poisson equation $\nabla^2 u = f$. In Chap. 6 the so-called pressure method, consisting in combining continuity and momentum equations governing a flow to derive a Poisson equation, is first introduced from the theoretical and practical viewpoints. Numerical methods are then explained in detail both for the Poisson equation and for a "Poisson-like" equation. It is implied that this equation is formally similar to the equation which would be obtained by discretizing a Poisson equation, apart from the fact that the coefficients are not constant.

Let a Laplace equation:

$$\nabla^2 u = 0 \tag{2.97}$$

be given in a two-dimensional domain D defined by $(0 \le x \le x_{\mathbf{max}}, 0 \le y \le y_{\mathbf{max}})$ with given boundary conditions $u = u_B$ on the boundary Γ of D. Equation (2.97) would describe, for instance, the temperature distribution in a body with constant heat conductivity and with given fixed temperature distributions at the boundaries.

In the following we give examples of numerical solutions of Eq. (2.97) in a square domain with $x_{\mathbf{max}} = y_{\mathbf{max}} = 4$ and with the following boundary conditions:

$$\begin{aligned} u(x,0) &= 0, \\ u(0,y) &= 0, \\ u(x,y_{\text{max}}) &= x^3, \\ u(x_{\text{max}},y) &= y^2_{\text{max}}y. \end{aligned} \tag{2.98}$$

The boundary condition varies as x^3 on the upper boundary of the domain D and varied linearly with y on the right boundary to reach the maximum value $x^3_{\text{max}} = y^3_{\text{max}} = 64$ at three right upper corner of D.

We describe briefly the three solution techniques coded in the following programs:

(i) **PROG_II_1_LAPLMC** with Subroutine LAPLMC, based on the Monte Carlo technique
(ii) **PROG_II _2_LAPLFI** with Subroutine LAPLFI, based on the 5-point formula
(iii) **PROG_II _3_LAPLUN** with Subroutine LAPLUN, based on under-relaxation.

2.2.4.1 Numerical Solution with Monte Carlo Technique

Let the definition domain D, in which Eq. (2.97) is given with boundary conditions (2.98), be partitioned in NI and NJ cells in the x and y directions, respectively. Letting $NI = NJ$, the cells have uniform lengths $\Delta x = x_{\mathbf{max}}/NI$ and $\Delta y = y_{\mathbf{max}}/NJ$. In the domain D is thus set a grid with a number of nodes given by NNODES $= (NI + 1)(NJ + 1)$. In the numerical example shown, we chose $NI = NJ = 19$; hence NNODES $= 400$. We denote the nodes in the grid with indices (i, j) with $(1 \le i \le NI + 1,\ \ 1 \le j \le NJ + 1)$.

The numerical solution with the Monte Carlo technique is based on the following procedure. Let us consider a node P_0 $(i = i0, j = j0)$ of the grid in the domain D and start a two-dimensional random walk from node P_0. Generating a random number R = 1/2/3/4, the point $P(i,j)$, representative of a "particle" making the random walk, moves as follows:

From $P(i,j)$ to $P(i-1, j)$ if $R = 1$
From $P(i,j)$ to $P(i, j + 1)$ if $R = 2$
From $P(i,j)$ to $P(i + 1, j)$ if $R = 3$

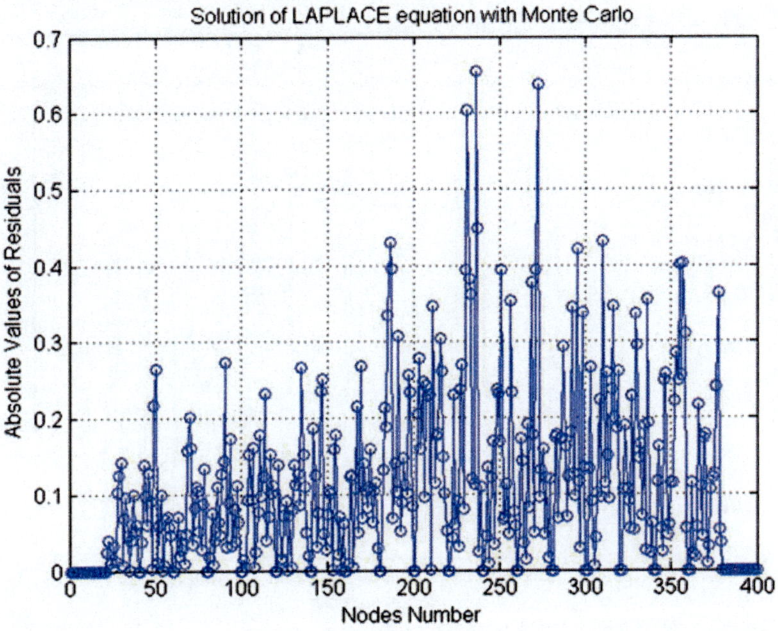

Fig. 2.3 Absolute value of residuals computed for the model Laplace equation (2.97) with boundary condition (2.98) with Monte Carlo technique with $N_w = 10^5$ sweeps over the full domain of definition

From $P(i,j)$ to $p(i, j-1)$ if $R = 4$

When the particle reaches a boundary at, say, $i = iB$, $j = jB$, the random walk stops and the boundary value $u(iB, jB)$ is assigned to the node$P_0 = (i0, j0)$. We repeat the random walk a number N_w of times (typically $N_w = 100,000$, but we present also results computed with $N_w = 10^6$) summing up the contributions of the ending boundary points to a counter C_0 for P_0. At the end the numerical approximation of the solution $u(P_0)$ will be $u(P_0) = u(\text{i}0, \text{j}0) = c_0/N_w$. Repeating the procedure for all inner points of the grid in the domain D, one obtains numerical approximations of the solution in the full domain.

The distribution of $u(i,j)$ computed with the Monte Carlo technique, with $N_w = 10^5$ sweeps for every node, is shown in Fig. 2.3 versus the node number. The nodes are numbered sequentially from left ($x = 0$) to right, till ($x = x_{\mathbf{max}}$), along the lines $x = (i - 1)\Delta x$ ($i = 1, 2, \ldots . NI$), from bottom ($y = 0$) to top ($y = y_{\mathbf{max}}$). The residuals are zero at the boundary points.

Figure 2.4 shows the residuals from a computation made with $N_w = 10^6$ sweeps for every node. In this second calculation, the residuals are reduced by a factor of three with respect to the previous calculation, at the expense of a tenfold computational time. In a further calculation with $N_w = 10^7$ sweeps, the maximum residual was 0.074. It is thus evident that the solution obtained with the Monte Carlo procedure converges very slowly to the correct solution with increasing number of

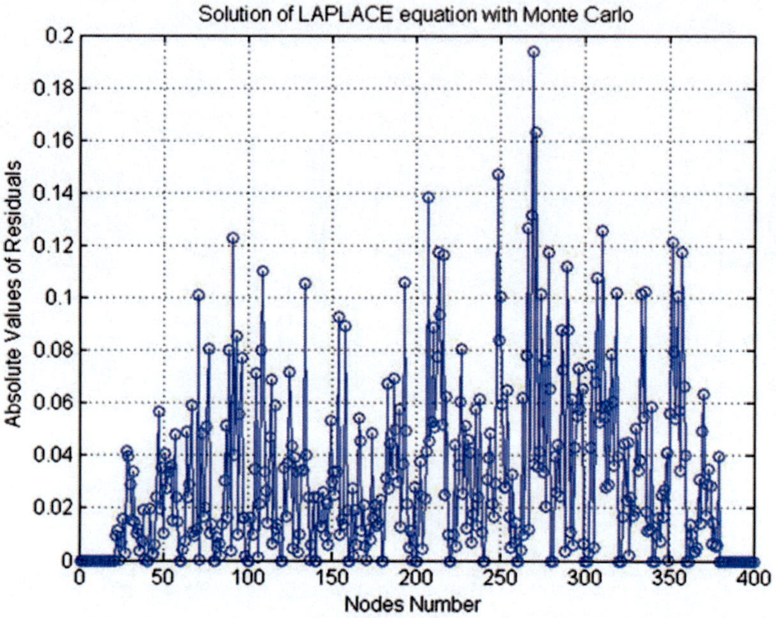

Fig. 2.4 Absolute value of residuals computed for the model Laplace equation (2.97) with boundary condition (2.98) with Monte Carlo technique with $N_w = 10^6$ sweeps over the full domain of definition

sweeps. The residuals are large with respect to those obtained with other algorithms; hence the numerical approximation that can be reached with the Monte Carlo procedure is poor. The method is however of interest from the didactic viewpoint because two facts relative to the Laplace equation become evident:

(i) Unlike the case of hyperbolic equations, in which the domain of influence of a boundary point P_B is a subdomain of D, bounded by the characteristics through P_B, in case of elliptic equations the domain of influence of a boundary point is the full domain D.
(ii) The solution $u(x,y)$ holds the "mini-max" principle, which can be stated as follows: the solution $u(x,y)$ in D is bounded by the minimum and maximum values assigned on the boundary Γ of D. Besides, the solution $u(x,y)$ has neither maxima nor minima in the domain D.

The numerical solution computed with the Monte Carlo method is shown in Fig. 2.5 versus the node number. The two-dimensional representation of the same results is shown in Fig. 2.6.

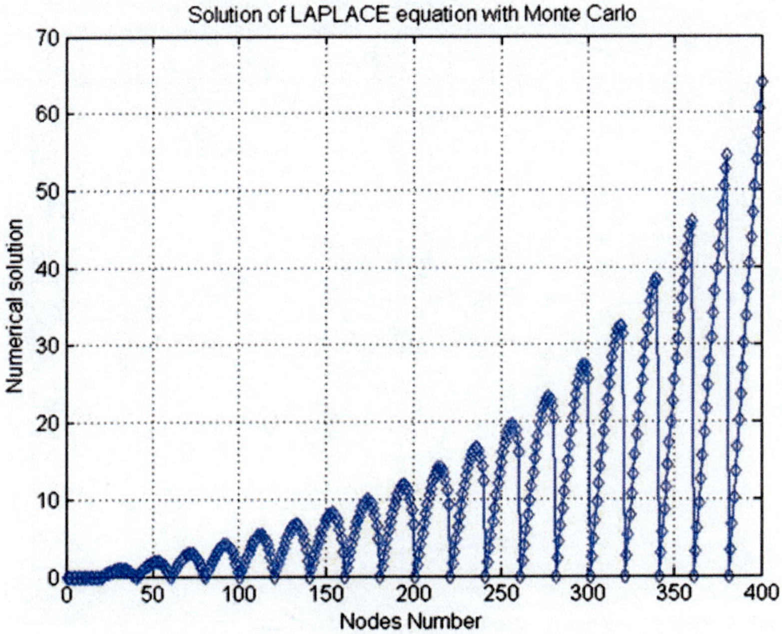

Fig. 2.5 Numerical solution of model Laplace equation (2.97) with boundary condition (2.98) computed with Monte Carlo technique with $N_w = 10^6$ sweeps versus node number

2.2.4.2 Numerical Solution with the 5-Point Formula

Discretization of Eq. (2.97) with central differences yields:

$$\frac{1}{\Delta x^2}\left(u_{i-1,j} - 2u_{i,j} + u_{i+1,j}\right) + \frac{1}{\Delta y^2}\left(u_{i,j-1} - 2u_{i,j} + u_{i,j+1}\right) = 0. \tag{2.99}$$

Let subscript 0 refer to the central node (i,j); subscripts 1, 2 refer to the neighboring nodes $(i-1,j)$, $(i+1,j)$ in the x coordinate direction; and subscripts 3, 4 refer to the neighboring nodes $(i,j-1)$, $(i,j+1)$ in the y direction. Hence from Eq. (2.99) we obtain:

$$(u_1 + u_2) + \frac{1}{\Delta y^2}(u_3 + u_4) - \left(\frac{2}{\Delta x^2} + \frac{2}{\Delta y^2}\right)u_0 = 0. \tag{2.100}$$

In case the meshes are uniform ($\Delta x = \Delta y = h$), one obtains:

$$u_1 + u_2 + u_3 + u_4 - 4u_0 = 0, \tag{2.101}$$

which is the so-called 5-point formula, characterized by the stencil:

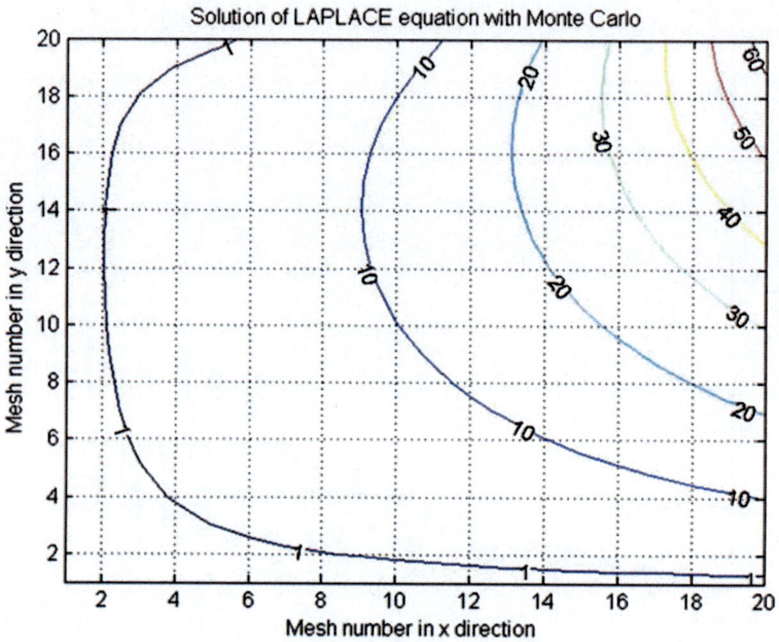

Fig. 2.6 Two-dimensional representation of the numerical solution of model Laplace equation (2.97) with boundary condition (2.98) computed with Monte Carlo technique with $N_w = 10^6$ sweeps

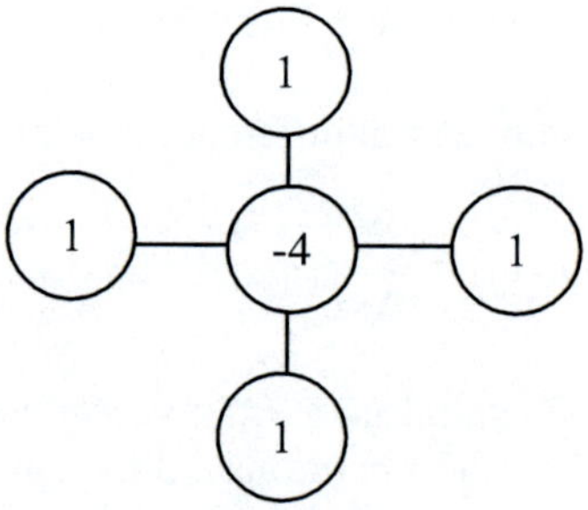

The coefficients of Eq. (2.101) are loaded in a matrix A, together with the boundary conditions. The matrix equation:

$$AU = B, \tag{2.102}$$

where the vector B collects the boundary conditions, is solved for $U = A^{-1}B$. The direct solution used in this case yields small residuals, of the order of magnitude of 10^{-14}, as shown in Fig. 2.7.

Results obtained with the 5-point formula are shown in Fig. 2.8 in one-dimensional representation, versus the node number.

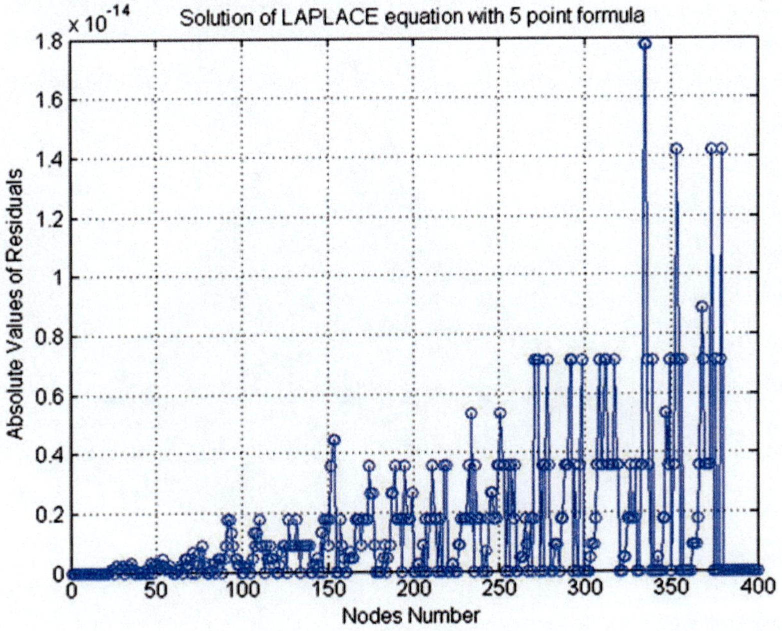

Fig. 2.7 Absolute value of residuals in the computational nodes from numerical solution of Laplace equation computed with the 5-point formula

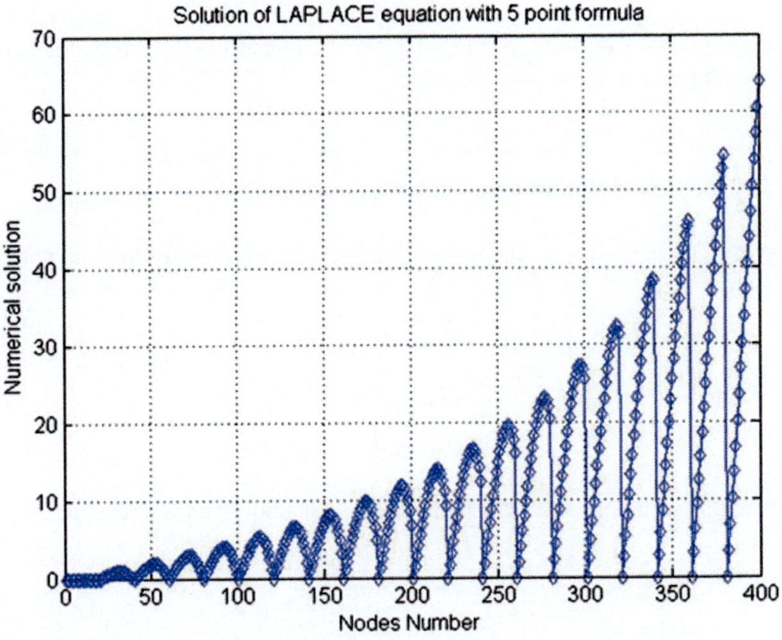

Fig. 2.8 Numerical solution of model Laplace equation computed with the 5-point formula versus node number

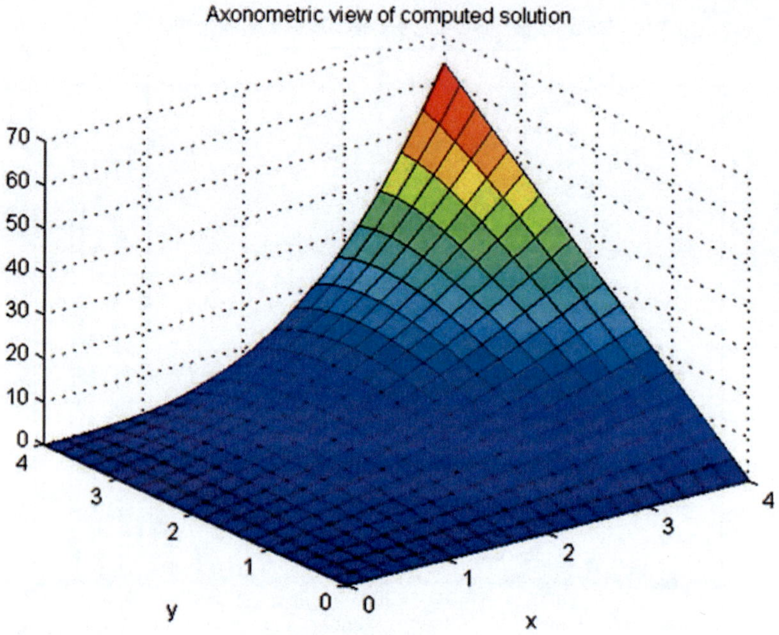

Fig. 2.9 Axonometric view of the numerical solution of model Laplace equation computed with the 5-point formula

Figure 2.9 shows an axonometric view of the numerical results obtained in the full domain with the five point formula. The level curves from the three-dimensional view of this figure are shown in Fig. 2.10.

2.2.4.3 Numerical Solution with Under-Relaxation

Equation (2.101) is used to evaluate the initial residuals for every internal node:

$$r_0 = u_1 + u_2 + u_3 + u_4 - 4u_0. \tag{2.103}$$

Equation (2.101), written for node 2 = $(i + 1, j)$, is:

$$u_0 + u_{i+1,j+1} + u_{i+1,j-1} + u_{i+2,j} - 4u_2 = 0. \tag{2.104}$$

Because this equation is not satisfied exactly, it actually yields a residual r_2 given by:

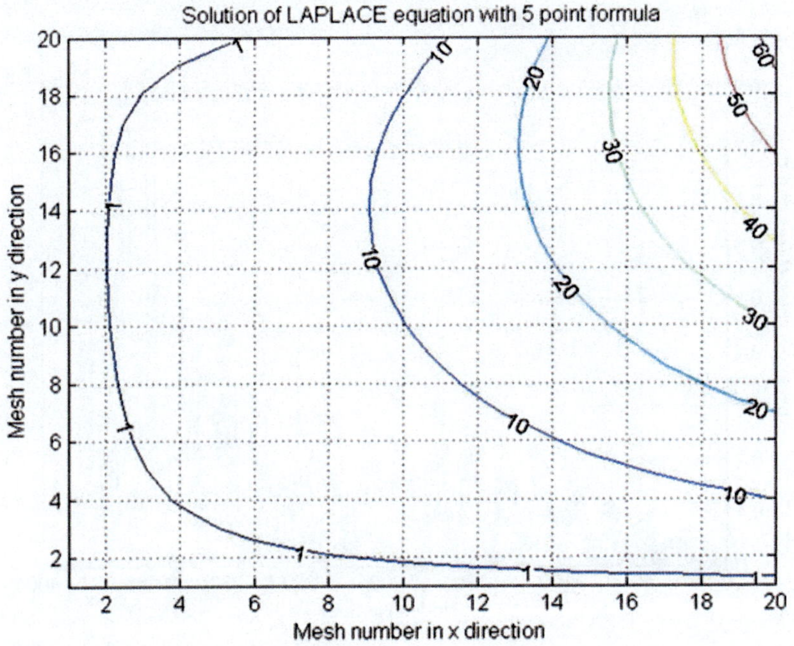

Fig. 2.10 Two-dimensional representation of the numerical solution of model Laplace equation computed with the five point formula

$$r_2 = u_0 + u_{i+1,j+1} + u_{i+1,j-1} + u_{i+2,j} - 4u_2. \tag{2.105}$$

Equations for the residuals in the other nodes surrounding node 0 can be written likewise. Equations (2.103) and (2.105) and consideration of the stencil shown in the previous section allow to deduce the following:

(i) A variation of the variable u_0 in node $0[=(i,j)]$ by one unity implies a reduction of the residual r_0 by four units.

(ii) A variation of the variable u_0 in node $0[=(i,j)]$ by one unity, positive or negative, implies a variation of the residuals in the neighboring nodes by one unit.

(iii) Variations of the residuals in node $0[=(i,j)]$ and in the neighboring nodes are of opposite signs.

These conclusions are taken into account in the under-relaxation method. In fact, using an under-relaxation parameter ω $(0 \leq \omega \leq 1)$, the residuals in node 0 and in the neighboring nodes are relaxed by:

$$\tilde{r}_0 = r_0 - \omega r_0 = r_0(1 - \omega), \tag{2.106a}$$

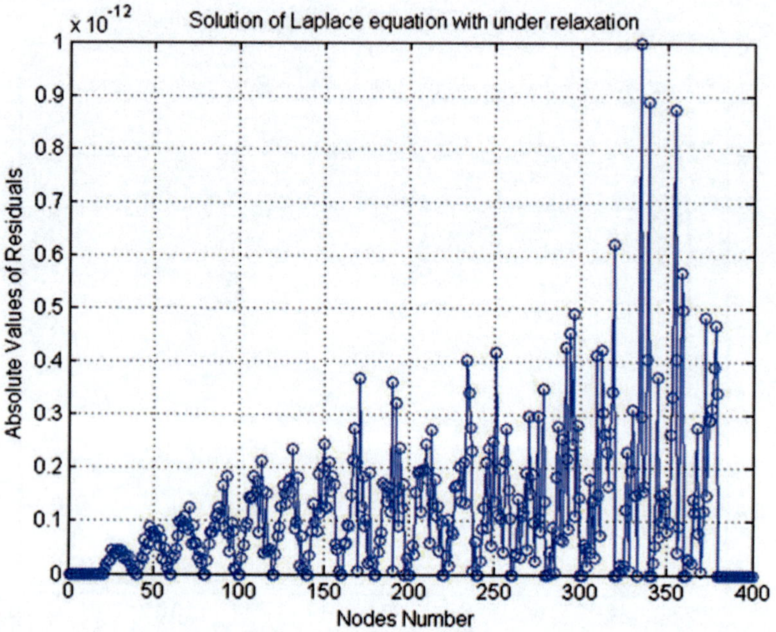

Fig. 2.11 Absolute value of residuals in the computational nodes from numerical solution of Laplace equation computed with under-relaxation

$$\widetilde{r}_1 = r_1 + \omega r_0/4, \tag{2.106b}$$

$$\widetilde{r}_2 = r_2 + \omega r_0/4, \tag{2.106c}$$

$$\widetilde{r}_3 = r_3 + \omega r_0/4, \tag{2.106d}$$

$$\widetilde{r}_4 = r_4 + \omega r_0/4, \tag{2.106e}$$

where the tilde denotes updated values.

The solution u_0 is updated by:

$$\widetilde{u}_0 = u_0 + \omega r_0/4. \tag{2.107}$$

Equation (2.106a) shows that for $\omega = 1$ the rest $\widetilde{r}_0$ is zeroed, or fully relaxed. Equations (2.106a, 2.106b, 2.106c, 2.106d, 2.106e) and (2.107) are applied to the full domain for one sweep. Sweeps are iterated till all residuals have been relaxed below a given threshold. Using $\omega = 0.7$ we needed 1900 iterations to reduce all residuals below 10^{-12}, as shown in Fig. 2.11.

Results obtained with the 5-point formula are shown in Fig. 2.12 in one-dimensional representation, versus the node number, and in Fig. 2.13 in the two-dimensional representation. Numerical results obtained with the 5-point formula and with under-relaxation are of the same quality and almost indistinguishable in practice.

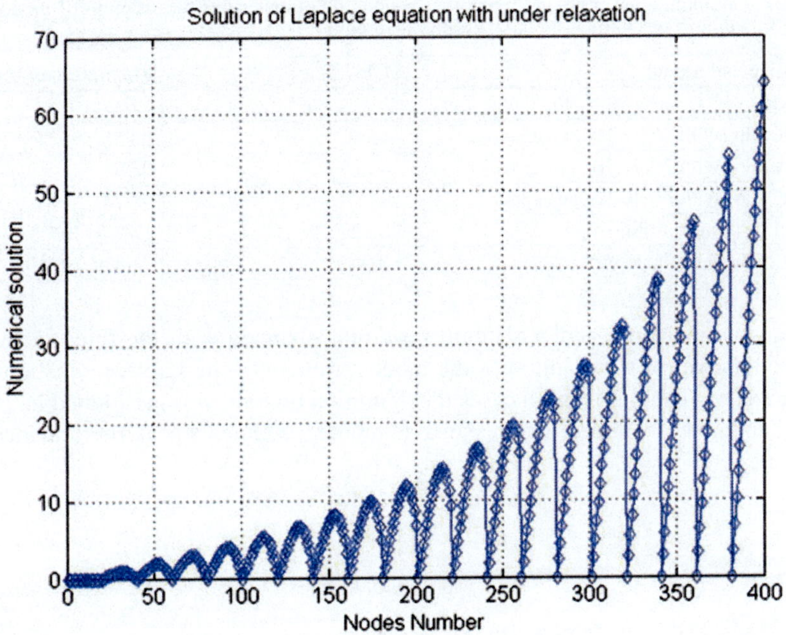

Fig. 2.12 Numerical solution of model Laplace equation computed with under-relaxation versus node number

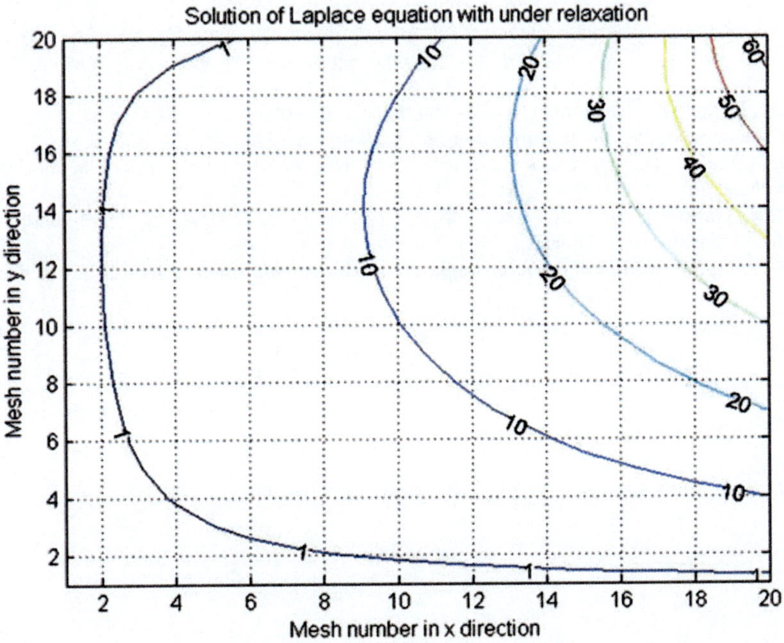

Fig. 2.13 Two-dimensional representation of the numerical solution of model Laplace equation computed with under-relaxation

Table 2.1 Computational time and maximum residuum computed for the solution of the model Laplace equation (2.97) with boundary conditions (2.98)

Subroutine and model	CPU time [s]	Maximum residuum
LAPLMC		
Monte Carlo (with $N_w = 10^5$ sweeps)	45.9	0.656
Monte Carlo (with $N_w = 10^6$ sweeps)	456.9	0.194
Monte Carlo (with $N_w = 10^7$ sweeps)	4467	0.074
LAPLFI (5-point formula)	0.187	1.77×10^{-14}
LAPLUN (under-relaxation)	0.047	9.98×10^{-13}

Computational time and maximum residuum obtained with the three methods explained above for the solution of the model problem of the Laplace equation are summarized in Table 2.1. In all cases the definition domain was partitioned in 20 × 20 cells and had 400 nodes. The personal computer used had a peak performance of 13 Gcycles.

References

Courant, R., & Hilbert, D. (1968). *Methoden der mathematischen Physik II*. Springer.

Forsythe, G. E., & Wasow, W. R. (1960). *Finite-difference methods for partial differential equations*. John Wiley & Sons.

Hadamard, J. (1923). *Lecures on Cauchy's problem*. Yale University Press.

Lax, P. D. (1954). Weak solutions of nonlinear hyperbolic equations and their numerical computation. *Communications on Pure and Applied Mathematics, 7*, 159–193.

Smith, G. D. (1969). *Numerical solution of partial differential equations*. Oxford University Press.

Stuhmiller, J. H. (July 1976). A review of the rational approach to two-phase flow modeling. *EPRI, NP-197*.

Stuhmiller, J. H. (1977). The influence of the interfacial pressure forces on the caracter of two-phase flow model equations. *International Journal of Multiphase Flow, 3*, 551–560.

Trapp, J. A. (1986). The mean flow character of two-phase flow equations. *International Journal Multiphase Flow, 12*(2), 263–276.

Travis, J. R., Harlow, F. H., & Amsden, A. A. (1976). Numerical calculations of two-phase flows. *Nuclear Science and Engineering, 61*, 1–10.

Zauderer, E. (1983). *Partial differential equations of applied mathematics*. John Wiley & Sons.

Chapter 3
Methods of Orthogonal Collocations (OC)

This chapter is dedicated to the method of orthogonal collocations (OC) which is a powerful alternative to finite differences methods, but far less used because of difficulties in its application.

In the first four sections of the chapter, we introduce the method, and some of its variants, showing their applications in case of a model problem consisting of a damped or un-damped wave. Section 3.5 gives some numerical applications of the explained variants. Section 3.6 explains the use of the OC method in case of a liquid-vapor flow problem, which, to the author's knowledge, was the first (or one of the first) applications of this technique to a complex two-phase flow problem.

Before going to the details of this application, we present a short historical overview of the mathematical and physical contest in which this class of methods evolved over some decades in the second half of the past century.

A strong impact on the development of mathematical methods for fluid dynamic problems has been given since the 1950s from several research centers around the world. Among these are worth mentioning the LANL (Los Alamos National Laboratory) in New Mexico, the ANL (Argonne National Laboratory) in Illinois, the Imperial College of London, the Centre d'Etudes Nucleaires de Grenoble (CENG), and the Gesellschaft für Reaktorsicherheit (GRS) (Reactor Safety Corporation) in Garching, Germany. Among a variety of thermal and fluid dynamic problems, a special place has been taken at these laboratories by the theoretical analysis of thermal-hydraulic behavior of nuclear reactors, both thermal and fast reactors. A large effort has been devoted to the development of refined numerical techniques for describing the rector coolant behavior in multidimensional geometrical configurations under both nominal and accident conditions. Among the most successful of these techniques is the implicit continuous-field Eulerian (ICE) method by Harlow and Amsden (1971) which is applicable for time-dependent viscous flows, for all Mach numbers from zero (incompressible limit) to infinity (hypersonic limit). A further improvement with respect to the ICE technique was documented by Harlow and Amsden (1975), in the implicit multifield (IMF) solution method. According to

M. Bottoni, *Physical Modeling and Computational Techniques for Thermal and Fluid-dynamics*, Mechanical Engineering Series,
https://doi.org/10.1007/978-3-030-79717-1_3

the quoted authors, the IMF methods present (among others) the following advantages with respect to previous techniques:

1. *Implicit treatments of mass convection and equation of state, so that flow speed can range from far subsonic (incompressible) to supersonic*
2. *Implicit coupling among the fields to allow forces that range from very week to those strong enough to completely tie the fields together*
3. *Energy equations that allow for heat production from condensation, nuclear reactions, exothermic chemistry, etc.*
4. *Interchange of material from one field to another, for example, through phase transition*

The most advanced application of the IMF method was probably the development of the SIMMER-I computer program (Bell et al., 1977) which has been since further improved in subsequent versions.

In parallel with the work at LANL, a new class of numerical methods, the asymmetric separated region weighted residual (ASWR) based on the orthogonal collocation technique, was developed for fluid dynamic calculations at the GRS. Pioneering work on these developments was documented in Romstedt and Werner (1977) and Werner (1979). The techniques proposed by these authors belong to the widespread category of the coarse-mesh methods which had been applied successfully for the numerical investigation of neutron diffusion and transport equations since the beginning of the 1960s. The state of the art of the ASWR methods was presented a few years later in Werner (1981) from the viewpoint of the treatment of discontinuities of the dependent variables traveling through the solution domain.

In the framework of the development at the Nuclear Research Centre of Karlsruhe (KfK) of computer programs describing transient two-phase flows for sodium-cooled fast reactors, under nominal and accident conditions, the capabilities of the OC method have been accessed in a one-dimensional case. For this purpose the BLOW-3A (Bottoni & Struwe, 1982) code was used. This code solves the coupled mass, momentum, and energy equations of the liquid and vapor phases on the basis of a heterogeneous multi-bubble slug ejection model. In the original code version, discretization of the conservation equations is done with finite differences both in space and time. An alternative version was developed solving the one-dimensional conservation equation for the sodium vapor phase with the residual method of orthogonal collocations. Section 3.6 of this chapter is dedicated to the development of the OC method linked to the BLOW-3A code. A synopsis of the code is given in Sect. 3.7.

3.1 Quasi-linear Damped Wave Model Equation

We consider as model equation the following damped wave equation for the dependent variable $u = u(z, t)$:

$$\frac{\partial u}{\partial t} + u\frac{\partial u}{\partial z} + cu = 0, \qquad (z, t \geq 0) \tag{3.1}$$

with c a given constant and with initial value:

$$u(z, 0) = F(z). \tag{3.2}$$

According to the definitions given in Sect. 2.1.1, Eq. (3.1) is quasi-linear. The solution of Eq. (3.1) with initial condition (3.2) is (Cfr. Eq. (2.23)):

$$u(z, t) = F(z - ut)e^{-ct}, \tag{3.3}$$

which represents a wave propagating in the direction of positive z with speed:

$$\frac{dz}{dt} = u, \tag{3.4}$$

and damped by the factor e^{-ct} with respect to the initial values (Eq. 3.2). In case $c = 0$ the wave is undamped.

The model Eq. (3.1) can be written in the form:

$$Au = 0, \tag{3.5}$$

with the definition of the operators:

$$A = \frac{\partial}{\partial t} + u\frac{\partial}{\partial z} + c = \frac{\partial}{\partial t} + L, \tag{3.6}$$

$$L = u\frac{\partial}{\partial z} + c. \tag{3.7}$$

We look for an approximation to the solution of Eq. (3.5) in the form:

$$u \approx u_N = \sum_{k=1}^{N} a_k \Phi_k, \tag{3.8}$$

where a_k are nodal values and $\mathbf{\Phi}_k$ are a set of chosen shape functions. The approximation (Eq. 3.8), when introduced into Eq. (3.5), gives the residuals:

$$A\sum_{k=1}^{N} a_k \Phi_k = R. \tag{3.9}$$

We want to minimize the residuals by imposing the orthogonality condition to a set of chosen weight functions Ψ_j ($j = 1, 2, \ldots, M$):

$$(R, \Psi_j) = \left(A \sum_{k=1}^{N} a_k \Phi_k, \Psi_j\right) = 0, \qquad (j = 1, 2, \ldots, M) \tag{3.10}$$

where the scalar product of any two functions f_1, f_2 is defined in the domain $\Omega(0 \leq z < \infty)$ by:

$$(f_1, f_2) = \int_{\Omega} f_1 f_2 d\Omega. \tag{3.11}$$

Using Eq. (3.6), Eq. (3.10) becomes:

$$\frac{\partial}{\partial t} \sum_{k=1}^{N} a_k (\Phi_k, \Psi_j) + \sum_{k=1}^{N} a_k (L\Phi_k, \Psi_j) = 0. \qquad (j = 1, 2, \ldots, M) \tag{3.12}$$

Discretizing the first term of Eq. (3.12) with respect to time and introducing a time discretization parameter θ $[0 \leq \theta \leq 1]$ to evaluate the second term at two levels t_n, t_{n+1}, one obtains:

$$\frac{1}{\Delta t_n}\left[\sum_{k=1}^{N} a_k^{n+1} (\Phi_k, \Psi_j) - \sum_{k=1}^{N} a_k^{n} (\Phi_k, \Psi_j)\right] + \theta \sum_{k=1}^{N} a_k^{n+1} (L\Phi_k, \Psi_j) + (1-\theta) \sum_{k=1}^{N} a_k^{n} (L\Phi_k, \Psi_j) = 0, \tag{3.13}$$

and, upon rearrangement:

$$\sum_{k=1}^{N} a_k^{n+1} (\Phi_k, \Psi_j) + \theta \Delta t_n \sum_{k=1}^{N} a_k^{n+1} (L\Phi_k, \Psi_j) = \sum_{k=1}^{N} a_k^{n} (\Phi_k, \Psi_j) - (1-\theta)\Delta t_n \sum_{k=1}^{N} a_k^{n} (L\Phi_k, \Psi_j). \tag{3.14}$$

3.2 Solution of Model Equation with the Subdomain Collocations Method

We partition the domain $\Omega(0 \leq z < \infty)$ into NE disjoint, but adjacent (not overlapping) subdomain $\Omega_e(e = 1, 2, \ldots, NE)$ with $\Omega = \ \cup \Omega_e$. The last domain Ω_{NE} extends to a point $z = z_{\max}$ which will not be reached by the wave represented by Eq. (3.1) in the problem time considered.

We choose the weight functions defined for every subdomain Ω_e by:

$$\Psi_j = \begin{cases} 1 & z \in \Omega_e, \\ 0 & z \in \Omega_e, \end{cases} \tag{3.15}$$

With this choice Eq. (3.14) yields for every subdomain (or "element") the following "element equation":

$$\begin{gathered} \text{Term A} \qquad\qquad\qquad \text{B} \\ \int_{\Omega_e} \sum_{k=1}^{N} a_k^{n+1} \Phi_k d\Omega_e + \theta \Delta t_n \int_{\Omega_e} \sum_{k=1}^{N} a_k^{n+1} L\Phi_k d\Omega_e \qquad (e = 1, 2, \ldots, NE) \\ = \int_{\Omega_e} \sum_{k=1}^{N} a_k^{n} \Phi_k d\Omega_e - (1-\theta)\Delta t_n \int_{\Omega_e} \sum_{k=1}^{N} a_k^{n} L\Phi_k d\Omega_e . \\ \text{D} \qquad\qquad\qquad \text{E} \end{gathered} \tag{3.16}$$

For every subdomain Ω_e of width Δz, we consider a local coordinate system $(z_0 = 0 \leq z \leq z + \Delta z = z_1)$ (not to be confused with the main coordinate system) and three nodes, the left boundary node, the center point, and the right boundary node, labeled 0, 1/2, and 1, respectively. Defined on this local coordinate system, we consider the following parabolic approximations:

$$u = a_0 + \frac{z}{\Delta z}\left(-3a_0 + 4a_{1/2} - a_1\right) + \frac{z^2}{\Delta z^2}\left(2a_0 - 4a_{1/2} + 2a_1\right). \tag{3.17}$$

It is easy to verify that for $z = 0$, $u = a_0$; for $z = \Delta z/2$, $u = a_{1/2}$ and for $z = \Delta z$, $u = a_1$. The coefficients a_0, $a_{1/2}$, a_1 represent the unknown function u in the respective nodes. Therefore in the following we may use the symbols u_0, $u_{1/2}$, u_1 instead of a_0, $a_{1/2}$, a_1.

Equation (3.17) can also be written as:

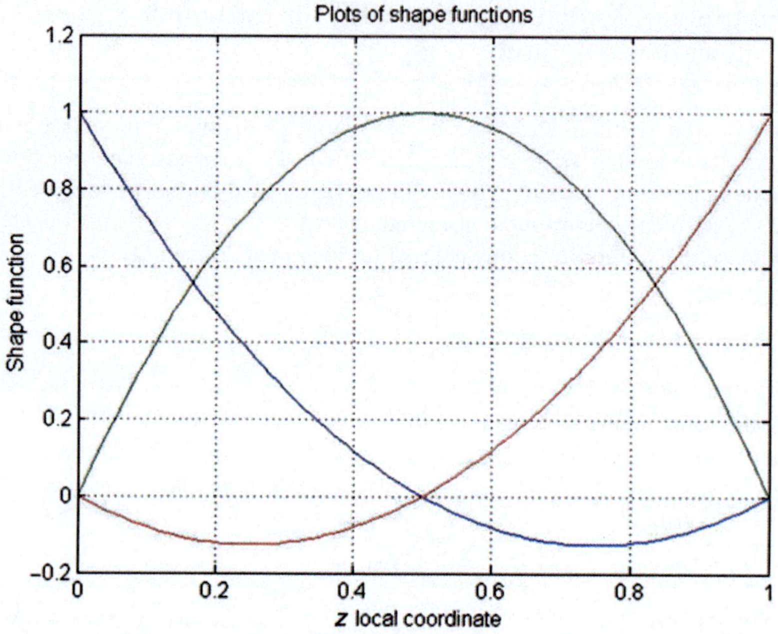

Fig. 3.1 Plots of shape functions in the local coordinate system ($0 \leq z \leq \Delta z = 1$): blue plot shows function (3.19a); green plot shows function (3.19b); red plot shows function (3.19c)

$$
\begin{aligned}
u =& a_0\left(1 - 3\frac{z}{\Delta z} + 2\frac{z^2}{\Delta z^2}\right) + a_{1/2}\left(4\frac{z}{\Delta z} - 4\frac{z^2}{\Delta z^2}\right) + a_1\left(-\frac{z}{\Delta z} + 2\frac{z^2}{\Delta z^2}\right) \\
=& a_0\Phi_0 + a_{1/2}\Phi_{1/2} + a_1\Phi_1,
\end{aligned} \tag{3.18}
$$

with the definition of the shape functions, plotted in Fig. 3.1:

$$\Phi_0 = 1 - 3\frac{z}{\Delta z} + 2\frac{z^2}{\Delta z^2}, \tag{3.19a}$$

$$\Phi_{1/2} = 4\frac{z}{\Delta z} - 4\frac{z^2}{\Delta z^2}, \tag{3.19b}$$

$$\Phi_1 = -\frac{z}{\Delta z} + 2\frac{z^2}{\Delta z^2}. \tag{3.19c}$$

The derivatives of the shape functions, with respect to z, are:

$$\frac{\partial \Phi_0}{\partial z} = -\frac{3}{\Delta z} + 4\frac{z}{\Delta z^2}, \tag{3.20a}$$

$$\frac{\partial \Phi_{1/2}}{\partial z} = \frac{4}{\Delta z} - 8\frac{z}{\Delta z^2}, \tag{3.20b}$$

$$\frac{\partial \Phi_1}{\partial z} = -\frac{1}{\Delta z} + 4\frac{z}{\Delta z^2}. \tag{3.20c}$$

From Eq. (3.8) we obtain:

$$\frac{\partial u}{\partial z} \approx \frac{\partial u_N}{\partial z} = \sum_{k=1}^{N} a_k \frac{\partial \Phi_k}{\partial z}. \qquad (k = 0, 1/2, 1) \tag{3.21}$$

Hence (using the symbols u_0, $u_{1/2}$, u_1 for the a_k):

$$\left(\frac{\partial u}{\partial z}\right)_0 = \left(\frac{\partial u}{\partial z}\right)_{z=0} = \frac{1}{\Delta z}\left(-3u_0 + 4u_{1/2} - u_1\right) = \frac{1}{\Delta z}\mathrm{DUDZN0}, \tag{3.22a}$$

$$\left(\frac{\partial u}{\partial z}\right)_{1/2} = \left(\frac{\partial u}{\partial z}\right)_{z=\Delta z/2} = \frac{1}{\Delta z}(u_1 - u_0) = \frac{1}{\Delta z}\mathrm{DUDZ12}, \tag{3.22b}$$

$$\left(\frac{\partial u}{\partial z}\right)_1 = \left(\frac{\partial u}{\partial z}\right)_{z=\Delta z} \frac{1}{\Delta z}\left(u_0 - 4u_{1/2} + 3u_1\right) = \frac{1}{\Delta z}\mathrm{DUDZN1}, \tag{3.22c}$$

with the definitions:

$$\mathrm{DUDZN0} = -3u_0 + 4u_{1/2} - u_1, \tag{3.23a}$$

$$\mathrm{DUDZ12} = u_1 - u_0, \tag{3.23b}$$

$$\mathrm{DUDZN1} = u_0 - 4u_{1/2} + 3u_1. \tag{3.23c}$$

In the following, these definitions, which have been introduced for reducing the length of the formulas, are used only for terms referring to time level n. When the analytical treatment is terminated, the definitions are applied in the backward direction to reintroduce the real variables u_k ($k = 0, 1/2, 1$).

The integrals of the shape functions over the elements are:

$$\int_0^{\Delta z} \Phi_0(z)dz = \frac{\Delta z}{6}, \tag{3.24a}$$

$$\int_0^{\Delta z} \Phi_{1/2}(z)dz = \frac{2}{3}\Delta z, \tag{3.24b}$$

$$\int_0^{\Delta z} \Phi_1(z)dz = \frac{\Delta z}{6}. \tag{3.24c}$$

Using Eq. (3.24) Term A of the elements equation (3.16) is computed by:

$$\text{Term A} = \sum_k a_k^{n+1} \int_{\Omega_e} \Phi_k d\Omega_e = u_0^{n+1} \frac{\Delta z}{6} + u_{1/2}^{n+1} \frac{2}{3} \Delta z + u_1^{n+1} \frac{\Delta z}{6} s. \tag{3.25}$$

Term B of the element equation (3.16) is computed as follows (using Eq. (3.8)):

$$\text{Term B} = \theta \Delta t_n \sum_k a_k^{n+1} \int_{\Omega_e} L\Phi_k d\Omega_e = \theta \Delta t_n \int_{\Omega_e} Lu^{n+1} d\Omega_e. \tag{3.26}$$

The integral of (Eq. 3.26) can be linearized as follows:

$$\begin{aligned}
Lu^{n+1} &= u^{n+1} \frac{\partial u^{n+1}}{\partial z} + cu^{n+1} \\
&= u^n \frac{\partial u^n}{\partial z} + \frac{\partial}{\partial t}\left(u^n \frac{\partial u^n}{\partial z}\right) \Delta t + cu^{n+1} \\
&= u^n \frac{\partial u^n}{\partial z} + \frac{\partial u^n}{\partial t} \frac{\partial u^n}{\partial z} \Delta t + u^n \frac{\partial}{\partial t}\left(\frac{\partial u^n}{\partial z}\right) \Delta t + cu^{n+1} \\
&= u^n \frac{\partial u^n}{\partial z} + \frac{\partial u^n}{\partial t} \frac{\partial u^n}{\partial z} \Delta t + u^n \frac{\partial}{\partial z}\left(\frac{u^{n+1} - u^n}{\Delta t}\right) \Delta t + cu^{n+1} \\
&= u^n \frac{\partial u^{n+1}}{\partial z} + \frac{u^{n+1} - u^n}{\Delta t} \frac{\partial u^n}{\partial z} \Delta t + cu^{n+1} \\
&= u^n \frac{\partial u^{n+1}}{\partial z} + \left(u^{n+1} - u^n\right) \frac{\partial u^n}{\partial z} + cu^{n+1}.
\end{aligned} \tag{3.27}$$

Inserting Eq. (3.27) into Eq. (3.26) yields:

$$\text{Term B} = \theta \Delta t_n \int_{\Omega_e} \Bigg[\underbrace{u^n \frac{\partial u^{n+1}}{\partial z}}_{\text{Term B1}} + \underbrace{\left(u^{n+1} - u^n\right) \frac{\partial u^n}{\partial z}}_{\text{B2}} + \underbrace{cu^{n+1}}_{\text{B3}} \Bigg] d\Omega_e. \tag{3.28}$$

By using the Simpson formula for evaluating definite integrals:

$$\int_a^b f(z) dz = \frac{b-a}{6}\left[f(a) + 4f\left(\frac{a+b}{2}\right) + f(b)\right], \tag{3.29}$$

the first two terms B1 and B2 of Eq. (3.28) are computed as follows:

$$
\begin{aligned}
\text{Term B1} &= \theta\Delta t_n \int_{\Omega_e} u^n \frac{\partial u^{n+1}}{\partial z} d\Omega_e = \theta\Delta t_n \int_0^{\Delta z} u^n \frac{\partial u^{n+1}}{\partial z} dz \\
&= \theta\Delta t_n \frac{\Delta z}{6}\left[u_0^n\left(\frac{\partial u^{n+1}}{\partial z}\right)_0 + 4u_{1/2}^n\left(\frac{\partial u^{n+1}}{\partial z}\right)_{1/2} + u_1^n\left(\frac{\partial u^{n+1}}{\partial z}\right)_1\right] \\
&= \frac{\theta\Delta t_n}{6}\left[u_0^n\left(-3u_0^{n+1} + 4u_{1/2}^{n+1} - u_1^{n+1}\right) + 4u_{1/2}^n\left(u_1^{n+1} - u_0^{n+1}\right) + u_1^n\left(u_0^{n+1} - 4u_{1/2}^{n+1} + 3u_1^{n+1}\right)\right] \\
&= \frac{\theta\Delta t_n}{6}\left[u_0^{n+1}\left(-3u_0^n - 4u_{1/2}^n + u_1^n\right) + 4u_{1/2}^{n+1}\left(u_0^n - u_1^n\right) + u_1^{n+1}\left(-u_0^n + 4u_{1/2}^n + 3u_1^n\right)\right],
\end{aligned}
\tag{3.30}
$$

having used Eq. (3.22).

$$
\begin{aligned}
\text{Term B2} &= \theta\Delta t_n \int_{\Omega_e} \left(u^{n+1} - u^n\right)\frac{\partial u^n}{\partial z} d\Omega_e = \theta\Delta t_n \int_0^{\Delta z} \left(u^{n+1} - u^n\right)\frac{\partial u^n}{\partial z} dz \\
&= \theta\Delta t_n \frac{\Delta z}{6}\frac{1}{\Delta z}\Big[\left(u_0^{n+1} - u_0^n\right)\text{DUDZN0} + 4\left(u_{1/2}^{n+1} - u_{1/2}^n\right)\text{DUDZ12} \\
&\qquad + \left(u_1^{n+1} - u_1^n\right)\text{DUDZN1}\Big] \\
&= \frac{\theta\Delta t_n}{6}\left(u_0^{n+1}\text{DUDZN0} + 4u_{1/2}^{n+1}\text{DUDZ12} + u_1^{n+1}\text{DUDZN1}\right) - \frac{\theta\Delta t_n}{6} \\
&\quad \times \left(u_0^n\text{DUDZN0} + 4u_{1/2}^n\text{DUDZ12} + u_1^n\text{DUDZN1}\right).
\end{aligned}
\tag{3.31}
$$

Using Eq. (3.8) Term B3 is computed directly as:

$$
\begin{aligned}
\text{Term B3} &= \theta\Delta t_n \int_{\Omega_e} c u^{n+1} d\Omega_e = c\theta\Delta t_n \int_0^{\Delta z} \sum\nolimits_k a_k^{n+1}\Phi_k dz \\
&= c\theta\Delta t_n\left(\frac{\Delta z}{6}u_0^{n+1} + \frac{2\Delta z}{3}u_{1/2}^{n+1} + \frac{\Delta z}{6}u_1^{n+1}\right).
\end{aligned}
\tag{3.32}
$$

Likewise one obtains:

$$
\begin{aligned}
\text{Term D} &= \sum\nolimits_k a_k^n \int_{\Omega_e} \Phi_k d\Omega_e = \sum\nolimits_k a_k^n \int_0^{\Delta z} \Phi_k dz \\
&= \frac{\Delta z}{6}u_0^n + \frac{2\Delta z}{3}u_{1/2}^n + \frac{\Delta z}{6}u_1^n.
\end{aligned}
\tag{3.33}
$$

$$\begin{aligned}
\text{Term E} &= -(1-\theta)\Delta t_n \sum\nolimits_k a_k^n \int_{\Omega_e} L\Phi_k d\Omega_e = -(1-\theta)\Delta t_n \int_{\Omega_e} Lu^n d\Omega_e \\
&= -(1-\theta)\Delta t_n \int_{\Omega_e} \left(u^n \frac{\partial u^n}{\partial z} + cu^n \right) d\Omega_e \\
&= -(1-\theta)\Delta t_n \frac{\Delta z}{6} \frac{1}{\Delta z} \left(u_0^n \text{DUDZN0} + 4u_{1/2}^n \text{DUDZ12} + u_1^n \text{DUDZN1} \right) \\
&\quad - (1-\theta)\Delta t_n c \left(\frac{\Delta z}{6} u_0^n + \frac{2\Delta z}{3} u_{1/2}^n + \frac{\Delta z}{6} u_1^n \right).
\end{aligned} \tag{3.34}$$

Inserting Eqs. (3.25) (Term A), (3.30), (3.31), (3.32) (Term B = B1 + B2 + B3), (3.33) (Term D), (3.34) (Term E) into the element equation (3.16) and rearranging one obtains:

$$b_0 u_0^{n+1} + b_{1/2} u_{1/2}^{n+1} + b_1 u_1^{n+1} = G_1^n, \tag{3.35}$$

with:

$$\begin{aligned}
b_0 &= \frac{\Delta z}{6} + \frac{\theta \Delta t_n}{6} \left[\left(-3u_0^n - 4u_{1/2}^n + u_1^n \right) + \text{DUDZN0} \right] + c\theta \Delta t_n \frac{\Delta z}{6} \\
&= \frac{\Delta z}{6} (1 + c\theta \Delta t_n) - \theta \Delta t_n u_0^n,
\end{aligned} \tag{3.36a}$$

$$\begin{aligned}
b_{1/2} &= \frac{2\Delta z}{3} + \frac{\theta \Delta t_n}{6} \left[4\left(u_0^n - u_1^n \right) + 4\text{DUDZ12} \right] + c\theta \Delta t_n \frac{2\Delta z}{3} \\
&= \frac{2\Delta z}{3} (1 + c\theta \Delta t_n)
\end{aligned} \tag{3.36b}$$

$$\begin{aligned}
b_1 &= \frac{\Delta z}{6} + \frac{\theta \Delta t_n}{6} \left[\left(-u_0^n + 4u_{1/2}^n + 3u_1^n \right) + \text{DUDZN1} \right] + c\theta \Delta t_n \frac{\Delta z}{6} \\
&= \frac{\Delta z}{6} (1 + c\theta \Delta t_n) + \theta \Delta t_n u_1^n,
\end{aligned} \tag{3.36c}$$

$$\begin{aligned}
G_1^n &= \frac{1}{6} [\theta \Delta t_n - (1-\theta)\Delta t_n] \left(u_0^n \text{DUDZN0}^n + u_{1/2}^n 4\text{DUDZ12}^n + u_1^n \text{DUDZN1}^n \right) \\
&\quad + \frac{\Delta z}{6} \left(u_0^n + 4u_{1/2}^n + u_1^n \right) [1 - c(1-\theta)\Delta t_n].
\end{aligned} \tag{3.36d}$$

Using the definitions (3.23), Eq. (3.36d), the second member of Eq. (3.35), can be written as:

$$
\begin{aligned}
G_1^n = &[\theta\Delta t_n - (1-\theta)\Delta t_n]\frac{1}{2}\left[-\left(u_0^n\right)^2 + u_0^n u_{1/2}^n - u_{1/2}^n u_1^n + \left(u_1^n\right)^2\right] \\
&+ [1 - c(1-\theta)\Delta t_n]\frac{\Delta z}{6}\left(u_0^n + 4u_{1/2}^n + u_1^n\right).
\end{aligned}
\tag{3.37}
$$

The first term of G_1^n vanishes if $\theta = 1 - \theta = 0.5$.

The total number of unknowns in the NE elements is 1 + 2NE. There are NE element equations and two boundary conditions. To close the system, it is necessary to use the $NE - 1$ continuity equations at the interface between the elements, given by:

$$
\left(\frac{\partial u^{n+1}}{\partial z}\right)_{\Omega_{e,1}} = \left(\frac{\partial u^{n+1}}{\partial z}\right)_{\Omega_{e+1,0}}. \tag{3.38}
$$

The subscript notation implies the right node of element Ω_e, which coincides with the left node of element Ω_{e+1}. Using Eqs. (3.22a) and (3.22c), one obtains the continuity condition:

$$
u_{e,0}^{n+1} - 4u_{e,1/2}^{n+1} + 6u_{e,1}^{n+1} - 4u_{e+1,1/2}^{n+1} + u_{e+1,1}^{n+1} = 0. \tag{3.39}
$$

The system of Eqs. (3.35) and (3.39) with two boundary conditions at $z = 0$ and $z = \Delta z \cdot NE$ can be solved for the unknown $u_{e\alpha}(e = 1, 2, \ldots, NE; \alpha = 0, 1/2, 1)$ with standard software for linear algebra.

As an example, we partition the domain $\Omega(0 \le z < \infty)$ in three elements Ω_1, Ω_2, and Ω_3 with seven unknown u_1 through u_7. The values u_1 and u_7 are determined by the boundary conditions $u_1 = \overline{u_1}$, $u_7 = \overline{u_7}$. These two boundary conditions, the coefficients of three element equations (3.35) and two continuity equations (3.39), are arranged in a matrix equation:

$$
AU = B, \tag{3.40}
$$

where A is a matrix of order 7, $U = (u_1, u_2, \ldots, u_7)$ is the vector of the unknowns, and B is the vector of the known right-hand sides of Eqs. (3.35) and (3.38) (the latter being zero).

The form of Eq. (3.39) is shown in Fig. 3.2.

3.3 Method of Point Collocations

In every subdomain Ω_e ($e = 1, 2, \ldots, NE$), a set of points $z = z_j$ ($j = 1, 2, \ldots, M$) is chosen, and the weight functions are defined using the delta function, by:

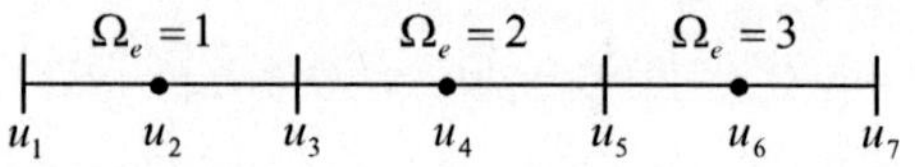

(a)

A **U** **B**

$$
\begin{bmatrix}
1 & & & & & & \\
b_0 & b_{1/2} & b_1 & & & & \\
1 & -4 & 6 & -4 & 1 & & \\
 & & b_0 & b_{1/2} & b_1 & & \\
 & & 1 & -4 & 6 & -4 & 1 \\
 & & & & b_0 & b_{1/2} & b_1 \\
 & & & & & & 1
\end{bmatrix}
\begin{bmatrix} u_1 \\ u_2 \\ u_3 \\ u_4 \\ u_5 \\ u_6 \\ u_7 \end{bmatrix}
=
\begin{bmatrix} \overline{u}_1 \\ G_1 \\ 0 \\ G_1 \\ 0 \\ G_1 \\ \overline{u}_7 \end{bmatrix}
$$

(b)

Fig. 3.2 (**a**) Partition of the definition domain **Ω** in three subdomains $\boldsymbol{\Omega}_e (e = 1, 2, 3)$; (**b**) form of the matrix equation $AU = B$

$$\psi_j = \delta(z - z_j). \tag{3.41}$$

In the following we use three collocation points defined on the element $\boldsymbol{\Omega}_e$ by $z_0 = 0$, $z_{1/2} = \Delta z/2$, $z_1 = \Delta z$, respectively. We use the same parabolic shape functions (3.19).

The shape functions, computed in the collocation points, are:

$$(\Phi_0)_{z0} = 1, \tag{3.42a}$$

$$(\Phi_0)_{z1/2} = 0, \tag{3.42b}$$

$$(\Phi_0)_{z1} = 0, \tag{3.42c}$$

$$(\Phi_{1/2})_{z0} = 0, \tag{3.42d}$$

$$(\Phi_{1/2})_{z1/2} = 1, \tag{3.42e}$$

$$(\Phi_{1/2})_{z1} = 0, \tag{3.42f}$$

$$(\Phi_1)_{z0} = 0, \tag{3.42g}$$

$$(\Phi_1)_{z1/2} = 0, \tag{3.42h}$$

$$(\Phi_1)_{z1} = 1. \tag{3.42i}$$

The derivatives of the shape functions, computed in the collocation points, are, from Eq. (3.20):

$$\left(\frac{\partial \Phi_0}{\partial z}\right)_{z0} = -\frac{3}{\Delta z}, \tag{3.43a}$$

$$\left(\frac{\partial \Phi_0}{\partial z}\right)_{z1/2} = -\frac{1}{\Delta z}, \tag{3.43b}$$

$$\left(\frac{\partial \Phi_0}{\partial z}\right)_{z1} = \frac{1}{\Delta z}, \tag{3.43c}$$

$$\left(\frac{\partial \Phi_{1/2}}{\partial z}\right)_{z0} = \frac{4}{\Delta z}, \tag{3.43d}$$

$$\left(\frac{\partial \Phi_{1/2}}{\partial z}\right)_{z1/2} = 0, \tag{3.43e}$$

$$\left(\frac{\partial \Phi_{1/2}}{\partial z}\right)_{z1} = -\frac{4}{\Delta z}, \tag{3.43f}$$

$$\left(\frac{\partial \Phi_1}{\partial z}\right)_{z0} = -\frac{1}{\Delta z}, \tag{3.43g}$$

$$\left(\frac{\partial \Phi_1}{\partial z}\right)_{z1/2} = \frac{1}{\Delta z}, \tag{3.43h}$$

$$\left(\frac{\partial \Phi_1}{\partial z}\right)_{z1} = \frac{3}{\Delta z}. \tag{3.43i}$$

Using the weight functions (3.41), the element equation (3.14) becomes:

$$\int_{\Omega_e} \sum_{k=1}^{N} a_k^{n+1} (\Phi_k)_{zj} d\Omega_e + \theta \Delta t_n \int_{\Omega_e} \sum_{k=1}^{N} a_k^{n+1} (L\Phi_k)_{zj} d\Omega_e, \tag{3.44}$$

for $(e = 1, 2, \ldots, NE)$ and $(j = 1, 2, 3 = M)$.

In this variant of the orthogonal collocation method, there are three element equations for every subdomain.

(i) *Element Equation for* $j = 1$, $(z = z_0 = 0)$

For the first collocation point $(z = = 0)$, the element equation (3.44) becomes:

$$\begin{aligned}
&\underbrace{\int_{\Omega_e}\sum_{k=1}^{N} a_k^{n+1}(\Phi_k)_{z0}d\Omega_e}_{\text{Term A}} + \underbrace{\theta\Delta t_n\int_{\Omega_e}\sum_{k=1}^{N} a_k^{n+1}(L\Phi_k)_{z0}d\Omega_e}_{\text{Term B}} \\
&= \underbrace{\int_{\Omega_e}\sum_{k=1}^{N} a_k^{n}(\Phi_k)_{z0}d\Omega_e}_{\text{Term D}} - \underbrace{(1-\theta)\Delta t_n\int_{\Omega_e}\sum_{k=1}^{N} a_k^{n}(L\Phi_k)_{z0}d\Omega_e}_{\text{Term E}}.
\end{aligned} \tag{3.45}$$

Term A is computed using (3.42) as:

$$\text{Term A} = \int_{\Omega_e}\sum_{k=1}^{N} a_k^{n+1}(\Phi_k)_{z0}d\Omega_e = \int_0^{\Delta z} a_0^{n+1}(\Phi_0)_{z0} = a_0^{n+1}\Delta z. \tag{3.46}$$

Term B is discretized with respect to time as shown in Eq. (3.27) and yields:

$$\begin{aligned}
\text{Term B} =&\theta\Delta t_n\int_{\Omega_e}\sum_{k=1}^{N} a_k^{n+1}(L\Phi_k)_{z0}d\Omega_e = \theta\Delta t_n\int_{\Omega_e}\left(Lu^{n+1}\right)_{z0}d\Omega_e \\
=&\theta\Delta t_n\int_{\Omega_e}\Bigg[\underbrace{u^n\frac{\partial u^{n+1}}{\partial z}}_{\text{Term B1}} + \underbrace{\left(u^{n+1}-u^n\right)\frac{\partial u^n}{\partial z}}_{\text{B2}} + \underbrace{cu^{n+1}}_{\text{B3}}\Bigg]_{z0} d\Omega_e.
\end{aligned} \tag{3.47}$$

Using (Eq. 3.42), Term B1 is computed as follows:

$$\begin{aligned}
\text{Term B1} =&\theta\Delta t_n\int_{\Omega_e}\left[u^n\frac{\partial u^{n+1}}{\partial z}\right]_{z0} d\Omega_e \\
=&\theta\Delta t_n\int_0^{\Delta z}\sum\nolimits_k a_k^n(\Phi_k)_{z0}\frac{\partial}{\partial z}\sum\nolimits_k a_k^{n+1}(\Phi_k)_{z0}dz \\
=&\theta\Delta t_n\int_0^{\Delta z} a_0^n\left[a_0^{n+1}\left(\frac{\partial\Phi_0}{\partial z}\right)_{z0} + a_{1/2}^{n+1}\left(\frac{\partial\Phi_{1/2}}{\partial z}\right)_{z0} + a_1^{n+1}\left(\frac{\partial\Phi_1}{\partial z}\right)_{z0}\right]dz \\
=&\theta\Delta t_n a_0^n\left(-3a_0^{n+1} + 4a_{1/2}^{n+1} - a_1^{n+1}\right).
\end{aligned} \tag{3.48}$$

Similarly, Term B2 yields:

$$\begin{aligned}
\text{Term B2} =& \theta\Delta t_n \int_{\Omega_e} \left[\left(u^{n+1} - u^n\right) \frac{\partial u^n}{\partial z} \right]_{z0} d\Omega_e \\
=& \theta\Delta t_n \int_0^{\Delta z} \sum\nolimits_k \left[a_k^{n+1} (\Phi_k)_{z0} - a_k^n (\Phi_k)_{z0} \right] \sum\nolimits_k a_k^n \left(\frac{\partial \Phi_k}{\partial z} \right)_{z0} dz \\
=& \theta\Delta t_n \int_0^{\Delta z} \left(a_0^{n+1} - a_0^n\right) \left[a_0^n \left(\frac{\partial \Phi_0}{\partial z} \right)_{z0} + a_{1/2}^n \left(\frac{\partial \Phi_{1/2}}{\partial z} \right)_{z0} + a_1^n \left(\frac{\partial \Phi_1}{\partial z} \right)_{z0} \right] dz \\
=& \theta\Delta t_n \left(a_0^{n+1} - a_0^n\right) \left(-3a_0^n + 4a_{1/2}^n - a_1^n \right) = \theta\Delta t_n \left(a_0^{n+1} - a_0^n\right).
\end{aligned} \tag{3.49}$$

Term B3 is computed by:

$$\begin{aligned}
\text{Term B3} &= \theta\Delta t_n \int_{\Omega_e} \left[c u^{n+1} \right]_{z0} d\Omega_e = \theta\Delta t_n c \int_0^{\Delta z} \sum\nolimits_k a_k^{n+1} (\Phi_k)_{z0} dz \\
&= \theta\Delta t_n c a_0^{n+1} \Delta z.
\end{aligned} \tag{3.50}$$

Summing up the contributions of Terms B1–B3 (Eqs. 3.48–3.50), one obtains:

$$\begin{aligned}
\text{Term B} =& \theta\Delta t_n \int_{\Omega_e} \sum_{k=1}^{N} a_k^{n+1} (L\Phi_k)_{z0} d\Omega_e \\
=& \theta\Delta t_n \left[a_0^{n+1} \left(-3a_0^n + \text{DUDZN0} + c\Delta z \right) + 4a_0^n a_{1/2}^{n+1} - a_0^n a_1^{n+1} - a_0^n \text{DUDZN0} \right].
\end{aligned} \tag{3.51}$$

Term D is computed as Term A, but referring to time level n, and yields:

$$\text{Term D} = \int_{\Omega_e} \sum_{k=1}^{N} a_k^n (\Phi_k)_{z0} d\Omega_e = a_0^n \Delta z. \tag{3.52}$$

Term E is computed, similarly to Eq. (3.34), by:

$$
\begin{aligned}
\mathrm{Term\,E} &= -(1-\theta)\Delta t_n \int_{\Omega_e} \sum_{k=1}^{N} a_k^n (L\Phi_k)_{z0} d\Omega_e = -(1-\theta)\Delta t_n \int_{\Omega e} (Lu^n)_{z0} d\Omega_e \\
&= -(1-\theta)\Delta t_n \int_{\Omega e} \left[\left(u^n \frac{\partial u^n}{\partial z} \right)_{z0} + c(u^n)_{z0} \right] d\Omega_e \\
&= -(1-\theta)\Delta t_n \int_0^{\Delta z} \left[\sum\nolimits_k a_k^n (\Phi_k)_{z0} \sum\nolimits_k a_k^n \left(\frac{\partial \Phi_k}{\partial z} \right)_{z0} + \sum\nolimits_k c a_k^n (\Phi_k)_{z0} \right] dz \\
&= -(1-\theta)\Delta t_n \left[a_0^n \left(-3a_0^n + 4a_{1/2}^n - a_1^n \right) + c a_0^n \Delta z \right] \\
&= -(1-\theta)\Delta t_n a_0^n [\mathrm{DUDZN0} + c\Delta z].
\end{aligned}
\tag{3.53}
$$

Collecting Term A, Eq. (3.46); Term B, Eq. (3.51); Term D, Eq. (3.52); and Term E, Eq. (3.53), one obtains the element equation for $j = 1$:

$$
\begin{aligned}
& a_0^{n+1} \left[\Delta z + \theta \Delta t_n \left(-3a_0^n + \mathrm{DUDZN0} + c\Delta z \right) \right] + a_{1/2}^{n+1} \left(4a_0^n \theta \Delta t_n \right) + a_1^{n+1} \left(-a_0^n \theta \Delta t_n \right) \\
&= a_0^n \Delta z + a_0^n \mathrm{DUDZN0} [\theta \Delta t_n - (1-\theta)\Delta t_n] - (1-\theta)\Delta t_n a_0^n c \Delta z.
\end{aligned}
\tag{3.54}
$$

(ii) *Element Equation for* $j = 2$, ($z = z_{1/2} = \Delta z/2$)

For the collocation point in the middle of the element ($z = z_{1/2} = \Delta z/2$), the element equation (3.44) becomes:

$$
\begin{aligned}
& \overbrace{\int_{\Omega_e} \sum_{k=1}^{N} a_k^{n+1} (\Phi_k)_{z1/2} d\Omega_e}^{\mathrm{Term\,A}} + \overbrace{\theta \Delta t_n \int_{\Omega_e} \sum_{k=1}^{N} a_k^{n+1} (L\Phi_k)_{z1/2} d\Omega_e}^{\mathrm{Term\,B}} \\
&= \underbrace{\int_{\Omega_e} \sum_{k=1}^{N} a_k^n (\Phi_k)_{z1/2} d\Omega_e}_{\mathrm{Term\,D}} - \underbrace{(1-\theta)\Delta t_n \int_{\Omega_e} \sum_{k=1}^{N} a_k^n (L\Phi_k)_{z1/2} d\Omega_e}_{\mathrm{Term\,E}}.
\end{aligned}
\tag{3.55}
$$

The four terms of Eq. (3.55) are computed with the same procedure followed for $j = 1$, also using Eqs. (3.42) and (3.43). This yields:

$$
\mathrm{Term\,A} = \int_{\Omega_e} \sum_{k=1}^{N} a_k^{n+1} (\Phi_k)_{z1/2} d\Omega_e = a_{1/2}^{n+1} \Delta z. \tag{3.56}
$$

$$\begin{aligned}\text{Term B} =&\theta\Delta t_n\int_{\Omega_e}\sum_{k=1}^{N}a_k^{n+1}(L\Phi_k)_{z1/2}d\Omega_e\\ =&\theta\Delta t_n\left[a_{1/2}^{n}\left(-a_0^{n+1}+a_1^{n+1}\right)+\left(a_{1/2}^{n+1}-a_{1/2}^{n}\right)\left(-a_0^{n}+a_1^{n}\right)+ca_{1/2}^{n+1}\Delta z\right].\end{aligned} \tag{3.57}$$

$$\text{Term D} = \int_{\Omega_e}\sum_{k=1}^{N}a_k^{n}(\Phi_k)_{z1/2}d\Omega_e = a_{1/2}^{n}\Delta z. \tag{3.58}$$

$$\begin{aligned}\text{Term E} =& -(1-\theta)\Delta t_n\int_{\Omega_e}\sum_{k=1}^{N}a_k^{n}(L\Phi_k)_{z1/2}d\Omega_e\\ =& -(1-\theta)\Delta t_n a_{1/2}^{n}\left(-a_0^{n}+a_1^{n}+c\Delta z\right).\end{aligned} \tag{3.59}$$

Using Eq. (3.23b) [$DUDZ12 = u_1 - u_0$], the element equation for $j = 2$ becomes:

$$\begin{aligned}&a_0^{n+1}\left(-\theta\Delta t_n a_{1/2}^{n}\right)+a_{1/2}^{n+1}[\Delta z+\theta\Delta t_n(\text{DUDZ12}+c\Delta z)]+a_1^{n+1}\left(\theta\Delta t_n a_{1/2}^{n}\right)\\ &= a_{1/2}^{n}\Delta z + a_{1/2}^{n}\text{DUDZ12}[\theta\Delta t_n-(1-\theta)\Delta t_n]-(1-\theta)\Delta t_n a_{1/2}^{n}c\Delta z.\end{aligned} \tag{3.60}$$

(iii) *Element Equation for j = 3, ($z = z_1 = \Delta z$)*

For the collocation point at the right boundary of the element ($z = z_1 = \Delta z$), the element equation 3.44 becomes:

$$\begin{aligned}&\overbrace{\int_{\Omega_e}\sum_{k=1}^{N}a_k^{n+1}(\Phi_k)_{z1}d\Omega_e}^{\text{Term A}}+\theta\Delta t_n\overbrace{\int_{\Omega_e}\sum_{k=1}^{N}a_k^{n+1}(L\Phi_k)_{z1}d\Omega_e}^{\text{Term B}}\\ &=\underbrace{\int_{\Omega_e}\sum_{k=1}^{N}a_k^{n}(\Phi_k)_{z1}d\Omega_e}_{\text{Term D}}-(1-\theta)\Delta t_n\underbrace{\int_{\Omega_e}\sum_{k=1}^{N}a_k^{n}(L\Phi_k)_{z1}d\Omega_e}_{\text{Term E}}.\end{aligned} \tag{3.61}$$

Terms of Eq. (3.61) are computed as follows:

$$\text{Term A} = \int_{\Omega_e}\sum_{k=1}^{N}a_k^{n+1}(\Phi_k)_{z1}d\Omega_e = a_1^{n+1}\Delta z. \tag{3.62}$$

$$\begin{aligned}\text{Term B} =&\theta\Delta t_n\int_{\Omega_e}\sum_{k=1}^{N}a_k^{n+1}(L\Phi_k)_{z1}d\Omega_e\\ =&\theta\Delta t_n\left[a_1^n\left(a_0^{n+1}-4a_{1/2}^{n+1}+3a_1^{n+1}\right)+\left(a_1^{n+1}-a_1^n\right)\left(a_0^n-4a_{1/2}^n+3a_1^n\right)+ca_1^{n+1}\Delta z\right]\\ =&\theta\Delta t_n\left[a_1^n\left(a_0^{n+1}-4a_{1/2}^{n+1}+3a_1^{n+1}\right)+\left(a_1^{n+1}-a_1^n\right)\text{DUDZN1}+ca_1^{n+1}\Delta z\right].\end{aligned} \tag{3.63}$$

$$\text{Term D} = \int_{\Omega_e}\sum_{k=1}^{N}a_k^n(\Phi_k)_{z1}d\Omega_e = a_1^n\Delta z. \tag{3.64}$$

$$\begin{aligned}\text{Term E} =& -(1-\theta)\Delta t_n\int_{\Omega_e}\sum_{k=1}^{N}a_k^n(L\Phi_k)_{z1}d\Omega_e\\ =& -(1-\theta)\Delta t_n a_1^n\left(a_0^n-4a_{1/2}^n+3a_1^n+c\Delta z\right)\\ =& -(1-\theta)\Delta t_n a_1^n(\text{DUDZN1}+c\Delta z).\end{aligned} \tag{3.65}$$

Collecting terms, the element equation for $j = 3$ becomes:

$$\begin{aligned}&a_0^{n+1}\left(\theta\Delta t_n a_1^n\right)+a_{1/2}^{n+1}\left(-4a_1^n\theta\Delta t_n\right)+a_1^{n+1}\left[\Delta z+\theta\Delta t_n\left(3a_1^n+\text{DUDZN1}+c\Delta z\right)\right]\\ &= a_1^n\Delta z+a_1^n\text{DUDZN1}[\theta\Delta t_n-(1-\theta)\Delta t_n]-(1-\theta)\Delta t_n c a_1^n\Delta z.\end{aligned} \tag{3.66}$$

The equations for $j = 1, 2, 3$, Eqs. (3.54), (3.60), and (3.66), respectively, in the unknowns $a_0^{n+1} = u_0^{n+1}$, $a_{1/2}^{n+1} = u_{1/2}^{n+1}$, $a_1^{n+1} = u_1^{n+1}$ must be summed to yield the combined element equation:

$$\begin{aligned}&u_0^{n+1}\left[\Delta z+\vartheta\Delta t_n\left(-6u_0^n+3u_{1/2}^n+c\Delta z\right)\right]+u_{1/2}^{n+1}\left[\Delta z+\vartheta\Delta t_n\left(3u_0^n-3u_1^n+c\Delta z\right)\right]\\ &\quad+u_1^{n+1}\left[\Delta z+\vartheta\Delta t_n\left(-3u_0^n+6u_1^n+c\Delta z\right)\right]\\ &=\left(u_0^n+u_{1/2}^n+u_1^n\right)\Delta z+\left(u_0^n\text{DUDZN0}+u_{1/2}^n\text{DUDZ12}+u_1^n\text{DUDZN1}\right)\left[\vartheta\Delta t_n+(1-\vartheta)\Delta t_n\right]\\ &\quad-(1-\vartheta)\Delta t_n\left(u_0^n+u_{1/2}^n+u_1^n\right)c\Delta z.\end{aligned} \tag{3.67}$$

The element equation (3.67) is formally identical with Eq. (3.35) obtained with the method of subdomain collocation and is treated likewise, combined with the continuity equation (3.39).

3.4 Galerkin Method of Orthogonal Collocations

In the variant of the Galerkin method, the weight functions are chosen identical with the shape functions:

$$\psi_j = \Phi_j. \qquad (j = 1, 2, \ldots, N) \tag{3.68}$$

With this choice the element equation (3.14) becomes:

$$\begin{aligned} &\int_{\Omega_e} \sum_{k=1}^{N} a_k^{n+1} (\Phi_k, \Phi_j) d\Omega_e + \theta \Delta t_n \int_{\Omega_e} \sum_{k=1}^{N} a_k^{n+1} (L\Phi_k, \Phi_j) d\Omega_e \\ &= \int_{\Omega_e} \sum_{k=1}^{N} a_k^{n} (\Phi_k, \Phi_j) d\Omega_e - (1-\theta) \Delta t_n \int_{\Omega_e} \sum_{k=1}^{N} a_k^{n} (L\Phi_k, \Phi_j) d\Omega_e. \end{aligned} \tag{3.69}$$

Using the parabolic shape functions (3.19) and Eq. (3.68) with $j = 1, 2, 3 = N$, three element equations are obtained. In deriving the analytical expressions of the element equations, the integrals shown in the following list are used:

(i) Group 1: Integrals of shape functions and of product of shape functions:

$$\int_0^{\Delta z} \Phi_0 dz = (1/6)\Delta z, \tag{3.69a}$$

$$\int_0^{\Delta z} \Phi_{1/2} dz = (2/3)\Delta z, \tag{3.69b}$$

$$\int_0^{\Delta z} \Phi_1 dz = (1/6)\Delta z, \tag{3.69c}$$

$$\int_0^{\Delta z} \Phi_0^2 dz = (2/15)\Delta z, \tag{3.69d}$$

$$\int_0^{\Delta z} \Phi_{1/2}^2 dz = (8/15)\Delta z, \tag{3.69e}$$

$$\int_0^{\Delta z} \Phi_1^2 dz = (2/15)\Delta z, \tag{3.69f}$$

$$\int_0^{\Delta z} \Phi_0 \Phi_{1/2} dz = \Delta z/15, \tag{3.69g}$$

$$\int_0^{\Delta z} \Phi_0 \Phi_1 dz = -\Delta z/30, \tag{3.69h}$$

$$\int_0^{\Delta z} \Phi_{1/2} \Phi_1 dz = \Delta z/15. \tag{3.69i}$$

(ii) Group 2: Integrands contain the derivative $\frac{d\Phi_0}{dz}$:

$$\int_0^{\Delta z} \Phi_0 \frac{d\Phi_0}{dz} dz = -1/2, \tag{3.70a}$$

$$\int_0^{\Delta z} \Phi_{1/2} \frac{d\Phi_0}{dz} dz = -2/3, \tag{3.70b}$$

$$\int_0^{\Delta z} \Phi_1 \frac{d\Phi_0}{dz} dz = 1/6, \tag{3.70c}$$

$$\int_0^{\Delta z} \Phi_0^2 \frac{d\Phi_0}{dz} dz = -1/3, \tag{3.70d}$$

$$\int_0^{\Delta z} \Phi_{1/2}^2 \frac{d\Phi_0}{dz} dz = -8/15, \tag{3.70e}$$

$$\int_0^{\Delta z} \Phi_1^2 \frac{d\Phi_0}{dz} dz = 1/15, \tag{3.70f}$$

$$\int_0^{\Delta z} \Phi_0 \Phi_{1/2} \frac{d\Phi_0}{dz} dz = -1/5, \tag{3.70g}$$

$$\int_0^{\Delta z} \Phi_0 \Phi_1 \frac{d\Phi_0}{dz} dz = 1/30, \tag{3.70h}$$

$$\int_0^{\Delta z} \Phi_{1/2} \Phi_1 \frac{d\Phi_0}{dz} dz = 1/15. \tag{3.70i}$$

(iii) Group 3: Integrands contain the derivative $\frac{d\Phi_{1/2}}{dz}$:

$$\int_0^{\Delta z} \Phi_0 \frac{d\Phi_{1/2}}{dz} dz = 2/3, \tag{3.71a}$$

$$\int_0^{\Delta z} \Phi_{1/2} \frac{d\Phi_{1/2}}{dz} dz = 0, \tag{3.71b}$$

$$\int_0^{\Delta z} \Phi_1 \frac{d\Phi_{1/2}}{dz} dz = -2/3, \tag{3.71c}$$

$$\int_0^{\Delta z} \Phi_0^2 \frac{d\Phi_{1/2}}{dz} dz = 2/5, \tag{3.71d}$$

$$\int_0^{\Delta z} \Phi_{1/2}^2 \frac{d\Phi_{1/2}}{dz} dz = 0, \tag{3.71e}$$

$$\int_0^{\Delta z} \Phi_1^2 \frac{d\Phi_{1/2}}{dz} dz = -2/5, \tag{3.71f}$$

$$\int_0^{\Delta z} \Phi_0 \Phi_{1/2} \frac{d\Phi_{1/2}}{dz} dz = 4/15, \tag{3.71g}$$

$$\int_0^{\Delta z} \Phi_0 \Phi_1 \frac{d\Phi_{1/2}}{dz} dz = 0, \tag{3.71h}$$

$$\int_0^{\Delta z} \Phi_{1/2} \Phi_1 \frac{d\Phi_{1/2}}{dz} dz = -4/15. \tag{3.71i}$$

(iv) Group 4: Integrands contain the derivative $\frac{d\Phi_1}{dz}$:

$$\int_0^{\Delta z} \Phi_0 \frac{d\Phi_1}{dz} dz = -1/6, \tag{3.72a}$$

$$\int_0^{\Delta z} \Phi_{1/2} \frac{d\Phi_1}{dz} dz = 2/3, \tag{3.72b}$$

$$\int_0^{\Delta z} \Phi_1 \frac{d\Phi_1}{dz} dz = 1/2, \tag{3.72c}$$

$$\int_0^{\Delta z} \Phi_0^2 \frac{d\Phi_1}{dz} dz = -1/15, \tag{3.72d}$$

$$\int_0^{\Delta z} \Phi_{1/2}^2 \frac{d\Phi_1}{dz} dz = 8/15, \tag{3.72e}$$

$$\int_0^{\Delta z} \Phi_1^2 \frac{d\Phi_1}{dz} dz = 1/3, \tag{3.72f}$$

$$\int_0^{\Delta z} \Phi_0 \Phi_{1/2} \frac{d\Phi_1}{dz} dz = -1/15, \tag{3.72g}$$

$$\int_0^{\Delta z} \Phi_0 \Phi_1 \frac{d\Phi_1}{dz} dz = -1/30, \tag{3.72h}$$

$$\int_0^{\Delta z} \Phi_{1/2} \Phi_1 \frac{d\Phi_1}{dz} dz = 1/5. \tag{3.72i}$$

(i) *Element equation for $j = 1$*

For $j = 1$ the element equation (3.69) becomes:

$$\begin{aligned}
&\overbrace{\int_{\Omega_e} \sum_{k=1}^{N} a_k^{n+1} \Phi_k \Phi_0 d\Omega_e}^{\text{Term A}} + \overbrace{\theta \Delta t_n \int_{\Omega_e} \sum_{k=1}^{N} a_k^{n+1} (L\Phi_k) \Phi_0 d\Omega_e}^{\text{Term B}} \\
&= \underbrace{\int_{\Omega_e} \sum_{k=1}^{N} a_k^{n} \Phi_k \Phi_0 d\Omega_e}_{\text{Term D}} - \underbrace{(1-\theta)\Delta t_n \int_{\Omega_e} \sum_{k=1}^{N} a_k^{n} (L\Phi_k) \Phi_0 d\Omega_e}_{\text{Term E}}.
\end{aligned} \tag{3.73}$$

Term A is computed, using the integrals (3.70) by:

$$\begin{aligned}
\text{Term A} = \int_{\Omega_e} \sum_{k=1}^{N} a_k^{n+1} \Phi_k \Phi_0 d\Omega_e &= \int_0^{\Delta z} \left(a_0^{n+1} \Phi_0^2 + a_{1/2}^{n+1} \Phi_{1/2} \Phi_0 + a_1^{n+1} \Phi_1 \Phi_0 \right) dz \\
&= a_0^{n+1} \frac{2}{15} \Delta z + a_{1/2}^{n+1} \frac{1}{15} \Delta z - a_1^{n+1} \frac{1}{30} \Delta z.
\end{aligned} \tag{3.74}$$

Term B is discretized with respect to time using Eq. (3.27) and yields:

$$\begin{aligned}
\text{Term B} &= \theta \Delta t_n \int_{\Omega_e} \sum_{k=1}^{N} a_k^{n+1} (L\Phi_k) \Phi_0 d\Omega_e = \theta \Delta t_n \int_{\Omega_e} \sum_{k=1}^{N} (Lu^{n+1}) \Phi_0 d\Omega_e \\
&= \theta \Delta t_n \int_{\Omega e} \Bigg[\underbrace{\left(u^n \frac{\partial u^{n+1}}{\partial z} \right) \Phi_0}_{\text{Term B1}} + \underbrace{\left(u^{n+1} - u^n \right) \left(\frac{\partial u^n}{\partial z} \right) \Phi_0}_{\text{Term B2}} + \underbrace{c u^{n+1} \Phi_0}_{\text{Term B3}} \Bigg] d\Omega_e.
\end{aligned} \tag{3.75}$$

Term B1 in Eq. (3.75) is computed, using integrals (3.71) through (3.73) by:

$$\begin{aligned}
\text{Term B1} =& \theta \Delta t_n \int_{\Omega e} \left[\left(u^n \frac{\partial u^{n+1}}{\partial z} \right) \Phi_0 \right] d\Omega_e \\
=& \theta \Delta t_n \int_0^{\Delta z} \left(a_0^n \Phi_0 + a_{1/2}^n \Phi_{1/2} + a_1^n \Phi_1 \right) \left(a_0^{n+1} \frac{d\Phi_0}{dz} + a_{1/2}^{n+1} \frac{d\Phi_{1/2}}{dz} + a_1^{n+1} \frac{d\Phi_1}{dz} + \right) \Phi_0 dz \\
=& \vartheta \Delta t_n \left[a_0^{n+1} \left(-\frac{1}{3} a_0^n - \frac{1}{5} a_{1/2}^n + \frac{1}{30} a_1^n \right) + a_{1/2}^{n+1} \left(\frac{2}{5} a_0^n + \frac{4}{15} a_{1/2}^n \right) \right] \\
&+ \vartheta \Delta t_n \left[a_1^{n+1} \left(-\frac{1}{15} a_0^n - \frac{1}{15} a_{1/2}^n - \frac{1}{30} a_1^n \right) \right].
\end{aligned} \tag{3.76}$$

Likewise Term B2 yields:

$$\begin{aligned}
\text{Term B2} =& \theta \Delta t_n \int_{\Omega e} \left[\left(u^{n+1} - u^n \right) \left(\frac{\partial u^n}{\partial z} \right) \Phi_0 \right] d\Omega_e \\
=& \theta \Delta t_n \int_{\Omega e} \left[\sum_k \left(a_k^{n+1} - a_k^n \right) \Phi_k \sum\nolimits_k a_k^n \frac{d\Phi_k}{dz} \right] \Phi_0 d\Omega_e \\
=& \theta \Delta t_n \int_0^{\Delta z} \left[\left(a_0^{n+1} - a_0^n \right) \Phi_0 + \left(a_{1/2}^{n+1} - a_{1/2}^n \right) \Phi_{1/2} + \left(a_1^{n+1} - a_1^n \right) \Phi_1 \right] \\
&\times \left[a_0^n \frac{d\Phi_0}{dz} + a_{1/2}^n \frac{d\Phi_{1/2}}{dz} + a_1^n \frac{d\Phi_1}{dz} \right] \Phi_0 dz \\
=& \theta \Delta t_n \left[a_0^{n+1} \left(-\frac{1}{3} a_0^n + \frac{2}{5} a_{1/2}^n - \frac{1}{15} a_1^n \right) + a_{1/2}^{n+1} \left(-\frac{1}{5} a_0^n + \frac{4}{15} a_{1/2}^n - \frac{1}{15} a_1^n \right) \right] \\
&+ \theta \Delta t_n \left[a_1^{n+1} \left(\frac{1}{30} a_0^n - \frac{1}{30} a_1^n \right) + \frac{1}{3} \left(a_0^n \right)^2 - \frac{4}{15} \left(a_{1/2}^n \right)^2 + \frac{1}{30} \left(a_1^n \right)^2 \right] \\
&+ \theta \Delta t_n \left[-\frac{1}{5} a_0^n a_{1/2}^n + \frac{1}{30} a_0^n a_1^n + \frac{1}{15} a_{1/2}^n a_1^n \right].
\end{aligned} \tag{3.77}$$

Using integrals (3.70) Term B3 is computed by:

$$\begin{aligned}
\text{Term B3} =& \theta \Delta t_n \int_{\Omega e} \left[c u^{n+1} \Phi_0 \right] d\Omega_e = c \theta \Delta t_n \int_0^{\Delta z} \left(\sum\nolimits_k a_k^{n+1} \Phi_k \right) \Phi_0 dz \\
=& c \theta \Delta t_n \left(a_0^{n+1} \frac{2}{15} \Delta z + a_{1/2}^{n+1} \frac{1}{15} \Delta z - a_1^{n+1} \frac{1}{30} \Delta z \right).
\end{aligned} \tag{3.78}$$

Collecting contributions from Eqs. (3.76)–(3.78), one obtains:

$$
\begin{aligned}
\text{Term B} =&\theta\Delta t_n \int_{\Omega_e} \sum_{k=1}^{N} a_k^{n+1} (L\Phi_k)\Phi_0 d\Omega_e \\
=&\theta\Delta t_n \left[a_0^{n+1}\left(-\frac{2}{3}a_0^n+\frac{1}{5}a_{1/2}^n-\frac{1}{30}a_1^n+c\frac{2}{15}\Delta z\right)\right] \\
&+\theta\Delta t_n \left[a_{1/2}^{n+1}\left(\frac{1}{5}a_0^n+\frac{8}{5}a_{1/2}^n-\frac{1}{15}a_1^n+c\frac{1}{15}\Delta z\right)\right] \\
&+\theta\Delta t_n \left[a_1^{n+1}\left(-\frac{1}{30}a_0^n-\frac{1}{15}a_{1/2}^n-\frac{1}{15}a_1^n-c\frac{1}{30}\Delta z\right)\right] \\
&+\theta\Delta t_n \left[\frac{1}{3}\left(a_0^n\right)^2-\frac{4}{15}\left(a_{1/2}^n\right)^2+\frac{1}{30}\left(a_1^n\right)^2-\frac{1}{5}a_0^n a_{1/2}^n+\frac{1}{30}a_0^n a_1^n+\frac{1}{15}a_{1/2}^n a_1^n\right].
\end{aligned}
\tag{3.79}
$$

Term D is computed like term A, but referring to time level n:

$$
\text{Term D} = \int_{\Omega_e} \sum\nolimits_k a_k^n \Phi_k \Phi_0 d\Omega_e = a_0^n \frac{2}{15}\Delta z + a_{1/2}^n \frac{\Delta z}{15} - a_1^n \frac{\Delta z}{30}. \tag{3.80}
$$

Term E is computed by:

$$
\begin{aligned}
\text{Term E} =& -(1-\theta)\Delta t_n \int_{\Omega_e} \sum\nolimits_k a_k^n (L\Phi_k)\Phi_0 d\Omega_e \\
=& -(1-\theta)\Delta t_n \int_{\Omega_e} Lu^n \Phi_0 d\Omega_e = -(1-\sigma)\Delta t_n \int_{\Omega_e}\left[u^n \frac{\partial u^n}{\partial z}+c\right]\Phi_0 d\Omega_e \\
=& -(1-\theta)\Delta t_n \int_0^{\Delta z}\left[\left(a_0^n\Phi_0^n + a_{1/2}^n\Phi_{1/2}^n + a_1^n\Phi_1^n\right)\right. \\
&\qquad \left.\times\left(a_0^n\frac{d\Phi_0}{dz}+a_{1/2}^n\frac{d\Phi_{1/2}}{dz}+a_1^n\frac{d\Phi_1}{dz}\right)+cu^n\right]\Phi_0 dz \\
=& -(1-\theta)\Delta t_n \left[-\frac{1}{3}\left(a_0^n\right)^2+\frac{4}{15}\left(a_{1/2}^n\right)^2-\frac{1}{30}\left(a_1^n\right)^2+\frac{1}{5}a_0^n a_{1/2}^n\right] \\
=& -(1-\theta)\Delta t_n \left[-\frac{1}{30}a_0^n a_1^n-\frac{1}{15}a_{1/2}^n a_1^n+a_0^n\frac{2}{15}c\Delta z+a_{1/2}^n\frac{c\Delta z}{15}-a_1^n\frac{c\Delta z}{30}\right].
\end{aligned}
\tag{3.81}
$$

Collecting all terms, namely, Term A (Eq. 3.74), Term B (Eq. 3.79), Term D (Eq. 3.80), and Term E (Eq. 3.81), one obtains the element equation for $j = 1$ (letting $a_k = u_k$, $k = 0$, 1/2, 1):

$$\begin{aligned}
&u_0^{n+1}\left[\frac{2}{15}\Delta z+\theta\Delta t_n\left(-\frac{2}{3}u_0^n+\frac{1}{5}u_{1/2}^n-\frac{1}{30}u_1^n+\frac{2}{15}c\Delta z\right)\right]\\
&+u_{1/2}^{n+1}\left[\frac{\Delta z}{15}+\theta\Delta t_n\left(\frac{1}{5}u_0^n+\frac{8}{15}u_{1/2}^n-\frac{1}{15}u_1^n+\frac{1}{15}c\Delta z\right)\right]\\
&+u_1^{n+1}\left[-\frac{\Delta z}{30}+\theta\Delta t_n\left(-\frac{1}{30}u_0^n-\frac{1}{15}u_{1/2}^n-\frac{1}{15}u_1^n-\frac{1}{30}c\Delta z\right)\right]\\
&=u_0^n\frac{2}{15}\Delta z+u_{1/2}^n\frac{\Delta z}{15}-u_1^n\frac{\Delta z}{30}+[\theta\Delta t_n-(1-\theta)\Delta t_n]\\
&\times\left[-\frac{1}{3}\left(u_0^n\right)^2+\frac{4}{15}\left(u_{1/2}^n\right)^2-\frac{1}{30}\left(u_1^n\right)^2+\frac{1}{5}u_0^nu_{1/2}^n-\frac{1}{30}u_0^nu_1^n-\frac{1}{15}u_{1/2}^nu_1^n\right]\\
&-(1-\theta)\Delta t_nc\Delta z\left(\frac{2}{15}u_0^n+\frac{1}{15}u_{1/2}^n-\frac{1}{30}u_1^n\right).
\end{aligned}\tag{3.82}$$

(ii) *Element equation for* $j = 2$

For $j = 2$ the element equation (3.69) becomes:

$$\begin{aligned}
&\overbrace{\int_{\Omega_e}\sum\nolimits_k a_k^{n+1}\boldsymbol{\Phi}_k\boldsymbol{\Phi}_{1/2}d\boldsymbol{\Omega}_e}^{\text{Term A}}+\overbrace{\theta\Delta t_n\int_{\Omega_e}\sum\nolimits_k a_k^{n+1}(L\boldsymbol{\Phi}_k)\boldsymbol{\Phi}_{1/2}d\boldsymbol{\Omega}_e}^{\text{Term B}}\\
&=\underbrace{\int_{\Omega_e}\sum\nolimits_k a_k^{n}\boldsymbol{\Phi}_k\boldsymbol{\Phi}_{1/2}d\boldsymbol{\Omega}_e}_{\text{Term D}}-\underbrace{(1-\theta)\Delta t_n\int_{\Omega_e}\sum\nolimits_k a_k^{n}(L\boldsymbol{\Phi}_k)\boldsymbol{\Phi}_{1/2}d\boldsymbol{\Omega}_e}_{\text{Term E}}.
\end{aligned}\tag{3.83}$$

The terms of Eq. (3.83) are computed in the same way as those of Eq. (3.73), just replacing the shape function $\boldsymbol{\Phi}_0$ with $\boldsymbol{\Phi}_{1/2}$. The results are as follows:

$$\text{Term A}=\int_{\Omega_e}\sum\nolimits_k a_k^{n+1}\boldsymbol{\Phi}_k\boldsymbol{\Phi}_{1/2}d\boldsymbol{\Omega}_e=a_0^{n+1}\frac{\Delta z}{15}+a_{1/2}^{n+1}\frac{8}{15}\Delta z+a_1^{n+1}\frac{\Delta z}{15},\tag{3.84}$$

$$\begin{aligned}
\text{Term B}=&\theta\Delta t_n\int_{\Omega_e}\sum\nolimits_k a_k^{n+1}(L\boldsymbol{\Phi}_k)\boldsymbol{\Phi}_{1/2}d\boldsymbol{\Omega}_e\\
=&\theta\Delta t_n\left\{a_0^{n+1}\left(-\frac{2}{5}a_0^n-\frac{4}{15}a_{1/2}^n+\frac{1}{15}c\Delta z\right)+a_{1/2}^{n+1}\left(-\frac{4}{15}a_0^n+\frac{4}{15}a_1^n+\frac{8}{15}c\Delta z\right)\right.\\
&\left.+a_1^{n+1}\left(\frac{4}{15}a_{1/2}^n+\frac{2}{5}a_1^n+\frac{1}{15}c\Delta z\right)\right\}\\
&+\theta\Delta t_n\left\{\left[+\frac{1}{5}\left(a_0^n\right)^2+\frac{4}{15}a_0^na_{1/2}^n-\frac{4}{15}a_{1/2}^na_1^n-\frac{1}{5}\left(a_1^n\right)^2\right]\right\},
\end{aligned}\tag{3.85}$$

$$\text{Term D}=\int_{\Omega_e}\sum\nolimits_k a_k^{n}\boldsymbol{\Phi}_k\boldsymbol{\Phi}_{1/2}d\boldsymbol{\Omega}_e=a_0^{n}\frac{\Delta z}{15}+a_{1/2}^{n}\frac{8}{15}\Delta z+a_1^{n}\frac{\Delta z}{15},\tag{3.86}$$

$$\begin{aligned}\text{Term E} &= -(1-\theta)\Delta t_n \int_{\Omega_e} \sum\nolimits_k a_k^n (L\Phi_k)\Phi_{1/2} d\Omega_e \\ &= -(1-\theta)\Delta t_n \left[-\frac{1}{5}\left(a_0^n\right)^2 - \frac{4}{15} a_0^n a_{1/2}^n + \frac{4}{15} a_{1/2}^n a_1^n + \frac{1}{5}\left(a_1^n\right)^2\right] \\ &\quad -(1-\theta)\Delta t_n \left[c\Delta z\left(\frac{1}{15} a_0^n + \frac{8}{15} a_{1/2}^n + \frac{1}{15} a_1^n\right)\right].\end{aligned} \tag{3.87}$$

Collecting all terms and letting $u_k = a_k$, $(k = 0, 1/2,\ \ 1)$, one obtains the element equation for $j = 2$:

$$\begin{aligned} & u_0^{n+1}\left[\frac{\Delta z}{15} + \theta\Delta t_n\left(-\frac{2}{5} u_0^n - \frac{4}{15} u_{1/2}^n + \frac{1}{15} c\Delta z\right)\right] \\ & + u_{1/2}^{n+1}\left[\frac{8}{15}\Delta z + \theta\Delta t_n\left(-\frac{4}{15} u_0^n + \frac{4}{15} u_1^n + \frac{8}{15} c\Delta z\right)\right] \\ & + u_1^{n+1}\left[\frac{\Delta z}{15} + \theta\Delta t_n\left(\frac{4}{15} u_{1/2}^n + \frac{2}{5} u_1^n + \frac{1}{15} c\Delta z\right)\right] \\ & = u_0^n \frac{\Delta z}{15} + u_{1/2}^n \frac{8}{15}\Delta z + u_1^n \frac{\Delta z}{15} + \left(\theta\Delta t_n - (1-\theta)\Delta t_n\right) \\ & \times\left[-\frac{1}{5}\left(a_0^n\right)^2 - \frac{4}{15} a_0^n a_{1/2}^n + \frac{4}{15} a_{1/2}^n a_1^n + \frac{1}{5}\left(a_1^n\right)^2\right] \\ & -(1-\theta)\Delta t_n c\Delta z\left(\frac{1}{15} a_0^n + \frac{8}{15} a_{1/2}^n + \frac{1}{15} a_1^n\right)\end{aligned} \tag{3.88}$$

(iii) *Element equation for* $j = 3$

For $j = 3$ the element equation (3.69) becomes:

$$\begin{aligned} & \overbrace{\int_{\Omega_e} \sum\nolimits_k a_k^{n+1} \Phi_k \Phi_1 d\Omega_e}^{\text{Term A}} + \theta\Delta t_n \overbrace{\int_{\Omega_e} \sum\nolimits_k a_k^{n+1} (L\Phi_k)\Phi_1 d\Omega_e}^{\text{Term B}} \\ & = \underbrace{\int_{\Omega_e} \sum\nolimits_k a_k^n \Phi_k \Phi_1 d\Omega_e}_{\text{Term D}} - (1-\theta)\Delta t_n \underbrace{\int_{\Omega_e} \sum\nolimits_k a_k^n (L\Phi_k)\Phi_1 d\Omega_e}_{\text{Term E}}.\end{aligned} \tag{3.89}$$

The terms of Eq. (3.89) are computed in the same way as those of Eq. (3.73), just replacing the shape function Φ_0 with Φ_1.

The results are as follows:

$$\begin{aligned}\text{Term A} &= \int_{\Omega_e} \sum\nolimits_k a_k^{n+1} \Phi_k \Phi_1 d\Omega_e \\ &= a_0^{n+1}\left(-\frac{\Delta z}{30}\right) + a_{1/2}^{n+1} \frac{\Delta z}{15} + a_1^{n+1} \frac{2}{15}\Delta z,\end{aligned} \tag{3.90}$$

$$\begin{aligned}
\text{Term B} =&\theta\Delta t_n\int_{\Omega_e}\sum\nolimits_k a_k^{n+1}(L\Phi_k)\Phi_1 d\Omega_e\\
=&\theta\Delta t_n\left\{\begin{aligned}&a_0^{n+1}\left(\frac{1}{15}a_0^n+\frac{1}{15}a_{1/2}^n+\frac{1}{30}a_1^n-\frac{1}{30}c\Delta z\right)\\&+a_{1/2}^{n+1}\left(\frac{1}{15}a_0^n-\frac{8}{15}a_{1/2}^n-\frac{1}{5}a_1^n+\frac{1}{15}c\Delta z\right)\end{aligned}\right\}\\
&+\theta\Delta t_n\left\{\begin{aligned}&+a_1^{n+1}\left(\frac{1}{30}a_0^n-\frac{1}{5}a_{1/2}^n+\frac{2}{3}a_1^n+\frac{2}{15}c\Delta z\right)\\&-\left[\frac{1}{30}\left(a_0^n\right)^2-\frac{4}{15}\left(a_{1/2}^n\right)^2+\frac{1}{3}\left(a_1^n\right)^2\right]\end{aligned}\right\}\\
&+\theta\Delta t_n\left\{\left[\frac{1}{15}a_0^na_{1/2}^n+\frac{1}{30}a_0^na_1^n-\frac{1}{5}a_{1/2}^na_1^n\right]\right\},
\end{aligned}\tag{3.91}$$

$$\text{Term D}=\int_{\Omega_e}\sum\nolimits_k a_k^n\Phi_k\Phi_1 d\Omega_e=-\frac{\Delta z}{30}a_0^n+\frac{\Delta z}{15}a_{1/2}^n+\frac{2}{15}\Delta z a_1^n,\tag{3.92}$$

$$\begin{aligned}
\text{Term E}=&-(1-\theta)\Delta t_n\int_{\Omega_e}\sum\nolimits_k a_k^n(L\Phi_k)\Phi_1 d\Omega_e\\
=&-(1-\theta)\Delta t_n\left[\frac{1}{30}\left(a_0^n\right)^2-\frac{4}{15}\left(a_{1/2}^n\right)^2+\frac{1}{3}\left(a_1^n\right)^2+\frac{1}{15}a_0^na_{1/2}^n+\frac{1}{30}a_0^na_1^n-\frac{1}{5}a_{1/2}^na_1^n\right]\\
&-(1-\theta)\Delta t_n\left[c\Delta z\left(-\frac{1}{30}a_0^n+\frac{1}{15}a_{1/2}^n+\frac{2}{15}a_1^n\right)\right].
\end{aligned}\tag{3.93}$$

Collecting all terms, and letting $u_k = a_k$, $(k = 0, 1/2,\ \ 1)$, one obtains the element equation for $j = 3$:

$$\begin{aligned}
&u_0^{n+1}\left[-\frac{\Delta z}{30}+\theta\Delta t_n\left(\frac{1}{15}a_0^n+\frac{1}{15}a_{1/2}^n+\frac{1}{30}a_1^n-\frac{1}{30}c\Delta z\right)\right]\\
&+u_{1/2}^{n+1}\left[\frac{\Delta z}{15}+\theta\Delta t_n\left(\frac{1}{15}a_0^n-\frac{8}{15}a_{1/2}^n-\frac{1}{5}a_1^n+\frac{1}{15}c\Delta z\right)\right]\\
&+u_1^{n+1}\left[\frac{2}{15}\Delta z+\theta\Delta t_n\left(\frac{1}{30}a_0^n-\frac{1}{5}a_{1/2}^n+\frac{2}{3}a_1^n+\frac{2}{15}c\Delta z\right)\right]\\
&=\frac{\Delta z}{30}a_0^n+\frac{\Delta z}{15}a_{1/2}^n+\frac{2}{15}\Delta z u_1^n+\left[\theta\Delta t_n-(1-\theta)\Delta t_n\right]\\
&\times\left[\frac{1}{30}\left(a_0^n\right)^2-\frac{4}{15}\left(a_{1/2}^n\right)^2+\frac{1}{3}\left(a_1^n\right)^2+\frac{1}{15}a_0^na_{1/2}^n+\frac{1}{30}a_0^na_1^n-\frac{1}{5}a_{1/2}^na_1^n\right]\\
&-(1-\theta)\Delta t_n c\Delta z\left(-\frac{1}{30}a_0^n+\frac{1}{15}a_{1/2}^n+\frac{2}{15}a_1^n\right).
\end{aligned}\tag{3.94}$$

Summing up the three element equations (Eqs. 3.82, 3.88, and 3.94), one obtains the combined element equation:

$$
\begin{aligned}
& u_0^{n+1}\left[\frac{\Delta z}{6}(1+c\theta\Delta t_n)-\theta\Delta t_n u_0^n\right]+u_{1/2}^{n+1}\left[\frac{2}{3}\Delta z(1+c\theta\Delta t_n)\right] \\
& \quad +u_1^{n+1}\left[\frac{\Delta z}{6}(1+c\theta\Delta t_n)+\theta\Delta t_n u_1^n\right] \\
& = [\theta\Delta t_n-(1-\theta)\Delta t_n]\frac{1}{2}\left[\left(u_1^n\right)^2-\left(u_0^n\right)^2\right]+[1-c(1-\theta)\Delta t_n]\frac{\Delta z}{6}\left(u_0^n+4u_{1/2}^n+u_1^n\right)
\end{aligned}
\tag{3.95}
$$

Notice that the first member of Eq. (3.95) is identical with the first member of Eq. (3.35) obtained with the method of subdomain collocation. The second member of these equations is however different. Equation (3.37) has two terms more than the second member of Eq. (3.95), namely, the contributions of $u_0^n u_{1/2}^n$ and $u_{1/2}^n u_1^n$. In case of $\theta = 1/2$, the second member of Eq. (3.37) and of Eq. (3.95) becomes identical.

A word of caution directed to young or inexperienced numerical analysts: more than 40 years ago, the author of this monograph used for the first time in a computer program the Galerkin method of orthogonal collocations, described in Sect. 3.4. The program was not long, but convoluted, and it took a few weeks to complete the coding. The integrals of the shape functions and of their products, listed in formulas (3.70) through (3.73), were computed by means of Gaussian integration which is easy and quick to code. After coding came the phase of testing, what in jargon is called "debugging." When running the code, it always terminated with underflows. The author suspected there was a trivial coding mistake and spent a couple of days to check the coding, without finding any error. A further suspect was that there might be an error in the Gaussian integration of the shape functions. To clear this point, the results of all integrations (36 values) were printed out. They seemed to be fine, apart from three of them which were of the order of magnitude $\pm 10^{-20}$. These integrals were then computed analytically, and it was found out that their exact values were zero. After inserting the zeros into the code, also replacing the Gaussian integrations with the exact values, the code started running properly, giving the expected result. The Gaussian integration, although very precise, was troublesome, because the integrals were multiplied with each other and caused underflow.

The lesson which could be learned from that work is as follows: "Numerical mathematics is a great achievement, but it should not supersede analytical mathematics. One should always compute analytically whatever analytically can be computed." This is also the reason for having introduced in Chap. 1 of this monograph several sections dedicated to the solution of ordinary differential equations for which analytical solutions are known.

3.5 Numerical Examples

First Test Case: Example 5.A

We consider the nonlinear equation:

$$\frac{\partial u}{\partial t} + \frac{\partial}{\partial z}\left(\frac{u^2}{2}\right) = 0, \tag{3.96}$$

defined for $0 \leq z \leq \infty$ with given initial condition $u(0, z) = F(z)$. We consider as initial function a step function defined by:

$$\begin{aligned} F(z) &= 1, \qquad 0 \leq z \leq 1 \\ F(z) &= 0, \qquad 1 < z \end{aligned} \tag{3.97}$$

and limit the computational domain at $z = 2$. As boundary conditions we take:

$$\begin{aligned} F(z = 0, t) &= u(0, t) = 1, \\ F(z = 2, t) &= u(\infty, t) = 0. \end{aligned} \tag{3.98}$$

With these initial and boundary conditions, Eq. (3.96) represents a shock front wave propagation toward the positive z coordinates with speed $dz/dt = 0.5$. In one second the front therefore propagates from $z = 1$ to $z = 1, 5$, without changing its shape.

In the computer program **PROG._III.1_FRONT1**, we solve Eq. (3.96) with three different variants of the method of orthogonal collocation, namely:

(i) *Subdomain collocation*, documented in Sect. 3.2, with weight functions:

$$\begin{aligned} \psi_i &= 1, \qquad z \in \boldsymbol{\Omega}_e \\ \psi_i &= 0. \qquad z \notin \boldsymbol{\Omega}_e \end{aligned} \tag{3.99}$$

(ii) *Point collocations*, with weight functions:

$$\psi_i = \delta(z - z_i). \tag{3.100}$$

(iii) *Galerkin method*, with weight functions:

$$\psi_i = \boldsymbol{\Phi}_i. \tag{3.101}$$

In all cases the shape functions are given by Eq. (3.19).

As an alternative to these variants of the method of orthogonal collocations, the program FRONT1 solves Eq. (3.96) also with finite differences using the following solution techniques:

(iv) *Point Jacobi iterations*

Discretizing Eq. (3.1) with FTCS (forward time and central space) differences obtains:

$$u_j^{n+1} = u_j^n - \frac{1}{2}\frac{\Delta t}{2\Delta z}\left[\left(u_{j+1}^2\right)^r - \left(u_{j-1}^2\right)^r\right] - \Delta t c u_j^r, \tag{3.102}$$

where r denotes an iteration step and at convergence $u_j^{n+1} = u_j^{r+1}$. In this test case, about ten iterations are needed at every time step to reach convergence, with a time step of 0.01 s. Convergence is reached when for every mesh it holds: $\left|u_j^r - u_j^{r-1}\right| / \left|u_j^r\right| \leq \varepsilon \ (\varepsilon = 10^{-8})$.

(v) *Point Gauss iterations*

Using the updated values for nodes 1, 2, . . ., $j - 1$, Eq. (3.102) becomes:

$$u_j^{r+1} = u_j^n - \frac{1}{2}\frac{\Delta t}{2\Delta z}\left[\left(u_{j+1}^2\right)^r - \left(u_{j-1}^2\right)^{r+1}\right] - \Delta t c u_j^r. \tag{3.103}$$

The second term in the brackets at the right-hand side of Eq. (3.103) is computed from already known values at iteration step $r + 1$, which makes the difference between the Jacobi and Gauss schemes. With the Gauss scheme, eight iterations are needed to reach convergence, with a time step of 0.01 s.

(vi) *Implicit scheme*

We linearize Eq. (3.96) as follows:

$$\frac{1}{\Delta t}\left(u_j^{n+1} - u_j^n\right) + \frac{1}{2}\frac{1}{2\Delta z}\left(u_{j+1}^{n+1} u_{j+1}^n - u_{j-1}^{n+1} u_{j-1}^n\right) + c u_j^{n+1} = 0, \tag{3.104}$$

hence:

$$u_j^{n+1}(1 + c\Delta t) + u_{j+1}^{n+1}\left(\frac{\Delta t}{4\Delta z} u_{j+1}^n\right) + u_{j-1}^{n+1}\left(-\frac{\Delta t}{4\Delta z} u_{j-1}^n\right) = u_j^n. \tag{3.105}$$

Equation (3.105) has a tri-diagonal matrix of coefficients and can be solved exactly with the Thomas algorithm (see Sect. 1.3).

Numerical algorithms simulating shock waves are often referred to as "shock-capturing" schemes. In a seminal article (Beam & Warming, 1976), it has been shown that the capturing capability of numerical algorithm, intended as the capability to represent well the shock front without unphysical distortions, can be substantially improved by adding a dissipation term in the form:

$$-\frac{\omega}{8}\left(u_{j-2}^n - 4u_{j-1}^n + 6u_j^n - 4u_{j+1}^n + u_{j+2}^n\right), \tag{3.106}$$

where ω is an under-relaxation parameter, which for stability reasons must be in the range $0 < \omega \leq 1$. Comprehensive treatments of calculations of shock fronts by means of simulated viscosity terms are given in Forsythe and Wasow (1960) and in Richtmyer and Morton (1967).

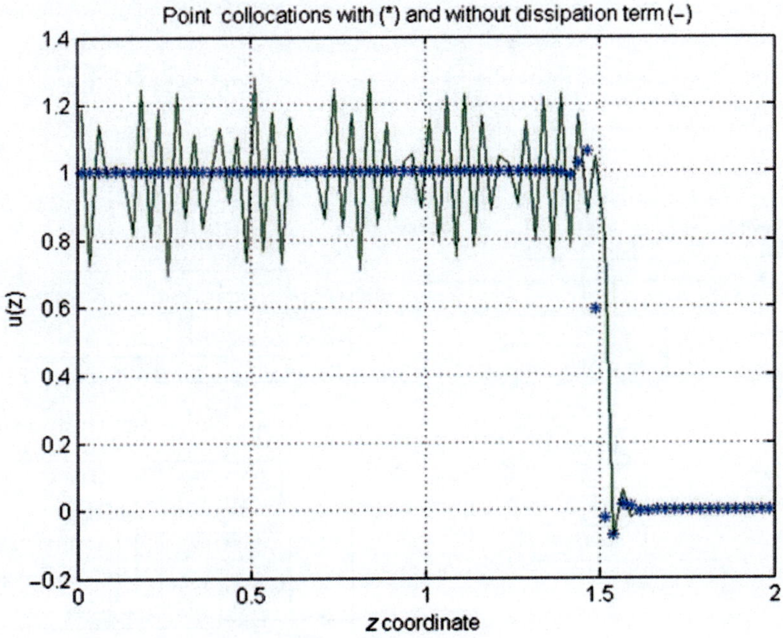

Fig. 3.3 Model problem, Eqs. (3.96)–(3.98). Comparison of results obtained with point collocations, with and without the dissipation term (3.106), at $t = 1$ s

Figure 3.3 shows a comparison between a calculation made with the point collocation method with and without the dissipation term (3.106). It is evident that without the dissipation term the amplitude of the shock wave cannot be simulated well.

Figure 3.4 gives a block diagram of the program **PROG_III.1 [FRONT1]**.

Figures 3.5, 3.6, and 3.7 show computational results at $t = 1.01$ s obtained with the subdomain collocations, the point collocations, and the Galerkin method, respectively. In each of the three figures, the results obtained with the finite difference (point Gauss, dotted) are also shown for comparison.

The code **PROG_III.1 [FRONT1]** is so organized that in a single run all six above mentioned solutions are computed. When applying the subdomain collocation method, Eq. (3.35) (with $c = 0$) is applied to compute the coefficients of the matrix [AM]; for the point collocations method, Eq. (3.67) is applied; for the Galerkin method, Eq. (3.95) is applied.

At the end of the calculations, a summary of all results is printed. Table 3.1 gives summary results of a calculation with 80 meshes and time step of 1 ms. (Meshes 1–50, where all wave values are $u = 1$, and meshes 71–80, where all wave values are $u = 0$, were eliminated.)

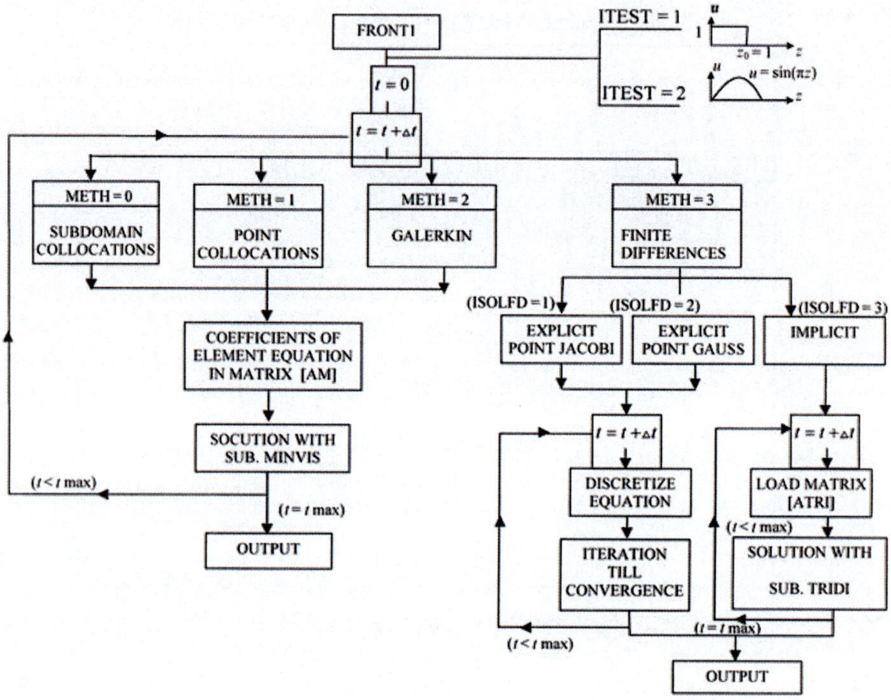

Fig. 3.4 Block diagram of program FRONT1

Second Test Case: Example 5.B
Consider the quasi-linear equation, already discussed in Sect. 2.1.4:

$$\frac{\partial u}{\partial t} + u\frac{\partial u}{\partial z} = 0, \tag{3.107}$$

with initial conditions:

$$F(z) = u(t = 0, z) = \sin(\pi z) \qquad 0 \leq z \leq 1 \tag{3.108a}$$

$$F(z) = 0, \qquad 1 < z \tag{3.108b}$$

and boundary condition:

$$F(z = 0, t) = u(0, t) = 0, \tag{3.109a}$$

$$F(z = 2, t) = u(\infty, t) = 0. \tag{3.109b}$$

As explained in Sect. 2.1.4, this wave breaks at $t_{B_r} = 1/\pi \cong 0.318\,\text{s}$.

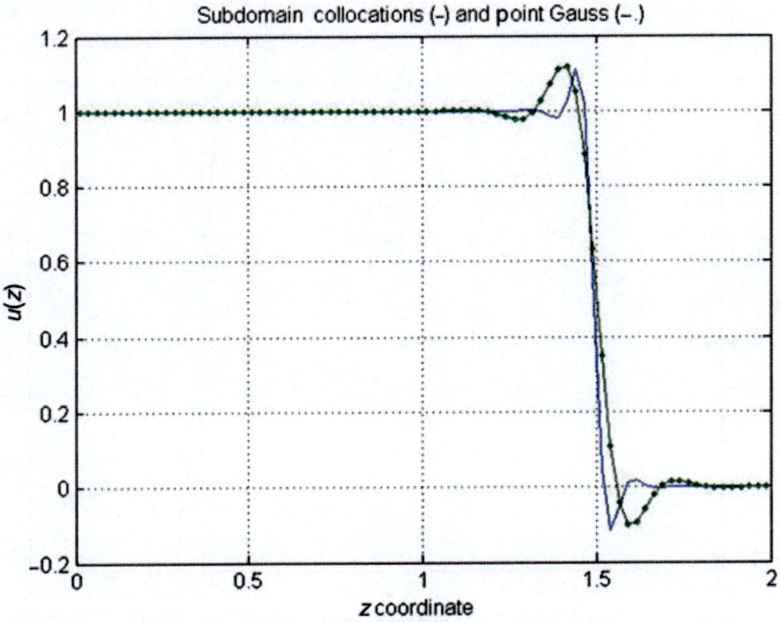

Fig. 3.5 Model problem, Eqs. (3.96)–(3.98). Comparison of results obtained with subdomain collocations and finite differences (point Gauss, dotted) at $t = 1$ s

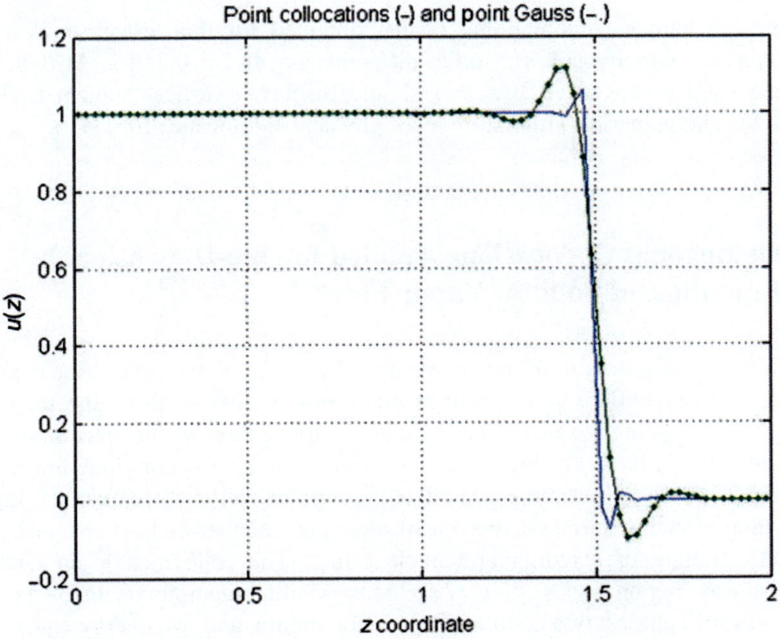

Fig. 3.6 Model problem, Eqs. (3.96)–(3.98). Comparison of results obtained with point collocations and finite differences (point Gauss, dotted) at $t = 1$ s

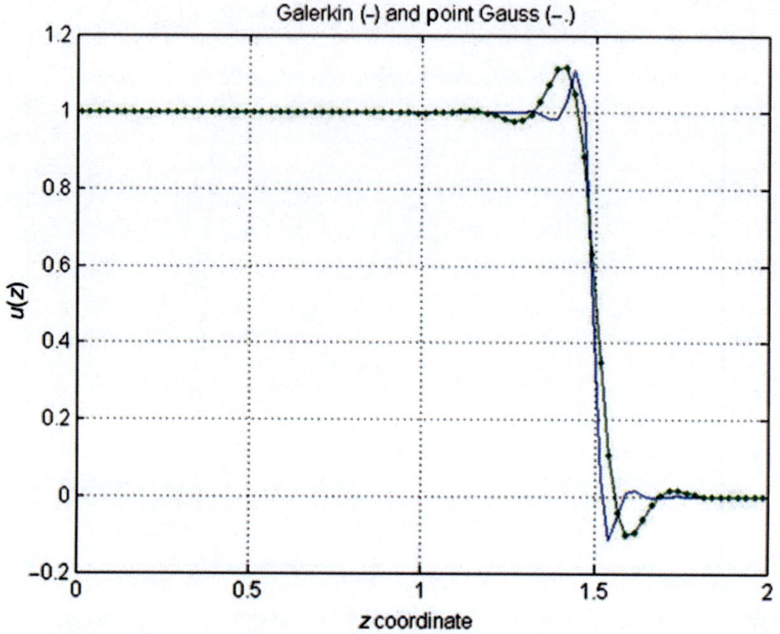

Fig. 3.7 Model problem, Eqs. (3.96)–(3.98). Comparison of results obtained with Galerkin method and finite differences (point Gauss, dotted) at $t = 1$ s

Figure 3.8 shows computational results obtained for this test case with the subdomain collocations and with finite differences at $t_{B_r} = 0.318$ s. At this time point, the front of the wave at $z = 1$ should display a vertical tangent (infinite derivative). The numerical simulation loses physical significance for $t > t_{B_r}$.

3.6 Orthogonal Collocations Applied to One-Dimensional Equations of Sodium Vapor Flow

The BLOW-3A program describes coolant boiling in sodium-cooled fast reactors under accident conditions like reactivity transients or loss of flow due to pump failure. Sodium boiling is described assuming a multiple bubble slug ejection model, which accounts for pressure and temperature distributions of vapor along the length of bubbles. Additional phenomena modeled are the behavior of the residual liquid film wetting cladding surfaces, dry-out of cladding, inlet mass flow reversal, and eventually melting of cladding and nuclear fuel. The code models an annular two-phase flow region by a system of six conservation equations for the separated liquid and vapor phases: two continuity, two momentum, and two energy equations.

Table 3.1 Solution obtained with the code PROG_III.1 [FRONT1] for test case of Example 5.A (front wave) for the six numerical methods explained in the text

Mesh Nr.	Z	Subd. CO	Point CO	Galerkin	Jacobi	Gauss	Implicit
51	0.1263E+01	0.1000E+01	0.1000E+01	0.1000E+01	0.9780E+00	0.9780E+00	0.9780E+00
52	0.1288E+01	0.1002E+01	0.1000E+01	0.1002E+01	0.9800E+00	0.9800E+00	0.9790E+00
53	0.1313E+01	0.1003E+01	0.1000E+01	0.1003E+01	0.9950E+00	0.9950E+00	0.9940E+00
54	0.1338E+01	0.9990E+00	0.1000E+01	0.9990E+00	0.1027E+01	0.1027E+01	0.1026E+01
55	0.1363E+01	0.9850E+00	0.1001E+01	0.9850E+00	0.1071E+01	0.1071E+01	0.1071E+01
56	0.1388E+01	0.9830E+00	0.9980E+00	0.9830E+00	0.1112E+01	0.1112E+01	0.1113E+01
57	0.1413E+01	0.1032E+01	0.9910E+00	0.1032E+01	0.1117E+01	0.1117E+01	0.1120E+01
58	0.1438E+01	0.1111E+01	0.1024E+01	0.1111E+01	0.1050E+01	0.1050E+01	0.1053E+01
59	0.1463E+01	0.1019E+01	0.1058E+01	0.1019E+01	0.8840E+00	0.8840E+00	0.8860E+00
60	0.1488E+01	0.5590E+00	0.5940E+00	0.5590E+00	0.6320E+00	0.6320E+00	0.6320E+00
61	0.1513E+01	0.4700E-01	−0.2200E-01	0.4700E-01	0.3490E+00	0.3490E+00	0.3480E+00
62	0.1538E+01	−0.1140E+00	−0.7400E-01	−0.1140E+00	0.1090E+00	0.1090E+00	0.1070E+00
63	0.1563E+01	−0.5000E-01	0.1800E-01	−0.5000E-01	−0.4200E-01	−0.4200E-01	−0.4400E-01
64	0.1588E+01	0.1100E-01	0.1200E-01	0.1100E-01	−0.1000E+00	−0.1000E+00	−0.1010E+00
65	0.1613E+01	0.1700E-01	−0.5000E-02	0.1700E-01	−0.9300E-01	−0.9300E-01	−0.9300E-01
66	0.1638E+01	0.4000E-02	−0.2000E-02	0.4000E-02	−0.5800E-01	−0.5800E-01	−0.5800E-01
67	0.1663E+01	−0.3000E-02	0.1000E-02	−0.3000E-02	−0.2100E-01	−0.2100E-01	−0.2100E-01
68	0.1688E+01	−0.2000E-02	0.0000E+00	−0.2000E-02	0.4000E-02	0.4000E-02	0.4000E-02
69	0.1713E+01	−0.0000E+00	−0.0000E+00	−0.0000E+00	0.1500E-01	0.1500E-01	0.1500E-01
70	0.1738E+01	0.1000E-02	−0.0000E+00	0.1000E-02	0.1500E-01	0.1500E-01	0.1500E-01

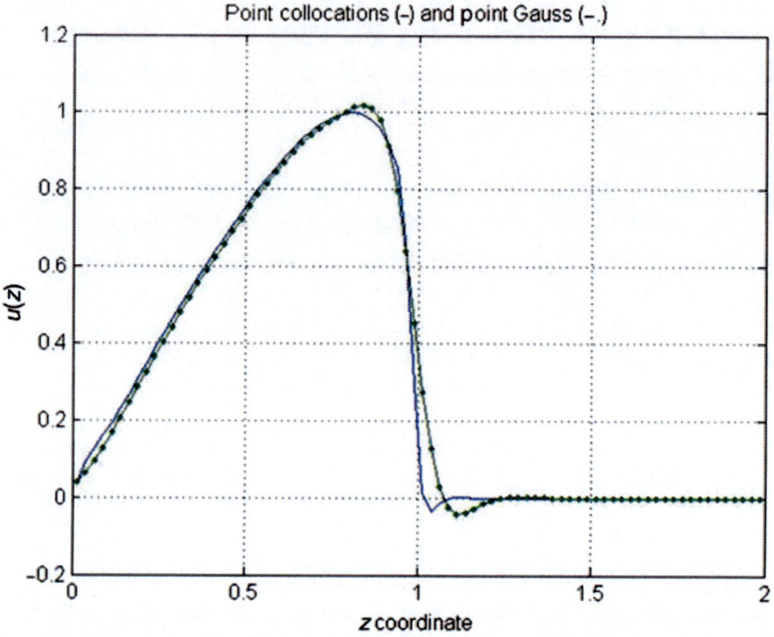

Fig. 3.8 Model problem, Eqs. (3.107)–(3.109). Comparison of results obtained with subdomain collocations and finite differences (point Gauss, dotted) at $t = 0.318$ s

The situation modelled in case of a well-developed vapor bubble is sketched in Fig. 3.9, which shows the bubble, the lower and upper liquid slugs, and the liquid film on the structural surface and, partially evaporated, on the surface of the cladding. Bubbles' displacements and growth or shrink are determined by the solution of the momentum equations for the liquid slugs between the bubbles, as well as for the inlet and outlet slugs. A bubble is considered "short" when it overlaps less than three computational meshes. In this case only mean values of temperature and pressure of the vapor inside the bubble are calculated. For "long" bubbles, overlapping at least partially three or more computational meshes, the axial temperature and pressure distributions of the vapor are computed solving the conservation equations of the vapor phase. In the standard BLOW-3A version, all conservation equations are discretized by finite differences and solved with an iterative scheme. In the OC (orthogonal collocations) variant of the code, the mass and momentum equations for the vapor phase inside the bubbles are solved with the algorithms explained in the following.

We refer to a z-axis coincident with the axis of a cylindrical fuel rod. The continuity, momentum, and energy equations for the vapor phase are:

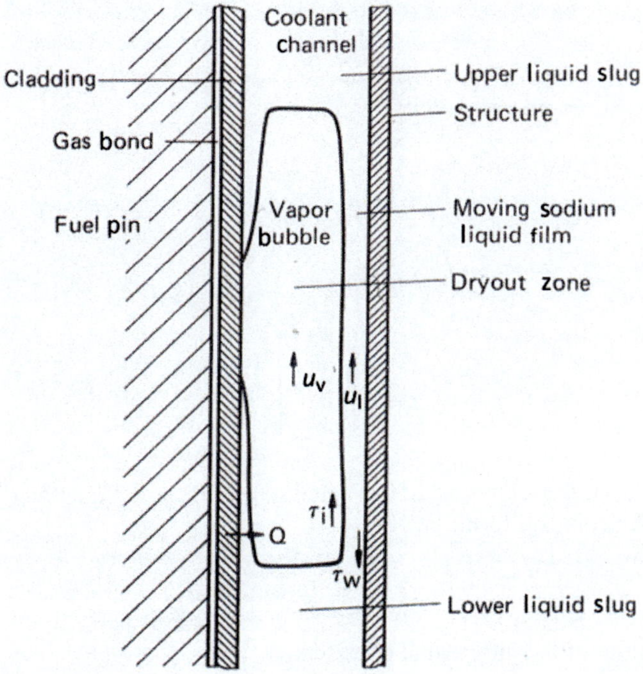

Fig. 3.9 Annular two-phase flow region as modelled in the BLOW-3A program

$$\frac{\partial \rho_v}{\partial t} + \frac{\partial (\rho_v u_v)}{\partial z} = \frac{U}{A_f} m, \tag{3.110}$$

$$\rho_v \frac{\partial u_v}{\partial t} + \rho_v u_v \frac{\partial u_v}{\partial z} = -\frac{\partial p_v}{\partial z} - \frac{U}{A_f}\left(\frac{\psi_i}{2}\rho_v u_v |u_v| + u_v \frac{m}{2}\right) - g\rho_v, \tag{3.111}$$

$$\frac{\partial (\rho_v h_v)}{\partial t} + \frac{\partial (\rho_v u_v h_v)}{\partial z} = Q_v. \tag{3.112}$$

In these equations A_f is the coolant channel cross section; U is the perimeter of wetted surfaces (cladding and structure); ψ_i is an interfacial (liquid-vapor) friction coefficient; m (kg/m^{2}s) is the mass flux by phase change, considered positive by evaporation; h_v (J/kg) is the vapor-specific enthalpy. In the following, we drop the subscript v for vapor.

Equations (3.110) and (3.111) can be written in matrix form as:

$$\frac{\partial Y}{\partial t} + F(Y)\frac{\partial Y}{\partial z} + E(Y) \cdot Y = B, \tag{3.113}$$

where Y is the unknown vector of the depending variables, density, and velocity of vapor:

$$Y = \begin{bmatrix} \rho \\ u \end{bmatrix}. \tag{3.114}$$

Matrices F and E and vector B are as follows:

$$F = \begin{pmatrix} u & \rho/\alpha \\ 1/\rho & u \end{pmatrix}, \tag{3.115}$$

$$E = \begin{pmatrix} 0 & 0 \\ 0 & \frac{U}{A_{\mathrm{f}}}\left(\frac{\psi_{\mathrm{i}}}{2}|u| + \frac{m}{2\rho}\right) \end{pmatrix}, \tag{3.116}$$

$$B = \begin{bmatrix} \frac{1}{\alpha}\frac{U}{A}m \\ -g \end{bmatrix}, \tag{3.117}$$

where $\alpha = \partial\rho/\partial p$ is the reciprocal of the sound speed in vapor. Equation (3.113) can be written in the compact form:

$$AY = B \tag{3.118}$$

with the definition of the differential operators:

$$A = \frac{\partial}{\partial t} + F(Y)\frac{\partial}{\partial z} + E(Y) = \frac{\partial}{\partial t} + L, \tag{3.119}$$

$$L = F(Y)\frac{\partial}{\partial z} + E(Y). \tag{3.120}$$

3.6.1 Subdomain Collocation Method

In the following we use the same shape functions, Eq. (3.19), as in the previous sections. With the weight functions (3.15), one obtains for every subdomain $\Omega_e = \Delta z$ an element equation in the form:

$$\begin{aligned} &\int_{\Omega_e} \sum_{k=1}^{N} a_k^{n+1}\Phi_k dz_e + \theta\Delta t_n \int_{\Omega_e} \sum_{k=1}^{N} a_k^{n+1} L\Phi_k dz - \theta\Delta t_n \int_{\Omega_e} B^{n+1} dz \\ &= \int_{\Omega_e} \sum_{k=1}^{N} a_k^n \Phi_k dz - (1-\theta)\Delta t_n \int_{\Omega_e} \sum_{k=1}^{N} a_k^n L\Phi_k dz + (1-\theta)\Delta t_n \int_{\Omega_e} \sum_{k=1}^{N} B^n dz. \end{aligned} \tag{3.121}$$

The first term of Eq. (3.121) yields:

$$\sum_{k=1}^{N} a_k^{n+1} \int_{\Omega_e} \sum_{k=1}^{N} \Phi_k dz = \frac{\Delta z}{6} \left[\begin{bmatrix} p_0^{n+1} \\ u_0^{n+1} \end{bmatrix} + 4 \begin{bmatrix} p_{1/2}^{n+1} \\ u_{1/2}^{n+1} \end{bmatrix} + \begin{bmatrix} p_1^{n+1} \\ u_1^{n+1} \end{bmatrix} \right]. \tag{3.122}$$

The second term of Eq. (3.121) can be linearized with respect to time by:

$$Lu^{n+1} = u^{n+1} \frac{\partial u^{n+1}}{\partial z} = u^n \frac{\partial u^n}{\partial z} + \frac{\partial}{\partial t} \left(u^n \frac{\partial u^n}{\partial z} \right) \Delta t. \tag{3.123}$$

After some calculations one derives:

$$\theta \Delta t_n \int_{\Omega_e} \left[F(Y^n) \frac{\partial Y^{n+1}}{\partial z} + \sum_{\ell=1}^{2} \frac{\partial F(Y^n)}{\partial q_\ell} \Delta q_\ell \frac{\partial Y^n}{\partial z} + E(Y^n) Y^{n+1} + \sum_{\ell=1}^{2} \frac{\partial E(Y^n)}{\partial q_\ell} \Delta q_\ell Y^n \right] dz \tag{3.124}$$

where $q_\ell (\ell = 1, 2)$ are the two dependent variables p, u and $\Delta q_\ell = q_\ell^{n+1} - q_\ell^n$.
The derivatives of the matrices $F(Y)$, $E(Y)$ in Eq. (3.124) are:

$$\frac{\partial F(Y)}{\partial p} = \begin{pmatrix} 0 & \dfrac{\partial (\rho/\alpha)}{\partial p} \\ \dfrac{\partial (1/\rho)}{\partial p} & 0 \end{pmatrix} = \begin{pmatrix} 0 & \beta \\ \gamma & 0 \end{pmatrix}, \tag{3.125}$$

$$\frac{\partial F(Y)}{\partial u} = \begin{pmatrix} 1 & 0 \\ 0 & 1 \end{pmatrix}, \tag{3.126}$$

$$\frac{\partial E(Y)}{\partial p} = \begin{pmatrix} 0 & 0 \\ 0 & \dfrac{U}{A_f} \dfrac{m}{2} \gamma \end{pmatrix}, \tag{3.127}$$

$$\frac{\partial E(Y)}{\partial u} = \begin{pmatrix} 0 & 0 \\ 0 & \pm \dfrac{U}{A_f} \dfrac{\psi_i}{2} \end{pmatrix}, \tag{3.128}$$

with:

$$\beta = \frac{\partial}{\partial p} \frac{\rho}{(\partial \rho / \partial p)} = 1 + \rho \frac{\frac{\partial}{\partial p} (\partial \rho / \partial p)}{\alpha^2}, \tag{3.129}$$

$$\gamma = \frac{\partial (1/\rho)}{\partial p}. \tag{3.130}$$

The sign in Eq. (3.128) is taken positive if $u > 0$, otherwise negative. The derivatives of the vector Y of the dependent variables, appearing in Eq. (3.124), computed in the nodes 0, 1/2, 1 of the element Δz, are (Cfr. Eq. (3.22)):

$$\left(\frac{\partial Y}{\partial z}\right)_0^{n+1} = \frac{1}{\Delta z}\left[-3\begin{bmatrix} p_0^{n+1} \\ u_0^{n+1}\end{bmatrix} + 4\begin{bmatrix} p_{1/2}^{n+1} \\ u_{1/2}^{n+1}\end{bmatrix} - \begin{bmatrix} p_1^{n+1} \\ u_1^{n+1}\end{bmatrix}\right], \tag{3.131a}$$

$$\left(\frac{\partial Y}{\partial z}\right)_{1/2}^{n+1} = \frac{1}{\Delta z}\left[\begin{bmatrix} p_1^{n+1} \\ u_1^{n+1}\end{bmatrix} - \begin{bmatrix} p_1^{n+1} \\ u_0^{n+1}\end{bmatrix}\right], \tag{3.131b}$$

$$\left(\frac{\partial Y}{\partial z}\right)_1^{n+1} = \frac{1}{\Delta z}\left[\begin{bmatrix} p_0^{n+1} \\ u_0^{n+1}\end{bmatrix} - 4\begin{bmatrix} p_{1/2}^{n+1} \\ u_{1/2}^{n+1}\end{bmatrix} + 3\begin{bmatrix} p_1^{n+1} \\ u_1^{n+1}\end{bmatrix}\right], \tag{3.131c}$$

and similarly at time level n.

The third term of Eq. (3.121) is evaluated with a parabolic formula for the integral and yields:

$$\begin{aligned} -\theta\Delta t_n \int_{\Omega_e} B^{n+1} dz &= -\theta\Delta t_n \frac{\Delta z}{6}\left(B_0^{n+1} + 4B_{1/2}^{n+1} + B_1^{n+1}\right) \\ &= -\theta\Delta t_n \Delta z \begin{bmatrix} \dfrac{1}{6}\dfrac{U}{A_f}\left(\dfrac{m_0^{n+1}}{\alpha_0^{n+1}} + 4\dfrac{m_{1/2}^{n+1}}{\alpha_{1/2}^{n+1}} + \dfrac{m_1^{n+1}}{\alpha_1^{n+1}}\right) \\ -g \end{bmatrix}. \end{aligned} \tag{3.132}$$

The terms at the right-hand side of Eq. (3.121) are treated likewise. Adding the contributions from all above terms, one derives for every subdomain $\Omega_e = \Delta z$ a system of two scalar equations in the six unknowns p_0^{n+1}, $p_{1/2}^{n+1}$, p_1^{n+1}, u_0^{n+1}, $u_{1/2}^{n+1}$, u_1^{n+1}, which can be written as:

$$b_1 p_0^{n+1} + b_2 u_0^{n+1} + b_3 p_{1/2}^{n+1} + b_4 u_{1/2}^{n+1} + b_5 p_1^{n+1} + b_6 u_1^{n+1} = N_1, \tag{3.133}$$

$$c_1 p_0^{n+1} + c_2 u_0^{n+1} + c_3 p_{1/2}^{n+1} + c_4 u_{1/2}^{n+1} + c_5 p_1^{n+1} + c_6 u_1^{n+1} = N_2, \tag{3.134}$$

with the coefficients:

$$b_1 = \frac{\Delta z}{6} + \frac{\theta\Delta t_n}{6}\left[-3u_0^n - 4u_{1/2}^n + u_1^n + \beta_0^n \Delta z\left(\frac{\partial u}{\partial z}\right)_0^n\right], \tag{3.135a}$$

$$b_2 = \frac{\theta\Delta t_n}{6}\left[-3\frac{\rho_0^n}{\alpha_0^n} - 4\frac{\rho_{1/2}^n}{\alpha_{1/2}^n} + \frac{\rho_1^n}{\alpha_1^n} + \Delta z\left(\frac{\partial p}{\partial z}\right)_0^n\right], \tag{3.135b}$$

$$b_3 = 4\frac{\Delta z}{6} + 4\frac{\theta\Delta t_n}{6}\left[u_0^n - u_1^n + \beta_{1/2}^n \Delta z\left(\frac{\partial u}{\partial z}\right)_{1/2}^n\right], \tag{3.135c}$$

$$b_4 = 4\frac{\theta\Delta t_n}{6}\left[\frac{\rho_0^n}{\alpha_0^n} - \frac{\rho_1^n}{\alpha_1^n} + \Delta z\left(\frac{\partial p}{\partial z}\right)_{1/2}^n\right], \tag{3.135d}$$

$$b_5 = \frac{\Delta z}{6} + \frac{\theta\Delta t_n}{6}\left[-u_0^n + 4u_{1/2}^n + 3u_1^n + \beta_1^n\Delta z\left(\frac{\partial u}{\partial z}\right)_1^n\right], \tag{3.135e}$$

$$b_6 = \frac{\theta\Delta t_n}{6}\left[-\frac{\rho_0^n}{\alpha_0^n} + 4\frac{\rho_{1/2}^n}{\alpha_{1/2}^n} + 3\frac{\rho_1^n}{\alpha_1^n} + \Delta z\left(\frac{\partial p}{\partial z}\right)_1^n\right], \tag{3.135f}$$

$$\begin{aligned}
N_1 =& \theta\Delta t_n\frac{\Delta z}{6}\left[u_0^n\left(\frac{\partial p}{\partial z}\right)_0^n + p_0^n\beta_0^n\left(\frac{\partial u}{\partial z}\right)_0^n + 4u_{1/2}^n\left(\frac{\partial p}{\partial z}\right)_{1/2}^n\right.\\
&\left.+4p_{1/2}^n\beta_{1/2}^n\left(\frac{\partial u}{\partial z}\right)_{1/2}^n + u_1^n\left(\frac{\partial p}{\partial z}\right)_1^n + p_1^n\beta_1^n\left(\frac{\partial u}{\partial z}\right)_1^n\right]\\
&+\theta\Delta t_n\frac{\Delta z}{6}\frac{U}{A_\mathrm{f}}\left(\frac{m_0^{n+1}}{\alpha_0^{n+1}} + 4\frac{m_{1/2}^{n+1}}{\alpha_{1/2}^{n+1}} + \frac{m_1^{n+1}}{\alpha_1^{n+1}}\right)\\
&+(1-\theta)\Delta t_n\frac{\Delta z}{6}\frac{U}{A_\mathrm{f}}\left(\frac{m_0^n}{\alpha_0^{n+1}} + 4\frac{m_{1/2}^n}{\alpha_{1/2}^{n+1}} + \frac{m_1^n}{\alpha_1^{n+1}}\right) + \frac{\Delta z}{6}\left(p_0^n + 4p_{1/2}^n + p_1^n\right)\\
&+(1-\theta)\Delta t_n\frac{\Delta z}{6}\left[u_0^n\left(\frac{\partial p}{\partial z}\right)_0^n + \frac{\rho_0^n}{\alpha_0^n}\left(\frac{\partial u}{\partial z}\right)_0^n + 4u_{1/2}^n\left(\frac{\partial p}{\partial z}\right)_{1/2}^n\right.
\end{aligned} \tag{3.135g}$$

$$c_1 = \frac{\theta\Delta t_n}{6}\left[-\frac{3}{\rho_0^n} - \frac{4}{\rho_{1/2}^n} + \frac{1}{\rho_1^n} + \gamma_0^n\Delta z\left(\frac{\partial p}{\partial z}\right)_0^n + \Delta z\gamma_0^n u_0^n\frac{U}{A_\mathrm{f}}\frac{m_0^n}{2}\right], \tag{3.136a}$$

$$\begin{aligned}
c_2 =& \frac{\Delta z}{6} + \frac{\theta\Delta t_n}{6}\\
&\times\left[-3u_0^n - 4u_{1/2}^n + u_1^n + \Delta z\left(\frac{\partial u}{\partial z}\right)_0^n + \Delta z\frac{\lambda_{i0}^n}{2D_\mathrm{h}}\left(u_0^n + |u_0^n|\right) + \frac{\Delta z}{\rho_0^n}\frac{U}{A_\mathrm{f}}\frac{m_0^n}{2}\right]
\end{aligned} \tag{3.136b}$$

$$c_3 = 4\frac{\theta\Delta t_n}{6}\left[\frac{1}{\rho_0^n} - \frac{1}{\rho_1^n} + \gamma_{1/2}^n\Delta z\left(\frac{\partial p}{\partial z}\right)_{1/2}^n + \Delta z\gamma_{1/2}^n u_{1/2}^n\frac{U}{A_\mathrm{f}}\frac{m_{1/2}^n}{2}\right], \tag{3.136c}$$

$$c_4 = \frac{4\Delta z}{6} + \frac{4\theta\Delta t_n}{6} \times \left[u_0^n - u_1^n + \Delta z \left(\frac{\partial u}{\partial z}\right)_{1/2}^n + \Delta z \frac{\lambda_{i1/2}^n}{2D_\mathrm{h}} \left(u_{1/2}^n + \left|u_{1/2}^n\right| \right) + \frac{\Delta z}{\rho_{1/2}^n} \frac{U}{A_\mathrm{f}} \frac{m_{1/2}^n}{2} \right] \tag{3.136d}$$

$$c_5 = \frac{\theta\Delta t_n}{6} \left[-\frac{1}{\rho_0^n} + \frac{4}{\rho_{1/2}^n} + \frac{3}{\rho_1^n} + \gamma_1^n \Delta z \left(\frac{\partial p}{\partial z}\right)_1^n + \Delta z \gamma_1^n u_1^n \frac{U}{A_\mathrm{f}} \frac{m_1^n}{2} \right], \tag{3.136e}$$

$$c_6 = \frac{\Delta z}{6} + \frac{\theta\Delta t_n}{6} \times \left[-u_0^n + 4u_{1/2}^n + 3u_1^n + \Delta z \left(\frac{\partial u}{\partial z}\right)_1^n + \Delta z \frac{\lambda_{i1}^n}{2D_\mathrm{h}} \left(u_1^n + |u_1^n| \right) + \frac{\Delta z}{\rho_1^n} \frac{U}{A_\mathrm{f}} \frac{m_1^n}{2} \right] \tag{3.136f}$$

$$\begin{aligned} N_2 = {} & \theta\Delta t_n \frac{\Delta z}{6} \left[u_0^n \left(\frac{\partial u}{\partial z}\right)_0^n + p_0^n \gamma_0^n \left(\frac{\partial p}{\partial z}\right)_0^n + 4u_{1/2}^n \left(\frac{\partial u}{\partial z}\right)_{1/2}^n + 4p_{1/2}^n \gamma_{1/2}^n \left(\frac{\partial p}{\partial z}\right)_{1/2}^n \right. \\ & \left. + u_1 \left(\frac{\partial u}{\partial z}\right)_1^n + p_1^n \gamma_1^n \left(\frac{\partial p}{\partial z}\right)_1^n \right] \\ & + \frac{\theta\Delta t_n}{6} \Delta z \left(p_0^n \gamma_0^n u_0^n \frac{U}{A_\mathrm{f}} \frac{m_0^n}{2} + 4p_{1/2}^n \gamma_{1/2}^n u_{1/2}^n \frac{U}{A_\mathrm{f}} \frac{m_{1/2}^n}{2} + p_1^n \gamma_1^n u_1^n \frac{U}{A_\mathrm{f}} \frac{m_1^n}{2} \right. \\ & \left. + \frac{\lambda_{i0}^n}{2D_\mathrm{h}} u_0^n |u_0^n| + 4 \frac{\lambda_{i1/2}^n}{2D_\mathrm{h}} u_{1/2}^n \left|u_{1/2}^n\right| + \frac{\lambda_{i1}^n}{2D_\mathrm{h}} u_1^n |u_1^n| \right) + \frac{\Delta z}{6} \left(u_0^n + 4u_{1/2}^n + u_1^n \right) \\ & - (1-\theta)\Delta t_n \frac{\Delta z}{6} \left[u_0^n \left(\frac{\partial u}{\partial z}\right)_0^n + \frac{1}{\rho_0^n} \left(\frac{\partial p}{\partial z}\right)_0^n + 4u_{1/2}^n \left(\frac{\partial u}{\partial z}\right)_{1/2}^n + \frac{4}{\rho_0^{1/2}} \left(\frac{\partial p}{\partial z}\right)_{1/2}^n \right. \\ & \left. + u_1^n \left(\frac{\partial u}{\partial z}\right)_1^n + \frac{1}{\rho_1^n} \left(\frac{\partial p}{\partial z}\right)_1^n \right] \\ & - \frac{(1-\theta)\Delta t_n}{6} \Delta z \left(\frac{u_0^n}{\rho_0^n} \frac{U}{A_\mathrm{f}} \frac{m_0^n}{2} + 4 \frac{u_{1/2}^n}{\rho_{1/2}^n} \frac{U}{A_\mathrm{f}} \frac{m_{1/2}^n}{2} + \frac{u_1^n}{\rho_1} \frac{U}{A_\mathrm{f}} \frac{m_1^n}{2} \right. \\ & \left. + \frac{\lambda_{i0}^n}{2D_\mathrm{h}} u_0^n |u_0^n| + 4 \frac{\lambda_{i1/2}^n}{2D_\mathrm{h}} u_{1/2}^n \left|u_{1/2}^n\right| + \frac{\lambda_{i1}^n}{2D_\mathrm{h}} u_1^n |u_1^n| \right) - g\Delta z \Delta t_n. \end{aligned} \tag{3.136g}$$

Because the definition domain of the matrix equation (3.118) has been partitioned in adjacent but disjoint elements, additional equations are necessary to provide the coupling between consecutive intervals. The coupling is provided by imposing the continuity of the dependent variables and of their space derivatives at the element interfaces. Letting Ω_e and Ω_{e+1} be two intervals of lengths Δz_j and Δz_{j+1}, the condition for the variable p at the interface $j + 1/2$ is:

$$\left(\frac{\partial p}{\partial z}\right)^{n+1}_{j+1/2} = \left(\frac{\partial p}{\partial z}\right)^{n+1}_{j+1-1/2}, \tag{3.137}$$

which yields:

$$\frac{1}{\Delta z_j}\left(p^{n+1}_{j-1/2} - 4p^{n+1}_j + 3p^{n+1}_{j+1/2}\right) = \frac{1}{\Delta z_{j+1}} \times \left(-3p^{n+1}_{j+1/2} + 4p^{n+1}_{j+1} - 1p^{n+1}_{j+3/2}\right), \tag{3.138}$$

and similarly for the variable u. One obtains therefore two additional equations for every interface. At the lower and upper phase boundaries, between the vapor bubbles and the liquid coolant slugs, boundary conditions are imposed for both pressure and velocity. All equations are assembled to form a system of linear algebraic equations of the form:

$$\{M\}\{Y^{n+1}\} = \{B^n\} \tag{3.139}$$

which can be solved by inverting the matrix $\{M\}$.

3.6.2 Point Collocation Method

The point collocation method can be used as alternative to the subdomain collocation method. With the weight functions (3.41), one obtains for every subdomain $\Omega_e = \Delta z$ an element equation in the form:

$$\int_{\Omega_e} \sum_{k=1}^{N} a_k^{n+1} (\Phi_k)_{zj} dz + \theta \Delta t_n \int_{\Omega_e} \sum_{k=1}^{N} a_k^{n+1} (L\Phi_k)_{zj} dz - \theta \Delta t_n \int_{\Omega_e} \sum_{k=1}^{N} (B^{n+1})_{zj} dz$$
$$= \int_{\Omega_e} \sum_{k=1}^{N} a_k^{n} (\Phi_k)_{zj} dz - (1-\theta) \Delta t_n \int_{\Omega_e} \sum_{k=1}^{N} a_k^{n} (L\Phi_k)_{zj} dz + (1-\theta) \Delta t_n \int_{\Omega_e} \sum_{k=1}^{N} (B^{n})_{zj} dz \tag{3.140}$$

The analytical development follows the same procedure explained above and leads likewise to two equations formally similar to Eqs. (3.133) and (3.134), with the coefficients:

$$b_1 = \Delta z + \theta \Delta t_n \left[-3u_0^n - u_{1/2}^n + u_1^n + \beta_0^n \Delta z \left(\frac{\partial u}{\partial z}\right)_0^n\right], \tag{3.141a}$$

$$b_2 = \theta \Delta t_n \left[-3\frac{\rho_0^n}{\alpha_0^n} - \frac{\rho_{1/2}^n}{\alpha_{1/2}^n} + \frac{\rho_1^n}{\alpha_1^n} + \Delta z \left(\frac{\partial p}{\partial z} \right)_0^n \right], \tag{3.141b}$$

$$b_3 = \Delta z + \theta \Delta t_n \left[4u_0^n - 4u_1^n + \beta_{1/2}^n \Delta z \left(\frac{\partial u}{\partial z} \right)_{1/2}^n \right], \tag{3.141c}$$

$$b_4 = \theta \Delta t_n \left[4\frac{\rho_0^n}{\alpha_0^n} - 4\frac{\rho_1^n}{\alpha_1^n} + \Delta z \left(\frac{\partial p}{\partial z} \right)_{1/2}^n \right], \tag{3.141d}$$

$$b_5 = \Delta z + \theta \Delta t_n \left[-u_0^n + u_{1/2}^n + 3u_1^n + \beta_1^n \Delta z \left(\frac{\partial u}{\partial z} \right)_1^n \right], \tag{3.141e}$$

$$b_6 = \theta \Delta t_n \left[-\frac{\rho_0^n}{\alpha_0^n} + \frac{\rho_{1/2}^n}{\alpha_{1/2}^n} + 3\frac{\rho_1^n}{\alpha_1^n} + \Delta z \left(\frac{\partial p}{\partial z} \right)_1^n \right], \tag{3.141f}$$

$$\begin{aligned} N_1 = {} & \theta \Delta t_n \Delta z \left[u_0^n \left(\frac{\partial p}{\partial z} \right)_0^n + p_0^n \beta_0^n \left(\frac{\partial u}{\partial z} \right)_0^n + u_{1/2}^n \left(\frac{\partial p}{\partial z} \right)_{1/2}^n + p_{1/2}^n \beta_{1/2}^n \left(\frac{\partial u}{\partial z} \right)_{1/2}^n \right. \\ & \left. + u_1^n \left(\frac{\partial p}{\partial z} \right)_1^n + p_1^n \beta_1^n \left(\frac{\partial u}{\partial z} \right)_1^n \right] \\ & + \theta \Delta t_n \Delta z \frac{U}{A_\mathrm{f}} \left(\frac{m_0^{n+1}}{\alpha_0^{n+1}} + \frac{m_{1/2}^{n+1}}{\alpha_{1/2}^{n+1}} + \frac{m_1^{n+1}}{\alpha_1^{n+1}} \right) \\ & + (1-\theta) \Delta t_n \Delta z \frac{U}{A_\mathrm{f}} \left(\frac{m_0^n}{\alpha_0^{n+1}} + \frac{m_{1/2}^n}{\alpha_{1/2}^{n+1}} + \frac{m_1^n}{\alpha_1^{n+1}} \right) + \Delta z \left(p_0^n + p_{1/2}^n + p_1^n \right) \\ & - (1-\theta) \Delta t_n \Delta z \left[u_0^n \left(\frac{\partial p}{\partial z} \right)_0^n + \frac{\rho_0^n}{\alpha_0^n} \left(\frac{\partial u}{\partial z} \right)_0^n + u_{1/2}^n \left(\frac{\partial p}{\partial z} \right)_{1/2}^n + \frac{\rho_{1/2}^n}{\alpha_{1/2}^n} \left(\frac{\partial u}{\partial z} \right)_{1/2}^n \right. \\ & \left. + u_1^n \left(\frac{\partial p}{\partial z} \right)_1^n + \frac{\rho_1^n}{\alpha_1^n} \left(\frac{\partial u}{\partial z} \right)_1^n \right], \end{aligned} \tag{3.142}$$

$$c_1 = \theta \Delta t_n \left[-\frac{3}{\rho_0^n} - \frac{1}{\rho_{1/2}^n} + \frac{1}{\rho_1^n} + \gamma_0^n \Delta z \left(\frac{\partial p}{\partial z} \right)_0^n + \Delta z \gamma_0^n u_0^n \frac{U}{A_\mathrm{f}} \frac{m_0^n}{2} \right], \tag{3.143a}$$

$$\begin{aligned} c_2 = {} & \Delta z \\ & + \theta \Delta t_n \left[-3u_0^n - u_{1/2}^n + u_1^n + \Delta z \left(\frac{\partial u}{\partial z} \right)_0^n + \Delta z \frac{\lambda_{i0}^n}{2D_\mathrm{h}} \left(u_0^n + |u_0^n| \right) + \frac{\Delta z}{\rho_0^n} \frac{U}{A_\mathrm{f}} \frac{m_0^n}{2} \right] \end{aligned} \tag{3.143b}$$

$$c_3 = \theta\Delta t_n\left[4\frac{1}{\rho_0^n} - 4\frac{1}{\rho_1^n} + \gamma_{1/2}^n\Delta z\left(\frac{\partial p}{\partial z}\right)_{1/2}^n + \Delta z\gamma_{1/2}^n u_{1/2}^n \frac{U}{A_\mathrm{f}}\frac{m_{1/2}^n}{2}\right], \tag{3.143c}$$

$$c_4 = \Delta z + \theta\Delta t_n\left[4u_0^n - 4u_1^n + \Delta z\left(\frac{\partial u}{\partial z}\right)_{1/2}^n + \Delta z\frac{\lambda_{i1/2}^n}{2D_\mathrm{h}}\left(u_{1/2}^n + \left|u_{1/2}^n\right|\right) + \frac{\Delta z}{\rho_{1/2}^n}\frac{U}{A_\mathrm{f}}\frac{m_{1/2}^n}{2}\right] \tag{3.143d}$$

$$c_5 = \theta\Delta t_n\left[-\frac{1}{\rho_0^n} + \frac{1}{\rho_{1/2}^n} + \frac{3}{\rho_1^n} + \gamma_1^n\Delta z\left(\frac{\partial p}{\partial z}\right)_1^n + \Delta z\gamma_1^n u_1^n\frac{U}{A_\mathrm{f}}\frac{m_1^n}{2}\right], \tag{3.143e}$$

$$c_6 = \Delta z + \theta\Delta t_n\left[-u_0^n + u_{1/2}^n + 3u_1^n + \Delta z\left(\frac{\partial u}{\partial z}\right)_1^n + \Delta z\frac{\lambda_{i1}^n}{2D_\mathrm{h}}\left(u_1^n + \left|u_1^n\right|\right) + \frac{\Delta z}{\rho_1^n}\frac{U}{A_\mathrm{f}}\frac{m_1^n}{2}\right] \tag{3.143f}$$

$$\begin{aligned}
N_2 = {} & \theta\Delta t_n\Delta z\left[u_0^n\left(\frac{\partial u}{\partial z}\right)_0^n + p_0^n\gamma_0^n\left(\frac{\partial p}{\partial z}\right)_0^n + u_{1/2}^n\left(\frac{\partial u}{\partial z}\right)_{1/2}^n + p_{1/2}^n\gamma_{1/2}^n\left(\frac{\partial p}{\partial z}\right)_{1/2}^n \right.\\
& \left. + u_1\left(\frac{\partial u}{\partial z}\right)_1^n + p_1^n\gamma_1^n\left(\frac{\partial p}{\partial z}\right)_1^n\right]\\
& + \theta\Delta t\Delta z\left(p_0^n\gamma_0^n u_0^n\frac{U}{A_\mathrm{f}}\frac{m_0^n}{2} + p_{1/2}^n\gamma_{1/2}^n u_{1/2}^n\frac{U}{A_\mathrm{f}}\frac{m_{1/2}^n}{2} + p_1^n\gamma_1^n u_1^n\frac{U}{A_\mathrm{f}}\frac{m_1^n}{2}\right.\\
& \left. + \frac{\lambda_{i0}^n}{2D_\mathrm{h}}u_0^n\left|u_0^n\right| + \frac{\lambda_{i1/2}^n}{2D_\mathrm{h}}u_{1/2}^n\left|u_{1/2}^n\right| + \frac{\lambda_{i1}^n}{2D_\mathrm{h}}u_1^n\left|u_1^n\right|\right)\\
& - (1-\theta)\Delta t_n\Delta z\left[u_0^n\left(\frac{\partial u}{\partial z}\right)_0^n + \frac{1}{\rho_0^n}\left(\frac{\partial p}{\partial z}\right)_0^n + u_{1/2}^n\left(\frac{\partial u}{\partial z}\right)_{1/2}^n + \frac{1}{\rho_{1/2}^n}\left(\frac{\partial p}{\partial z}\right)_{1/2}^n \right.\\
& \left. + u_1^n\left(\frac{\partial u}{\partial z}\right)_1^n + \frac{1}{\rho_1^n}\left(\frac{\partial p}{\partial z}\right)_1^n\right]\\
& + (1-\theta)\Delta t_n\Delta z\left(\frac{u_0^n}{\rho_0^n}\frac{U}{A_\mathrm{f}}\frac{m_0^n}{2} + \frac{u_{1/2}^n}{\rho_{1/2}^n}\frac{U}{A_\mathrm{f}}\frac{m_{1/2}^n}{2} + \frac{u_1^n}{\rho_1}\frac{U}{A_\mathrm{f}}\frac{m_1^n}{2}\right.\\
& \left. + \frac{\lambda_{i0}^n}{2D_\mathrm{h}}u_0^n\left|u_0^n\right| + \frac{\lambda_{i1/2}^n}{2D_\mathrm{h}}u_{1/2}^n\left|u_{1/2}^n\right| + \frac{\lambda_{i1}^n}{2D_\mathrm{h}}u_1^n\left|u_1^n\right|\right) + \Delta z\left(u_0^n + u_{1/12}^n + u_1^n\right) - g\Delta z\Delta t_n.
\end{aligned} \tag{3.143g}$$

The quality of results obtained with orthogonal collocation methods is generally better than that of results obtained with finite differences, at the expense of more analytical and programming work.

The coupling of the above explained methods of orthogonal collocations to the computer code BLOW-3-A is explained in Bottoni (1982).

3.7 Synopsis of the BLOW-3A Code

The computer program BLOW-3A describes sodium boiling phenomena in fast breeder reactor subassemblies as consequence of different reactor failure conditions. These include mass flow reduction due to pump failure, single subassembly inlet blockage, and mild or rapid reactivity transients. The program is based on a theoretical approach assuming a multiple bubble slug ejection model as documented in Wirtz (1973), and in its final version in Bottoni and Struwe (1982). Some of the most important modelling features of the code are:

- Arbitrary (input data left to the user) number of radial meshes (usually 8–10) in fuel pin and arbitrary axial discretization.
- Allowance of negligible (smaller than 0.5 K) or large superheat of coolant prior to boiling onset.
- Simulation of the behavior of liquid film on structural surfaces after boiling initialization.
- Calculation of vapor pressure distribution inside continuous two-phase flow regions, overlapping at least three discretization meshes in axial direction.
- Application of different approximations for the calculation of two-phase pressure drop coefficients at the liquid-vapor interface.
- Calculation of single-phase coolant compressibility in case of rapid coolant heat-up.
- Calculation of heat and mass exchanges between upper liquid slugs or vapor regions and upper reactor plenum.
- Allowance for cladding melting after dry-out.
- An independent function library for materials properties is linked to the code.

Easy comparison of computed and experimental results is made possible by several plotting capabilities directly linked to the code.

The code, written in FORTRAN IV, consists of about 14,000 statements organized in 104 subroutines. Out of these, 39 subroutines compute physical properties of materials inside the reactor core. The standard code version is based on finite differences. The variant using the orthogonal collocation techniques for the calculation of the vapor flow in large coolant bubbles has been coded in a separate version.

From the historical viewpoint, it is worthwhile to remark that the theoretical background and physical modelling of the BLOW-3A code are similar to those of the worldwide known SAS2A program (Dunn et al., 1974).

References

Beam, R. M., & Warming, R. F. (1976). An implicit finite-difference algorithm for hyperbolic systems in conservation-law form. *Journal of Computational Physics, 22*, 87–110.

Bell, C. R., Bleiweis, P. B., Boudreau, S. E., Parker, F. R., & Smith, L. L. (1977). *SIMMER-I: An S_n, implicit, multifield, multicomponent, Eulerian, recriticality code for LMFBR disruptive core analysis* (NRC Report LA-NUREG-6467-MS). Los Alamos National Laboratory.

Bottoni, M. (1982). Evaluation of the residual method of orthogonal collocations and solution of one-dimensional equations of sodium vapour flow. *Nuclear Science and Engineering, 82*, 1–18.

Bottoni, M., & Struwe, D. (1982, December). *BLOW-3A, a theoretical model to describe transient two-phase flow conditions in LMFBR coolant channels* (Report No. KfK 3317).

Dunn, F. E., Fischer, G. J., Heames, T. J., Pizzica, P. A., McNeal, M. A., Bohl, W. R., & Prastein, S. M. (1974, October). *The SAS2A LMFBR accident-analysis computer code* (Report No. ANL-8138).

Forsythe, G. E., & Wasow, W. R. (1960). *Finite-difference methods for partial differential equations*. Wiley.

Harlow, F. H., & Amsden, A. A. (1971). A numeric fluid dynamics calculation method for all flow speeds. *Journal of Computational Physics, 8*(2), 197–213.

Harlow, F. H., & Amsden, A. A. (1975). Numerical calculation of multiphase fluid flow. *Journal of Computational Physics Volume, 17*, 19–52.

Richtmyer, R. D., & Morton, K. W. (1967). *Difference methods for initial-value problems*. Interscience.

Romstedt, P., & Werner, W. (1977). Efficient high-order method for the solution of fluid dynamics equations. *Nuclear Science and Engineering, 64*, 208–218.

Werner, W. (1979). High order methods in fluid dynamics. In *Proceedings of the Topical Meeting on Computational Methods in Nuclear Engineering*, Williamsburg, Virginia, April 23–25, 1979. American Nuclear Society.

Werner, W. (1981). Weighted residuals methods for the solution of fluid dynamics problems. In *Proceedings of the Topical Meeting on Advances in Mathematical Methods for the Solution of Nuclear Engineering Problems* (Vol. I), München, April 27–29, 1981. American Nuclear Society.

Wirtz, P. (1973, October). *Ein Beitrag zur theoretischen Beschreibung des Siedens unter Störfallbedingungen in natriumgekühlten schnellen Reaktoren* (Report No. KFK 1858).

Chapter 4
Numerical Methods for the Solution of the Convection-Diffusion Equation and QUICK Algorithm

In this chapter we discussed several variants of the algorithms applied for the numerical solution of convective-diffusion equations and also presents several methods to analyze the stability of the numerical schemes. The chapter is organized as follows.

In Sect. 4.1 we show how statistical models of Brownian motion allow to derive a convection-diffusion equation which then can be used to verify the correctness of algorithms for random number generation. Sections 4.2 and 4.3 introduce the model convective-diffusion equation used throughout the chapter and give a more precise definition of the concepts of convergence, consistency, stability, and truncation error of a numerical algorithm. Section 4.4 is dedicated to the diffusion equation which is analyzed in several solutions variants. Section 4.5 deals with the convection equation and Sect. 4.6 with the convection-diffusion equation. Eventually, Sect. 4.7 is dedicated to an application of the QUICK (Quadratic Upstream Interpolation for Convective Kinematics) algorithm to basic equations of two-phase flow simulation. The reason for introducing this section into this chapter is that the original reference for the QUICK scheme, Leonard (1979), deals mainly with the convection-diffusion equation, which is the topic of the chapter. We show however the application of the method to more complex conservation equations of two-phase fluid flow.

4.1 A Statistical Model of Brownian Motion Leading to the Convection-Diffusion Equation (One- and Two-Dimensional Random Walks)

In this chapter we examine some discretization methods to solve numerically the convection-diffusion equation. As an introduction to this topic, we show in this preface how the convection-diffusion equation arises from a statistical model of the

M. Bottoni, *Physical Modeling and Computational Techniques for Thermal and Fluid-dynamics*, Mechanical Engineering Series,
https://doi.org/10.1007/978-3-030-79717-1_4

Brownian motion of a small particle suspended in a fluid. We consider a one-dimensional model of the Brownian motion in which a particle, starting from a position $x = 0$, undergoes a sequence of collisions. Every collision displaces it along an x-axis by a step δ, with probability p, or by a step $-\delta$ (in the negative x-direction) with probability q. To every collision we associate a random variable x_i which can take the values $a_1 = \delta$ or $a_2 = -\delta$, with probabilities p and q, respectively. Therefore we can write formally:

$$P_{\text{rob}}(x_i = \delta) = p, \tag{4.1a}$$

$$P_{\text{rob}}(x_i = -\delta) = q, \tag{4.1b}$$

with

$$p + q = 1. \tag{4.2}$$

Let us consider n consecutive collisions of the particle and define the new variable:

$$X_n = x_1 + x_2 + \ldots\ldots + x_n = \sum_{i=1}^{n} x_i. \tag{4.3}$$

X_n represents the position of the particle along the x-axis after n collisions and is itself a random variable.

We recall that if a discrete valued random variable x can assume M values a_1, $a_1 \ldots\ldots a_m \ldots\ldots a_M$ each one with probability p_m, its mean value is defined as:

$$\overline{x} = \sum_{m=1}^{M} a_m p_m. \tag{4.4}$$

The variance of x is given by:

$$V(x) = \overline{(x - \overline{x})^2} = \overline{x^2 - 2x\overline{x} + (\overline{x})^2} = \overline{x^2} - 2\overline{x}\overline{x} + (\overline{x})^2 = \overline{x^2} - (\overline{x})^2. \tag{4.5}$$

Therefore mean value and variance of the random variable x_i are:

$$\overline{x_i} = \sum_{m=1}^{2} \alpha_m p_m = \delta p + (-\delta)q = (p - q)\delta, \tag{4.6}$$

$$V(x_i) = \overline{x_i^2} - (\overline{x_i})^2. \tag{4.7}$$

The mean value of X_n is:

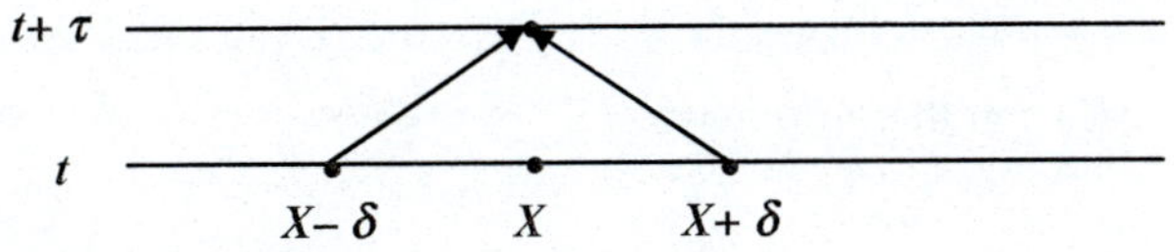

Fig. 4.1 Position of the particle at time $t + \tau$

$$\overline{X_n} = \overline{\sum_{i=1}^{n} x_i} = \sum_{i=1}^{n} \overline{x_i} = \sum_{i=1}^{n} \delta(p - q) = (p - q)n\delta. \tag{4.8}$$

The variance of X_n is:

$$V(X_n) = V\left(\sum_{i=1}^{n} x_i\right) = \sum_{i=1}^{n} V(x_i) = \sum_{i=1}^{n} \left[\overline{x_i^2} - (\overline{x_i})^2\right]. \tag{4.9}$$

The value of x_i^2 can be δ^2, with probability p or $(-\delta)^2$ with probability q; therefore we can write:

$$\overline{x_i^2} = \overline{\delta^2 p + (-\delta)^2 q} = \overline{\delta^2 (p + q)} = \delta^2. \tag{4.10}$$

Using Eqs. (4.10) and (4.6), Eq. (4.9) becomes:

$$\begin{aligned} V(X_n) &= \sum_{i=1}^{n} \left[\delta^2 - \delta^2 (p - q)^2\right] = \sum_{i=1}^{n} \delta^2 \left[1 - (p - q)^2\right] \\ &= \sum_{i=1}^{n} \delta^2 \left[1 - (p + q)^2 + 4pq\right] = 4pq\delta^2 n \end{aligned} \tag{4.11}$$

Let us call C the mean displacement of the particle per unit of time and D its variance. Let r be the collision rate (number of collisions per unit of time) and $\tau = 1/r$ the mean time interval between collisions. Using formulas (4.8) and (4.11) (mean value and variance of X_n), we can write that the state of the particle after r collisions is represented by:

$$C = (p - q)r\delta, \tag{4.12}$$

$$D = 4pq\delta^2 r. \tag{4.13}$$

We now introduce a probability distribution function formally defined by $v(x, t) = P_{\mathbf{rob}}(x = X_n, t)$, namely, the probability that at time t the particle is at the position x. The particle can be at position x at time $t + \tau$ if:

(i) It was at position $x - \delta$ at time t (see Fig. 4.1) and underwent a collision which moved it a step δ to the right (event with probability p).

(ii) It was at position $x + \delta$ at time t and moved a step $-\delta$ (to the left) (event with probability q).

We can therefore write:

$$v(x, t + \tau) = pv(x - \delta, t) + qv(x + \delta, t). \tag{4.14}$$

Using a Taylor series expansion for time around point (x, t), we have:

$$v(x, t + \tau) = v(x, t) + \tau v_t'(x, t) + O(\tau^2), \tag{4.15a}$$

and likewise using a Taylor series expansion for the space variable around the same point (x, t):

$$v(x + \delta, t) = v(x, t) + \delta v_x'(x, t) + {}^{\delta^2}\!/_{2} v_{xx}''(x, t) + O(\delta^3), \tag{4.15b}$$

$$v(x - \delta, t) = v(x, t) - \delta v_x'(x, t) + {}^{\delta^2}\!/_{2} v_{xx}''(x, t) + O(\delta^3). \tag{4.15c}$$

Introducing Eqs. (4.15) into Eq. (4.14) yields:

$$\begin{aligned} v(x, t) + \tau v_t'(x, t) = p\left[v(x, t) - \delta v_x'(x, t) + \frac{\delta^2}{2} v_{xx}''(x, t)\right] \\ + q\left[v(x, t) + \delta v_x'(x, t) + \frac{\delta^2}{2} v_{xx}''(x, t)\right] \end{aligned} \tag{4.16}$$

Simplifying and multiplying by $r = 1/\tau$, Eq. (4.16) yields:

$$v_t'(x, t) = -\delta r(p - q) v_x'(x, t) + r\frac{\delta^2}{2} v_{xx}''(x, t). \tag{4.17}$$

The coefficient which multiplies v_x' is $-C$ [cfr. Eq. (4.12)], while the coefficient which multiplies v_{xx}'' is $D/(8pq)$ [cfr. Eq. (4.13)]. Hence Eq. (4.17) becomes:

$$\frac{\partial v(x, t)}{\partial t} + C\frac{\partial v(x, t)}{\partial x} = \frac{D}{8pq} \frac{\partial^2 v(x, t)}{\partial x^2}. \tag{4.18}$$

Equation (4.18) has the form of a convection-diffusion equation. The convective velocity C is the displacement of the particle per unit of time, while the diffusion coefficient $D/(8pq)$ is related to the variance of the position of the particle around the mean value C and is a measure of the diffusion of particles, in case of the random motion of several particles originating from the source point.

Let us consider now a two-dimensional random walk assuming that the probabilities of one step in positive or negative x- or y-directions be the same, namely, $p = ¼$. Let us introduce the two-dimensional probability distribution function σ *(x, y, t)* as the probability that a particle is in the position *P (x, y)* at time *t*. Thus a particle will be at *P (x, y)* at time $(t + \tau)$ if at time *t* it was at $P\ (x \pm \delta, y)$ or at $P\ (x, y \pm \delta)$. We can therefore write, in analogy with Eq. (4.14):

$$v(x,y,t+\tau)=\frac{1}{4}[v(x+\sigma,y,t)+v(x-\sigma,y,t)+v(x,y+\sigma,t)+v(x,y-\sigma,t)]. \quad (4.19)$$

Using a Taylor series expansion around *(x, y, t)*, one has:

$$v(x,y,t+\tau) = v(x,y,t) + \tau v_t'(x,y,t) + O(\tau^2), \quad (4.20a)$$

$$v(x+\delta,y,t) = v(x,y,t) + \delta v_x'(x,y,t) + \frac{\delta^2}{2} v_{xx}''(x,y,t) + O(\delta^3), \quad (4.20b)$$

$$v(x-\delta,y,t) = v(x,y,t) - \delta v_x'(x,y,t) + \frac{\delta^2}{2} v_{xx}''(x,y,t) + O(\delta^3), \quad (4.20c)$$

$$v(x,y+\delta,t) = v(x,y,t) + \delta v_y'(x,y,t) + \frac{\delta^2}{2} v_{yy}''(x,y,t) + O(\delta^3), \quad (4.20d)$$

$$v(x,y-\delta,t) = v(x,y,t) - \delta v_y'(x,y,t) + \frac{\delta^2}{2} v_{yy}''(x,y,t) + O(\delta^3). \quad (4.20e)$$

Inserting Eq. (4.20) into Eq. (4.19) and simplifying, one derives:

$$v_t'(x,y,t) = \frac{\delta^2}{4\tau}\left[v_{xx}''(x,y,t) + v_{yy}''(x,y,t)\right], \quad (4.21)$$

which represents a two-dimensional diffusion equation with diffusion coefficient

$$D = \delta^2/4\tau\left[m^2/s\right].$$

The above formalism can be easily applied to obtain a three-dimensional diffusion equation.

The unconstrained random walk which we have considered allows for a check of the performance of a random number generator, based upon Einstein's law. In a one- or more-dimensional random walk, let R_N be the distance of a particle from the origin after *N* steps. Let us repeat this *N*-step walk a large number *M* of times (f.i., some thousand times) and take the mean value of $R^2{}_N$ over all walks:

$$\overline{R_N^2} = \frac{1}{M}\sum_1^M mR_{N,m}^2. \quad (4.22)$$

Einstein's law states that, in the limit for $N \rightarrow \infty$, $\overline{R_N^2}$ is proportional to the length N of the walk:

$$\overline{R_N^2} \equiv \mathrm{N}. \tag{4.23}$$

Because τ is the mean time interval between subsequent steps, the time which elapses during a walk of length N is $t = \tau N$; hence Einstein's law can be formulated in terms of time as:

$$\overline{R_N^2} \equiv t. \tag{4.24}$$

Performing M random walks for every random walk length $N(N = 1, 2, \ldots N_{\mathbf{max}})$ (where $N_{\mathbf{max}}$ may be $\approx 5000 \div 10{,}000$ steps) and plotting $\overline{R_N^2}$ versus N allow for checking whether a random number generator performs well or not.

A listing of a small computer program which performs this test in two dimensions, called **PROG_IV_1 [TEST_RNG]**, is given in Appendix A.IV. This program checks two random number generators: the first (Function RANDOM) is based on the modulus function; the second (Function RF) is more sophisticated. The first check of these two functions is made by computing the number π in the following way: Let us consider the two-dimensional domain $D = (0 < x < 1, 0 < y < 1))$ and consider a sequence of couples of random numbers (x_i, y_i) interpreted as coordinates of a point in the domain D. If the distance $r_i = \left(x_i^2 + y_i^2\right)^{1/2}$ of the points from the origin of the coordinate system is smaller than 1, the points are inside the circle with center at the origin and radius = 1. The frequency with which the points fall inside the circle is $\pi/4$. In this way π can be computed. The precision of the computations gives information about the quality of the random number generator. Using imax = 10^9 couples of points, the results obtained with the two functions are, in terms of the relative errors defined by $error = |\pi - \pi_c|/\pi$, where π_c is the computed value, as follows:

Function RANDOM: error = 1.58×10^{-4}; Function RF: error = 5.72×10^{-6}.

Therefore the function RF performs much better than the function RANDOM.

Results of a further test based on Einstein's law are shown in the following two figures. Figure 4.2 shows the performance of the RANDOM function which departs from Einstein's law (straight line) already for $N > 2000$. The performance is poor and implies that the random number generator has a rather short cycle.

Figure 4.3 shows the performance of the RF number generator which satisfies well with Einstein's law till $N = 10,000$.

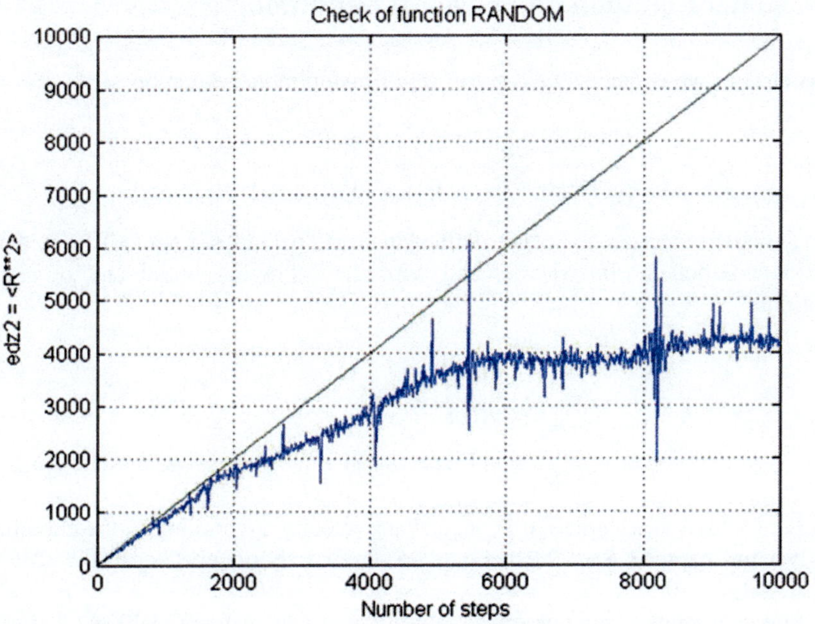

Fig. 4.2 Check of RANDOM function for random number generation, against Einstein's law

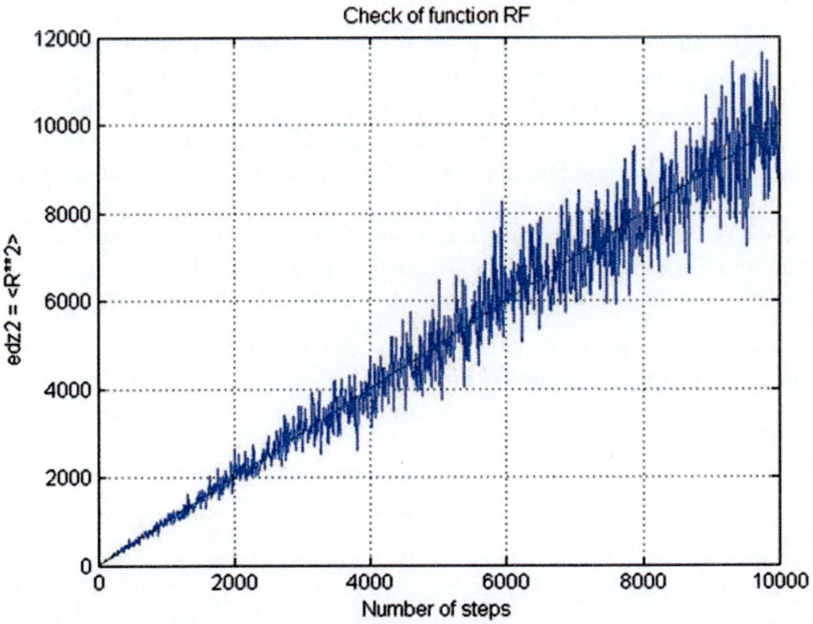

Fig. 4.3 Check of RF function for random number generation, against Einstein's law

4.2 Model Equation and General Definitions

In this chapter we consider the convection-diffusion model equation:

$$\frac{\partial u}{\partial t} + V\frac{\partial u}{\partial x} = a\frac{\partial^2 u}{\partial x^2}. \tag{4.25}$$

for an unknown function $u = u(x, t)$ defined in the domain $(0 \leq x \leq L, t \geq 0)$, with the diffusion coefficient $a[m^2/s] > 0$ and with the following initial and boundary conditions:

$$u(x,0) = F(x), \tag{4.26}$$

$$u(0,t) = u_0(t), \tag{4.27a}$$

$$u(L,t) = u_L(t). \tag{4.27b}$$

In Eq. (4.25) V is assumed to be a constant velocity. According to the classification given in Chapter 2, Sect. 2.2.1, Eq. (4.25) is of parabolic type, because $b = c = 0$, $\Delta = b^2 - 4ac = 0$.

Before explaining some numerical solution methods of the model Eq. (4.25) and of its subcases, the diffusion equation (obtained by dropping the convective term) and the convective equation (obtained by dropping the diffusion term), we recall the following general definitions and concepts.

(i) "Well-posedeness." According to the definition given by **Hadamard** (1923), a problem of mathematical physics is "well posed" if its solution exists, is unique, and depends continuously on the data (initial data and boundary conditions).

(ii) "Convergence." We assume that the partial differential Eq. (4.25) has been replaced, for its numerical solution, by a finite difference equation, obtained discretizing every term of (4.25). The solution of the finite difference equation is convergent if, in every point $u(x, t)$ of the definition domain, it approaches the solution of the partial differential equation when the grid sizes Δx, Δt tend to zero.

Let $U = U(x, t)$ be the exact solution of the differential equation and $u_{i,\,n}$ be the solution of the difference equation in the mesh modes $(x_i = i\Delta x, t_n = n\Delta t)$. The "order of convergence" of the difference equation is given by the exponents m_x and m_t in a formula of the type:

$$u_{i,n} - U(x_i, t_n) = O(\Delta x^{m_x}) + O(\Delta t^{m_t}). \tag{4.28}$$

(iii) Consistency. Let $L(u) = 0$ be a differential equation with solution $U(x, t)$; let $F(u) = 0$ be a difference equation (approximating the given differential equation) with the exact (apart from round-off errors) solution $u_{i,\ n}$. Let v be a continuous and differentiable function for which $L(v)$ and $F(v)$ can be computed.

The "truncation error" is defined by:

$$T_{i,n}(v) = F[v(x_i, t_n)] - L[v(x_i, t_n)]. \tag{4.29}$$

In the following we will show, more in detail, that the truncation error represents the discrepancy between the finite difference form of the differential equation and the differential equation itself.

A finite difference approximation to a differential equation is consistent with the differential equation if the truncation error $[T_{i,\ n}(v)]$ tends to zero, when the grid sizes Δx, Δt tend to zero.

The "order of consistency" is given by the exponents k_x and k_t in a formula of the type:

$$T_{i,n} = O\left(\Delta x^{k_x}\right) + O\left(\Delta t^{k_t}\right). \tag{4.30}$$

(iv) Stability.

Let:

- U be the exact (analytical) solution of a differential equation,
- u be the exact solution of the approximating difference equation (i.e., the solution which would be obtained if there were no round-off errors).
- N be the numerical solution of the difference equation (affected by round-off errors).

The "total error" $e = U - N$ can be written as:

$$e = U - N = \underset{(a)}{(U - u)} + \underset{(b)}{(u - N)}. \tag{4.31}$$

Term (a) in Eq. (4.31) represents the discretization error; Term (b) represents the "round-off" error. Let e_{ij}^n be the error in the mesh node $\rho_{ij} = \rho(x_i, y_j) = \rho(i\Delta x, j\Delta y)$ [in case of a two-dimensional problem] at time point $t_n = n\Delta t$.

A difference scheme is stable if the cumulative effect of all rounding-off errors is negligible, so that $\left|e_{ij}^n\right| < \delta$ in all mesh points, as time progresses or, in case of time-independent problems, as the numerical solution progresses.

(v) Lax's equivalence theorem.

Let us consider a well-posed, linear, initial value problem. Lax's equivalence theorem states that "If a finite difference approximation is consistent, then stability is the necessary and sufficient condition for convergence."

A shorter formulation of the theorem is: “Consistency and stability imply convergence.”

In the course of this chapter, we will first consider a model diffusion equation, obtained from Eq. (4.25) dropping the convective term. On the basis of this diffusion equation, we will introduce and analyze several methods of stability analysis (Sects. 4.4.1 thorough 4.4.3). Then, in Sect. 4.5, we will consider the convection equation obtained from Eq. (4.25) by dropping the diffusion term. Eventually, in Sect. 4.6, we will consider the full model Eq. (4.25) and analyze some methods of differencing the derivatives.

4.3 Analysis of Convergence and Truncation Error for the Diffusion Equation

(i) **Convergence**

We consider the one-dimensional heat diffusion equation:

$$\frac{\partial u}{\partial t} = a\frac{\partial^2 u}{\partial x^2}, (0 \leq x \leq 1), (a \geq 0), (t \geq 0) \tag{4.32}$$

for a normalized temperature $u(x, t)$, with initial and boundary conditions:

$$u(x, 0) = F(x), \tag{4.33}$$

$$u(0, t) = g_0(t), \tag{4.34a}$$

$$u(1, t) = g_1(t). \tag{4.34b}$$

Equation (4.32) is obtained from the model Eq. (4.25) with $V = 0$. We discretize Eq. (4.32) with the explicit FTCS (forward time, central space) scheme (also referred to as Euler scheme):

$$\frac{1}{\Delta t}\left(u_j^{n+1} - u_j^n\right) = \frac{a}{\Delta x^2}\left(u_{j+1}^n - 2u_j^n + u_{j-1}^n\right). \tag{4.35}$$

Introducing the dimensionless parameter

$$s = \frac{a\Delta t}{\Delta x^2}. \tag{4.36}$$

and solving with respect to u_j^{n+1}, one derives:

$$u_j^{n+1} = su_{j-1}^n + (1 - 2s)u_j^n + su_{j+1}^n. \tag{4.37}$$

Let $\overline{u}(x,t)$ be the exact (analytical) solution of the diffusion Eq. (4.32) and u_j^n be the exact solution of the difference Eq. (4.37) in the node $(x_j = j\Delta x, t_n = n\Delta t)$ (without considering the influence of round-off errors).

We define the "discretization error" as:

$$e_j^n = \overline{u}(x_j, t_n) - u_j^n = \overline{u}_j^n - u_j^n, \tag{4.38}$$

hence:

$$u_j^n = \overline{u}_j^n - e_j^n. \tag{4.39}$$

Inserting Eq. (4.39) into Eq. (4.37) yields:

$$\overline{u}_j^{n+1} - e_j^{n+1} = s\left(\overline{u}^n_{j-1} - e^n_{j-1}\right) + (1-2s)\left(\overline{u}_j^n - e_j^n\right) + s\left(\overline{u}^n_{j+1} - e^n_{j+1}\right). \tag{4.40}$$

Solving Eq. (4.40) with respect to e_j^{n+1} yields:

$$\begin{aligned} e_j^{n+1} &= se^n_{j-1} + (1-2s)e_j^n + se^n_{j+1} + \overline{u}_j^{n+1} - \overline{u}_j^n \\ &\quad - s\left(\overline{u}^n_{j-1} + \overline{u}^n_{j+1} - 2\overline{u}_j^n\right). \end{aligned} \tag{4.41}$$

We use a Taylor series expansion of the values $\overline{u}^n_{j+1}, \overline{u}_j^n, \overline{u}^n_{j-1}$ around the grid point (j, n) with the remainder term expressed in Lagrange's form:

$$\overline{u}^n_{j+1} = \overline{u}_j^n + \Delta x\left(\frac{\partial \overline{u}}{\partial x}\right)_j^n + \frac{\Delta x^2}{2}\left(\frac{\partial^2 \overline{u}}{\partial x^2}\right)_{j+\theta_1}^n + \ldots\ldots \tag{4.42}$$

$$\overline{u}^n_{j-1} = \overline{u}_j^n - \Delta x\left(\frac{\partial \overline{u}}{\partial x}\right)_j^n + \frac{\Delta x^2}{2}\left(\frac{\partial^2 \overline{u}}{\partial x^2}\right)_{j-\theta_2}^n + \ldots\ldots \tag{4.43}$$

$$\overline{u}_j^{n+1} = \overline{u}_j^n + \Delta t\left(\frac{\partial \overline{u}}{\partial t}\right)_j^{n+\theta_3} + \ldots\ldots \tag{4.44}$$

where, for instance, $j+\theta_1$ is a short form for $(j\Delta x + \theta_1\Delta x, n\Delta t)$ and $n+\theta_3$ is a short form for $(j\Delta x, n\Delta t + \theta_3\Delta t)$ with $(0 < \theta_1, \theta_2, \theta_3 < 1)$.

Inserting Eqs. (4.42), (4.43), and (4.44) into (4.41), one obtains:

$$\begin{aligned} e_j^{n+1} &= se^n_{j-1} + (1-2s)e_j^n + se^n_{j+1} + \Delta t\left(\frac{\partial \overline{u}}{\partial t}\right)_j^{n+\theta_3} \\ &\quad - s\left\{\frac{\Delta x^2}{2}\left[\left(\frac{\partial^2 \overline{u}}{\partial x^2}\right)_{j+\theta_1}^n + \left(\frac{\partial^2 \overline{u}}{\partial x^2}\right)_{j-\theta_2}^n\right]\right\} \end{aligned} \tag{4.45}$$

In the following we consider $s \leq 1/2$ so that $1 - 2s \geq 0$. In this case, taking the absolute values of the right- and left-hand sides of Eq. (4.45), and using $s\Delta x^2 = a\Delta t$, yields:

$$\left|e_j^{n+1}\right| \leq s\left|e^n_{j-1}\right| + (1-2s)\left|e_j^n\right| + s\left|e^n_{j+1}\right| + \Delta t\left|\left(\frac{\partial \overline{u}}{\partial t}\right)_j^{n+\theta_3} - \frac{a}{2}\left[\left(\frac{\partial^2 \overline{u}}{\partial x^2}\right)^n_{j+\theta_1} + \left(\frac{\partial^2 \overline{u}}{\partial x^2}\right)^n_{j-\theta_2}\right]\right|. \tag{4.46}$$

Let us call ΔtD the last term of Eq. (4.46), call M the maximum of the D values in all grid points and call E^n the maximum of all $\left|e_j^n\right|$:

$$M = \max(D), \ (j = 1, 2, \ldots\ldots J-1) \tag{4.47}$$

$$E^n = \max\left|e_j^n\right|.(j = 0, 1, 2\ldots\ldots)\ (n = 0, 1, 2\ldots\ldots) \tag{4.48}$$

With these definitions, from Eq. (4.46) one obtains:

$$\begin{aligned}\left|e_j^{n+1}\right| &\leq sE^n + (1-2s)E^n + sE^n + \Delta tM,\\ \left|e_j^{n+1}\right| &\leq E^n + \Delta tM.\end{aligned} \tag{4.49}$$

Applying Eq. (4.49) subsequently for $n = 1, 2, 3\ldots$ yields:

$$\left|e_j^1\right| \leq E^0 + \Delta tM, \tag{4.50a}$$

$$\left|e_j^2\right| \leq E^1 + \Delta tM \leq \left(E^0 + \Delta tM\right) + \Delta tM = E^0 + 2\Delta tM \tag{4.50b}$$

……..

$$\left|e_j^N\right| \leq E^0 + N\Delta tM = E^0 + Mt_N \tag{4.50c}$$

using $t_N = N\Delta t$. Because for $t = 0$ we use the initial condition (4.33), the error E^0 is zero. For $\Delta x \to 0$ the two second-order derivatives in Eq. (4.46) tend to the same value, because both $j + \theta_1$ and $j - \theta_2$ tend to j. Hence the last two terms of Eq. (4.46), within the sign of absolute value, are the exact Eq. (4.32) and $D \to 0$; therefore $M \to 0$ and $\left|e_j^N\right| \to 0$. Hence the solution of the difference scheme (4.37) converges to the solution of the differential equation, and the scheme is convergent.

(ii) **Truncation Error**

We recall Eq. (4.37) written for the exact solution $\overline{u}$, replace the term u_j^{n+1} by a Taylor series expansion up to the second-order time derivative, and replace the terms u^n_{j+1}, u^n_{j-1} by a series expansion up to the fourth-order derivative in space:

$$\begin{aligned}
&\overline{u}_j^n + \Delta t\left(\frac{\partial \overline{u}}{\partial t}\right)_j^n + \frac{\Delta t^2}{2}\left(\frac{\partial^2 \overline{u}}{\partial t^2}\right)_j^n + O(\Delta t^3)\\
&= s\left[\overline{u}_j^n + \Delta x\left(\frac{\partial \overline{u}}{\partial x}\right)_j^n + \frac{\Delta x^2}{2}\left(\frac{\partial^2 \overline{u}}{\partial x^2}\right)_j^n + \frac{\Delta x^3}{3!}\left(\frac{\partial^3 \overline{u}}{\partial x^3}\right)_j^n + \frac{\Delta x^4}{4!}\left(\frac{\partial^4 \overline{u}}{\partial x^4}\right)_j^n + O(\Delta x^5)\right].\\
&+s\left[\overline{u}_j^n - \Delta x\left(\frac{\partial \overline{u}}{\partial x}\right)_j^n + \frac{\Delta x^2}{2}\left(\frac{\partial^2 \overline{u}}{\partial x^2}\right)_j^n - \frac{\Delta x^3}{3!}\left(\frac{\partial^3 \overline{u}}{\partial x^3}\right)_j^n + \frac{\Delta x^4}{4!}\left(\frac{\partial^4 \overline{u}}{\partial x^4}\right)_j^n + O(\Delta x^5)\right]\\
&+(1-2s)\overline{u}_j^n.
\end{aligned} \tag{4.51}$$

Using $a = s\Delta x^2/\Delta t$ [cfr. Eq. (4.36)], one obtains:

$$\left(\frac{\partial \overline{u}}{\partial t}\right)_j^n = a\left(\frac{\partial^2 \overline{u}}{\partial x^2}\right)_j^n + \left[-\frac{\Delta t}{2}\left(\frac{\partial^2 \overline{u}}{\partial t^2}\right)_j^n + a\frac{\Delta x^2}{12}\left(\frac{\partial^4 \overline{u}}{\partial x^4}\right)_j^n\right] + O\left(\Delta t^2, \Delta x^5\right). \tag{4.52}$$

The last terms in Eq. (4.52) represent the truncation error, or the discrepancy between the finite difference form of the differential equation and the differential Eq. (4.32). Thus we define:

$$T_{j,n} = -\frac{\Delta t}{2}\left(\frac{\partial^2 \overline{u}}{\partial t^2}\right)_j^n + a\frac{\Delta x^2}{12}\left(\frac{\partial^4 \overline{u}}{\partial x^4}\right)_j^n + O\left(\Delta t^2, \Delta x^5\right). \tag{4.53}$$

The second-order derivative with respect to time can be also expressed by means of a fourth-order derivative with respect to space:

$$\begin{aligned}
\frac{\partial^2 \overline{u}}{\partial t^2} &= \frac{\partial}{\partial t}\left(\frac{\partial \overline{u}}{\partial t}\right) = \frac{\partial}{\partial t}\left(a\frac{\partial^2 \overline{u}}{\partial x^2}\right) = a\frac{\partial^2}{\partial x^2}\left(\frac{\partial \overline{u}}{\partial t}\right) = a\frac{\partial^2}{\partial x^2}\left(a\frac{\partial^2 \overline{u}}{\partial x^2}\right)\\
&= a^2\frac{\partial^4 \overline{u}}{\partial x^4}.
\end{aligned} \tag{4.54}$$

Inserting Eq. (4.54) into Eq. (4.53) yields:

$$
\begin{aligned}
T_{j,n} &= -\frac{\Delta t}{2}a^2\left(\frac{\partial^4\overline{u}}{\partial x^4}\right)_j^n + a\frac{\Delta x^2}{12}\left(\frac{\partial^4\overline{u}}{\partial x^4}\right)_j^n + O(\Delta t^2, \Delta x^5) \\
&= -\frac{\Delta t}{2}a^2\left(\frac{\partial^4\overline{u}}{\partial x^4}\right)_j^n + \frac{a}{12}\frac{a\Delta t}{s}\left(\frac{\partial^4\overline{u}}{\partial x^4}\right)_j^n + O(\Delta t^2, \Delta x^5) \qquad (4.55) \\
&= -\frac{a^2\Delta t}{2}\left(1-\frac{1}{6s}\right)\left(\frac{\partial^4\overline{u}}{\partial x^4}\right)_j^n + O(\Delta t^2, \Delta x^5),
\end{aligned}
$$

having used $\Delta x^2 = a\Delta t/s$.

The first term at the right-hand side of Eq. (4.55) vanishes for $s = 1/6$. This value of s corresponds to the fastest decreasing of $T_{j,\,n}$ as Δt and Δx tend to zero. Anyway, even for $s \neq 1/6$, as Δt and Δx tend to zero, $T_{j,\,n}$ also tends to zero, therefore the difference scheme (4.37) is consistent with the differential Eq. (4.32).

An alternative calculation of the truncation error can be made, based on definition (4.29). In case of the diffusion Eq. (4.32), we define:

$$
L(u) = \frac{\partial u}{\partial t} - a\frac{\partial^2 u}{\partial x^2} = 0, \qquad (4.56)
$$

$$
F(u) = \frac{1}{\Delta t}\left(u_j^{n+1} - u_j^n\right) - \frac{a}{\Delta x^2}\left(u_{j+1}^n - 2u_j^n + u_{j-1}^n\right) = 0. \qquad (4.57)
$$

Let v be a continuous and differentiable function for which $L(v)$ and $F(v)$ can be computed. Then the truncation error is defined [Cfr. Eq. (4.29)] by:

$$
\begin{aligned}
T_{j,n}(v) &= F[v(x_j, t_n)] - L[v(x_j, t_n)] \\
&= \frac{1}{\Delta t}\left(v_j^{n+1} - v_j^n\right) - \frac{a}{\Delta x^2}\left(v_{j+1}^n - 2v_j^n + v_{j-1}^n\right) - \left(v_t'\right)_j^n + a\left(v_{xx}''\right)_j^n.
\end{aligned} \qquad (4.58)
$$

Inserting the Taylor series expansions:

$$
v_j^{n+1} = v_j^n + \Delta t\left(v_t'\right)_j^n + \frac{\Delta t^2}{2}\left(v_{tt}''\right)_j^n + \cdots, \qquad (4.59a)
$$

$$
v_{j+1}^n = v_j^n + \Delta x\left(v_x'\right)_j^n + \frac{\Delta x^2}{2}\left(v_{xx}''\right)_j^n + \frac{\Delta x^3}{6}\left(v_{xxx}'''\right)_j^n + \frac{\Delta x^4}{24}\left(v''''_{xxxx}\right)_j^n + \cdots, \qquad (4.59b)
$$

$$
v_{j-1}^n = v_j^n - \Delta x\left(v_x'\right)_j^n + \frac{\Delta x^2}{2}\left(v_{xx}''\right)_j^n - \frac{\Delta x^3}{6}\left(v_{xxx}'''\right)_j^n + \frac{\Delta x^4}{24}\left(v''''_{xxxx}\right)_j^n + \cdots, \qquad (4.59c)
$$

into Eq. (4.58), one obtains:

$$T_{j,n}(v) = \frac{1}{\Delta t}\left[\Delta t\left(v_t'\right)_j^n + \frac{\Delta t^2}{2}\left(v_{tt}''\right)_j^n\right] - \frac{a}{\Delta x^2}$$
$$\times\left[2v_j^n + \Delta x^2\left(v_{xx}''\right)_j^n + \frac{\Delta x^4}{12}\left(v''''_{xxxx}\right)_j^n - 2v_j^n\right] - \left(v_t'\right)_j^n + a\left(v_{xx}''\right)_j^n. \quad (4.60)$$

Simplifying one obtains:

$$T_{j,n}(v) = \frac{\Delta t}{2}\left(v_{tt}''\right)_j^n - \frac{a\Delta x^2}{12}\left(v''''_{xxxx}\right)_j^n, \quad (4.61)$$

which shows that $T_{j,\,n}(v)$ tends to zero as Δt and Δx tend to zero.

4.4 Stability Analysis for Diffusion Equation

In this section we consider different methods for the stability analysis of a difference scheme, namely, (i) the matrix method (also called "eigenvalue criterion"); (ii) the Fourier stability criterion after von Neumann (1944); and (iii) the Hirt's heuristic criterion. In the following we show applications to the diffusion equation and to the convection-diffusion equation.

4.4.1 The Matrix Method (Eigenvalue Criterion)

4.4.1.1 General Considerations

Let us consider a general difference scheme represented by the matrix equation

$$DU^{n+1} = BU^n + Q^n, \quad (4.62)$$

where the matrix D of rank N is assumed to be non-singular ($\det D \neq 0$) and the superscript n refers to time discretization. U^n and U^{n+1} are vectors of unknowns at time levels t_n and t_{n+1}, respectively, and Q^n is a source term.

For the exact unknown solution $\overline{u}$, it also holds:

$$D\overline{u}^{n+1} = B\overline{u}^n + Q^n. \quad (4.63)$$

Subtracting Eq. (4.62) from Eq. (4.63), one has:

$$D\left(\overline{u}^{n+1} - U^{n+1}\right) = B\left(\overline{u}^n - U^n\right). \quad (4.64)$$

or

$$De_{n+1} = Be_n, \tag{4.65}$$

with the definition of the error vector:

$$e_n = \overline{u}_n - U_n. \tag{4.66}$$

From Eq. (4.65) one obtains:

$$e_{n+1} = D^{-1}Be_n = He_n, \tag{4.67}$$

with the definition $H = D^{-1}B$. Let $\lambda_i (i = 1, 2.\ldots..N)$ be the eigenvalues of the matrix H [i.e., the solution of the algebraic equation $\det(H - \lambda I) = 0$, where I is the unit matrix] and let us assume that they are all distinct. In this case the corresponding N eigenvectors $\overline{x_i}$ form a system of linearly independent vectors, and it holds:

$$H\overline{x_i} = \lambda_i \overline{x_i}. \quad (i = 1, 2\ldots\ldots N) \tag{4.68}$$

The error vector e_0 can be represented as a linear combination of the eigenvectors, which build a "basis" in the N-dimensional vector space, namely

$$e_0 = \sum_{i=1}^{N} c_i \overline{x_i}, \tag{4.69}$$

where the c_i are constants.

Using Eqs. (4.67), (4.68) and (4.69) we can therefore write:

$$\begin{aligned} e_1 &= He_0 = H\sum_i c_i\overline{x_i} = \sum_i c_i H\overline{x_i} = \sum_i c_i\lambda_i\overline{x_i}, \\ e_2 &= He_1 = H\sum_i c_i\lambda_i\overline{x_i} = \sum_i c_i\lambda_i H\overline{x_i} = \sum_i c_i\lambda_i^2\overline{x_i}, \\ e_n &= \sum_i c_i\lambda_i^n\overline{x_i}. \end{aligned} \tag{4.70}$$

The error vector e_n remains limited and vanishes with increasing time if all eigenvalues λ_i are in modulus smaller than 1. The stability condition of the difference scheme (4.62) is therefore $|\lambda_i| < 1$, $(i = 1, 2.\ldots..N)$.

4.4.1.2 Leverrier Method for the Determination of the Eigenvalue of Maximum Modulus

In Chap. 5 we will show that a matrix equation:

$$AY = B, \tag{4.71}$$

can be solved iteratively by means of a numerical scheme given by the equation:

$$Y^r = H(\omega)Y^{r-1} + MB. \tag{4.72}$$

In Eq. (4.72) the matrix $H(\omega)$, called "iteration matrix," depends on a relaxation parameter ω $(0 < \omega < 2)\langle\,|\,\rangle$. The matrix M is not relevant for the analysis made in this section. The superscript r refers to the r-th iteration step in an iterative method to solve Eq. (4.71).

The exact (unknown) solution $\overline{Y}$ of Eq. (4.71) also satisfies an equation formally identical with (4.72):

$$\overline{Y} = H(\omega)\overline{Y} + MB. \tag{4.73}$$

Subtracting Eq. (4.73) from Eq. (4.72) yields:

$$E^r = HE^{r-1}, \tag{4.74}$$

with the definition of the error vector:

$$E^r = Y^r - \overline{Y}. \tag{4.75}$$

We now define "substitution vector" δ^r the difference between two consecutive solutions of Eq. (4.71):

$$\delta^r = Y^{r+1} - Y^r = \left(Y^{r+1} - \overline{Y}\right) + \left(\overline{Y} - Y^r\right) = E^{r+1} - E^r. \tag{4.76}$$

Using Eq. (4.74), Eq. (4.76) yields:

$$\begin{aligned} \delta^r &= HE^r - HE^{r-1} = H\delta^{r-1} \\ &= H^2E^{r-1} - H^2E^{r-2} = H^2\delta^{r-2} \\ \delta^r &= H^rE^1 - H^rE^0 = H^r\delta^0 \end{aligned} \tag{4.77}$$

Let us call $\eta_i (i = 1, 2. \ldots .. N)$ the eigenvalues of the iteration matrix H, and let us assume that they are distinct and real and labeled in order of increasing magnitude: $\eta_1 < \eta_2 < \ldots \ldots < \eta_N$. Let $\overline{x_i}$ be the corresponding eigenvectors of the matrix H, so that one has:

$$H\overline{x_i} = \eta_i\overline{x_i}. \tag{4.78}$$

Because the eigenvectors $\overline{x_i}$ are linearly independent and build a basis of the N-dimensional space, the substitution vector δ^0 can be written as a linear combination of the eigenvectors:

$$\delta^0 = \sum_{i=1}^{N} c_i\overline{x_i} = c_1\overline{x_1} + c_2\overline{x_2} + \ldots\ldots + c_N\overline{x_N}, \tag{4.79}$$

where c_i are constants.

Introducing Eq. (4.79) in Eq. (4.77) and using Eq. (4.78), one has subsequently:

$$\begin{aligned}
\delta^r &= H^r\delta^0 = H^r\sum_{i=1}^{N} c_i\overline{x_i} = H^{r-1}\sum_{i=1}^{N} c_iH\overline{x_i} = H^{r-1}\sum_{i=1}^{N} c_i\eta_i\overline{x_i} = H^{r-2}\sum_{i=1}^{N} c_i\eta_iH\overline{x_i} \\
&= H^{r-2}\sum_{i=1}^{N} c_i\eta_i^2\overline{x_i} = \ldots\ldots H^0\sum_{i=1}^{N} c_i\eta_i^{r-1}\overline{x_i} = \sum_{i=1}^{N} c_i\eta_i^r\overline{x_i}, \\
&= c_1\eta_1^r\overline{x_1} + c_2\eta_2^r\overline{x_2} + \ldots\ldots + c_N\eta_N^r\overline{x_N} \\
&= \eta_N^r\left[c_1\left(\frac{\eta_1}{\eta_N}\right)^r\overline{x_1} + c_2\left(\frac{\eta_2}{\eta_N}\right)^r\overline{x_2} + \ldots\ldots + c_{N-1}\left(\frac{\eta_{N-1}}{\eta_N}\right)^r\overline{x_{N-1}} + c_N\overline{x_N}\right] \\
&= \eta_N^r\left[c_N\overline{x_N} + O\left(\frac{\eta_{N-1}}{\eta_N}\right)^r\right].
\end{aligned} \tag{4.80}$$

Equation (4.80) shows that the vector δ^r is parallel to the vector $\overline{x_N}$ apart from an error of order of magnitude $(\eta_{N-1}/\eta_N)^r$ which becomes smaller and smaller, as r increases. The limit, as $r \to \infty$, of the norms of two consecutive values of the vector δ is:

$$\lim_{r\to\infty}\frac{\|\delta^{r+1}\|}{\|\delta^r\|} = \lim_{r\to\infty}\frac{\|\eta_N^{r+1}c_N\overline{x_N}\|}{\|\eta_N^r c_N\overline{x_N}\|} = \eta_N. \tag{4.81}$$

If $|\eta_N| < 1$ the iteration scheme (4.72) converges.

4.4.1.3 Matrix Method for the Diffusion Equation

We recall the diffusion Eq. (4.32), partition the definition domain $(0 \leq x \leq 1)$ in N intervals of length $\Delta x = 1/N$ and set $x_0 = 0$, $x_N = N\Delta x = 1$. Discretization with the FTCS (forward time, central space) scheme leads to [Cfr. Eq. (4.37)]:

$$u_j^{n+1} = su^n_{j-1} + (1 - 2s)u_j^n + su^n_{j+1}, \tag{4.82}$$

with $s = a\Delta t/\Delta x^2$.

Equation (4.82) can be written in matrix form

$$U^{n+1} = AU^n,, \tag{4.83}$$

or:

$$\begin{vmatrix} u_1^{n+1} \\ u_2^{n+1} \\ \cdots\cdots \\ u_{N-1}^{n+1} \end{vmatrix} = \begin{vmatrix} 1-2s & s & & & \\ s & 1-2s & s & & \\ & & \cdots\cdots & & \\ & & & s & 1-2s \end{vmatrix} \begin{vmatrix} u_1^n \\ u_2^n \\ \cdots\cdots \\ u_{N-1}^n \end{vmatrix}. \tag{4.84}$$

We recall from the matrix analysis that the eigenvalues of a tri-diagonal matrix of rank N:

$$\begin{vmatrix} a & b & & & & \\ c & a & b & & & \\ & c & a & b & & \\ & & & \cdots\cdots & & \\ & & & c & a & b \\ & & & & c & a \end{vmatrix}$$

are given by:

$$\lambda_i = a + 2b\sqrt{\frac{c}{b}}\cos\left(\frac{i\pi}{N+1}\right) \quad (i = 1, 2\ldots\ldots N). \tag{4.85}$$

Therefore the eigenvalues of the matrix A in Eq. (4.84) are:

$$\begin{aligned} \lambda_i &= 1 - 2s + 2s\cos\left(\frac{i\pi}{N}\right) = 1 - 2s\left[1 - \cos\left(\frac{i\pi}{N}\right)\right] \\ &= 1 - 4s\sin^2\left(\frac{i\pi}{2N}\right), \quad (i = 1, 2\ldots\ldots N). \end{aligned} \tag{4.86}$$

having used the identity $1 - \cos\alpha = 2\sin^2(\alpha/2)$.

The stability condition for the numerical scheme (4.82) is therefore:

$$\left|1 - 4s\sin^2\left(\frac{i\pi}{2N}\right)\right| \le 1, \tag{4.87}$$

or:

$$-1 \leq 1 - 4s\sin^2\left(\frac{i\pi}{2N}\right) \leq 1. \tag{4.88}$$

Disequation (4.88) implies $-4s\sin^2(i\pi/2N) \leq 0$, which is always satisfied, and:

$$\begin{aligned} -2 &\leq -4s\sin^2(i\pi/2N), \\ 1 &\geq 2s\sin^2(i\pi/2N). \end{aligned} \tag{4.89}$$

Because the maximum value of $\sin^2(i\pi/2N)$ is 1, Eq. (4.89) is satisfied when:

$$s = \frac{a\Delta t}{\Delta x^2} \leq \frac{1}{2}, \tag{4.90}$$

hence:

$$\Delta t \leq \frac{\Delta x^2}{2a}. \tag{4.91}$$

The ratio $\tau = \Delta x^2/a$ is referred to a "diffusion time" and represents the time that the physical diffusion process considered takes to travel through a mesh of length Δx. Hence the stability condition requires that the time step be smaller than half of the diffusion time.

Remark One obtains the same result $s \leq 1/2$ by considering the row-norm of the matrix A, defined by:

$$\begin{aligned} \|A\| &= \max{}_i \sum_{j=1}^{N} |a_{ij}| = |s| + |1 - 2s| + |s| = s + |1 - 2s| + s \\ &= 2s + |1 - 2s|, \end{aligned} \tag{4.92}$$

because $s > 0$.

We recall that every eigenvalue of a matrix is in modulus smaller or equal to every norm of the matrix. Therefore $|\lambda_i| \leq 1$, if $\|A\| \leq 1$, or:

$$\begin{aligned} &2s + |1 - 2s| \leq 1, \\ &|1 - 2s| \leq 1 - 2s, \\ &-(1 - 2s) \leq 1 - 2s \leq 1 - 2s. \end{aligned} \tag{4.93}$$

The disequation at the right-hand side is obviously satisfied. That at the left-hand side yields:

$$\begin{aligned} -1+2s &\leq 1-2s, \\ -2 &\leq -4s, \\ 1 &\geq 2s, \end{aligned} \tag{4.94}$$

which coincides with Eq. (4.90).

4.4.2 Fourier Stability Criterion After von Neumann

The Fourier stability criterion was first applied in 1944 by von Neumann to analyze the stability of the fluid dynamic equations describing shock waves. The method consists in assuming that the numerical solution is affected by errors, distributed in the full definition domain ($0 \leq x \leq L$), which can be represented by a Fourier series with M terms:

$$e(x) = \sum_{p=1}^{M} A_p^n e^{ip\frac{\pi x}{L}}, \tag{4.95}$$

where $i = \sqrt{-1}$ is the imaginary unit and A_p^n represents the amplitude of the error component p at time level n. Discretizing the definition domain in N + 1 nodes $(0, 1, 2.\ldots..N)$, the error distribution in the node j, for $x = x_j = j\Delta x$, ($\Delta x = L/N$), is

$$e(x_j) = \sum_{p=1}^{M} A_p^n e^{ip\frac{\pi x_j}{L}}. \tag{4.96}$$

We define the wave number of the component p by:

$$k_p = \frac{p\pi}{L} = \frac{p\pi}{N\Delta x}, \tag{4.97}$$

and the error $e(x_j)$ can be written as:

$$e(x_j) = \sum_{p=1}^{M} A_p^n e^{ik_p x_j}. \tag{4.98}$$

Because we consider linear equations, for which linear combinations of solutions are still a solution, we need not consider all p component of Eq. (4.98) but can restrict ourselves to analyzing the behavior of only one component, say $A^n e^{ikx_j} = A^n e^{ikj}$, using the short notation j for x_j.

We apply this technique to the one-dimensional heat diffusion Eq. (4.32) which we rewrite:

$$\frac{\partial u}{\partial t} = a\frac{\partial^2 u}{\partial x^2}.\ (a \geq 0),\ (a \geq 0),\ (0 \leq x \leq L) \tag{4.99}$$

4.4.2.1 Explicit and Semi-Implicit Schemes

We discretize the time-dependent term in Eq. (4.99) with forward differences and the space-dependent second derivative at two time levels, by means of the time-discretization parameter $\theta(0 \leq \theta \leq 1)$, obtaining:

$$\frac{1}{\Delta t}\left(u_j^{n+1} - u_j^n\right) = \frac{a}{\Delta x^2} \times \left[\theta\left(u_{j-1}^{n+1} - 2u_j^{n+1} + u_{j+1}^{n+1}\right) + (1-\theta)\left(u_{j-1}^n - 2u_j^n + u_{j+1}^n\right)\right]. \tag{4.100}$$

If $\theta = 1/2$ the scheme (4.100) is referred to as Crank-Nicolson scheme; if $\theta = 0$ it is referred to as Euler scheme.

Introducing into Eq. (4.100) the error component $A^n e^{ikj}$ yields:

$$\frac{1}{\Delta t}\left(A^{n+1} - A^n\right)e^{ikj} = \frac{a}{\Delta x^2}\begin{bmatrix} \theta A^{n+1}\left(e^{ik(j-1)} - 2e^{ikj} + e^{ik(j+1)}\right) \\ +(1-\theta)A^n\left(e^{ik(j-1)} - 2e^{ikj} + e^{ik(j+1)}\right)\end{bmatrix}. \tag{4.101}$$

Eliminating from Eq. (4.101) the common factor e^{ikj} yields:

$$\frac{1}{\Delta t}\left(A^{n+1} - A^n\right) = \frac{a}{\Delta x^2}\left[\theta A^{n+1}\left(e^{-ik} - 2 + e^{ik}\right) + (1-\theta)A^n\left(e^{-ik} - 2 + e^{ik}\right)\right]. \tag{4.102}$$

Using the identity $e^{ix} + e^{-ix} = 2\cos x$ yields:

$$\frac{1}{\Delta t}\left(A^{n+1} - A^n\right) = \frac{a}{\Delta x^2}\left[\theta A^{n+1}(2\cos k - 2) + (1-\theta)A^n(2\cos k - 2)\right]. \tag{4.103}$$

Using the identity $\cos x - 1 = -2\sin^2(x/2)$ and multiplying both members by Δt yields:

$$A^{n+1} - A^n = -\frac{4a\Delta t}{\Delta x^2}\left[\theta A^{n+1}\sin^2(k/2) + (1-\theta)A^n\sin^2(k/2)\right]. \tag{4.104}$$

Solving Eq. (4.104) with respect to A^{n+1}, and using $s = a\Delta t/\Delta x^2$, yields:

$$A^{n+1} = \frac{1 - 4s(1-\theta)\sin^2(k/2)}{1 + 4s\theta\sin^2(k/2)} A^n = GA^n, \tag{4.105}$$

with the definition of the amplification factor:

$$G = \frac{1 - 4s(1-\theta)\sin^2(k/2)}{1 + 4s\theta\sin^2(k/2)}. \tag{4.106}$$

In the course of the numerical solution of Eq. (4.100), as n increases, the errors remain limited or vanish if $|G| \leq 1$.

For $\theta = 0$ (explicit Euler scheme), Eq. (4.106) yields $G = 1 - 4s\sin^2(k/2)$ and the stability condition becomes:

$$\begin{aligned} &\left|1 - 4s\sin^2(k/2)\right| \leq 1, \\ &-1 \leq 1 - 4s\sin^2(k/2) \leq 1, \\ &-2 \leq -4s\sin^2(k/2) \leq 0. \end{aligned} \tag{4.107}$$

While the disequation at the right-hand side is always satisfied, that at the left side yields:

$$1 \geq 2s\sin^2(k/2). \tag{4.108}$$

Because the maximum value of $\sin^2(k/2)$ is 1, Eq. (4.108) implies $s \leq 1/2$, which coincides with the result obtained from the matrix method [see Eq. (4.90)].

For $\theta = 1/2$ (Crank-Nicolson scheme), the amplification factor becomes:

$$G = \frac{1 - 2s\sin^2(k/2)}{1 + 2s\sin^2(k/2)}, \tag{4.109}$$

and because $|G| \leq 1$ for all positive s, the scheme is unconditionally stable.

4.4.2.2 Leapfrog Scheme (Richardson's Method)

In the leapfrog scheme, the time derivative of the diffusion Eq. (4.99) is discretized with central differences:

$$\frac{1}{2\Delta t}\left(u_j^{n+1} - u_j^{n-1}\right) = \frac{a}{\Delta x^2}\left(u_{j-1}^n - 2u_j^n + u_{j+1}^n\right). \tag{4.110}$$

The leapfrog scheme differs from the Euler scheme [Cfr. Eq. (4.35)] because the time derivative is discretized with reference to two time intervals ($t_{n-1} - t_{n+1}$).

Introducing the harmonic $A^n e^{ikj}$ yields:

$$\frac{1}{2\Delta t}\left(A^{n+1} - A^{n-1}\right)e^{ikj} = \frac{a}{\Delta x^2}A^n\left[e^{ik(j+1)} - 2e^{ikj} + e^{ik(j-1)}\right]. \tag{4.111}$$

Multiplying by $2\Delta t/e^{ikj}$ and using $s = a\Delta t/\Delta x^2$ yields:

$$A^{n+1} = A^{n-1} + 2sA^n\left(e^{ik} - 2 + e^{-ik}\right). \tag{4.112}$$

Using the identity $(e^{ik} - 2 + e^{-ik}) = 2(\cos k - 1) = -2[2\sin^2(k/2)] = -4\sin^2(k/2)$ yields:

$$A^{n+1} = A^{n-1} - 8sA^n \sin^2(k/2). \tag{4.113}$$

Defining the amplification factor $G = A^{n+1}/A^n$ and using $A^{n-1}/A^n = 1/G$, one obtains:

$$\frac{A^{n+1}}{A^n} = G = \frac{1}{G} - 8s\sin^2(k/2), \tag{4.114}$$

hence:

$$G^2 + G8s\sin^2(k/2) - 1 = 0. \tag{4.115}$$

Letting $\alpha = 4s\sin^2(k/2)$, Eq. (4.115) has the two roots:

$$G_{1,2} = -\alpha \pm \sqrt{\alpha^2 + 1} = -4s\sin^2(k/2) \pm \sqrt{16s^2\sin^4(k/2) + 1}. \tag{4.116}$$

Because α is real and positive, at least one of the two roots is smaller than -1; therefore the stability condition $|G| \leq 1$ cannot be satisfied and the scheme (4.110) is unstable.

4.4.2.3 Dufort-Frankel Scheme

The Dufort-Frankel scheme is obtained from the leapfrog scheme (4.110) by replacing at the right-hand side the term $2u_j^n$ with $\left(u_j^{n+1} + u_j^{n-1}\right)$:

$$\frac{1}{2\Delta t}\left(u_j^{n+1} - u_j^{n-1}\right) = \frac{a}{\Delta x^2}\left[u^n_{j-1} - 2\left(u_j^{n+1} + u_j^{n-1}\right) + u^n_{j+1}\right]. \tag{4.117}$$

Multiplying by Δt and using $s = a\Delta t/\Delta x^2$ yield:

$$u_j^{n+1}(1+2s) = (1-2s)u_j^{n-1} + 2s\left(u^n_{j+1} + u^n_{j-1}\right), \tag{4.118}$$

hence:

$$u_j^{n+1} = \frac{1-2s}{1+2s}u_j^{n-1} + \frac{2s}{1+2s}\left(u^n_{j+1} + u^n_{j-1}\right). \tag{4.119}$$

Introducing the harmonic $A^n e^{ikj}$ yields:

$$A^{n+1}e^{ikj} = \frac{1-2s}{1+2s}A^{n-1}e^{ikj} + \frac{2s}{1+2s}A^n\left[e^{ik(j+1)} + e^{ik(j-1)}\right]. \tag{4.120}$$

Dividing by e^{ikj} one obtains:

$$\begin{aligned} A^{n+1} &= \frac{1-2s}{1+2s}A^{n-1} + \frac{2s}{1+2s}A^n\left(e^{ik} + e^{-ik}\right) \\ &= \frac{1-2s}{1+2s}A^{n-1} + \frac{2s}{1+2s}A^n 2\cos k. \end{aligned} \tag{4.121}$$

Defining the amplification factor $G = A^{n+1}/A^n$ and using $A^{n-1}/A^n = 1/G$, one obtains:

$$\frac{A^{n+1}}{A^n} = G = \frac{1-2s}{1+2s}\frac{1}{G} + \frac{4s}{1+2s}\cos k, \tag{4.122}$$

hence:

$$G^2 - \frac{4s}{1+2s}G\cos k - \frac{1-2s}{1+2s} = 0. \tag{4.123}$$

Equation (4.123) has the two roots:

$$\begin{aligned} G_{1,2} &= \frac{2s}{1+2s}\cos k \pm \sqrt{\frac{4s^2}{(1+2s)^2}\cos^2 k + \frac{1-2s}{1+2s}} \\ &= \frac{1}{1+2s}\left[2s\cos k \pm \sqrt{4s^2\cos^2 k + 1 - 4s^2}\right] \\ &= \frac{1}{1+2s}\left[2s\cos k \pm \sqrt{1 - 4s^2\sin^2 k}\right]. \end{aligned} \tag{4.124}$$

Case (i) If $1 - 4s^2\sin^2 k \geq 0$, then both roots G_1 and G_2 are the sum of two real numbers. Because $|2s\cos k| \leq 2s$ and $\left|\sqrt{1-4s^2\sin^2 k}\right| \leq 1$, we can write:

$$
\begin{aligned}
|G_{1,2}| &\leq \frac{1}{1+2s}\left|2s\cos k \pm \sqrt{1-4s^2\sin^2 k}\right| \\
&\leq \frac{1}{1+2s}\left[|2s\cos k| + \left|\sqrt{1-4s^2\sin^2 k}\right|\right] \leq \frac{2s+1}{1+2s} = 1. \qquad (4.125)
\end{aligned}
$$

Case (ii) If $1 - 4s^2\sin^2 k < 0$, then both roots G_1 and G_2 are complex and given by:

$$
G_{1,2} = \frac{1}{1+2s}\left[2s\cos k \pm i\sqrt{4s^2\sin^2 k - 1}\right]. \qquad (4.126)
$$

Hence the square of the modulus of the roots is:

$$
\begin{aligned}
|G_{1,2}|^2 &= \frac{1}{(1+2s)^2}\left(4s^2\cos^2 k + 4s^2\sin^2 k - 1\right) = \frac{4s^2-1}{(1+2s)^2} \\
&= \frac{4s^2-1}{4s^2+4s+1} < 1. \qquad (4.127)
\end{aligned}
$$

Therefore the Dufort-Frankel scheme is always stable.

4.4.2.4 Implicit Scheme

The implicit scheme for the diffusion Eq. (4.32) is given by:

$$
\frac{1}{\Delta t}\left(u_j^{n+1} - u_j^n\right) = \frac{a}{\Delta x^2}\left(u_{j-1}^{n+1} - 2u_j^{n+1} + u_{j+1}^{n+1}\right). \qquad (4.128)
$$

Using $s = a\Delta t/\Delta x^2$ and rearranging, one obtains:

$$
-su_{j+1}^{n+1} + (1+2s)u_j^{n+1} - su_{j-1}^{n+1} = u_j^n. \qquad (4.129)
$$

The matrix of coefficients of the system of Eq. (4.129) ($j = 1, 2. \ldots. .N - 1$) is tri-diagonal, and therefore Eq. (4.129) can be solved directly by means of the Thomas algorithm (see Sect. 1.3).

Introducing the harmonic $A^n e^{ikj}$ yields:

$$
-sA^{n+1}e^{ik(j+1)} + (1+2s)A^{n+1}e^{ikj} - sA^{n+1}e^{ik(j-1)} = A^n e^{ikj}. \qquad (4.130)
$$

Dividing by e^{ikj} one obtains:

$$
A^{n+1}\left[-se^{ik} + (1+2s) - se^{-ik}\right] = A^n. \qquad (4.131)
$$

Using the identity $2\cos k = e^{ik} + e^{-ik}$ and introducing the amplification factor $G = A^{n+1}/A^n$ yield:

$$\frac{A^{n+1}}{A^n} = G = \frac{1}{1 + 2s - 2s\cos k} = \frac{1}{1 + 2s(1 - \cos k)}. \tag{4.132}$$

Because $|G| < 1$ for every s, the implicit scheme is unconditionally stable. Equation (4.132) is equal to Eq. (4.106) for $\vartheta = 1$.

4.4.3 Hirt's Stability Criterion (For Diffusion Equation)

The heuristic stability criterion introduced by Hirt consists of the following three steps (Noye, 1978):

1. The given differential equation is discretized and terms at time level $n + 1$ and at space points $j \pm 1$ are expanded by means of Taylor series.
2. From the discretized equation obtained following the procedure 1), a new continuum differential equation is derived.
3. The stability condition is obtained from the condition that the new differential equation is (in some sense) compatible with the original differential equation.

We show the application of the Hirt's method to the diffusion equation:

$$\frac{\partial u}{\partial t} = a\frac{\partial^2 u}{\partial x^2}. \quad (0 \le x \le 1), (a > 0), (t > 0) \tag{4.133}$$

Step 1: Discretization of the differential equation and expansion in Taylor series
Discretizing Eq. (4.133) with the FTCS (forward time, central space) scheme yields [Cfr. Eq. (4.35)]:

$$\frac{1}{\Delta t}\left(u_j^{n+1} - u_j^n\right) - \frac{a}{\Delta x^2}\left(u^n_{j+1} - 2u_j^n + u^n_{j-1}\right) = 0. \tag{4.134}$$

Using Taylor series expansions:

$$u_j^{n+1} = u_j^n + \Delta t\left(\frac{\partial u}{\partial t}\right)_j^n + \frac{\Delta t^2}{2}\left(\frac{\partial^2 u}{\partial t^2}\right)_j^n + \ldots\ldots, \tag{4.135a}$$

$$u^n_{j+1} = u_j^n + \Delta x\left(\frac{\partial u}{\partial x}\right)_j^n + \frac{\Delta x^2}{2}\left(\frac{\partial^2 u}{\partial x^2}\right)_j^n + \ldots\ldots, \tag{4.135b}$$

$$u^n_{j-1} = u_j^n - \Delta x\left(\frac{\partial u}{\partial x}\right)_j^n + \frac{\Delta x^2}{2}\left(\frac{\partial^2 u}{\partial x^2}\right)_j^n + \ldots\ldots, \tag{4.135c}$$

and introducing in Eq. (4.134), one obtains:

$$\frac{1}{\Delta t}\left[\Delta t\left(\frac{\partial u}{\partial t}\right)_j^n+\frac{\Delta t^2}{2}\left(\frac{\partial^2 u}{\partial t^2}\right)_j^n\right]-\frac{a}{\Delta x^2}\left[\Delta x^2\left(\frac{\partial^2 u}{\partial x^2}\right)_j^n\right]=0. \tag{4.136}$$

Step 2: Obtain a new differential equation

From Eq. (4.136) we derive the following new partial differential equation:

$$\frac{\partial u}{\partial t}+\frac{\Delta t}{2}\frac{\partial^2 u}{\partial t^2}-a\frac{\partial^2 u}{\partial x^2}=0. \tag{4.137}$$

In Eq. (4.137) there is one term more than in the original Eq. (4.133). According to the classification of second-order partial differential equations given in Chap. 2 [Cfr. Eq. (2.50)], the coefficients of the second-order derivatives in Eq. (4.137) are $<a> = -a, <b> = 0, <c> = \Delta t/2$; therefore $\Delta = <b>^2 - 4<a><c> = 4a\Delta t/2 > 0$, and the equation is of hyperbolic type.

The equation of the characteristics of Eq. (4.137) is [Cfr. (2.65)]:

$$\left[<a>\left(\frac{dy}{dx}\right)^2-<b>\left(\frac{dy}{dx}\right)+<c>\right]=-a\left(\frac{dt}{dx}\right)^2+\frac{\Delta t}{2}=0. \tag{4.138}$$

Equation (4.138) has the two solutions:

$$\frac{dt}{dx}=\pm\sqrt{\frac{\Delta t}{2a}}, \tag{4.139}$$

which are two straight lines originating from a point $P(x, t)$ in the plane (t, x), as shown in Fig. 4.4:

The shaded area in the plane (x, t) shows the domain of influence of $P(x, t)$. This is the subdomain in which the solution is affected by the value of the unknown in point P.

Step 3: Physical constraint imposed on the new differential equation

From the discretized Eq. (4.134) it is clear that the three points $(j-1, n)$, (j, n), $(j+1, n)$ in the discretized (t, x) plane yield the solution at point $(j, n+1)$. The

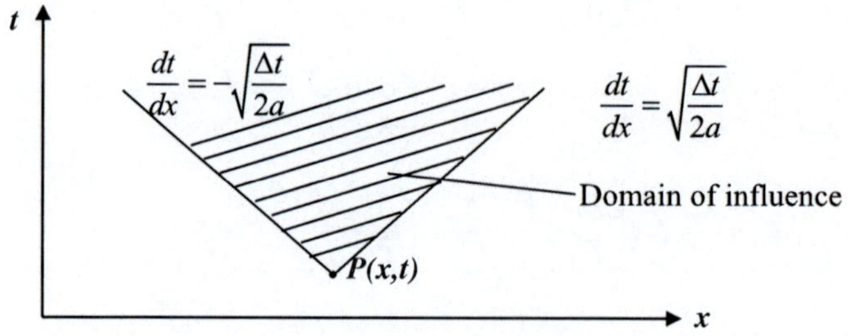

Fig. 4.4 Domain of influence of the differential Eq. (4.133)

domain of influence in the discretized (t, x) plane is within the characteristics of slope $\pm\Delta t/\Delta x$. An error made numerically at point (j, n) propagates to the points $(j - 1, n + 1)$, $(j, n + 1)$, $(j + 1, n + 1)$; therefore it propagates along the vertical line and along the characteristic lines. The errors propagating along the characteristic lines should remain outside of the domain of influence of the differential equation. This implies that the slope of the characteristic lines $\Delta t/\Delta x$ must be smaller than the slope of the characteristic lines dt/dx in Fig. 4.4. Therefore the stability requires that:

$$\frac{\Delta t}{\Delta x} \leq \sqrt{\frac{\Delta t}{2a}}, \tag{4.140}$$

or

$$\Delta t \leq \frac{\Delta x^2}{2a}, \tag{4.141}$$

which coincides with the results obtained with other methods of stability analysis [Cfr. Eq. (4.90) and Eq. (4.94)].

Table 4.1 summarizes the main discretization schemes and stability conditions discussed for the model diffusion equation.

4.5 The Convection Equation

In this section we analyze the stability of the convection equation

$$\frac{\partial u}{\partial t} + V\frac{\partial u}{\partial x} = 0, \tag{4.142}$$

where $u = u(x, t)$ is defined in a domain $(0 \leq x \leq L)$ and represents a physical quantity (temperature, concentration of transported particles, etc.) transported with velocity V. At $t = 0$ the initial condition is given as $u(x, 0) = F(x)$. This leads to a general solution [Cr. Eq. (2.23) with $\lambda = 0$] of the type:

$$u(x, t) = F(x - Vt). \tag{4.143}$$

Equation (4.143) implies that the initial distribution $u(x, 0) = F(x)$ is translated along the x-axis at the constant speed V, without changing amplitude and shape.

In the following subsections, we examine the numerical solution of Eq. (4.142) with the convective term discretized with central differences and with unwind differences.

Table 4.1 Summary of discretization schemes and stability conditions for the diffusion equation $\frac{\partial u}{\partial t} = a\frac{\partial^2 u}{\partial x^2}$

(i) Forward time central space (FTCS) difference (Euler scheme)

$\frac{1}{\Delta t}\left(u_j^{n+1} - u_j^n\right) = \frac{a}{\Delta x^2}\left(u_{j+1}^n - 2u_j^n + u_{j-1}^n\right)$. Eq. (4.35)

Amplification factor $G = 1 - 4s\sin^2(k/2)$. From Eq. (4.106) for $\theta = 0$

Stability condition: $s = \frac{a\Delta t}{\Delta x^2} \leq \frac{1}{2}$.

(ii) Semi-implicit scheme $(0 \leq \theta \leq 1)$ **[if $\theta = 1/2$, Crank-Nicolson scheme]**

$\frac{1}{\Delta t}\left(u_j^{n+1} - u_j^n\right) = \frac{a}{\Delta x^2}\left[\theta\left(u_{j-1}^{n+1} - 2u_j^{n+1} + u_{j+1}^{n+1}\right) + (1-\theta)\left(u_{j-1}^n - 2u_j^n + u_{j+1}^n\right)\right]$. Eq. (4.100)

Amplification factor $G = \frac{1-4s(1-\theta)\sin^2(k/2)}{1+4s\theta\sin^2(k/2)}$. Eq. (4.106)

For $\theta = 1/2$, $G = \frac{1-2s\sin^2(k/2)}{1+2s\sin^2(k/2)}$. Unconditionally stable. Eq. (4.109)

(iii) Hirt's stability criterion (using FTCS scheme)

Stability condition: $\Delta t \leq \frac{\Delta x^2}{2a}$. Eq. (4.141)

(iv) Leapfrog scheme (Richardson's method)

$\frac{1}{2\Delta t}\left(u_j^{n+1} - u_j^{n-1}\right) = \frac{a}{\Delta x^2}\left(u_{j-1}^n - 2u_j^n + u_{j+1}^n\right)$. Eq. (4.110)

Amplification factor:

$G_{1,2} = -\alpha \pm \sqrt{\alpha^2 + 1} = -4s\sin^2(k/2) \pm \sqrt{16s^2\sin^4(k/2) + 1}$. Eq. (4.116)

($\alpha = 4s\sin^2(k/2)$). Unstable.

(v) Dufort-Frankel scheme

$\frac{1}{2\Delta t}\left(u_j^{n+1} - u_j^{n-1}\right) = \frac{a}{\Delta x^2}\left[u_{j-1}^n - 2\left(u_j^{n+1} + u_j^{n-1}\right) + u_{j+1}^n\right]$. Eq. (4.117)

Amplification factor $G = \frac{1}{1+2s}\left[2s\cos k \pm \sqrt{1 - 4s^2\sin^2 k}\right]$ Eq. (4.124)

Unconditionally stable.

(vi) Implicit scheme

$\frac{1}{\Delta t}\left(u_j^{n+1} - u_j^n\right) = \frac{a}{\Delta x^2}\left(u_{j-1}^{n+1} - 2u_j^{n+1} + u_{j+1}^{n+1}\right)$. Eq. (4.128)

Amplification factor $G = \frac{1}{1+2s(1-\cos k)}$. Eq. (4.132)

Unconditionally stable.

4.5.1 *Central Differences for the Convective Term*

Discretization of Eq. (4.142) with central differences for the convective term yields:

$$\frac{1}{\Delta t}\left(u_j^{n+1} - u_j^n\right) = -\frac{V}{2\Delta x}\left(u_{j+1}^n - u_{j-1}^n\right). \tag{4.144}$$

Letting the Courant number C be defined by:

$$C = \frac{V\Delta t}{\Delta x}, \tag{4.145}$$

one obtains:

$$u_j^{n+1} = u_j^n - \frac{C}{2}\left(u_{j+1}^n - u_{j-1}^n\right). \tag{4.146}$$

Using the Fourier method by introducing in Eq. (4.144) the harmonic $A^n e^{ikx_j} = A^n e^{ikj}$ yields:

$$A^{n+1}e^{ikj} = A^n e^{ikj} - \frac{C}{2}A^n\left[e^{ik(j+1)} - e^{ik(j-1)}\right]. \tag{4.147}$$

Dividing by e^{ikj} and using the identity $e^{ik} - e^{-ik} = 2i \sin k$, one obtains:

$$A^{n+1} = A^n(1 - iC\sin k) = A^n G, \tag{4.148}$$

with the definition of the amplification factor:

$$G = (1 - iC\sin k). \tag{4.149}$$

Because $|G|^2 = 1 + C^2\sin^2 k$ is always ≥ 1, the stability condition $|G|^2 \leq 1$ is never satisfied and therefore the discretization scheme (4.144) is unconditionally unstable. For this reason one generally makes recourse to upwind differences for the discretization of convective terms, as shown in the next subsection.

4.5.2 *Upwind Differences for the Convective Term*

Equation (4.142) is discretized with upwind differences for the convective term by:

$$\frac{1}{\Delta t}\left(u_j^{n+1} - u_j^n\right) = -\frac{V}{\Delta x}\left(u_j^n - u_{j-1}^n\right) \quad (V \geq 0), \tag{4.150a}$$

$$\frac{1}{\Delta t}\left(u_j^{n+1} - u_j^n\right) = -\frac{V}{\Delta x}\left(u_{j+1}^n - u_j^n\right) \quad (V < 0). \tag{4.150b}$$

Let us refer to the equation for $V > 0$. With the definition (4.145) of the Courant number, one obtains:

$$u_j^{n+1} = (1 - C)u_j^n + Cu_{j-1}^n. \tag{4.151}$$

Introducing the harmonic $A^n e^{ikj}$ yields:

$$A^{n+1}e^{ikj} = A^n e^{ikj} - CA^n\left[e^{ikj} - e^{ik(j-1)}\right]. \tag{4.152}$$

Dividing by e^{ikj} one obtains:

$$
\begin{aligned}
A^{n+1} &= A^n\left[1 - C\left(1 - e^{ik}\right)\right] = A^n[1 - C + C(\cos k - i\sin k)] \\
&= A^n(1 - C + C\cos k - iC\sin k) = A^n G,
\end{aligned}
\tag{4.153}
$$

with the definition of the amplification factor:

$$
G = 1 - C + C\cos k - iC\sin k. \tag{4.154}
$$

The square of the modulus of the amplification factor is:

$$
\begin{aligned}
|G|^2 &= (1 - C + C\cos k)^2 + C^2\sin^2 k \\
&= (1 - C)^2 + 2(1 - C)C\cos k + C^2\cos^2 k + C^2\sin^2 k \\
&= 1 - 2C + 2C^2 + 2(1 - C)C\cos k \\
&= 1 - 2C(1 - C) + 2C(1 - C)\cos k \\
&= 1 - 2C(1 - C)(1 - \cos k).
\end{aligned}
\tag{4.155}
$$

If $C > 1$, $1 - C < 0$ and $|G|^2 \geq 1$; hence the method is unstable. If $C < 1$, $|G|^2$ has a maximum value of one when $\cos k = 1$, and therefore the scheme (4.148) is stable. The stability condition can therefore be written as:

$$
C = \frac{V\Delta t}{\Delta x} \leq 1, \tag{4.156}
$$

which is the very well-known Courant-Friedrichs-Lewy (CFL) condition. This condition generally applies to explicit finite difference schemes approximating differential equations of hyperbolic type. It implies that:

$$
V\Delta t \leq \Delta x, \tag{4.157}
$$

and therefore that the fluid, with velocity V, should not cross one mesh of width Δx during one time step of length Δt.

4.5.3 The Problem of Numerical Diffusion

The drawback of the upwind difference scheme is that it introduces into the difference equation a truncation error which is proportional to a second-order derivative of the dependent variable and therefore is generally referred to as numerical diffusion.

The concept of numerical diffusion is introduced as follows. After discretizing the original differential equation and obtaining a system of difference equations, one makes a consistency check to evaluate how closely the difference equation approximates the differential equation. The consistency is checked by expanding in a

Taylor series every term of the difference equation about the central node for which the equation is written. Summing up the contributions of every term in the difference equation, we obtain the discrete equation equivalent to the original differential equation.

In the case of upwind differences, the contributions of the second derivatives in the space Taylor expansions of the convective term do not cancel each other, as they do with central differences, and therefore represent an undesired discrepancy between the difference equation and the differential equation. The coefficient which multiplies the second-order derivatives is interpreted physically as a false diffusion coefficient.

We recall Eq. (4.151):

$$u_j^{n+1} = (1 - C)u_j^n + Cu^n_{j-1}, \tag{4.158}$$

and use the Taylor series expansions:

$$u_j^{n+1} = u_j^n + \Delta t\left(\frac{\partial u}{\partial t}\right)_j^n + \frac{\Delta t^2}{2}\left(\frac{\partial^2 u}{\partial t^2}\right)_j^n + O(\Delta t^3), \tag{4.159}$$

$$u^n_{j-1} = u_j^n - \Delta x\left(\frac{\partial u}{\partial x}\right)_j^n + \frac{\Delta x^2}{2}\left(\frac{\partial^2 u}{\partial x^2}\right)_j^n + O(\Delta x^3), \tag{4.160}$$

obtaining:

$$\begin{aligned} u_j^n + \Delta t\left(\frac{\partial u}{\partial t}\right)_j^n + \frac{\Delta t^2}{2}\left(\frac{\partial^2 u}{\partial t^2}\right)_j^n &= (1 - C)u_j^n \\ &+ C\left[u_j^n - \Delta x\left(\frac{\partial u}{\partial x}\right)_j^n + \frac{\Delta x^2}{2}\left(\frac{\partial^2 u}{\partial x^2}\right)_j^n\right]. \end{aligned} \tag{4.161}$$

Simplifying and dividing by Δt yield:

$$\left(\frac{\partial u}{\partial t}\right)_j^n + \frac{\Delta t}{2}\left(\frac{\partial^2 u}{\partial t^2}\right)_j^n = \frac{C}{\Delta t}\left[-\Delta x\left(\frac{\partial u}{\partial x}\right)_j^n + \frac{\Delta x^2}{2}\left(\frac{\partial^2 u}{\partial x^2}\right)_j^n\right]. \tag{4.162}$$

Using $C\Delta x/\Delta t = V$ yields:

$$\left(\frac{\partial u}{\partial t}\right)_j^n + V\left(\frac{\partial u}{\partial x}\right)_j^n = \frac{1}{2}\left[V\Delta x\left(\frac{\partial^2 u}{\partial x^2}\right)_j^n - \Delta t\left(\frac{\partial^2 u}{\partial t^2}\right)_j^n\right]. \tag{4.163}$$

The second-order time derivative can be eliminated by deriving Eq. (4.142) with respect to time:

$$\frac{\partial^2 u}{\partial t^2} = -V\frac{\partial}{\partial t}\left(\frac{\partial u}{\partial x}\right) = -V\frac{\partial}{\partial x}\left(\frac{\partial u}{\partial t}\right) = -V\frac{\partial}{\partial x}\left(-V\frac{\partial u}{\partial x}\right) = V^2\frac{\partial^2 u}{\partial x^2}. \quad (4.164)$$

Introducing Eq. (4.164) in Eq. (4.163) yields:

$$\begin{aligned}\left(\frac{\partial u}{\partial t}\right)_j^n + V\left(\frac{\partial u}{\partial x}\right)_j^n &= \frac{1}{2}(V\Delta x - V^2\Delta t)\left(\frac{\partial^2 u}{\partial x^2}\right)_j^n \\ &= \frac{1}{2}V\Delta x\left(1 - \frac{V\Delta t}{\Delta x}\right)\left(\frac{\partial^2 u}{\partial x^2}\right)_j^n \\ &= \frac{1}{2}V\Delta x(1 - C)\left(\frac{\partial^2 u}{\partial x^2}\right)_j^n = \alpha\left(\frac{\partial^2 u}{\partial x^2}\right)_j^n, \end{aligned} \quad (4.165)$$

with the definition of the numerical (or false) diffusion coefficient:

$$\alpha = \frac{1}{2}V\Delta x(1 - C). \quad (4.166)$$

Notice that for $C \to 1$, $\alpha \to 0$, but $C = 1$ represents the stability limit for the scheme (4.150).

Several methods have been proposed for minimizing the numerical diffusion. In Sect. 4.7 we consider the QUICK scheme and explain its implementation into a major computer program for CFD (computational fluids dynamics).

4.6 The Convection-Diffusion Equation

In this section we analyze the stability of the convection-diffusion equation

$$\frac{\partial u}{\partial t} + V\frac{\partial u}{\partial x} = a\frac{\partial^2 u}{\partial x^2}, \quad (4.167)$$

by discretizing the convective term with central differences and with upwind differences. In both cases we use the Fourier analysis. Then, also using central differences, we consider Hirt's stability criterion for this case.

4.6.1 *Central Differences for the Convective Term*

Discretizing Eq. (4.167) with forward time and central space differences, we obtain:

$$\frac{1}{\Delta t}\left(u_j^{n+1}-u_j^n\right)+\frac{V}{2\Delta x}\left(u^n_{j+1}-u^n_{j-1}\right)=\frac{a}{\Delta x^2}\left(u^n_{j+1}-2u_j^n+u^n_{j-1}\right). \quad (4.168)$$

Using the definitions $C = V\Delta t/\Delta x$ and $s = a\Delta t/\Delta x^2$, one obtains:

$$u_j^{n+1}=u_j^n-\frac{C}{2}\left(u^n_{j+1}-u^n_{j-1}\right)+s\left(u^n_{j+1}-2u_j^n+u^n_{j-1}\right). \quad (4.169)$$

Introducing the harmonic $A^n e^{ikx_j} = A^n e^{ikj}$ yields:

$$\begin{aligned} A^{n+1}e^{ikj} &= A^n e^{ikj}-\frac{C}{2}A^n\left[e^{ik(j+1)}-e^{ik(j-1)}\right] \\ &\quad + sA^n\left[e^{ik(j+1)}-2e^{ikj}+e^{ik(j-1)}\right]. \end{aligned} \quad (4.170)$$

Dividing by e^{ikj} one obtains:

$$\begin{aligned} A^{n+1} &= A^n - A^n\frac{C}{2}\left(e^{ik}-e^{-ik}\right)+sA^n\left(e^{ik}-2+e^{-ik}\right) \\ &= A^n\left[1-\frac{C}{2}2i\sin k+s(2\cos k-2)\right] \\ &= A^n[1-2s(1-\cos k)-Ci\sin k]=A^nG, \end{aligned} \quad (4.171)$$

with the definition of the amplification factor:

$$G = 1-2s(1-\cos k)-Ci\sin k. \quad (4.172)$$

The square of the modulus of the amplification factor is:

$$\begin{aligned} |G|^2 &= [1-2s(1-\cos k)]^2+C^2\sin^2 k \\ &= [1-2s(1-\cos k)]^2+C^2\left(1-\cos^2 k\right) \end{aligned} \quad (4.173)$$

and the stability condition is $|G|^2 \le 1$.

For $\cos k = -1$ the stability condition is:

$$\begin{aligned} &|G|^2=(1-4s)^2<1, \\ &|1-4s|<1, \\ &-1<1-4s<1, \\ &-2<-4s<0, \end{aligned} \quad (4.174)$$

and is always satisfied for:

$$\frac{a\Delta t}{\Delta x^2} = s < \frac{1}{2}, \tag{4.175}$$

or

$$\Delta t < \frac{\Delta x^2}{2a}. \tag{4.176}$$

For $\cos k = 0$ the stability condition is:

$$|G|^2 = (1 - 2s)^2 + C^2 = 1 - 4s + 4s^2 + C^2 \leq 1. \tag{4.177}$$

hence:

$$\begin{aligned} -4s + 4s^2 + C^2 &\leq 0, \\ 4s - 4s^2 &\geq C^2, \\ s - s^2 &\geq C^2/4. \end{aligned} \tag{4.178}$$

Because $s < 1/2$, according to condition (4.175), $s^2 < 1$ and $s > s - s^2$; hence (4.178) can be replaced "a fortiori" by:

$$\begin{aligned} s &> \frac{C^2}{4}, \\ \frac{a\Delta t}{\Delta x^2} &> \frac{1}{4}\left(V\frac{\Delta t}{\Delta x}\right)^2, \end{aligned} \tag{4.179}$$

which finally yields:

$$\Delta t < \frac{4a}{V^2}. \tag{4.180}$$

For $\cos k = 1$, Eq. (4.173) yields $|G|^2 = 1$ and represents a stability limit.

If we assume $V = 1$ m/s, $\Delta x = 10^{-2}$ m, $a = 1.5 \times 10^{-5}$ m^2/s (approximately the kinematic viscosity of air at room temperature), the two stability conditions (4.176) and (4.180) are:

- From Eq. (4.176):
  $\Delta t < \frac{\Delta x^2}{2\nu} = \frac{10^{-4}}{3\times 10^{-5}} = 3.33 s$,
- From Eq. (4.180):
  $\Delta t < \frac{4\nu}{V^2} = 4 \times 1.5 \times 10^{-5} = 6 \times 10^{-5} s$.

One therefore sees that the second condition is very restrictive, and this depends on the fact that the scheme (4.168) is explicit and also central differences have been used. In the next section, we show that the use of upwind differences leads to a less restrictive stability condition.

4.6.2 Upwind Differences for the Convective Term

Assuming $V > 0$, discretizing the convective term in Eq. (4.167) with upwind difference yields:

$$\frac{1}{\Delta t}\left(u_j^{n+1} - u_j^n\right) + \frac{V}{\Delta x}\left(u_j^n - u_{j-1}^n\right) = \frac{\nu}{\Delta x^2}\left(u_{j+1}^n - 2u_j^n + u_{j-1}^n\right). \tag{4.181}$$

Using the definitions $C = V\Delta t/\Delta x$ and $s = a\Delta t/\Delta x^2$ yields:

$$\begin{aligned} u_j^{n+1} &= u_j^n - C\left(u_j^n - u_{j-1}^n\right) + s\left(u_{j+1}^n - 2u_j^n + u_{j-1}^n\right) \\ &= u_{j-1}^n(C+s) + u_j^n(1 - C - 2s) + u_{j+1}^n s. \end{aligned} \tag{4.182}$$

Introducing the harmonic $A^n e^{ikx_j} = A^n e^{ikj}$ one obtains:

$$A^{n+1}e^{ikj} = A^n\left[(C+s)e^{ik(j-1)} + (1 - C - 2s)e^{ikj} + se^{ik(j+1)}\right]. \tag{4.183}$$

Dividing by e^{ikj} and using the identities $2\cos k = e^{ik} + e^{-ik}$, $e^{-ik} = \cos k - i\sin k$, one obtains:

$$\begin{aligned} A^{n+1} &= A^n\left[(C+s)e^{-ik} + 1 - C - 2s + se^{ik}\right] \\ &= A^n[C(\cos k - i\sin k) + 2s\cos k + 1 - C - 2s] \\ &= A^n[1 - C - 2s + (C+2s)\cos k - Ci\sin k] \\ &= A^n[1 - (C+2s)(1 - \cos k) - iC\sin k] \\ &= A^n G, \end{aligned} \tag{4.184}$$

with the definition of the amplification factor:

$$G = 1 - (C+2s)(1 - \cos k) - iC\sin k. \tag{4.185}$$

The square of the modulus of the amplification factor is:

$$\begin{aligned} |G|^2 &= [1 - (C+2s)(1 - \cos k)]^2 + C^2\sin^2 k \\ &= 1 - 2(C+2s)(1 - \cos k) + (C+2s)^2(1 - \cos k)^2 + C^2\sin^2 k \\ &= 1 - 2C(1 - \cos k) - 4s(1 - \cos k) + C^2(1 - \cos k)^2 \\ &\quad +4Cs(1 - \cos k)^2 + 4s^2(1 - \cos k)^2 + C^2\sin^2 k. \end{aligned} \tag{4.186}$$

The last two terms of Eq. (4.186) yield:

$$\begin{aligned}4s^2(1-\cos k)^2+C^2\sin^2 k &= 4s^2-8s^2\cos k+4s^2\cos^2 k+C^2\sin^2 k\\ &= 4s^2-8s^2\cos k+4s^2-4s^2\sin^2 k+C^2\sin^2 k\\ &= 8s^2(1-\cos k)-4s^2\sin^2 k+C^2\sin^2 k.\end{aligned} \tag{4.187}$$

Using Eq. (4.187), Eq. (4.186) becomes:

$$\begin{aligned}|G|^2 &= 1-2C(1-\cos k)-4s(1-\cos k)+4Cs(1-\cos k)^2\\ &\quad+8s^2(1-\cos k)-4s^2\sin^2 k+C^2(1-\cos k)^2+C^2\sin^2 k.\end{aligned} \tag{4.188}$$

The last two terms in Eq. (4.188) yield:

$$\begin{aligned}C^2(1-\cos k)^2+C^2\sin^2 k &= C^2\left(1-2\cos k+\cos^2 k+\sin^2 k\right)\\ &= 2C^2(1-\cos k).\end{aligned} \tag{4.189}$$

Using Eq. (4.189), Eq. (4.188) becomes:

$$\begin{aligned}|G|^2 &= 1-2C(1-\cos k)-4s(1-\cos k)+2C^2(1-\cos k)\\ &\quad+4Cs(1-\cos k)^2+8s^2(1-\cos k)-4s^2\sin^2 k,\end{aligned} \tag{4.190}$$

hence:

$$\begin{aligned}|G|^2 &= 1-2C(1-\cos k)(1-C)-4s(1-\cos k)(1-2s)\\ &\quad+4\mathrm{Cs}(1-\cos k)^2-4s^2\sin^2 k.\end{aligned} \tag{4.191}$$

The stability condition $|G|^2 \leq 1$ yields:

$$\begin{aligned}&1-2C(1-\cos k)(1-C)-4s(1-\cos k)(1-2s)+4\mathrm{Cs}(1-\cos k)^2-4s^2\sin^2 k\leq 1,\\ &4\mathrm{Cs}(1-\cos k)^2\leq 2C(1-\cos k)(1-C)+4s(1-\cos k)(1-2s)+4s^2\sin^2 k,\\ &4\mathrm{Cs}(1-\cos k)\leq 2C(1-C)+4s(1-2s)+4s^2\sin^2 k/(1-\cos k).\end{aligned} \tag{4.192}$$

For $\cos k = -1$ the left-hand side of the disequation (4.192) assumes its maximum value $8Cs$, while at the same time $\sin k = 0$; hence the right-hand side assumes the minimum value $2C(1–C) + 4s(1–2s)$. Therefore the stability condition becomes:

$$\begin{aligned}
&8Cs \le 2C(1-C) + 4s(1-2s), \\
&4Cs - C + C^2 - 2s + 4s^2 \le 0, \\
&(2s+C)^2 - (2s+C) \le 0, \\
&2s + C - 1 \le 0, \\
&2s + C \le 1.
\end{aligned} \tag{4.193}$$

Recalling the definitions of C and s, condition (4.193) becomes:

$$\frac{V\Delta t}{\Delta x} + 2\frac{a\Delta t}{\Delta x^2} \le 1, \tag{4.194}$$

or:

$$\Delta t \le \frac{1}{\frac{V}{\Delta x} + \frac{2a}{\Delta x^2}}. \tag{4.195}$$

Because $V/\Delta x > > 2a/\Delta x^2$, condition (4.195) can be approximated by:

$$\Delta t \le \frac{\Delta x}{V}, \tag{4.196}$$

which is far less restrictive than condition (4.180).

In the example given in the previous section with $V = 1$ m/s, $\Delta x = 10^{-2}$ m, $a = 1.5 \times 10^{-5}$ m^2/s, one has:

$$\Delta t \le \frac{1}{\frac{1}{10^{-2}} + 2\frac{1.5\times10^{-5}}{10^{-4}}}(s) = \frac{1}{10^2 + 0.3}(s) \approx 10^{-2}(s). \tag{4.197}$$

4.6.3 *Hirt's Stability Criterion (Central Differences)*

We apply Hirt's stability criterion to the convection-diffusion Eq. (4.167) proceeding in the three steps as explained in Sect. 4.4.3.

Step 1: Discretization of the differential equation and expansion in Taylor series

Using forward time and central space differences for both convective and diffusive terms. we obtain the difference equation:

$$\frac{1}{\Delta t}\left(u_j^{n+1} - u_j^n\right) + \frac{V}{2\Delta x}\left(u^n_{j+1} - u^n_{j-1}\right) - \frac{a}{\Delta x^2}\left(u^n_{j+1} - 2u_j^n + u^n_{j-1}\right) = 0. \tag{4.198}$$

We use the Taylor series expansions:

$$u_j^{n+1} = u_j^n + \Delta t\left(\frac{\partial u}{\partial t}\right)_j^n + \frac{\Delta t^2}{2}\left(\frac{\partial^2 u}{\partial t^2}\right)_j^n + O(\Delta t^3), \tag{4.199}$$

$$u^n_{j+1} = u_j^n + \Delta x\left(\frac{\partial u}{\partial x}\right)_j^n + \frac{\Delta x^2}{2}\left(\frac{\partial^2 u}{\partial x^2}\right)_j^n + O(\Delta x^3), \tag{4.200}$$

$$u^n_{j-1} = u_j^n - \Delta x\left(\frac{\partial u}{\partial x}\right)_j^n + \frac{\Delta x^2}{2}\left(\frac{\partial^2 u}{\partial x^2}\right)_j^n + O(\Delta x^3), \tag{4.201}$$

thus obtaining, from Eq. (4.198):

$$\frac{1}{\Delta t}\left[\Delta t\left(\frac{\partial u}{\partial t}\right)_j^n + \frac{\Delta t^2}{2}\left(\frac{\partial^2 u}{\partial t^2}\right)_j^n\right] + \frac{V}{2\Delta x}\left[2\Delta x\left(\frac{\partial u}{\partial x}\right)_j^n\right] - \frac{a}{\Delta x^2}\left[\Delta x^2\left(\frac{\partial^2 u}{\partial x^2}\right)_j^n\right] = 0. \tag{4.202}$$

Step 2: Obtain a new differential equation
From Eq. (4.202) the following new differential equation is obtained:

$$\frac{\partial u}{\partial t} + \frac{\Delta t}{2}\left(\frac{\partial^2 u}{\partial t^2}\right) + V\frac{\partial u}{\partial x} - a\frac{\partial^2 u}{\partial x^2} = 0, \tag{4.203}$$

which has one term more than the original differential Eq. (4.167).

This additional term can be expressed by means of spatial derivatives, by using the original differential equation, as follows:

$$\begin{aligned}
\frac{\partial^2 u}{\partial t^2} &= \frac{\partial}{\partial t}\left(-V\frac{\partial u}{\partial x} + a\frac{\partial^2 u}{\partial x^2}\right) = -V\frac{\partial}{\partial x}\left(\frac{\partial u}{\partial t}\right) + a\frac{\partial^2}{\partial x^2}\left(\frac{\partial u}{\partial t}\right) \\
&= -V\frac{\partial}{\partial x}\left(-V\frac{\partial u}{\partial x} + a\frac{\partial^2 u}{\partial x^2}\right) + \nu\frac{\partial^2}{\partial x^2}\left(-V\frac{\partial u}{\partial x} + a\frac{\partial^2 u}{\partial x^2}\right) \\
&= V^2\frac{\partial^2 u}{\partial x^2} - 2aV\frac{\partial^3 u}{\partial x^3} + \nu^2\frac{\partial^4 u}{\partial x^4} \approx V^2\frac{\partial^2 u}{\partial x^2}.
\end{aligned} \tag{4.204}$$

Because in the Taylor series expansions (4.199) through (4.201) we have used only terms up to the second-order derivatives, so we do consistently also in Eq. (4.204).

Inserting Eq. (4.204) in Eq. (4.203) yields:

$$\frac{\partial u}{\partial t} + V\frac{\partial u}{\partial x} - \left(a - \frac{\Delta t V^2}{2}\right)\frac{\partial^2 u}{\partial x^2} = 0. \tag{4.205}$$

Step 3: Physical constraint imposed on the new differential equation

The coefficient which multiplies the second-order derivative of Eq. (4.205) has the physical meaning of an effective viscosity, which must be positive; hence:

$$\nu_{eff} = a - \frac{\Delta t V^2}{2} \geq 0, \tag{4.206}$$

which leads to the time step constraint:

$$\Delta t \leq \frac{2a}{V^2}. \tag{4.207}$$

The condition (4.207) is in close agreement with the condition (4.180) obtained by the Fourier analysis. In case of small kinematic viscosity a, this condition is very restrictive.

4.7 Numerical Treatment of Conservation Equations with the QUICK Scheme

4.7.1 Introduction

In this section we show the application of one method of minimization of numerical diffusion, not from the theoretical viewpoint but from the applicative one. We explain all details necessary to understand the implementation of such a method into an existing computer program, well established in the field of CFD (computational fluid dynamics). The program to which the QUICK method has been linked is the COMMIX-2 Code, documented in Bottoni (1985).

The method of minimization of numerical diffusion chosen for this application example is the QUICK (Quadratic Upstream Interpolation for Convective Kinematics) algorithm by Leonard (1979). This method has gained wide popularity in the applications and the article by Leonard has become one of the most celebrated and quoted articles on CFD.

Before starting with the details of the implementation of QUICK in COMMIX-2, we give some introductory explanation about the numerical QUICK scheme and about the computer program involved.

(i) The QUICK Scheme

The first sentences of the introduction of the above referred article by Leonard summarized clearly the problems experienced by the numerical analysts in computations involving convection. We quote these first sentences, omitting only the given references:

> Satisfactory numerical modelling of convection presents a well-known dilemma to the computational fluid dynamicist. On the one hand, central differencing may lead to unphysical oscillatory behaviour in an implicit solution or to disastrous non convergence

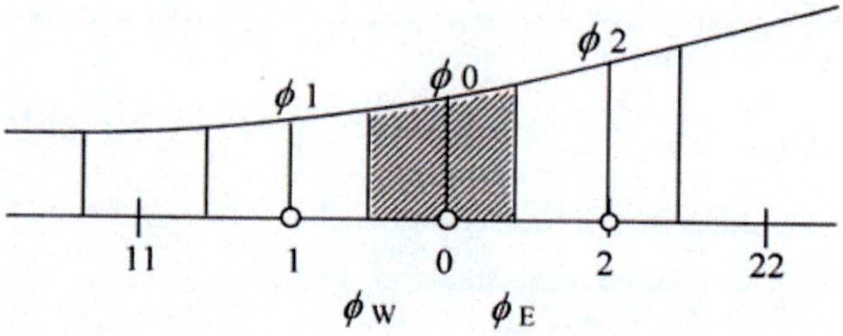

Fig. 4.5 Sketch showing the transported variable **Φ**, defined in the center nodes of the cells (labeled 0) and at the West (W) and East (E) boundaries

in an explicit computation in regions were convection strongly dominates diffusion. On the other hand, computations based on the classical alternative of upstream differencing often suffer from severe inaccuracies due to truncation error or streamline-to-grid skewness. Each of the latter error mechanism can be associated with equivalent artificial numerical diffusion terms introduced by the one-sided upstream differencing of convection. Although, in principle, grid refinement can alleviate all these problems, the necessary degree of refinement is often totally impracticable for engineering purposes, especially if one is attempting to model high-speed unsteady three-dimensional turbulent flows.

The alternative to central differencing and upwind differencing of convective terms, proposed and justified in Leonard's article, is given by the following scheme which involves (see Fig. 4.5), with reference to the x coordinate direction, a central node (labeled <0>), two nodes forward (<2> and <22>), and two nodes backward (<1> and <11>); besides, the surfaces bounding the central cell, around node <0>, are labeled <E> (East) and <W> (West). Denoting **Φ** a transported property, the QUICK scheme is given by:

$$\Phi_E = \frac{1}{2}(\Phi_0 + \Phi_2) - \frac{1}{8}(\Phi_1 - 2\Phi_0 + \Phi_2) \quad (u_E \geq 0) \tag{4.208a}$$

$$= \frac{1}{2}(\Phi_0 + \Phi_2) - \frac{1}{8}(\Phi_0 - 2\Phi_2 + \Phi_{22}) \quad (u_E < 0) \tag{4.208b}$$

This discretization corresponds to a linear interpolation, at the surface <E>, between the adjacent nodes <0> and <2>, corrected by an upstream term involving the additional node <1>. Similarly, for the West surface one has:

$$\Phi_W = \frac{1}{2}(\Phi_0 + \Phi_1) - \frac{1}{8}(\Phi_0 - 2\Phi_1 + \Phi_{11}) \quad (u_W \geq 0) \tag{4.209a}$$

$$= \frac{1}{2}(\Phi_0 + \Phi_1) - \frac{1}{8}(\Phi_1 - 2\Phi_0 + \Phi_2) \quad (u_W < 0) \tag{4.209b}$$

A von Neumann stability analysis of the QUICK scheme is given in the Leonard's article.

(ii) The Computer Program COMMIX-2

The program COMMIX-2 belongs to the family of COMMIX (COMponent MIXing) codes, developed over decades at the Argonne National Laboratory

(ANL), which describe steady-state and transient three-dimensional fluid flows with heat transfer in engineering systems, including nuclear reactor components (reactor tank, pipes, containment, etc.). The method of modelling the geometrical configuration with the fluid flow relies upon the concepts of volume porosity, directional surface porosity, distributes resistances, and distributed heat sources.

Within the COMMIX code family, COMMIX-2 is the variant which describes two-phase flows in the most general sense: two immiscible liquids, liquid and vapor, particles transported by liquids or gases, etc. The most common application of COMMIX-2 in the frame of reactor safety analysis has concerned the flow of water and steam, in case of thermal reactors, and the flow of sodium and its vapor, for the safety analysis of sodium-cooled fast reactors.

Regarding the case of the flow of a liquid and its vapor, COMMIX-2 has two modelling variants to describe this two-phase flow.

The first variant is based on a three-equation slip model (SM) in which the relative velocity between the liquid and vapor phases (slip velocity) is either constant or depends on the mixture velocity in each of the coordinate directions. The limiting case of equal phase velocities, known as homogeneous equilibrium model (HEM) is a subcase of the slip model. In the SM code version, the conservation equation for mass, momentum, and enthalpy of the two-phase mixture is formally identical to the equations for a single-phase flow, but with the inclusion of additional terms which take into account the so-called momentum slip and enthalpy slip between the phases.

The second variant is a five-equation unequal phase velocity, equal phase temperature (UVET) seriated two-phase continuum, which we refer to as the separated phase model (SPM). Its theoretical foundation was given in Solbrig (1975) and in Lyczkowski and Solbrig (1980).

For both two-phase flow models, the pressure method is applied in COMMIX-2 by deriving, through a combination of the momentum and continuity equations, a Poisson-like equation which describes the pressure distribution in the coolant. This procedure is shown in detail in Chap. 8 of this monograph for the transport equations of turbulence models. The numerical solution of the Poisson equation is obtained either with iterative methods (SOR, ADI, etc.) or by direct matrix inversion which uses a decomposition of the matrix of coefficients. The iterative methods are most suitable for large-size problems, while the direct matrix inversion method is applicable for relatively small problems (with modern computers up to about 50,000 cells).

A key feature used in the SPM variant consists in describing the space and time variation of the thickness of a coolant liquid film wetting the structural surfaces. The liquid phase is thus split in two components, liquid film and liquid droplets entrained by the flowing vapor from the film surface. The calculation of the liquid film thickness is essential for the evaluation of the structure to vapor heat fluxes.

Some more details about the historical development of the COMMIX-2 code at ANL and at KfK are given in Sect. 7.6.

In the following we recall the basic equations of both two-phase models mentioned above and then show the application of the QUICK method to the slip model.

4.7.2 Basic Equations of the Slip Model: Homogeneous Equilibrium Model [SM(HEM)]

The governing equations for mass, momentum and energy of the coolant for the slip model have been adopted in the form derived in Ishii (1977):

(i) Continuity equation:

$$\frac{\partial \rho_m}{\partial t} + \nabla \cdot (\rho_m \overline{v}_m) = 0. \tag{4.210}$$

(ii) Momentum equation:

$$\frac{\partial (\rho_m \overline{v}_m)}{\partial t} + \nabla \cdot (\rho_m \overline{v}_m \overline{v}_m) + \nabla \cdot \left[x(1-x)\rho_m \overline{V}_{SL} \overline{V}_{SL}\right] + \nabla p = \nabla \cdot (\mu \nabla \overline{v}_m) + \rho_m \overline{g} - \overline{R} \tag{4.211}$$

(iii) Energy equation (in terms of the mixture enthalpy h_m):

$$\frac{\partial (\rho_m h_m)}{\partial t} + \nabla \cdot (\rho_m \overline{v}_m h_m) + \nabla \cdot \left[x(1-x)\rho_m (h_g - h_l) \overline{V}_{SL}\right] = \\ = \frac{\partial p}{\partial t} + \overline{v}_m \cdot \nabla p + \left[\frac{1}{\rho_m} \alpha(1-\alpha)(\rho_l - \rho_g) \overline{V}_{SL}\right] \cdot \nabla p - \nabla \cdot \overline{q} + Q. \tag{4.212}$$

In these equation indices g, l, m refer to vapor (gas), liquid, and coolant mixture, respectively. Other symbols are defined as follows:

- $\overline{g}\,[m/s^2]$ = gravity acceleration;
- $h_m\,[J/kg]$ = specific enthalpy of mixture,
- p $[N/m^2]$ = pressure,
- $\overline{q} = -\Gamma_h \nabla h_m\,[W/m^2]$= heat flux,
- Q $[W/m^3]$ = specific power generation in the coolant,
- $\overline{R}\,[kg/m^2 s^2]$ = friction force,
- t $[s]$ = time,
- $\overline{v}_m\,[m/s]$ = mixture velocity,
- $\overline{V}_{SL}(= \overline{v}_g - \overline{v}_l)\,[m/s]$ = slip velocity,
- $x = [(h_m - h_l)/](h_g - h_l)$ = thermodynamic quality,
- α = void fraction,
- $\Gamma_h\,[kg/ms]$ = thermal diffusivity,
- $\rho_m\,[kg/m^3]$ = density of mixture.

In terms of contributions from both phases, the mixture enthalpy is given by:

$$h_m = \frac{1}{\rho_m}\left(\alpha_l \rho_l h_l + \alpha_g \rho_g h_g\right). \tag{4.213}$$

The continuity equation is formally identical with the equation for single-phase flow. The energy and momentum equations are formally similar to the equations for single-phase flow, but have additional terms depending on the slip velocity.

4.7.3 *Basic Equations of the Separated Phases Model (SPM)*

From the viewpoint of physical modelling, the strongest limitation of the slip model (SM) is the imposition of a slip ratio between the phase velocities that is kept constant during a calculation. In reality, the slip ratio depends on the flow regime and, in general, increases with increasing void fraction. This shortcoming is overcome in the more refined separated phase model (SPM) in which the momentum and continuity equations are treated separately for the vapor and liquid phases.

The governing equations for mass, momentum and energy of the coolant used in the SPM are as follows.

(i) Continuity equations:

$$\frac{\partial \rho'_g}{\partial t} + \nabla \cdot \left(\rho'_g \overline{u}_g\right) = M, \tag{4.214}$$

$$\frac{\partial \rho'_l}{\partial t} + \nabla \cdot \left(\rho'_l \overline{u}_l\right) = -M. \tag{4.215}$$

These equations are strictly coupled to each other by means of the source M[kg/sm^3] which is the coolant mass evaporating or condensing per unit time and volume, considered positive by net mass evaporation. Denoting S_{me} and S_{mc} the masses evaporating and condensing, respectively, one has:

$$M = S_{\mathrm{me}} - S_{\mathrm{mc}}. \tag{4.216}$$

ρ'_g and ρ'_l are macroscopic densities defined by

$$\rho'_g = \alpha \rho_g, \tag{4.217}$$

$$\rho'_l = (1 - \alpha)\rho_l. \tag{4.218}$$

(ii) Momentum equations:

$$\begin{aligned}\frac{\partial\left(\rho_g'\overline{u}_g\right)}{\partial t}+\nabla\cdot\left(\rho_g'\overline{u}_g\overline{u}_g\right) &= [0,M]\overline{u}_l-[0,-M]\overline{u}_g+\\ &-\alpha\nabla p+\overline{V}_g+\rho_g'\overline{g}+K\left(\overline{u}_l-\overline{u}_g\right)-\alpha\overline{R}_g,\end{aligned} \tag{4.219}$$

$$\begin{aligned}\frac{\partial\left(\rho_l'\overline{u}_l\right)}{\partial t}+\nabla\cdot\left(\rho_l'\overline{u}_l\overline{u}_l\right) &= -[0,M]\overline{u}_l-[0,-M]\overline{u}_g+\\ &-(1-\alpha)\nabla p+\overline{V}_l+\rho_l'\overline{g}-K\left(\overline{u}_l-\overline{u}_g\right)-(1-\alpha)\overline{R}_l,\end{aligned} \tag{4.220}$$

where $[a,b]=\max\,(a,b)$; $\overline{V}_g$, $\overline{V}_l[kg/ms^2]$ are the momentum density sources in vapor and liquid arising from viscous dissipation, K$[kg/m^2s]$ is the interface (liquid/vapor) drag force and $\overline{D}_g$ and $\overline{D}_l[kg/ms^2]$ are drag forces. The two momentum equations are strictly coupled to each other by means of the first, second and fifth terms at the right-hand side (terms depending on M and K, respectively).

(iii) Energy equation: For the coolant mixture, the same Eq. (4.212) is used as for the slip model.

4.7.4 Implementation of the QUICK Method in the Enthalpy Equation of the Slip Model

The equations for the SPM were given for the sake of clarity in the enunciation of the two-phase models used in COMMIX-2. In the following, however, we describe the implementation of the QUICK method in the slip model, only applied to the terms independent of the slip velocity. This does not mean that in the computer program these terms are neglected; it only means that these terms are not treated with the QUICK methodology but with standard finite difference schemes.

Dropping for simplicity the subscript m, which refers to the coolant mixture, the enthalpy equation reduces to:

$$\frac{\partial(\rho h)}{\partial t}+\nabla\cdot(\rho h\overline{v})=\frac{\partial p}{\partial t}+\overline{v}\cdot\nabla p-\nabla\cdot\overline{q}+Q. \tag{4.221}$$

In the following we denote with Φ the dependent variable ($\Phi=$ h) and introduce the definitions:

$$\overline{q} = -\Gamma_{\Phi} \nabla \Phi, \tag{4.222}$$

$$J_{\Phi j} = \rho h u_j - \Gamma_h \frac{\partial h}{\partial x_j} = \rho \Phi u_j - \Gamma_h \frac{\partial \Phi}{\partial x_j}, \tag{4.223}$$

$$S_{\Phi} = \frac{\partial p}{\partial t} + \overline{v} \cdot \nabla p + Q = \frac{dp}{dt} + Q. \tag{4.224}$$

The second term at the left side and the third term at the right side of Eq. (4.221) give:

$$\nabla \cdot \rho h \overline{v} + \nabla \cdot \overline{q} = \nabla \cdot (\rho h \overline{v} - \Gamma_{\Phi} \nabla \Phi) = \frac{\partial}{\partial x_j} \left(\rho h u_j - \Gamma_{\Phi} \frac{\partial \Phi}{\partial x_j} \right)$$
$$= \frac{\partial J_{\Phi j}}{\partial x_j} \tag{4.225}$$

Subscript <*j*>, ($j = 1, 2, 3$) refers to the coordinate directions. Using Eqs. (4.223), (4.224), and (4.225), Eq. (4.221) can be written in the compact form:

$$\frac{\partial (\rho \Phi)}{\partial t} + \frac{\partial J_{\Phi j}}{\partial x_j} = S_{\Phi}. \tag{4.226}$$

Let us assume that we have partitioned the definition domain of the problem in regular (equally spaced) cells and that the fluid volume in a cell is V_f. Integrating Equation (4.226) over the volume V_f of the fluid in a control cell, one obtains:

$$\int_{V_f} \frac{\partial (\rho \Phi)}{\partial t} dV + \int_{V_f} \frac{\partial J_{\Phi j}}{\partial x_j} dV = \int_{V_f} S_{\Phi} dV. \tag{4.227}$$

Using the Gauss theorem, we replace the volume integral of the divergence of vector J with the fluxes through the boundary surfaces of the cell. We denote with S2, S1 the surfaces of a cell normal to the x-axis, in the positive and negative directions, respectively; similarly S4, S3 bound the cell in the y direction and S6, S5 in the z direction. Thus Eq. (4.226) becomes:

$$\int_{V_f} \frac{\partial (\rho \Phi)}{\partial t} dV + \int_{S2} J_{\phi x} dS - \int_{S1} J_{\phi x} dS + \int_{S4} J_{\phi y} dS - \int_{S3} J_{\phi y} dS + \int_{S6} J_{\phi z} dS - \int_{S5} J_{\phi z} dS$$
$$= \int_{V_f} S_{\Phi} dV \tag{4.228}$$

Denoting with $A_\alpha(\alpha = 1, 2, ...6)$ the areas of the bounding surfaces, Eq. (4.228) can be approximated by

$$\frac{V_{F0}}{\Delta t}\left[(\rho_0 h_0)^{n+1} - (\rho_0 h_0)^n\right] + A_2\langle J_\phi\rangle_2 - A_1\langle J_\phi\rangle_1 + A_4\langle J_\phi\rangle_4 - A_3\langle J_\phi\rangle_3$$
$$+ A_6\langle J_\phi\rangle_6 - A_5\langle J_\phi\rangle_5$$
$$= V_{f0}\langle S_h\rangle_0, \tag{4.229}$$

where the symbol $\langle\rangle$ denotes average over a surface or volume, as appropriate, and superscript n refers to the time discretization. The fluxes of the vector J_ϕ through the boundary surfaces are evaluated by:

$$A_2\langle J_\phi\rangle_2 = \langle\rho U\Phi A\rangle_2 = A_2\left(\rho\Phi u - \Gamma_\phi\frac{\partial\Phi}{\partial x}\right)_2$$
$$= A_2(\rho\Phi u)_2 - D_2(\Phi_2 - \Phi_0), \tag{4.230}$$

$$A_1\langle J_\phi\rangle_1 = \langle\rho U\Phi A\rangle_1 = A_1\left(\rho\Phi u - \Gamma_\phi\frac{\partial\Phi}{\partial x}\right)_1$$
$$= A_1(\rho\Phi u)_1 - D_1(\Phi_0 - \Phi_1), \tag{4.231}$$

with:

$$D_2 = D_{i+1/2} = \frac{A_{f,i+1/2}}{\frac{1}{2}\left(\frac{\Delta x_i}{\Gamma_{\phi,0}} + \frac{\Delta x_{1+1}}{\Gamma_{\phi,2}}\right)}, \tag{4.232a}$$

$$D_1 = D_{i-1/2} = \frac{A_{f,i-1/2}}{\frac{1}{2}\left(\frac{\Delta x_i}{\Gamma_{\phi,0}} + \frac{\Delta x_{1-1}}{\Gamma_{\phi,1}}\right)}. \tag{4.232b}$$

In Eq. (4.232) subscript i refers to the x coordinate direction.

Referring to Fig. 4.5, the fluxes through the right (East = E) and left (West = W) surfaces bounding the cell in the x coordinate direction are given by Eqs. (4.208) and (4.209) which are repeated here for the sake of clarity:

$$\Phi_E = \frac{1}{2}(\Phi_0 + \Phi_2) - \frac{1}{8}(\Phi_1 - 2\Phi_0 + \Phi_2) \quad (u_E \geq 0) \tag{4.233a}$$

$$= \frac{1}{2}(\Phi_0 + \Phi_2) - \frac{1}{8}(\Phi_0 - 2\Phi_2 + \Phi_{22}) \quad (u_E < 0) \tag{4.233b}$$

$$\Phi_W = \frac{1}{2}(\Phi_0 + \Phi_1) - \frac{1}{8}(\Phi_0 - 2\Phi_1 + \Phi_{11}) \quad (u_W \geq 0) \tag{4.234a}$$

$$= \frac{1}{2}(\Phi_0 + \Phi_1) - \frac{1}{8}(\Phi_1 - 2\Phi_0 + \Phi_2) \quad (u_W < 0) \tag{4.234b}$$

Letting:

$$F = A\rho u, \tag{4.235}$$

the convective terms, for both positive and negative velocities u_E and u_W, can be written as:

$$\begin{aligned} F_E\Phi_E = {} & [F_E, 0]\left(\frac{6}{8}\Phi_0 + \frac{3}{8}\Phi_2 - \frac{1}{8}\Phi_1\right) - [-F_E, 0] \\ & \times \left(\frac{6}{8}\Phi_2 + \frac{3}{8}\Phi_0 - \frac{1}{8}\Phi_{22}\right), \end{aligned} \tag{4.236}$$

$$\begin{aligned} F_W\Phi_W = {} & [F_W, 0]\left(\frac{6}{8}\Phi_1 + \frac{3}{8}\Phi_0 - \frac{1}{8}\Phi_{11}\right) - [-F_W, 0] \\ & \times \left(\frac{6}{8}\Phi_0 + \frac{3}{8}\Phi_1 - \frac{1}{8}\Phi_2\right). \end{aligned} \tag{4.237}$$

The sign of the velocity component u is taken into account by means of the definition $[a, b] = \max(a, b)$ for any two real numbers, a and b.

In the most general case of unequal mesh lengths, Eq. (4.208) is replaced with the following formulae obtained with reference to the local coordinate systems sketched in Fig. 4.6.

For the flux Φ_E one has, for $u_E \geq 0$:

$$\begin{aligned} \Phi_E &= \Phi_E(\Phi_1, \Phi_0, \Phi_2) = \\ &= \frac{x_0x_1(x_0 - x_1)\Phi_2 - x_2x_1(x_2 - x_1)\Phi_0 + x_0x_2(x_2 - x_0)\Phi_1}{(x_2 - x_0)(x_0 - x_1)(x_2 - x_1)} \\ &= -C1AX \cdot \Phi_1 + C0AX \cdot \Phi_0 + C2AX \cdot \Phi_2, \end{aligned} \tag{4.238}$$

with:

$$C1AX = -x_0x_2(x_2 - x_0)/DENA, \tag{4.239a}$$

$$C0AX = -x_1x_2(x_2 - x_1)/DENA, \tag{4.239b}$$

$$C2AX = x_0x_1(x_0 - x_1)/DENA, \tag{4.239c}$$

$$DENA = (x_2 - x_0)(x_0 - x_1)(x_2 - x_1), \tag{4.239d}$$

and for $u_E < 0$:

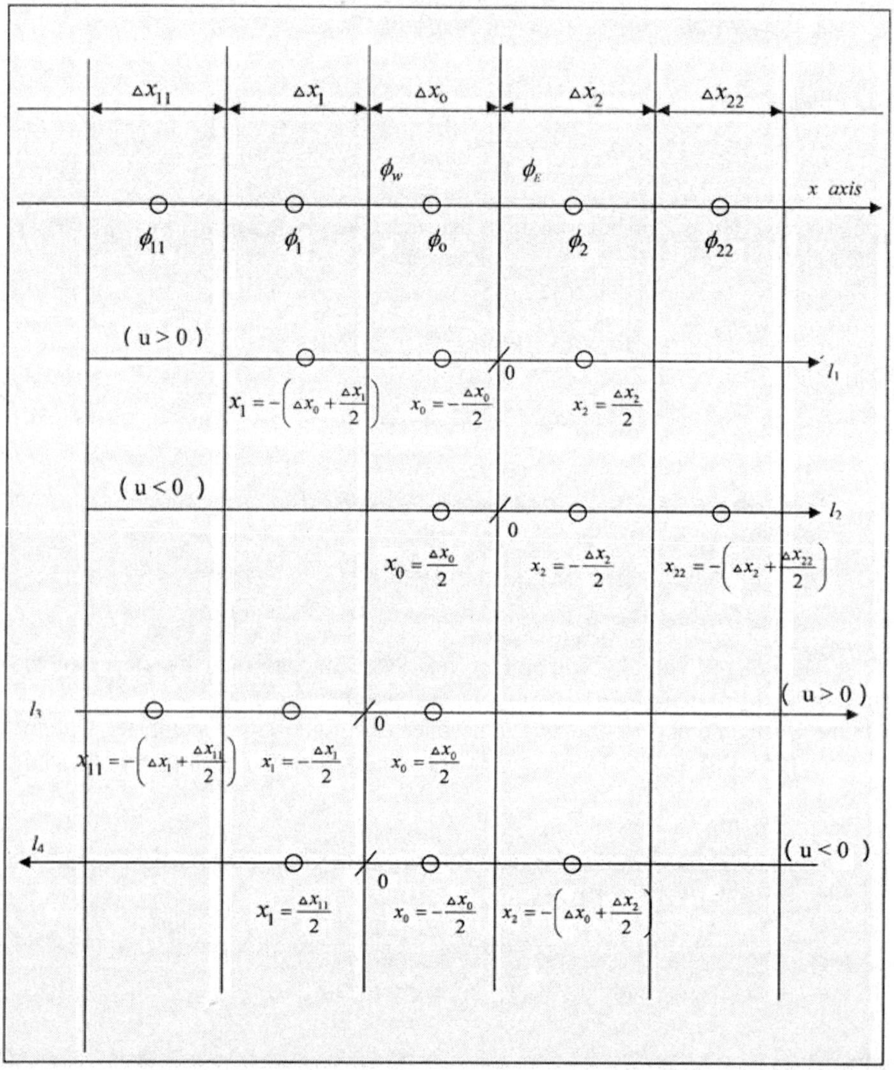

Fig. 4.6 Sketches showing the definition of the transported variable Φ in the central cell and in the neighboring cells, the discretization used for the x coordinate direction, and the definition of the coefficients in case of positive or negative velocity component u

$$\begin{aligned}
\Phi_E &= \Phi_E(\Phi_0, \Phi_2, \Phi_{22}) = \\
&= \frac{x_2 x_{22}(x_2 - x_{22})\Phi_0 - x_0 x_{22}(x_0 - x_{22})\Phi_2 - x_0 x_2(x_0 - x_2)\Phi_{22}}{(x_0 - x_2)(x_2 - x_{22})(x_0 - x_{22})} \\
&= C0BX \cdot \Phi_o + C2BX \cdot \Phi_2 - C22BX \cdot \Phi_{22}
\end{aligned} \tag{4.240}$$

with:

$$COBX = x_2 x_{22}(x_2 - x_{22})/DENB, \tag{4.241a}$$

$$C2BX = -x_0 x_{22}(x_0 - x_{22})/DENB, \tag{4.241b}$$

$$C22BX = -x_0 x_2(x_0 - x_2)/DENB, \tag{4.241c}$$

$$DENB = (x_0 - x_2)(x_2 - x_{22})(x_0 - x_{22}), \tag{4.241d}$$

For the flux Φ_W one has, for $u_W \geq 0$:

$$\begin{aligned}
\Phi_W &= \Phi_W(\Phi_1, \Phi_0, \Phi_{11}) = \\
&= \frac{-x_0 x_1(x_0 - x_1)\Phi_1 + x_1 x_{11}(x_1 - x_{11})\Phi_0 - x_0 x_1(x_0 - x_1)\Phi_{11}}{(x_0 - x_1)(x_1 - x_{11})(x_0 - x_{11})} \\
&= -C1CX \cdot \Phi_1 + C0CX \cdot \Phi_0 + C11CX \cdot \Phi_2,
\end{aligned} \tag{4.242}$$

with:

$$C11CX = -x_0 x_1(x_0 - x_1)/DENC, \tag{4.243a}$$

$$C1CX = -x_0 x_1(x_0 - x_{11})/DENC, \tag{4.243b}$$

$$COCX = x_1 x_{11}(x_1 - x_{11})/DENC, \tag{4.243c}$$

$$DENC = (x_0 - x_1)(x_1 - x_{11})(x_0 - x_{11}), \tag{4.243d}$$

and for $u_W < 0$:

$$\begin{aligned}
\Phi_W &= \Phi_W(\Phi_1, \Phi_0, \Phi_2) = \\
&= \frac{x_0 x_2(x_0 - x_2)\Phi_1 - x_1 x_2(x_1 - x_2)\Phi_0 - x_1 x_0(x_1 - x_0)\Phi_2}{(x_1 - x_0)(x_0 - x_2)(x_1 - x_2)} \\
&= C1DX \cdot \Phi_1 + C0DX \cdot \Phi_0 - C2DX \cdot \Phi_2
\end{aligned} \tag{4.244}$$

with:

$$C1DX = x_0 x_2(x_0 - x_2)/DEND, \tag{4.245a}$$

$$CODX = -x_1 x_2(x_1 - x_2)/DEND, \tag{4.245b}$$

$$C2DX = -x_1 x_0 (x_1 - x_0)/DEND, \tag{4.245c}$$

$$DEND = (x_1 - x_0)(x_0 - x_2)(x_1 - x_2), \tag{4.245d}$$

The coefficients in Eq. (4.238) through Eq. (4.241) are so defined that they are positive.

In the case of equal mesh lengths, the numerical values of the coefficients become:

$$\begin{array}{llllllll}
\text{C1AX} & = 1/8, & \text{C0BX} & = 3/8, & \text{C11CX} & = 1/8, & \text{C1DX} & = 3/8,\\
\text{C0AX} & = 6/8, & \text{C2BX} & = 6/8, & \text{C1CX} & = 6/8, & \text{C0DX} & = 6/8,\\
\text{C2AX} & = 3/8, & \text{C22BX} & = 1/8, & \text{C0CX} & = 3/8, & \text{C2DX} & = 1/8,
\end{array}$$

which were used in Eqs. (4.236) and (4.237).

Figure 4.6 shows the discretization used for the x direction. The uppermost line represents the x-axis with a central cell (labeled 0) and two neighboring cells in the positive (cells 2, 22) and negative (cells 1, 11) directions. Lines ℓ_1 and ℓ_2 show the local coordinate systems used for computing the value ϕ_E of the transported physical quantity, with origin on the right boundary of the central cell, when the velocity component u is positive or negative, respectively. Likewise, lines ℓ_3 and ℓ_4 show the local coordinate systems used for the computation of the value ϕ_W of the transported physical quantity.

For unequal mesh lengths, Eqs. (4.236) and (4.237) are replaced by:

$$\begin{aligned} F_E\Phi_E = & [F_E, 0](-C1AX \cdot \Phi_1 + C0AX \cdot \Phi_0 + C2AX \cdot \Phi_2) \\ & -[-F_E, 0](+C0BX \cdot \Phi_0 + C2BX \cdot \Phi_2 - C22BX \cdot \Phi_{22}) \end{aligned} \tag{4.246}$$

$$\begin{aligned} F_W\Phi_W = & [F_W, 0](-C11CX \cdot \Phi_{11} + C1CX \cdot \Phi_1 + C0CX \cdot \Phi_0) \\ & -[-F_W, 0](+C1DX \cdot \Phi_1 + C0DX \cdot \Phi_0 - C2DX \cdot \Phi_2) \end{aligned} \tag{4.247}$$

Not all terms of the two above equations are treated implicitly: in Eq. (4.246) only the central terms with the unknowns Φ_0 and Φ_2; in Eq. (4.247) only the central terms with the unknowns Φ_0 and Φ_1. Let superscript r denote an iteration index in a numerical scheme to solve the system of equations obtained for all cells. Terms with superscript $r + 1$ are treated implicitly because they belong to the vector of unknowns in the numerical solution of the system of equations; terms with superscript r are treated explicitly. In the system they are brought to the right-hand side and their current value is the value from previous iteration step. Thus Eqs. (4.246) and (4.247) are treated according to the iterative scheme:

$$\begin{aligned} F_E\Phi_E &= [F_E,0]\left(-C1AX\cdot\Phi_1^r + C0AX\cdot\Phi_0^{r+1} + C2AX\cdot\Phi_2^r\right) \\ &\quad -[-F_E,0]\left(+C0BX\cdot\Phi_0^r + C2BX\cdot\Phi_2^{r+1} - C22BX\cdot\Phi_{22}^r\right) \end{aligned} \tag{4.248}$$

$$\begin{aligned} F_W\Phi_W &= [F_W,0]\left(-C11CX\cdot\Phi_{11}^r + C1CX\cdot\Phi_1^{r+1} + C0CX\cdot\Phi_0^r\right) \\ &\quad -[-F_W,0]\left(+C1DX\cdot\Phi_1^r + C0DX\cdot\Phi_0^{r+1} - C2DX\cdot\Phi_2^r\right) \end{aligned} \tag{4.249}$$

Using the identities:

$$[F_E,0] = [-F_E,0] + F_E, \tag{4.250}$$

$$-[-F_W,0] = [-F_W,0] + F_W, \tag{4.251}$$

Equations (4.248) and (4.249) become:

$$\begin{aligned} F_E\Phi_E = {} & [-F_E,0]\left(C0AX\cdot\Phi_0^{r+1} - C2BX\cdot\Phi_2^{r+1}\right) + F_E\cdot C0AX\cdot\Phi_0^{r+1} \\ & +[F_E,0]\left(C2AX\cdot\Phi_2^r - C1AX\cdot\Phi_1^r\right) - [-F_E,0]\left(C0BX\cdot\Phi_0^r - C22BX\cdot\Phi_{22}^r\right) \end{aligned} \tag{4.252}$$

$$\begin{aligned} F_W\Phi_W = {} & [F_W,0]\left(C1CX\cdot\Phi_1^{r+1} - C0DX\cdot\Phi_0^{r+1}\right) + F_W\cdot C0DX\cdot\Phi_0^{r+1} \\ & +[F_W,0]\left(C0CX\cdot\Phi_0^r - C11CX\cdot\Phi_{11}^r\right) - [-F_W,0]\left(C1DX\cdot\Phi_1^r - C2DX\cdot\Phi_2^r\right) \end{aligned} \tag{4.253}$$

Similar procedures hold for the other coordinate directions. For the y direction, we denote the boundary surfaces of a cell with indexes N (North) and S (South); for the z direction with T (Top) and B (Bottom). With these convections one obtains:

$$\begin{aligned} F_N\Phi_N = {} & [-F_N,0]\left(C0AY\cdot\Phi_0^{r+1} - C4BY\cdot\Phi_4^{r+1}\right) + F_N\cdot C0AY\cdot\Phi_0^{r+1} \\ & +[F_N,0]\left(C4AY\cdot\Phi_4^r - C3AY\cdot\Phi_3^r\right) - [-F_N,0]\left(C0BY\cdot\Phi_0^r - C44BY\cdot\Phi_{44}^r\right), \end{aligned} \tag{4.254}$$

$$\begin{aligned} F_S\Phi_S = {} & [F_S,0]\left(C3CY\cdot\Phi_3^{r+1} - C0DY\cdot\Phi_0^{r+1}\right) + F_S\cdot C0DY\cdot\Phi_0^{r+1} \\ & +[F_S,0]\left(C0CY\cdot\Phi_0^r - C33CY\cdot\Phi_{33}^r\right) - [-F_S,0]\left(C3DY\cdot\Phi_3^r - C4DY\cdot\Phi_4^r\right), \end{aligned} \tag{4.255}$$

$$\begin{aligned} F_T\Phi_T = {} & [-F_T,0]\left(C0AZ\cdot\Phi_0^{r+1} - C6BZ\cdot\Phi_6^{r+1}\right) + F_T\cdot C0AZ\cdot\Phi_0^{r+1} \\ & +[F_T,0]\left(C6AZ\cdot\Phi_0^r - C5AZ\cdot\Phi_5^r\right) - [-F_T,0]\left(C0BZ\cdot\Phi_0^r - C66BZ\cdot\Phi_{66}^r\right), \end{aligned} \tag{4.256}$$

$$
\begin{aligned}
F_B\Phi_B = {} & [F_B,0]\left(C5CZ\cdot\Phi_5^{r+1}-C0DZ\cdot\Phi_0^{r+1}\right)+F_B\cdot C0DZ\cdot\Phi_0^{r+1} \\
& +[F_B,0]\left(C0CZ\cdot\Phi_0^{r}-C55CZ\cdot\Phi_{55}^{r}\right)-[-F_B,0]\left(C5DZ\cdot\Phi_5^{r}-C6DZ\cdot\Phi_6^{r}\right).
\end{aligned}
\tag{4.257}
$$

Inserting Eqs. (4.252)–(4.257) into Eq. (4.229), taking also into account the diffusive terms and rearranging (setting at the left-hand side the implicit terms with superscript $r+1$ and at the right-hand side the explicit terms with superscripts r) one obtains:

$$
\begin{aligned}
& \frac{V_{F0}}{\Delta t}\left[(\rho_0\Phi_0)^{r+1}\right]+\left\{[-F_E,0]\left(C0AX\cdot\Phi_0^{r+1}-C2BX\cdot\Phi_2^{r+1}\right)+F_E\cdot C0AX\cdot\Phi_0^{r+1}\right\} \\
& -D_2(\Phi_2-\Phi_0)^{r+1} \\
& -\left\{[F_W,0]\left(C1CX\cdot\Phi_1^{r+1}-C0DX\cdot\Phi_0^{r+1}\right)+F_W\cdot C0DX\cdot\Phi_0^{r+1}\right\}+D_1(\Phi_0-\Phi_1)^{r+1} \\
& +\left\{[-F_N,0]\left(C0AY\cdot\Phi_0^{r+1}-C4BY\cdot\Phi_4^{r+1}\right)+F_N\cdot C0AY\cdot\Phi_0^{r+1}\right\}-D_4(\Phi_4-\Phi_0)^{r+1} \\
& -\left\{[F_S,0]\left(C3CY\cdot\Phi_3^{r+1}-C0DY\cdot\Phi_0^{r+1}\right)+F_S\cdot C0DY\cdot\Phi_0^{r+1}\right\}+D_3(\Phi_0-\Phi_3)^{r+1} \\
& +\left\{[-F_T,0]\left(C0AZ\cdot\Phi_0^{r+1}-C6BZ\cdot\Phi_6^{r+1}\right)+F_T\cdot C0AZ\cdot\Phi_0^{r+1}\right\}-D_6(\Phi_6-\Phi_0)^{r+1} \\
& -\left\{[F_B,0]\left(C5CZ\cdot\Phi_5^{r+1}-C0DZ\cdot\Phi_0^{r+1}\right)+F_B\cdot C0DZ\cdot\Phi_0^{r+1}\right\}+D_5(\Phi_0-\Phi_5)^{r+1} \\
= {} & -[F_E,0]\left(C2AX\cdot\Phi_2^{r}-C1AX\cdot\Phi_1^{r}\right)-[-F_E,0]\left(C0BX\cdot\Phi_0^{r}-C22BX\cdot\Phi_{22}^{r}\right) \\
& +[F_W,0]\left(C0CX\cdot\Phi_0^{r}-C11CX\cdot\Phi_{11}^{r}\right)-[-F_W,0]\left(C1DX\cdot\Phi_1^{r}-C2DX\cdot\Phi_2^{r}\right) \\
& -[F_N,0]\left(C4AY\cdot\Phi_4^{r}-C3AY\cdot\Phi_3^{r}\right)-[-F_N,0]\left(C0BY\cdot\Phi_0^{r}-C44BY\cdot\Phi_{44}^{r}\right) \\
& +[F_S,0]\left(C0CY\cdot\Phi_0^{r}-C33CY\cdot\Phi_{33}^{r}\right)-[-F_S,0]\left(C3DY\cdot\Phi_3^{r}-C4DY\cdot\Phi_4^{r}\right) \\
& -[F_T,0]\left(C6AZ\cdot\Phi_0^{r}-C5AZ\cdot\Phi_5^{r}\right)-[-F_T,0]\left(C0BZ\cdot\Phi_0^{r}-C66BZ\cdot\Phi_{66}^{r}\right) \\
& +[F_B,0]\left(C0CZ\cdot\Phi_0^{r}-C55CZ\cdot\Phi_{55}^{r}\right)-[-F_B,0]\left(C5DZ\cdot\Phi_5^{r}-C6DZ\cdot\Phi_6^{r}\right) \\
& +\frac{V_{F0}}{\Delta t}[(\rho_0\Phi_0)^{r}]+V_{f0}\langle S_\Phi\rangle
\end{aligned}
\tag{4.258}
$$

We now rearrange Eq. (4.258) to write it as a 7-point formula for the unknowns Φ_α^{r+1} $(\alpha=0,1,2,\ldots 6)$ leaving at the right-hand side all terms at iteration level r. Thus one obtains:

$$\begin{aligned}
&\Phi_0^{r+1}\Bigg\{\frac{V_{F0}\rho_0^{r+1}}{\Delta t}+[-F_E,0]\cdot C0AX+[-F_N,0]\cdot C0AY+[-F_T,0]\cdot C0AZ\\
&\quad+[F_W,0]\cdot C0DX+[F_S,0]\cdot C0DY+[F_B,0]\cdot C0DZ+F_E\cdot C0AX\\
&\quad+F_N\cdot C0AY+F_T\cdot C0AZ\\
&\quad-F_W\cdot C0DX-F_S\cdot C0DY-F_B\cdot C0DZ+D_1+D_2+D_3+D_4+D_5+D_6\Bigg\}\\
&-\Phi_1^{r+1}\{C1CX\cdot[F_W,0]+D_1\}-\Phi_2^{r+1}\{C2BX\cdot[-F_E,0]+D_2\}\\
&-\Phi_3^{r+1}\{C3CY\cdot[F_S,0]+D_3\}-\Phi_4^{r+1}\{C4BY\cdot[-F_N,0]+D_4\}\\
&-\Phi_5^{r+1}\{C5CZ\cdot[F_B,0]+D_5\}-\Phi_6^{r+1}\{C6BZ\cdot[-F_T,0]+D_6\}\\
&=\Phi_0^{r}\{[-F_E,0]\cdot C0BX+[-F_N,0]\cdot C0BY+[-F_T,0]\cdot C0BZ\\
&\quad+[F_W,0]\cdot C0CX+[F_S,0]\cdot C0CY+[F_B,0]\cdot C0CZ\}\\
&-\Phi_{11}^{r}\{C11CX\cdot[F_w,0]\}+\Phi_1^{r}\{C1AX\cdot[F_E,0]-C1DX\cdot[-F_W,0]\}\\
&\Phi_2^{r}\{-C2AX\cdot[F_E,0]+C2DX\cdot[-F_W,0]\}-\Phi_{22}^{r}\{C22BX\cdot[-F_E,0]\}\\
&-\Phi_{33}^{r}\{C33CY\cdot[F_S,0]\}+\Phi_3^{r}\{C3AY\cdot[F_N,0]-C3DY\cdot[-F_S,0]\}\\
&\Phi_4^{r}\{-C4AY\cdot[F_N,0]+C4DY\cdot[-F_S,0]\}-\Phi_{44}^{r}\{C44BY\cdot[-F_N,0]\}\\
&-\Phi_{55}^{r}\{C55CZ\cdot[F_B,0]\}+\Phi_5^{r}\{C5AZ\cdot[F_T,0]-C5DZ\cdot[-F_B,0]\}\\
&\Phi_6^{r}\{-C6AZ\cdot[F_T,0]+C6DZ\cdot[-F_B,0]\}-\Phi_{66}^{r}\{C66BZ\cdot[-F_T,0]\}\\
&\quad+\frac{V_{F0}}{\Delta t}[(\rho_0\Phi_0)^r]+V_{f0}\langle S_\Phi\rangle.
\end{aligned} \tag{4.259}$$

Equation (4.259) is a Poisson-like equation describing the distribution of the fluid enthalpy in the cells of the definition domain of the problem. It can be written in the compact form (superscript E refers to the enthalpy equation):

$$a_0^E\Phi_0^{r+1}-\sum_{\beta=1}^{6}a_\beta^E\Phi_\beta^{r+1}=b_0^E, \tag{4.260}$$

with the following definitions of the coefficients:

$$
\begin{aligned}
a_0^E \quad = \quad & \left\{ \frac{V_{F0}\rho_0^{r+1}}{\Delta t} + [-F_E, 0] \cdot C0AX + [-F_N, 0] \cdot C0AY + [-F_T, 0] \cdot C0AZ \right. \\
& + [F_W, 0] \cdot C0DX + [F_S, 0] \cdot C0DY + [F_B, 0] \cdot C0DZ \\
& + F_E \cdot C0AX + F_N \cdot C0AY + F_T \cdot C0AZ \\
& \left. - F_W \cdot C0DX - F_S \cdot C0DY - F_B \cdot C0DZ + D_1 + D_2 + D_3 + D_4 + D_5 + D_6 \right\}
\end{aligned}
\tag{4.261a}
$$

$$a_1^E = C1CX \cdot [F_W, 0] + D_1, \tag{4.261b}$$

$$a_2^E = C2BX \cdot [-F_E, 0] + D_2, \tag{4.261c}$$

$$a_3^E = C3CY \cdot [F_S, 0] + D_3, \tag{4.261d}$$

$$a_4^E = C4BY \cdot [-F_N, 0] + D_4, \tag{4.261e}$$

$$a_5^E = C5CZ \cdot [F_B, 0] + D_5, \tag{4.261f}$$

$$a_6^E = C6BZ \cdot [-F_T, 0] + D_6, \tag{4.261g}$$

$$
\begin{aligned}
b_0^E \quad = \quad & \Phi_0^r \{ [-F_E, 0] \cdot C0BX + [-F_N, 0] \cdot C0BY + [-F_T, 0] \cdot C0BZ + [F_W, 0] \cdot C0CX \\
& \qquad + [F_S, 0] \cdot C0CY + [F_B, 0] \cdot C0CZ \} \\
& - \Phi_{11}^r \{ C11CX \cdot [F_w, 0] \} + \Phi_1^r \{ C1AX \cdot [F_E, 0] - C1DX \cdot [-F_W, 0] \} \\
& \Phi_2^r \{ -C2AX \cdot [F_E, 0] + C2DX \cdot [-F_W, 0] \} - \Phi_{22}^r \{ C22BX \cdot [-F_E, 0] \} \\
& - \Phi_{33}^r \{ C33CY \cdot [F_S, 0] \} + \Phi_3^r \{ C3AY \cdot [F_N, 0] - C3DY \cdot [-F_S, 0] \} \\
& \Phi_4^r \{ -C4AY \cdot [F_N, 0] + C4DY \cdot [-F_S, 0] \} - \Phi_{44}^r \{ C44BY \cdot [-F_N, 0] \} \\
& - \Phi_{55}^r \{ C55CZ \cdot [F_B, 0] \} + \Phi_5^r \{ C5AZ \cdot [F_T, 0] - C5DZ \cdot [-F_B, 0] \} \\
& \Phi_6^r \{ -C6AZ \cdot [F_T, 0] + C6DZ \cdot [-F_B, 0] \} - \Phi_{66}^r \{ C66BZ \cdot [-F_T, 0] \} \\
& + \frac{V_{F0}}{\Delta t} [(\rho_0 \Phi_0)^r] + V_{f0} \langle S_\Phi \rangle .
\end{aligned}
\tag{4.261h}
$$

The Poisson-like Eq. (4.260) can be solved with the methods explained in Chap. 7.

The reader interested in more details about the application of the QUICK scheme to boundary cells is referred to Maekawa et al. (1987).

4.7.5 FRAM-Correction of the QUICK solution

It is well known from the computational praxis that the solution of a transport equation obtained with the QUICK method displays over- and under-shoots at the locations of large gradients of the transported quantity, in this case of the specific enthalpy of the fluid. To cope with this problem and smooth unphysical oscillations a correction of the numerical solution obtained with the QUICK method can be applied following the guidelines of the FRAM (filtering remedy and methodology) method, explained in Chapman (1981). Application of the FRAM methodology implies combining the QUICK solution with another solution of the transport equation written in Lagrangian form. We present in this section a simplified variant of the FRAM method which was applied with success in the COMMIX-2 code. The problem is however complex and offers opportunities of more refined investigations at research level.

Using the identities:

$$\nabla \cdot (\rho h\overline{v}) = \rho h(\nabla \cdot \overline{v}) + \overline{v} \cdot \nabla(\rho h), \tag{4.262}$$

$$\frac{D(\rho h)}{Dt} = \frac{\partial(\rho h)}{\partial t} + \overline{v} \cdot \nabla(\rho h), \tag{4.263}$$

the Eulerian form of the enthalpy equation, Eq. (4.221) can be written in the Lagrangian form:

$$\frac{D(\rho h)}{Dt} + \rho h(\nabla \cdot \overline{v}) = \frac{dp}{dt} - \nabla \cdot \overline{q} + Q. \tag{4.264}$$

We further make the limiting assumption that, by solving the Lagrangian form of the equation, the fluid can be considered as incompressible, so that the divergence of the velocity field is zero ($\nabla \cdot \overline{v} = 0$). This is obviously a crude assumption in case of heated fluids.

Again we denote with $\boldsymbol{\Phi}$ the dependent variable ($\boldsymbol{\Phi}$ = h) and introduce the definitions:

$$\overline{q} = -\Gamma_h \nabla h = -\Gamma_{\Phi} \nabla \Phi, \tag{4.265}$$

$$J_{\phi j} = -\Gamma_h \frac{\partial h}{\partial x_j} = -\Gamma_{\Phi} \frac{\partial \Phi}{\partial x_j}, \tag{4.266}$$

$$S_{\Phi} = \frac{\partial p}{\partial t} + \overline{v} \cdot \nabla p + Q = \frac{dp}{dt} + Q. \tag{4.267}$$

With these definitions and using subscript j for the coordinate directions ($j = 1, 2, 3$), Eq. (4.264) can be written in the compact form:

$$\frac{D(\rho\Phi)}{Dt} + \frac{\partial J_{\phi j}}{\partial x_j} = S_{\Phi}. \tag{4.268}$$

As explained in the previous section, Eq. (4.268) is integrated over the fluid volume in a cell; the volume integral of the divergence of the vector J_ϕ is replaced with the fluxes through the boundary surfaces of the cell. Thus we obtain an equation formally identical with (4.229):

$$\begin{aligned}&\frac{V_{F0}}{\Delta t}\left[(\rho_0 h_0)^{n+1} - (\rho_0 h_0)^n\right] + A_2\langle J_\phi\rangle_2 - A_1\langle J_\phi\rangle_1 + A_4\langle J_\phi\rangle_4 - A_3\langle J_\phi\rangle_3\\&+ A_6\langle J_\phi\rangle_6 - A_5\langle J_\phi\rangle_5\\&= V_{f0}\langle S_\phi\rangle,\end{aligned} \tag{4.269}$$

We use Eqs. (4.230) and (4.231) and the same definitions of the diffusion coefficients $D_\alpha (\alpha = 1, 2, ...6)$ given in Eq. (4.232) for the x coordinate direction. Similar equations apply for the y and z coordinate directions. Thus Eq. (4.269) becomes

$$\begin{aligned}&\frac{V_{F0}}{\Delta t}\left[(\rho_0\Phi_0)^{n+1} - (\rho_0\Phi_0)^n\right]\\&\quad -D_2(\Phi_2 - \Phi_0) + D_1(\Phi_0 - \Phi_1) - D_4(\Phi_4 - \Phi_0) + D_3(\Phi_0 - \Phi_3)\\&\quad -D_6(\Phi_6 - \Phi_0) + D_5(\Phi_0 - \Phi_5) =\\&= V_{f0}\langle S_\phi\rangle.\end{aligned} \tag{4.270}$$

Equation (4.270) can be rearranged as:

$$\begin{aligned}&\frac{V_{F0}}{\Delta t}\left[(\rho_0\Phi_0)^{n+1} + D_1 + D_2 + D_3 + D_4 + D_5 + D_5\right]\\&\quad -\Phi_1 D_1 - \Phi_2 D_2 - \Phi_3 D_3 - \Phi_4 D_4 - \Phi_5 D_5 - \Phi_6 D_6\\&\qquad = \frac{V_{f0}}{\Delta t}(\rho_0\Phi_0)^n + V_{f0}\langle S_\phi\rangle.\end{aligned} \tag{4.271}$$

Equation (4.271) is again a Poisson equation which can be solved with the methods explained in Chap. 7. Let us refer to its solution as $\mathbf{\Phi_{LAG}}$ (Lagrangian solution) for every cell (i, j, k) in the definition domain, while we call $\mathbf{\Phi_{QUICK}}$ the solution obtained with the QUICK method. Furthermore we compute a third solution discretizing the convective terms of Eq. (4.221) with the upwind scheme and call it $\mathbf{\Phi_{UPW}}$.

We now compute for every cell the maximum and minimum of the Lagrangian solution $\mathbf{\Phi_{LAG}}$ in the cell itself and in its six neighboring cells in the three coordinate directions and define:

$$L_{\max} = \max\left(\Phi_{\mathrm{LAG},\beta}\right)\ (\beta = 0, 1, 2, \ldots 6), \tag{4.272}$$

$$L_{\min} = \min\left(\Phi_{\mathrm{LAG},\beta}\right)\ (\beta = 0, 1, 2, \ldots 6), \tag{4.273}$$

$$\delta_L = L_{\max} - L_{\min} \tag{4.274}$$

The QUICK solution Φ_{QUICK} is compared with these maximum and minimum values to ascertain whether it is in the range:

$$L_{\min} + c\delta_L \leq \Phi_{\mathrm{QUICK}} \leq L_{\max} - c\delta_L \tag{4.275}$$

with

- $c = 0$, for a central cell (cell with index $\beta = 0$)
- $c = 0.05$, for the cells adjacent to the central cell (cells wit index $\beta = 1, 2, \ldots 6$)
- $c = 0.1$, for boundary cells

The values of the coefficient c are rather arbitrary and have been suggested by the computational experience. If the solution Φ_{QUICK} lies in the above range, it is accepted without correction. If conversely Φ_{QUICK} lies outside of the above range, a correction is applied by replacing it with a value weighted between Φ_{QUICK} itself and Φ_{UPW}:

$$\Phi_{\mathrm{QUICK}}^{\mathrm{Updated}} = f \cdot \Phi_{\mathrm{UPW}} + (1 - f) \cdot \Phi_{\mathrm{QUICK}}, \tag{4.276}$$

with:

$$f = (L_{\min} + c\delta_L - \Phi_{\mathrm{QUICK}})/\delta_L,\ [\text{if } \Phi_{\mathrm{QUICK}} \leq L_{\min} + c\delta_L], \tag{4.277a}$$

$$f = \Phi_{\mathrm{QUICK}} - (L_{\max} - c\delta_L)/\delta_L\ \ [\text{if } \Phi_{\mathrm{QUICK}} \geq L_{\max} - c\delta_L]. \tag{4.277b}$$

Furthermore the condition is imposed that the weighting factor be in the range

$$0.01 < f < 1.$$

With the FRAM methodology, unphysical oscillations in the space distribution of the transported quantity are damped. One has however to be careful about the numerical damping because it must not overshadow the physical diffusion inherent in the diffusion terms of the equation to be solved. The normal praxis consists in testing a code in cases where an analytical solution is known and calibrate the free parameters c and f, i.e., find the best values of the free parameters which fit the numerical solution to the known analytical solution. Following the calibration, the parameters should not be changed any more. In many cases however experience in *the art of scientific computing* may be the best help.

4.7.6 *Implementation of the QUICK Method in the Momentum Equation of the Slip Model*

We recall the momentum Eq. (4.211) dropping the subscript m and the slip term:

$$\frac{\partial(\rho\overline{v})}{\partial t} + \nabla\cdot(\rho\overline{v}\overline{v}) = \nabla\cdot(\mu\nabla\overline{v}) + \rho\overline{g} - \overline{R} - \nabla p. \tag{4.278}$$

The component of the momentum equation in the x coordinate direction is:

$$\frac{\partial(\rho u)}{\partial t} + \frac{\partial(\rho u v_j)}{\partial x_j} = \frac{\partial}{\partial x_j}\left(\mu\frac{\partial u}{\partial x_j}\right) + \rho g_x - R_x - \frac{\partial p}{\partial x}, \tag{4.279}$$

and similarly for the y and z coordinate directions. Equation (4.279) can be written in the compact form:

$$\frac{\partial(\rho u)}{\partial t} + \frac{\partial(J_{u,j})}{\partial x_j} = S_u, \tag{4.280}$$

with the definitions:

$$J_{u,j} = \rho u v_j - \mu\frac{\partial u}{\partial x_j}, \tag{4.281}$$

$$S_u = \rho g_x - R_x - \frac{\partial p}{\partial x}. \tag{4.282}$$

Equation (4.280) is formally identical with Eq. (4.226) and is treated in a similar way to obtain a Poisson-like equation similar to Eq. (4.260) (superscript M refers to the momentum equation and $\Phi = u$):

$$a_0^M \Phi_0^{r+1} - \sum_{\beta=1}^{6} a_\beta^M \Phi_\beta^{r+1} = b_0^M. \tag{4.283}$$

Equation (4.283) is solved in the same way as Eq. (4.260). Similar Poisson-like equations are obtained for the velocity components v and w in the y and z coordinate directions, respectively.

References

Bottoni, M., et al. (1985, Sept). *COMMIX-2: a three-dimensional transient computer program for thermal-hydraulic analysis of two-phase flow*, Report NUREG/CR-4371, ANL-85-47.

Chapman, M. (1981). FRAM-non linear damping algorithms for the continuity equation. *Journal of Computational Physics, 44*.

Hadamard, J. (1923). *Lectures on Cauchy's Problem*, New Haven, 1923.

Ishii M. (1977). One-dimensional drift-flux model and constitutive equations for relative motion between phases in various two-phase flow regimes, ANL-77-47, October 1977.

Leonard, B. P. (1979). A stable and accurate convective modeling procedure based on quadratic interpolation. *Computer Methods in Applied Mechanics and Engineering, 19*.

Lyczkowski, R. W., & Solbrig, C. W. (1980). Calculation of the governing equations for a seriated unequal velocity, equal temperature two-phase continuum. *AICHE Journal*, (1).

Maekawa, T., Muramatsu, M., Matsumoto, H., Ninokata, M., & Bottoni (1987). *Cooperation on the development of AQUA and COMMIX-KfK. Information exchange and numerical models discussions*. Power Reactor and Nuclear Fuel Development Corporation (PNC), O-Arai Engineering Center, O-Arai, Japan, Report PNC SN9410 87-167, 1987.

Noye, J. (Ed.). (1978). *Numerical simulation of fluid motion*. North-Holland Publishing Company.

Solbrig C. W. (1975). Governing equations for a Seriated Continuum: An Unequal Velocity Model for Two-Phase Flow, Aerojet Nuclear Company, ANCR-1193, UC-78b, May 1975.

Von Neumann, J. (1944). *Proposal and analysis of a numerical method for the treatment of hydrodynamical shock problems*, Nat. Def. Report AM-551, March 1944.

Chapter 5
Numerical Solution of Large Linear Systems

The numerical methods explained in this chapter are usually applied to large linear systems arising from the discretization of partial differential equations of elliptic type. They can however be applied to systems of linear equations arising from different types of differential equations, provided some conditions are satisfied by the matrix of the coefficients of the system. In the following section, we introduce some basic concepts and properties of the matrix of coefficients and explain some sufficient conditions for the convergence of the numerical iterative schemes explained later.

5.1 Basic Concepts and Definitions

(i) *Definition of T-matrix.* A matrix A is defined as T-matrix if it is block tri-diagonal and the blocks of the main diagonal are diagonal matrices. The form of a T-matrix is as follows:

$$A = \begin{bmatrix} D_1 & H_1 & & & & & \\ K_1 & D_2 & H_2 & & & & \\ & K_2 & D_3 & H_3 & & & \\ & & & \cdots\cdots & & & \\ & & & K_{s-2} & D_{s-1} & H_{s-1} \\ & & & & K_{s-1} & D_s \end{bmatrix}, \tag{5.1}$$

where the D matrices are diagonal matrices.

M. Bottoni, *Physical Modeling and Computational Techniques for Thermal and Fluid-dynamics*, Mechanical Engineering Series,
https://doi.org/10.1007/978-3-030-79717-1_5

(ii) *Definition of consistently ordered matrix.* A matrix A of order N is consistently ordered if the set of natural numbers $W = \{1, 2. \ldots . . N\}$ can be divided in subsets S_1, $S_2 . \ldots . . S_t$ with $W = \ \cup S_k (k = 1, 2. \ldots . . t)$, such that for every off-diagonal element $a_{ij} \neq 0$ of the matrix A with $i \in S_k$ it holds:

$$\left(a_{ij} \neq 0, i \in S_k\right)\begin{cases} j \in S_{k+1} & \text{if } j > i, \\ j \in S_{k-1} & \text{if } j < i. \end{cases} \tag{5.2}$$

Theorem 5.1: *If the matrix A is a T-matrix, it is a consistently ordered matrix.*

The inverse of the statement of Theorem 5.1 does not hold: a matrix can be consistently ordered without being a T-matrix.

(iii) *Definition of ordering vector.* Let A be a matrix of order N. The vector $Y_o = |Y_1, Y_2 . \ldots . . Y_N|^T$ of natural numbers Y_i (not necessarily different from each other) is an ordering vector for the matrix A, if for every element $a_{ij} \neq 0$, $(i \neq j)$ of A it holds $|Y_i - Y_j| = 1$. It can be proved (Young, 1971) that the ordering vector has at most two different values.

(iv) *Definition of suitable ordering vector.* Let A be a matrix of order N. The vector $Y_{so} = |Y_1, Y_2 . \ldots . . Y_N|^T$ is a suitable ordering vector for A, if for every $a_{ij} \neq 0$, $(i \neq j)$ it holds:

$$\left(a_{ij} \neq 0\right), (i \neq j)\begin{cases} Y_i - Y_j = 1 & \text{if } i > j, \\ Y_i - Y_j = -1 & \text{if } i < j. \end{cases} \tag{5.3}$$

Theorem 5.2: *A matrix A of order N is a consistently ordered matrix, if and only if there is a suitable ordering vector for A.*

(v) *Definition of "Property A."* Let A be a matrix of order N. The matrix A has the "Property A" if the set $W = \{1, 2. \ldots . . N\}$ of natural numbers can be split in two subsets S_1,S_2 with $W = \cup S_k (k = 1, 2)$, such that for every element $a_{ij} \neq 0$, $(i \neq j)$ of A it holds either $i \in S_1, j \in S_2$ or $i \in S_2, j \in S_1$.

If the matrix A has "Property A," there is an ordering vector for A.

Theorem 5.3: *If the matrix A has "Property A," there is a permutation matrix P such that the matrix $M' = P^{-1}AP$ has the form*

$$M' = P^{-1}AP = \begin{vmatrix} D_1 & H \\ K & D_2 \end{vmatrix}, \tag{5.4}$$

where the D_1,D_2 matrices are diagonal matrices, and the matrix M' is consistently ordered.

(vi) *Alternative definition of "Property A,"* A matrix A has the "Property A" if there is a permutation matrix P such that the matrix PAP^T is block tri-diagonal and has therefore the structure:

$$PAP^T = \begin{vmatrix} D_1 & F_1 & & & \\ E_2 & D_2 & F_2 & & \\ & & \cdots\cdots & & \\ & & E_{m-1} & D_{m-1} & F_{m-1} \\ & & & E_m & D_m \end{vmatrix} = T, \tag{5.5}$$

where the D_i matrices are diagonal matrices.

It is always possible to reduce the structure of the matrix T in Eq. (5.5) to a matrix with $m = 2$:

$$PAP^T = \begin{vmatrix} D_1 & F \\ E & D_2 \end{vmatrix}. \tag{5.6}$$

For instance, a matrix with the structure of Eq. (5.5) can be obtained in the following case: let us consider a problem defined on a rectangular domain partitioned in cells with a grid parallel to the coordinate axes (x, y) and numbered sequentially with the natural numbers 1,2,3. . . . . .N. If the equations for the problem considered are written first for the cells with odd numbers (1,3,5. . . . .) and then for the cells with even numbers (2,4,6. . . . . .), the matrix of the coefficients of the system thus obtained has the structure shown in Eq. (5.5).

A third less formal definition of "Property A" can be given in the following way:

(vii) *Less formal definition of "Property A."* Let us assume that the cells which discretize the definition domain of a two-dimensional problem are divided into two classes, black (B) and white (W), as on a chessboard:

$$\begin{matrix} B & W & B & W & B \\ W & B & W & B & W \\ B & W & B & W & B \\ W & B & W & B & W \end{matrix}$$

If the difference equation written for one cell, for instance, a cell of class W, includes only other cells of class B (or vice versa), then the matrix of coefficients of the resulting system of equations has the "Property A."

The sequence in which the unknowns $x_i (i = 1, 2. \ldots\ldots N)$ of a system of difference equations are computed during an iteration step which brings the solution from step r to step $r + 1$ is crucial for the convergence of the iterative scheme used.

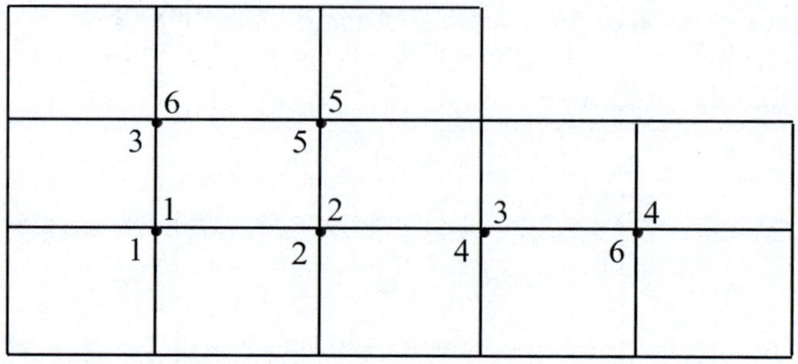

Fig. 5.1 Domain D

This sequence, or order, must satisfy some conditions to be consistent with the block tri-diagonal structure of Eq. (5.5).

(viii) *Definition of consistent order*. Let the matrix A of the equation $AX = B$ be reduced to the block tri-diagonal structure shown in Eq. (5.5). Let S_j be the set of unknown x_i which belong to the block D_j: $S_j = \{x_i; x_i \in D_j\}$. An order (sequence) of solution of the equations of the system $AX = B$ is consistent with the block tri-diagonal structure (5.5) of matrix A if, within every step of an iteration scheme, for every $j = 1,2.\ldots.m$, all unknowns x_i of S_{j-1} are computed (updated to step $r + 1$) before every unknown x_k of S_j with which they are coupled. (Two components x_i,x_k of the vector X of the unknowns are said to be "coupled" if either $a_{ik} \neq 0$ or $a_{ki} \neq 0$).

Practical rule to decide whether an order of nodes is consistent. We consider the inner nodes of a discretization mesh of a two-dimensional domain, and number these nodes sequentially: 1,2,3... Considering couples of neighboring nodes, like P and Q, we set an arrow toward the node with higher index. We circulate around the mesh, for instance, from node 1 to 3, 4, 2 back to 1. If we move along the arrow, add +1, to a counter (initially set to zero), otherwise (if we move against the arrow) add −1. If the "circulation" (content of the counter) at the end is zero, the ordering is consistent.

Example 5.1
Let us consider a two-dimensional domain D discretized with 13 cells as shown in Fig. 5.1 and number the inner nodes as shown by the natural number 1,2.....6 written right above the nodes. In the boundary nodes, boundary conditions are given, and therefore these nodes do not need to be considered, because we do not need to write equations corresponding to these nodes. It can be proved that with the numbering just given (sequential numbering) the matrix A of coefficients of the system of Eqs. (5.1) through (5.6) has "Property A" (Cfr. sets $S_1\{1, 3, 5\}$ and $S_2\{2, 6, 4\}$), but is not consistently ordered. There is however a permutation P such that the matrix $M' = P^{-1}AP$ is consistently ordered (according to Definition ii). The permutation corresponds to a renumbering of the nodes as shown by the natural numbers 1, 2.....6 written left below the nodes ("diagonal numbering").

It is easy to verify that the first numbering of nodes yields a matrix with the following structure:

$$A = \begin{bmatrix} a_{11} & a_{12} & 0 & 0 & 0 & a_{16} \\ a_{21} & a_{22} & a_{23} & 0 & a_{25} & 0 \\ 0 & a_{32} & a_{33} & a_{34} & 0 & 0 \\ 0 & 0 & a_{43} & a_{44} & 0 & 0 \\ 0 & a_{52} & 0 & 0 & a_{55} & a_{56} \\ a_{61} & 0 & 0 & 0 & a_{65} & a_{66} \end{bmatrix} \quad \text{(Sequential numbering)}$$

After building the sets $S_1\{1\}$, $S_2\{2,6\}$, $S_3\{3,5\}$, $S_4\{4\}$, corresponding to the elements $a_{12};a_{16}$, $a_{21};a_{23},a_{25}$, $a_{32};a_{34}$, a_{52}, of the matrix A, there is a conflict when treating element a_{56} because the number 6 should belong to S_4, but it has already been set in S_2. In this case the ordering vector is $Y_o = |1,0,1,0,1,0|^T$.

With the second ("diagonal") numbering of nodes, the matrix has the following structure:

$$A = \begin{bmatrix} a_{11} & a_{12} & a_{13} & 0 & 0 & 0 \\ a_{21} & a_{22} & 0 & a_{24} & a_{25} & 0 \\ a_{31} & 0 & a_{33} & 0 & a_{35} & 0 \\ 0 & a_{42} & 0 & a_{44} & 0 & a_{46} \\ 0 & a_{52} & a_{53} & 0 & a_{55} & 0 \\ 0 & 0 & 0 & a_{64} & 0 & a_{66} \end{bmatrix} \quad \text{(Diagonal numbering)}$$

The sets S_k (according to Definition ii) are $S_1\{1,\}$, $S_2\{2,3,\}$, $S_3\{4,5,\}$, $S_4\{6,\}$.

With the diagonal numbering, the ordering vector is $Y_o = |1,0,0,1,1,0|^T$and the suitable ordering vector is $Y_{so} = |1,2,2,3,3,4|^T$.

It is easy to verify that for $a_{ij} \neq 0$, $(i \neq j)$ condition (5.2) is verified.

The relationships between "Property A," consistent order, and T-form of a matrix are illustrated in Fig. 5.2, which must be interpreted in topological sense: properties of the innermost case are also properties of the case drawn around it, but not vice-versa.

Example 5.2

First example of consistently ordered matrix. Given the matrix A of order $N = 4$:

$$A = \begin{vmatrix} 1 & 1 & 1 & 0 \\ 1 & 1 & 0 & 1 \\ 1 & 0 & 1 & 1 \\ 0 & 1 & 1 & 1 \end{vmatrix},$$

Matrix A with "**Property A**": There is an ordering vector $Y_o = |Y_1, Y_2 ...... Y_N|^T$ for

A such that for $a_{ij} \neq 0 (i \neq j) \rightarrow |Y_i - Y_j| = 1$,

$$W = \{1, 2 ...... N\} = \bigcup S_k \ (k = 1, 2).$$

It holds either $[i \in S_1, j \in S_2]$ or $[i \in S_2, j \in S_1]$.

Permutation P gives: $P^{-1} AP = M'$

↓

Consistently ordered matrix M': There is a **suitable ordering vector**

$Y_{so} = |Y_1, Y_2 ...... Y_N|^T$ for M' such that for

$m'_{ij} \neq 0 \ (i \neq j)$ it holds

$$Y_i - Y_j = \begin{cases} +1, i > j, \\ -1, i < j. \end{cases}$$

If there is a suitable ordering vector, A is a **consistently ordered matrix**:

$$W = \{1, 2 ..... N\} = \bigcup S_k$$

$$a_{ij} \neq 0 (i \neq j), i \in S_k \begin{cases} j \in S_{k+1} (i < j) \\ j \in S_{k-1} (i > j) \end{cases}$$

T-matrix

(Block tri-diagonal form)

$$\begin{vmatrix} D_1 & H_1 & & & \\ K_1 & D_2 & H_2 & & \\ & K_2 & D_3 & H_3 & ..... \\ & & & ...... & ..... \end{vmatrix}$$

(D_i :square diagonal)

Fig. 5.2 Relationships between "Property A," consistent order and T-form of a matrix

let us consider the two subsets $S_a\{1, 4\}$, $S_b\{2, 3\}$ of the set $W = \cup S_k (k = 1, 2) = \{1, 2, 3, 4\}$. For every off-diagonal, non-zero element a_{ij} it holds $i \in S_a$ and $j \in S_b$, or vice-versa. For example for the element a_{34}, $i = 3 \in S_b$ and $j = 4 \in S_a$. Therefore, according to definition (v), the matrix A has "Property A."

In this case the ordering vector is $Y_o = |1, 0, 0, 1|^T$.

Let us now consider the vector: $Y_{so} = |1, 2, 2, 3|^T$; for the off-diagonal, non-zero elements of the matrix, it holds:

$$
\begin{array}{llllll}
i=1 & a_{12} \rightarrow Y_1 - Y_2 & = 1-2 & = -1 & (i<j) \\
 & a_{13} \rightarrow Y_1 - Y_3 & = 1-2 & = -1 & (i<j) \\
i=2 & a_{21} \rightarrow Y_2 - Y_1 & = 2-1 & = +1 & (i>j) \\
 & a_{24} \rightarrow Y_2 - Y_4 & = 2-3 & = -1 & (i<j) \\
i=3 & a_{31} \rightarrow Y_3 - Y_1 & = 2-1 & = +1 & (i>j) \\
 & a_{34} \rightarrow Y_3 - Y_4 & = 2-3 & = -1 & (i<j) \\
i=4 & a_{42} \rightarrow Y_4 - Y_2 & = 3-2 & = +1 & (i>j) \\
 & a_{43} \rightarrow Y_4 - Y_3 & = 3-2 & = +1 & (i>j)
\end{array}
$$

According to definition (iv) the above given vector Y_{so} is a suitable ordering vector.

Corresponding to the following off-diagonal and non-zero elements of the matrix A a_{12},a_{13}/a_{21},a_{24}/a_{31},a_{34}/a_{42},a_{43} one can build the following sets: $S_1\{1\}$, $S_2\{2, 3\}$, $S_3\{4\}$.

Consider, for instance, the element a_{31}; $i = 3 \in S_2 = S_k, j = 1 \in S_1 = S_{k-1}$. For a_{12}; $i = 1 \in S_1 = S_k, j = 2 \in S_2 = S_{k+1}$, etc. Therefore the matrix A not only has Property "A" but (according to Definition ii) is also consistently ordered, which is a more restrictive condition.

Example 5.3

Second example of consistently ordered matrix. Given the matrix A of order $N = 9$:

$$
\begin{array}{c}
\qquad\qquad j = \; 1 \;\; 2 \;\; 3 \;\; 4 \;\; 5 \;\; 6 \;\; 7 \;\; 8 \;\; 9 \\
\begin{array}{c} \\ \\ \\ \\ i = \\ \\ \\ \\ \\ \end{array}
\begin{array}{c} 1 \\ 2 \\ 3 \\ 4 \\ 5 \\ 6 \\ 7 \\ 8 \\ 9 \end{array}
\begin{array}{c} \\ \\ \\ \\ A = \\ \\ \\ \\ \\ \end{array}
\begin{bmatrix}
1 & 0 & 0 & 0 & 0 & 0 & 0 & 0 & 0 \\
0 & 0 & 1 & 0 & 0 & 0 & 0 & 0 & 0 \\
0 & 1 & 0 & 0 & 0 & 0 & 0 & 0 & 0 \\
0 & 0 & 0 & 0 & 0 & 0 & 1 & 0 & 0 \\
0 & 0 & 0 & 0 & 0 & 1 & 0 & 1 & 0 \\
0 & 0 & 0 & 0 & 1 & 0 & 0 & 0 & 1 \\
0 & 0 & 0 & 1 & 0 & 0 & 0 & 0 & 0 \\
0 & 0 & 0 & 0 & 1 & 0 & 0 & 0 & 1 \\
0 & 0 & 0 & 0 & 0 & 1 & 0 & 1 & 0
\end{bmatrix},
\end{array}
$$

let us consider the two subsets $S_a\{1, 2, 4, 5, 9\}$, $S_b\{3, 7, 6, 8\}$ of the set $W = \cup S_k(k = 1, 2) =$ $W = \{1, 2, 3, ...9\}$. For every off-diagonal, non-zero element a_{ij} it holds $i \in S_a$ and $j \in S_b$, or vice-versa. For example, for the element a_{23}, $i = 2 \in S_a$

and $j = 3 \in S_b$; for the element a_{65}, $i = 6 \in S_b$ and $j = 5 \in S_a$. Therefore, according to definition (v) the matrix A has "Property A."

In this case the ordering vector is $Y_o = |1, 1, 2, 1, 1, 2, 2, 2, 3|^T$.

Consider now the vector $Y_{so} = |1, 1, 2, 2, 2, 3, 3, 3, 4|^T$. For the off-diagonal, non-zero coefficients of the matrix, it holds:

$$
\begin{aligned}
a_{23} &\rightarrow Y_2 - Y_3 = 1 - 2 = -1 \quad (i < j)\\
a_{32} &\rightarrow Y_3 - Y_2 = 2 - 1 = 1 \quad (i > j)\\
a_{47} &\rightarrow Y_4 - Y_7 = 2 - 3 = -1 \quad (i < j)\\
a_{56} &\rightarrow Y_5 - Y_6 = 2 - 3 = -1 \quad (i < j)\\
a_{58} &\rightarrow Y_5 - Y_8 = 2 - 3 = -1 \quad (i < j)\\
a_{65} &\rightarrow Y_6 - Y_5 = 3 - 2 = 1 \quad (i > j)\\
a_{69} &\rightarrow Y_6 - Y_9 = 2 - 3 = -1 \quad (i < j)\\
a_{74} &\rightarrow Y_7 - Y_4 = 3 - 2 = 1 \quad (i > j)\\
a_{85} &\rightarrow Y_8 - Y_5 = 3 - 2 = 1 \quad (i > j)\\
a_{89} &\rightarrow Y_8 - Y_9 = 2 - 3 = -1 \quad (i < j)\\
a_{96} &\rightarrow Y_9 - Y_6 = 4 - 3 = 1 \quad (i > j)\\
a_{98} &\rightarrow Y_9 - Y_8 = 4 - 3 = 1 \quad (i > j)
\end{aligned}
$$

According to definition (iv), the above given vector Y_{so} is a suitable ordering vector.

Consider now the set $S_1\{1, 2\}$, $S_2\{3, 4, 5\}$, $S_3\{7, 6, 8\}$, $S_4\{9\}$.

For the element a_{23}; $i = 2 \in S_1 = S_k$, $j = 3 \in S_2 = S_{k+1}$.

For a_{56}; $i = 5 \in S_2 = S_k$, $j = 6 \in S_3 = S_{k+1}$, etc.

Therefore the matrix A not only has "Property A" but it is also consistently ordered.

A computer program testing a given matrix for "Property A," given in **PROG_V_1 [TEST_PROA]**, is based on the logical diagram documented in Young (1971). This program is applied to several examples of matrices, including the matrices of Examples 5.1, 5.2, 5.3, 5.5, and 5.6.

5.2 The Young-Frankel Theory of Successive Over-Relaxation (SOR)

5.2.1 The Numerical Schemes

Let us consider a system of equations:

$$AX = B, \tag{5.7}$$

where A is a square non-singular matrix of order N, $X = (x_1, x_2 \ldots\ldots x_N)^T$ is a vector of unknowns, and B is a vector of known quantities. Equation (5.7) consists of N linear algebraic equations:

$$\begin{aligned}
&a_{11}x_1 + a_{12}x_2 + && && +a_{1N}x_N = b_1,\\
&a_{21}x_1 + a_{22}x_2 + && && +a_{2N}x_N = b_2,\\
&a_{i1}x_1 + a_{i2}x_2 + && +a_{ii}x_i + && +a_{iN}x_N = b_i,\\
&a_{N1}x_1 + a_{N2}x_2 + && && +a_{NN}x_N = b_N.
\end{aligned} \tag{5.8}$$

Let us assume we want to solve Eq. (5.8) iteratively and, starting from an arbitrary initial vector $X^0 = \left(x_1^0, x_2^0 \ldots\ldots x_N^0\right)^T$, we found an approximate solution at iteration step r, $X^r = \left(x_1^r, x_2^r \ldots\ldots x_N^r\right)^T$. From this value X^r we want to find a next better approximation X^{r+1}.

The method of Jacobi, or "simultaneous displacements" method, consists in solving the equations of the system (5.8) sequentially one after the other, without using, when solving the i-th equation, the values previously found for the unknowns $x_1^{r+1}, x_2^{r+1} \ldots\ldots x_{i-1}^{r+1}$. The solution sequence is therefore:

$$\begin{aligned}
x_1^{r+1} &= \frac{1}{a_{11}}\left(b_1 - \sum_{j=2}^{N} a_{1j}x_j^r\right),\\
x_2^{r+1} &= \frac{1}{a_{22}}\left(b_2 - a_{21}x_1^r - \sum_{j=3}^{N} a_{2j}x_j^r\right),\\
x_i^{r+1} &= \frac{1}{a_{ii}}\left(b_i - \sum_{j=1}^{i-1} a_{ij}x_j^r - \sum_{j=i+1}^{N} a_{ij}x_j^r\right),\\
x_N^{r+1} &= \frac{1}{a_{NN}}\left(b_N - \sum_{j=1}^{N-1} a_{Nj}x_j^r\right).
\end{aligned} \tag{5.9}$$

An improvement of the convergence rate is obtained by the Gauss-Seidel method, or method of "successive displacements" in which the previously found values $x_1^{r+1}, x_2^{r+1} \ldots\ldots x_{i-1}^{r+1}$ are used when solving the i-th equation for x_i^{r+1}. The solution sequence becomes:

$$\begin{aligned}
x_1^{r+1} &= \frac{1}{a_{11}}\left(b_1 - \sum_{j=2}^{N} a_{1j}x_j^r\right),\\
x_2^{r+1} &= \frac{1}{a_{22}}\left(b_2 - a_{21}x_1^{r+1} - \sum_{j=3}^{N} a_{2j}x_j^r\right),\\
x_i^{r+1} &= \frac{1}{a_{ii}}\left(b_i - \sum_{j=1}^{i-1} a_{ij}x_j^{r+1} - \sum_{j=i+1}^{N} a_{ij}x_j^r\right),\\
x_N^{r+1} &= \frac{1}{a_{NN}}\left(b_N - \sum_{j=1}^{N-1} a_{Nj}x_j^{r+1}\right).
\end{aligned} \tag{5.10}$$

A further improvement with respect to the Gauss-Seidel method is generally achieved when the vector X^{r+1} obtained from Eq. (5.10) is considered only as a provisional value, say $X*$, and a better value is searched for by using the equation:

$$X^{r+1} = X^r + \omega(X^* - X^r). \tag{5.11}$$

Equation (5.11) means that a better value X^{r+1} is searched for by adding to the previous solution X^r an increment $(X^* - X^r)$ multiplied by a parameter ω, called relaxation parameter. For reasons of stability, ω must be in the range $(0 \leq \omega \leq 2)$ and the scheme (5.11) is called successive under-relaxation if $\omega \leq 1$ and successive over-relaxation (SOR) if $1 < \omega \leq 2$. If $\omega = 1$, $X^{r+1} = X^*$, which means the Gauss-Seidel method (5.10) is used.

A justification of the fact that the relaxation parameter must be smaller than 2 is given, based on different arguments, in Sect. (5.3.1).

The matrix A can always be written as:

$$A = E + D + F, \tag{5.12}$$

as sum of a diagonal matrix D, of a matrix E with the elements of A below the main diagonal, and a matrix F with the elements of A above the main diagonal.

The method of successive relaxation is represented by Eq. (5.11) together with the matrix equation:

$$EX^{r+1} + DX^* + FX^r = B. \tag{5.13}$$

Equation (5.13) implies that, as in the Gauss-Seidel method (5.10), the values x_l^{r+1} $(l = 1, 2 \ldots\ldots i-1)$ are used when solving for the i–th unknown x_i^{r+1}.

The provisional value X^* can be eliminated between Eqs. (5.11) and (5.13). From Eq. (5.11) we obtain:

$$\begin{aligned} \omega X^* &= X^{r+1} - X^r + \omega X^r, \\ X^* &= \omega^{-1}\left(X^{r+1} - X^r\right) + X^r, \end{aligned} \tag{5.14}$$

hence from Eq. (5.13):

$$\begin{aligned} EX^{r+1} + D\left[\omega^{-1}\left(X^{r+1} - X^r\right) + X^r\right] + FX^r &= B, \\ \left(E + D\omega^{-1}\right)X^{r+1} + \left(-\omega^{-1}D + D + F\right)X^r &= B, \\ X^{r+1} = -\left(E + D\omega^{-1}\right)^{-1}&\left[\left(1 - \omega^{-1}\right)D\right. \\ \left.+F\right]X^r + &\left(E + D\omega^{-1}\right)^{-1}B, \\ X^{r+1} &= H(\omega)X^r + M(\omega)B, \end{aligned} \tag{5.15}$$

having defined the matrices:

$$H(\omega) = -\left(E + D\omega^{-1}\right)^{-1}\left[\left(1 - \omega^{-1}\right)D + F\right], \tag{5.16}$$

$$M(\omega) = \left(E + D\omega^{-1}\right)^{-1}. \tag{5.17}$$

The matrix $H(\omega)$, function of the relaxation parameter ω, is called the iteration matrix. For the Gauss-Seidel method, $\omega = 1$ and the iteration matrix becomes:

$$H_{GS} = H(\omega = 1) = -(D + E)^{-1}F, \tag{5.18}$$

which can be obtained directly from Eq. (5.13) with $X^* = X^{r+1}$.

Equation (5.13) written for the Gauss-Seidel method, with $\omega = 1$, implies that at every iteration step we subtract, from the previous solution X^r, the residuals in order to get an updated solution. In fact from Eq. (5.13) (with $X^* = X^{r+1}$) we obtain, adding and subtracting DX^r:

$$\begin{aligned} DX^{r+1} &= -EX^{r+1} - FX^r + B = DX^r - \left(DX^r + EX^{r+1} + FX^r - B\right) \\ &= DX^r - R^r, \end{aligned} \tag{5.19}$$

having defined:

$$R^r = DX^r + EX^{r+1} + FX^r - B, \tag{5.20}$$

the residual vector at iteration step r.

Without loss of generality, the diagonal matrix D can be assumed as the identity matrix, because every equation of the system (5.8) can be divided by a_{ii}. Thus Eq. (5.19) shows that the residuals R^r are subtracted from the previous solution X^r to obtain the updated solution X^{r+1}. This is expressed by saying that at every iteration step the solution is "relaxed" by subtracting the residuals.

The Jacobi method is represented by the matrix equation:

$$EX^r + DX^{r+1} + FX^r = B, \tag{5.21}$$

which yields:

$$X^{r+1} = -D^{-1}(E + F)X^r + D^{-1}B = H_J X^r + M_J B. \tag{5.22}$$

with:

$$H_J = -D^{-1}(E + F) = -D^{-1}(A - D) = -D^{-1}A + I = -M_J A + I, \tag{5.23}$$

$$M_J = D^{-1}. \tag{5.24}$$

From Eq. (5.23) we also have:

$$M_J = -(H_J - I)A^{-1}. \tag{5.25}$$

5.2.2 *Relationship Between SOR, Gauss-Seidel, and Jacobi Methods*

Apart from the different presentation, this section relies upon the treatment given in Forsythe and Wasow (1960). We aim at analyzing the relationship between the eigenvalues of the iteration matrices of the SOR, Gauss-Seidel, and Jacobi methods. We recall that these schemes are given by the vector equations:

$$X^{r+1} = H(\omega)X^r + M(\omega)B, \qquad \text{(SOR and Gauss-Seidel)} \tag{5.26}$$

$$X^{r+1} = H_J X^r + M_J B, \qquad \text{(Jacobi)} \tag{5.27}$$

with the iteration matrices defined by:

$$H(\omega) = -\left(E + D\omega^{-1}\right)^{-1}\left[\left(1 - \omega^{-1}\right)D + F\right], \qquad \text{(SOR)} \tag{5.28}$$

$$H(\omega = 1) = H_{GS} = -(D + E)^{-1}F, \qquad \text{(Gauss-Seidel)} \tag{5.29}$$

$$\begin{aligned} H_J &= -D^{-1}(E + F) = -D^{-1}(A - D) = -D^{-1}A + I \\ &= -M_J A + I. \quad \text{(Jacobi)} \end{aligned} \tag{5.30}$$

The eigenvalues of the iteration matrix $H(\omega)$ are the solutions of the algebraic equation:

$$\det[H(\omega) - \eta I] = 0, \tag{5.31a}$$

or

$$Q(\eta) = 0. \tag{5.31b}$$

We recall that for any non-singular matrices A, B, C, it holds:

$$\text{If} A \quad = \quad BC, \quad \det A \quad = \quad \det B \det C. \tag{5.32}$$

Equation (5.31b) can be written as:

$$Q(\omega) = \det[H(\omega) - \eta I] = 0. \tag{5.33}$$

Multiplying for the constant $\det(\omega^{-1}D + E)$, using Eq. (5.32) and the definition (5.28), one obtains:

$$
\begin{aligned}
Q(\omega) &= \det\left(\omega^{-1}D+E\right)\det[H(\omega)-\eta I]=0,\\
&= \det\left\{\left(\omega^{-1}D+E\right)[H(\omega)-\eta I]\right\}=0,\\
&= \det\left\{\left(\omega^{-1}D+E\right)\left(\omega^{-1}D+E\right)^{-1}\left[\left(1-\omega^{-1}\right)D+F\right]+\eta\left(\omega^{-1}D+E\right)\right\}=0,\\
&= \det\left[\left(1-\omega^{-1}\right)D+F+\eta\left(\omega^{-1}D+E\right)\right]=0.
\end{aligned}
\tag{5.34}
$$

The eigenvalues of the iteration matrix H_J are the roots of the characteristic equation:

$$
\begin{aligned}
P(\lambda) &= \det(H_J-\lambda I)=0,\\
&= \det D(H_J-\lambda I)=0,\\
&= \det D\left[-D^{-1}(E+F)-\lambda I\right]=0,\\
&= \det(E+\lambda D+F)=0.
\end{aligned}
\tag{5.35}
$$

It can be proved (Forsythe & Wasow, 1960):

(i) For an integer r $(r\geq 0)$ the polynomial $P(\lambda)$ can be written in the form:

$$
P(\lambda)=\begin{cases}\lambda^{2r}p_{N/2-r}\left(\lambda^2\right) & \textit{if } N \textit{ is even},\\ \lambda^{2r+1}p_{(N-1)/2-r}\left(\lambda^2\right) & \textit{if } N \textit{ is odd},\end{cases}
\tag{5.36}
$$

where $p_\alpha(\rho)$ denotes a polynomial of degree α in ρ with $p_\alpha(0)\neq 0$ (hence $\rho=0$ is not a root of p_α). Consequently the equation $P(\lambda)=0$ has a root $\lambda=0$ with order of multiplicity $2r$ (if N is even) or $2r+1$ (if N is odd) and has $N/2-r$ [or $(N-1)/2-r$] non-zero pairs of roots $\pm\lambda_i$, each pair corresponding to one root of $p(\lambda^2)$.

(ii) If, and only if, the matrix A has "Property A," the roots of the characteristic equations $Q(\eta)=0$ and $P(\lambda)=0$ are related by the equation:

$$
\frac{(\eta+\omega-1)^2}{\eta}=\lambda^2\omega^2.
\tag{5.37}
$$

Hence, from Eq. (5.37) one deduces:

(iii) If $\omega=1$ (Gauss-Seidel method):

(a) To every root $\lambda_i=0$ of $P(\lambda)=0$ corresponds a root $\eta_i=0$ of $Q(\eta)=0$.
(b) To every couple of roots $\pm\lambda_i$ of $P(\lambda)=0$ corresponds the couple $\eta_i=0$, $\eta_i=\lambda_i^2$ of $Q(\eta)=0$.

(iv) There remains to analyze the situation for $\omega\neq 1$. In this case $\eta=0$ cannot be a solution of $Q(\eta)=0$.

Let λ_i be a root of $P(\lambda) = 0$. The corresponding root η_i of $Q(\eta) = 0$ is obtained from Eq. (5.37), as follows:

$$(\eta_i + \omega - 1)^2 = \lambda_i^2 \omega^2 \eta_i, \tag{5.38a}$$

$$\eta_i + \omega - 1 = \lambda_i \omega \eta_i^{1/2}, \tag{5.38b}$$

$$\eta_i - \omega \lambda_i \eta_i^{1/2} + \omega - 1 = 0, \tag{5.38c}$$

$$\eta_i^{1/2} = \frac{\omega \lambda_i \pm \sqrt{\omega^2 \lambda_i^2 - 4(\omega - 1)}}{2}. \tag{5.38d}$$

We consider the positive sign in Eq. (5.38d) because we want to analyze and minimize the maximum value of η_i.

For $\lambda_i = 0$, from Eq. (5.38b) we obtain $\eta_i = -(\omega - 1)$.

If $\lambda_i \neq 0$, the maximum of the two values of $\eta_i^{1/2}$ is:

$$\eta_i^{1/2} = \frac{1}{2}\omega \lambda_i + \frac{1}{2}\sqrt{\omega^2 \lambda_i^2 - 4(\omega - 1)}. \tag{5.39}$$

From Eq. (5.39) one derives also the useful relation:

$$\sqrt{\omega^2 \lambda_i^2 - 4(\omega - 1)} = 2\eta_i^{1/2} - \omega \lambda_i. \tag{5.40}$$

Let us consider the case of over-relaxation ($1 \leq \omega < 2$) and examine the behavior of $\eta_i^{1/2}$ as ω increases from 1. The derivative of $\eta_i^{1/2}$ with respect to ω is.

$$\frac{d\left(\eta_i^{1/2}\right)}{d\omega} = \frac{1}{2}\left[\lambda_i + \frac{2\omega\lambda_i^2 - 4}{2\sqrt{\omega^2\lambda_i^2 - 4(\omega - 1)}}\right] = \frac{1}{2}\left[\lambda_i + \frac{\omega\lambda_i^2 - 2}{2\eta_i^{1/2} - \omega\lambda_i}\right]$$

[using Eq. (5.40)]

$$= \frac{1}{2}\left[\frac{2\lambda_i\eta_i^{1/2} - \omega\lambda_i^2 + \omega\lambda_i^2 - 2}{2\eta_i^{1/2} - \omega\lambda_i}\right] = \frac{1}{2}\left[\frac{\lambda_i\eta_i^{1/2} - 1}{\eta_i^{1/2} - \frac{\omega\lambda_i}{2}}\right] = -\frac{1}{2}\left[\frac{1 - \lambda_i\eta_i^{1/2}}{\eta_i^{1/2} - \frac{\omega\lambda_i}{2}}\right] \tag{5.41}$$

As ω increases (from $\omega = 1$) the derivative $d\left(\eta_i^{1/2}\right)/d\omega$ remains negative, and therefore $\eta_i^{1/2}$ decreases, as long as $\eta_i^{1/2} > \omega\lambda_i/2$. Eventually a value $\omega = \omega_i$ is reached such that the denominator of the right-hand side of Eq. (5.41) becomes zero, hence [from Eq. (5.40)]:

$$\omega_i^2\lambda_i^2 - 4(\omega_i - 1) = 0. \tag{5.42}$$

From Eq. (5.38c) one obtains (using $2\eta_i^{1/2} = \omega\lambda_i$) the minimum value of η_i:

$$\eta_i - 2\eta_i^{1/2}\eta_i^{1/2} + \omega - 1 = 0,$$

$$\eta_i = \omega - 1, \tag{5.43a}$$

$$\eta_i^{1/2} = \sqrt{\omega - 1}. \tag{5.43b}$$

The value of ω which makes $\eta_i^{1/2}$ (hence η_i) minimum is obtained from Eq. (5.42) ($\omega_i^2\lambda_i^2 - 4\omega_i + 4 = 0$) and is:

$$\omega_i = \frac{2 \pm \sqrt{4 - 4\lambda_i^2}}{\lambda_i^2}. \tag{5.44}$$

In Eq. (5.44) we take the negative sign because ω_i must not exceed the value 2. Hence, from Eq. (5.44) we obtain:

$$\omega_i = \frac{2\left(1 - \sqrt{1 - \lambda_i^2}\right)}{\lambda_i^2} = \frac{2(1 - 1 + \lambda_i^2)}{\lambda_i^2\left(1 + \sqrt{1 - \lambda_i^2}\right)} = \frac{2}{1 + \sqrt{1 - \lambda_i^2}}. \tag{5.45}$$

To this value of ω_i corresponds the minima:

$$\eta_{\min}^{1/2} = \sqrt{\omega_i - 1}, \tag{5.46}$$

$$\eta_{\min} = \omega_i - 1 = \frac{2}{1 + \sqrt{1 - \lambda_i^2}} - 1 = \frac{1 - \sqrt{1 - \lambda_i^2}}{1 + \sqrt{1 - \lambda_i^2}}. \tag{5.47}$$

Let us now assume that the roots of $P(\lambda) = 0$ are ordered in the decreasing sequence. To the maximum value λ_1 corresponds the maximum $\eta = \eta_1\left(= \lambda_1^2, \text{if } \omega = 1\right)$. The value of ω which minimizes this maximum value of η is:

Table 5.1 Relationship between roots λ_i of $P(\lambda) = 0$ and roots η_i of $Q(\eta) = 0$

If $\omega = 1$: To every $\lambda_i = 0$ corresponds $\rightarrow \eta_i = 0$	
To every couple $\pm\lambda_i \neq 0 \rightarrow$ the couple	$\begin{cases} \eta_i = 0 \\ \eta_i = \lambda_i^2 \end{cases}$
If $\omega \neq 1$: To every $\lambda_i = 0 \rightarrow \eta_i = -(\omega - 1)$	
To every couple $\pm\lambda_i \neq 0 \rightarrow$ 2 solutions of Eq. (5.37)	

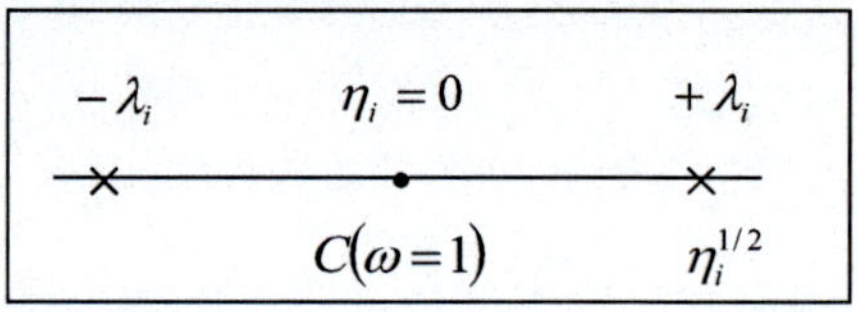

Fig. 5.3 Relationship between couples of roots $\pm\lambda_i$ and $\eta_i = 0$, $\eta_i = \lambda_i^2$ for the circle $C(\omega = 1)$

$$\omega_{\text{opt}} = \omega_1 = \frac{2}{1 + \sqrt{1 - \lambda_1^2}}, \tag{5.48}$$

to which corresponds:

$$\eta_{\min} = \{\max{}_i |\eta_i(\omega)|\}_{\min} = \omega_{opt} - 1 = \frac{1 - \sqrt{1 - \lambda_1^2}}{1 + \sqrt{1 - \lambda_1^2}}. \tag{5.49}$$

The following Table 5.1 summarizes the results obtained:

The behavior of the roots of the equation $Q(\eta) = 0$ can be examined in the complex plane, in which we define a circle $C_\omega = \{\eta : |\eta| = \omega - 1\}$. For $\omega = 1$, the circle C_ω collapses to the origin of the complex plane, and the situation of the roots is shown in Fig. 5.3:

The figure shows that the couple $\pm\lambda_i$ of roots of $P(\lambda) = 0$ corresponds the couple $\eta_i = 0$, $\eta_i = \lambda_i^2$ of roots of $Q(\eta) = 0$.

In case $1 < \omega < \omega_{\mathbf{opt}}$, there are some couples of real roots (η_1, η_1'), $(\eta_2, \eta_2')\ldots$, but the other couples are complex conjugate. To the roots $\lambda_i = 0$ of $P(\lambda) = 0$ correspond roots with moduli $\eta_i = (\omega - 1)$. The situation is sketched in Fig. 5.4a.

When ω increases the points which represent real roots, like (η_i, η_i'), move toward each other (toward the circle C_ω), and when ω reaches the value given by Eq. (5.45), the couple meets on the circle C_{ω_i}, because [cfr. Eq. (5.47)] then it holds $\eta_i = \eta_i' = \omega_i - 1$.

When $\omega = \omega_{\mathbf{opt}}$, all non-zero roots η_i are complex, but the couple η_1, η_1' which coincide on the circle $C_{\omega_{\text{opt}}} = \{\eta : |\eta| = \eta_1 = \eta_1' = \omega_{\text{opt}} - 1\}$. In this case all eigenvalues of the matrix $H(\omega = \omega_{\mathbf{opt}})$ have the same modulus $\omega_{\mathbf{opt}} - 1$. The situation is sketched in Fig. 5.4b.

The dependence of the maximum of the modulus of $\eta_i(\omega)$ [$\max_i|\eta_i(\omega)|$] on the parameter ω is shown in Fig. 5.5.

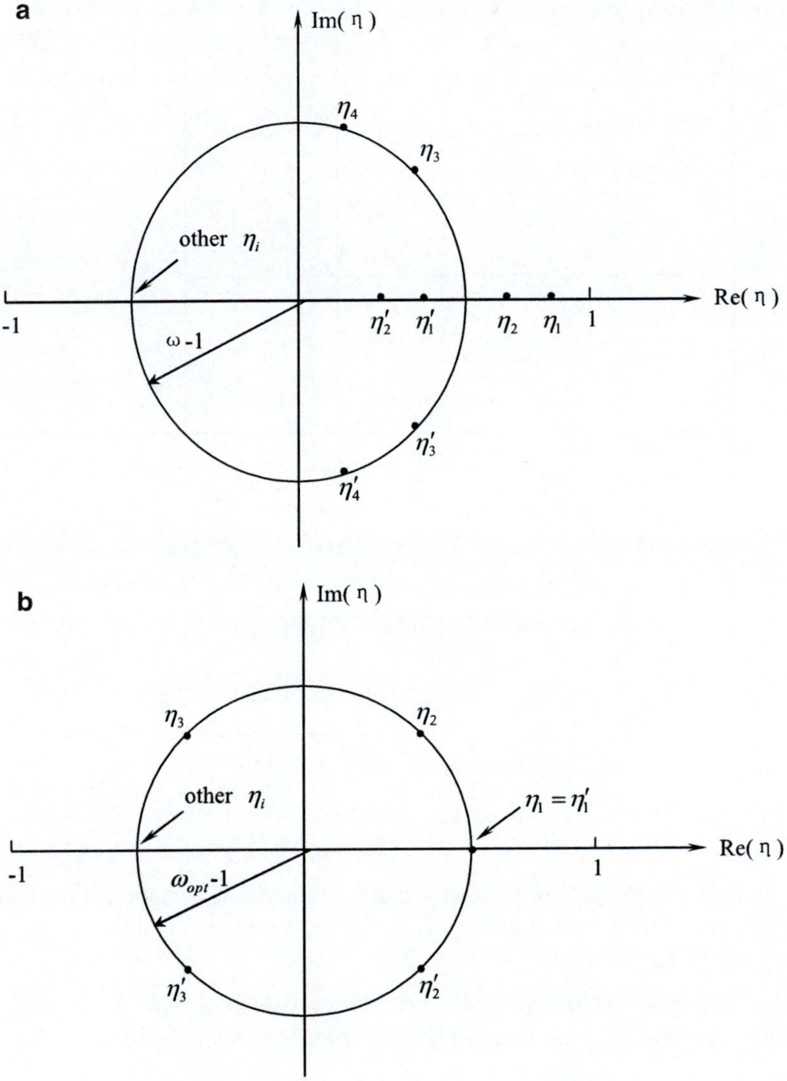

Fig. 5.4 (**a**) Loci of the roots of $Q(\eta) = 0$ for $1 < \omega < \omega_{\text{opt}}$. (**b**) Loci of the roots of $Q(\eta) = 0$ for $\omega = \omega_{\text{opt}}$

The convergence of the method of successive over-relaxations requires that all eigenvalues λ_i of the Jacobi method be smaller than one. In fact if one eigenvalue λ_i is greater than one, from Eq. (5.38d) follows that for one of the two roots $\eta^{1/2}$, it holds:

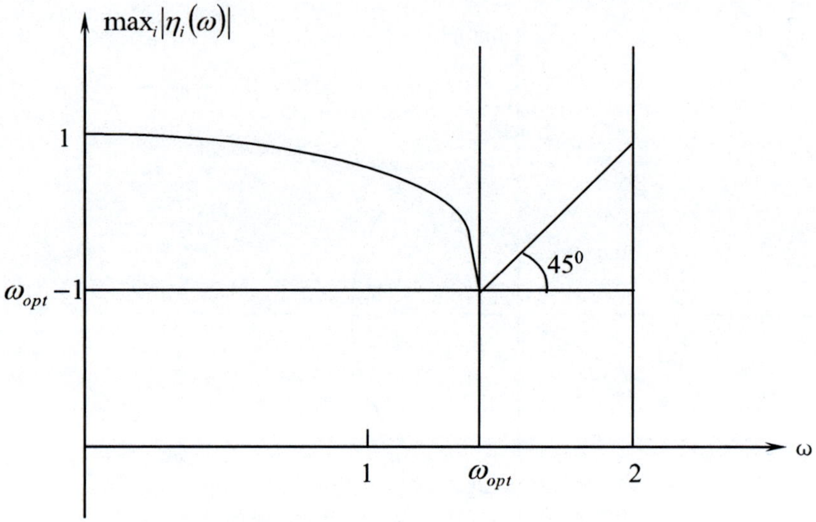

Fig. 5.5 Spectral radius of the SOR method versus relaxation parameter ω

$$\begin{aligned}
\left|\eta_i^{1/2}\right| &> \frac{1}{2}|\omega| + \frac{1}{2}\left|\sqrt{\omega^2\lambda_i^2 - 4(\omega-1)}\right| \\
&> \frac{1}{2}|\omega| + \frac{1}{2}\sqrt{\omega^2 - \omega^2 - 4\omega + 4 + \omega^2\lambda_i^2} \\
&> \frac{1}{2}|\omega| + \frac{1}{2}\sqrt{(\omega-2)^2 + \omega^2\left(\lambda_i^2 - 1\right)} \\
&> \frac{1}{2}|\omega| + \frac{1}{2}|\omega - 2| \geq 1.
\end{aligned} \tag{5.50}$$

The following theorem is important because it establishes a relationship between the eigenvalues of the iteration matrix H_J of the Jacobi method and the matrix A of the given Eq. (5.7).

Theorem 5.4: *If the matrix* $A = E + D + F$ *is symmetric and* $\det D \neq 0$, *the eigenvalues* λ_i *of the iteration matrix* H_J *of the Jacobi method are real.*

Proof: The eigenvalues of H_J are given by Eq. (5.51):

$$P(\lambda) = \det(H_J - \lambda I) = \det(E + F + \lambda D) = 0. \tag{5.51}$$

Because, by assumption $\det D \neq 0$, also $\det D^{-1/2} \neq 0$. Multiplying Eq. (5.51) at the left side by ($-\det D^{-1/2}$), and at the right side by ($\det D^{-1/2}$), yields:

$$\begin{aligned}
-\det\left(D^{-1/2}\right)\det[(E+F) + \lambda D]\det\left(D^{-1/2}\right) &= 0, \\
\det\left[-D^{-1/2}(E+F)D^{-1/2} - \lambda D^{-1/2}DD^{-1/2}\right] &= 0, \\
\det\left[-D^{-1/2}(E+F)D^{-1/2} - \lambda I\right] = 0 &= 0,
\end{aligned} \tag{5.52}$$

having used the identity $D^{-1/2}DD^{-1/2} = D^{-1/2}D^{1/2}D^{1/2}D^{-1/2} = I \cdot I = I$. Equation (5.52) is equivalent to Eq. (5.51) and gives the eigenvalues of the matrix $-D^{-1/2}(E+F)$ $D^{-1/2}$. This matrix is symmetric (because by hypothesis A is symmetric), and its eigenvalues, hence the eigenvalues of H_J, are real.

From the practical viewpoint, the most important result of the theory of SOR is the following theorem:

Theorem 5.5: *The SOR method for the solution of the matrix equation $AX = B$ converges for $0 < \omega < 2$ if the matrix A is symmetric and positive definite.*

In practice this condition is often satisfied. It is however a sufficient, not necessary, condition.

Example 5.4

(i) *Statement of the problem.*

Given a symmetric matrix, calculate the eigenvalues of the iteration matrices of the Jacobi, SOR, and Gauss-Seidel methods, and prove the theoretical relationships among them.

Let us consider the symmetric matrix:

$$A = \begin{bmatrix} 1 & -\frac{1}{2} & \frac{1}{6} \\ -\frac{1}{2} & 2 & 0 \\ \frac{1}{6} & 0 & 3 \end{bmatrix}. \tag{5.53}$$

According to Eq. (5.12) we define the matrices:

$$E + F = \begin{bmatrix} 0 & 0 & 0 \\ -\frac{1}{2} & 0 & 0 \\ \frac{1}{6} & 0 & 0 \end{bmatrix} + \begin{bmatrix} 0 & -\frac{1}{2} & \frac{1}{6} \\ 0 & 0 & 0 \\ 0 & 0 & 0 \end{bmatrix} = \begin{bmatrix} 0 & -\frac{1}{2} & \frac{1}{6} \\ -\frac{1}{2} & 0 & 0 \\ \frac{1}{6} & 0 & 0 \end{bmatrix}, \tag{5.54}$$

$$D = \begin{bmatrix} 1 & 0 & 0 \\ 0 & 2 & 0 \\ 0 & 0 & 3 \end{bmatrix}, \tag{5.55}$$

hence $\det(D) = 6 \neq 0$ and:

$$D^{-1} = \begin{bmatrix} 1 & 0 & 0 \\ 0 & \frac{1}{2} & 0 \\ 0 & 0 & \frac{1}{3} \end{bmatrix}, \tag{5.56}$$

$$D^{-1/2} = \begin{bmatrix} 1 & 0 & 0 \\ 0 & \frac{1}{\sqrt{2}} & 0 \\ 0 & 0 & \frac{1}{\sqrt{3}} \end{bmatrix}. \tag{5.57}$$

With the usual "rows times columns" matrix multiplication, from (5.54) and (5.57) one obtains:

$$(E+F)D^{-1/2} = \begin{bmatrix} 0 & -\frac{1}{2} & \frac{1}{6} \\ -\frac{1}{2} & 0 & 0 \\ \frac{1}{6} & 0 & 0 \end{bmatrix} \begin{bmatrix} 1 & 0 & 0 \\ 0 & \frac{1}{\sqrt{2}} & 0 \\ 0 & 0 & \frac{1}{\sqrt{3}} \end{bmatrix} = \begin{bmatrix} 0 & -\frac{1}{2\sqrt{2}} & \frac{1}{6\sqrt{3}} \\ -\frac{1}{2} & 0 & 0 \\ \frac{1}{6} & 0 & 0 \end{bmatrix}, \tag{5.58}$$

hence, from (5.57) and (5.58)

$$\begin{aligned} D^{-1/2}(E+F)D^{-1/2} &= \begin{bmatrix} 1 & 0 & 0 \\ 0 & \frac{1}{\sqrt{2}} & 0 \\ 0 & 0 & \frac{1}{\sqrt{3}} \end{bmatrix} \begin{bmatrix} 0 & -\frac{1}{2\sqrt{2}} & \frac{1}{6\sqrt{3}} \\ -\frac{1}{2} & 0 & 0 \\ \frac{1}{6} & 0 & 0 \end{bmatrix} \\ &= \begin{bmatrix} 0 & -\frac{1}{2\sqrt{2}} & \frac{1}{6\sqrt{3}} \\ -\frac{1}{2\sqrt{2}} & 0 & 0 \\ \frac{1}{6\sqrt{3}} & 0 & 0 \end{bmatrix}. \end{aligned} \tag{5.59}$$

(ii) *Eigenvalues of the iteration matrix of the Jacobi method.*

The eigenvalues of the iteration matrix H_J of the Jacobi method are given by the equation:

$$P(\lambda) = \det(H_J - \lambda I) = \det(E + F + \lambda D) = 0, \tag{5.60}$$

which was shown to be equivalent to:

$$\det\left[-D^{-1/2}(E + F)D^{-1/2} - \lambda I\right] = 0. \tag{5.61}$$

In this example, using (5.59), one has:

$$\det\left[-D^{-1/2}(E + F)D^{-1/2} - \lambda I\right] = \begin{bmatrix} -\lambda & -\frac{1}{2\sqrt{2}} & \frac{1}{6\sqrt{3}} \\ -\frac{1}{2\sqrt{2}} & -\lambda & 0 \\ \frac{1}{6\sqrt{3}} & 0 & -\lambda \end{bmatrix} = 0. \tag{5.62}$$

With a few calculations, from Eq. (5.62) one obtains the equation:

$$\lambda\left(\lambda^2 - \frac{29}{216}\right) = 0, \tag{5.63}$$

which has the solutions:

$$\lambda_1 = 0, \tag{5.64a}$$

$$\lambda_2 = \sqrt{\frac{29}{216}}, \tag{5.64b}$$

$$\lambda_3 = -\sqrt{\frac{29}{216}}. \tag{5.64c}$$

It is easy to verify that these solutions are also obtained straightly from Eq. (5.60), which in this case becomes:

$$P(\lambda) = \det(H_J - \lambda I) = \det(E + F + \lambda D) = \begin{bmatrix} \lambda & -\frac{1}{2} & \frac{1}{6} \\ -\frac{1}{2} & 2\lambda & 0 \\ \frac{1}{6} & 0 & 3\lambda \end{bmatrix} = 0. \tag{5.65}$$

Thus Theorem 5.4 is verified: because the matrix A is symmetric and $\det(D) \neq 0$, the eigenvalues of the iteration matrix of the Jacobi method are real.

(iii) *Eigenvalues of the iteration matrix of the SOR method.*

The eigenvalues of the iteration matrix $H(\omega)$ of the SOR method are given by:

$$H(\omega) = -\left(E + \frac{D}{\omega}\right)^{-1}\left[F + \frac{\omega - 1}{\omega}D\right]. \tag{5.66}$$

Let us define the auxiliary matrix:

$$Y = E + \frac{D}{\omega} = \begin{bmatrix} \frac{1}{\omega} & 0 & 0 \\ -\frac{1}{2} & \frac{2}{\omega} & 0 \\ \frac{1}{6} & 0 & \frac{3}{\omega} \end{bmatrix}, \tag{5.67}$$

which has determinant $\det(Y) = 6/\omega^3$ and cofactors:

$$\begin{aligned} Y_{11} &= 6/\omega^2, & Y_{12} &= -3/2\omega, & Y_{13} &= -1/3\omega, \\ Y_{21} &= 0, & Y_{22} &= 3/\omega^2, & Y_{23} &= 0, \\ Y_{31} &= 0, & Y_{32} &= 0, & Y_{33} &= 2/\omega^2, \end{aligned} \tag{5.68}$$

Letting:

$$X = Y^{-1} = \left(E + \frac{D}{\omega}\right)^{-1}, \tag{5.69}$$

the elements of the matrix X are given by $X_{rs} = (-1)^{r+s}\frac{Y_{sr}}{\det(Y)}$, which yields:

$$X = Y^{-1} = \left(E + \frac{D}{\omega}\right)^{-1} = \begin{bmatrix} \omega & 0 & 0 \\ \frac{\omega^2}{4} & \frac{\omega}{2} & 0 \\ \frac{-\omega^2}{18} & 0 & \frac{\omega}{3} \end{bmatrix}. \tag{5.70}$$

From (5.54) and (5.55), one obtains:

$$F + \frac{\omega - 1}{\omega}D = \begin{bmatrix} \frac{\omega - 1}{\omega} & -\frac{1}{2} & \frac{1}{6} \\ 0 & 2\frac{\omega - 1}{\omega} & 0 \\ 0 & 0 & 3\frac{\omega - 1}{\omega} \end{bmatrix}. \tag{5.71}$$

From (5.70) and (5.71), one obtains:

$$
\begin{aligned}
H(\omega) &= -\left(E+\frac{D}{\omega}\right)^{-1}\left[F+\frac{\omega-1}{\omega}D\right] = \\
&= -\begin{bmatrix} \omega & 0 & 0 \\ \frac{\omega^2}{4} & \frac{\omega}{2} & 0 \\ \frac{-\omega^2}{18} & 0 & \frac{\omega}{3} \end{bmatrix}\begin{bmatrix} \frac{\omega-1}{\omega} & -\frac{1}{2} & \frac{1}{6} \\ 0 & 2\frac{\omega-1}{\omega} & 0 \\ 0 & 0 & 3\frac{\omega-1}{\omega} \end{bmatrix} = \\
&= -\begin{bmatrix} \omega-1 & -\frac{\omega}{2} & \frac{\omega}{6} \\ \frac{\omega}{4}(\omega-1) & -\frac{\omega^2}{8}+(\omega-1) & \frac{\omega^2}{24} \\ -\frac{\omega}{18}(\omega-1) & \frac{\omega^2}{36} & -\frac{\omega^2}{108}+(\omega-1) \end{bmatrix}.
\end{aligned}
\tag{5.72}
$$

The eigenvalues of the iteration matrix $H(\omega)$ are the solutions of the algebraic equation:

$$
\begin{aligned}
Q(\eta) &= \det(H(\omega)-\eta I) = \\
&= \begin{bmatrix} -(\omega-1)-\eta & \frac{\omega}{2} & -\frac{\omega}{6} \\ -\frac{\omega}{4}(\omega-1) & \frac{\omega^2}{8}-(\omega-1)-\eta & -\frac{\omega^2}{24} \\ \frac{\omega}{18}(\omega-1) & -\frac{\omega^2}{36} & \frac{\omega^2}{108}-(\omega-1)-\eta \end{bmatrix} = 0
\end{aligned}
\tag{5.73}
$$

Developing the determinant, with a rather lengthy calculation, one obtains:

$$
Q(\eta) = (\eta+\omega-1)\left[(\eta+\omega-1)^2 - \frac{29}{216}(\eta+\omega-1)\omega^2 + \frac{29}{216}(\omega-1)\omega^2\right] = 0.
\tag{5.74}
$$

Equation (5.74) has one solution:

$$
\eta_1 = -(\omega-1),
\tag{5.75}
$$

and two further solutions from the second-degree algebraic equation obtained setting to zero the terms in square brackets, which yields:

$$\eta^2 + \eta\left[2(\omega - 1) - \frac{29}{216}\omega^2\right] + (\omega - 1)^2 = 0. \tag{5.76}$$

Solutions of Eq. (5.76) are:

$$\eta_2 = -(\omega - 1) + \frac{29}{216}\frac{\omega^2}{2} - \frac{1}{2}\sqrt{\left(\frac{29}{216}\right)^2\omega^4 - 4\frac{29}{216}\omega^2(\omega - 1)}, \tag{5.77a}$$

$$\eta_3 = -(\omega - 1) + \frac{29}{216}\frac{\omega^2}{2} + \frac{1}{2}\sqrt{\left(\frac{29}{216}\right)^2\omega^4 - 4\frac{29}{216}\omega^2(\omega - 1)}. \tag{5.77b}$$

Taking into considerations the eigenvalues of the iteration matrix H_J of the Jacobi method, given by formula (5.64a, 5.64b, 5.64c), and the eigenvalues of the iterations matrix of the SOR method, given by (5.75) and (5.77a, 5.77b), one find the correlations which were summarized in Table 5.1 for $\omega \neq 1$:

To the eigenvalue $\lambda_1 = 0$ corresponds the eigenvalue $\eta_1 = -(\omega - 1)$.

To the couples of eigenvalues $\lambda_i = \pm\sqrt{\frac{29}{216}}$ (i = 2, 3) correspond the eigenvalues η_2 and η_3 which satisfy Eq. (5.37), as it is easy to verify.

The optimum relaxation parameter for the SOR method is given by Eq. (5.48):

$$\omega_{\text{opt}} = \frac{2}{1 + \sqrt{1 - \lambda_1^2}} = \frac{2}{1 + \sqrt{1 - \frac{29}{216}}} = 1.036. \tag{5.78}$$

(iv) *Eigenvalues of the iteration matrix of the Gauss-Seidel method.*

With the relaxation parameter $\omega = 1$, Eq. (5.74) becomes:

$$\eta^2(216\eta - 29) = 0), \tag{5.79}$$

which has one solution:

$$\eta = 0, \tag{5.80a}$$

with order of multiplicity equal to 2, and one third solution:

$$\eta_3 = 29/216 = \lambda_i^2, (i = 2, 3). \tag{5.80b}$$

Example 5.5

(i) *Statement of the problem.*

Given the matrix:

$$A = \begin{bmatrix} 1 & a & a \\ a & 1 & 0 \\ a & 0 & 1 \end{bmatrix}, \tag{5.81}$$

show that it has "Property A," and there is an ordering vector for A, and find the eigenvalues of the iteration matrices of the Jacobi and Gauss-Seidel methods and proof of their relationship.

Matrix A may represent a physical problem concerning three bodies related to each other as shown in the following sketch: Body 1 is linked to both bodies 2 and 3; bodies 2 and 3 are linked only to body 1. The link is represented by the number <a> in the matrix.

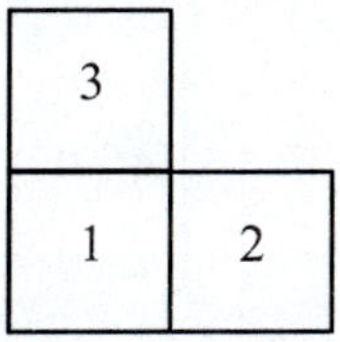

Consider the two sets of natural numbers $S_1\{1\}$, $S_2\{2, 3\}$. For the set $W = \{1, 2, 3\}$ it holds $W = \cup S_k (k = 1, 2)$. For every element $a_{ij} \neq 0$, $(i \neq j)$ of A, it holds either $i \in S_1, j \in S_2$ or $i \in S_2, j \in S_1$. Therefore, according to Definition (v), the matrix A has "Property A."

Consider now the vector $Y_o = |1, 0, 0|^T$. For the off-diagonal, non-zero elements of the matrix A, one has:

$$\begin{aligned} a_{12} &= |Y_1 - Y_2| = |1 - 0| = 1, \\ a_{13} &= |Y_1 - Y_3| = |1 - 0| = 1, \\ a_{21} &= |Y_2 - Y_1| = |0 - 1| = 1, \\ a_{31} &= |Y_3 - Y_1| = |0 - 1| = 1, \end{aligned}$$

therefore, according to Definition (iii), Y_o is an ordering vector.

Matrix (5.81) is also consistently ordered with suitable ordering vector $Y_{so} = |1, 2, 2|^T$. In fact:

$$\begin{aligned} a_{12} &= Y_1 - Y_2 = 1 - 2 = -1, \quad (i < j) \\ a_{13} &= Y_1 - Y_3 = 1 - 2 = -1, \quad (i < j) \\ a_{21} &= Y_2 - Y_1 = 2 - 1 = 1, \quad (i > j) \\ a_{31} &= Y_3 - Y_1 = 2 - 1 = 1. \quad (i > j) \end{aligned}$$

According to Eq. (5.12), we define:

$$D = D^{-1} = \begin{bmatrix} 1 & 0 & 0 \\ 0 & 1 & 0 \\ 0 & a & 1 \end{bmatrix} = I, \tag{5.82}$$

$$E = \begin{bmatrix} 0 & 0 & 0 \\ a & 0 & 0 \\ a & 0 & 0 \end{bmatrix}, \tag{5.83}$$

$$F = \begin{bmatrix} 0 & a & a \\ 0 & 0 & 0 \\ 0 & 0 & 0 \end{bmatrix}, \tag{5.84}$$

$$E + F = \begin{bmatrix} 0 & a & a \\ a & 0 & 0 \\ a & 0 & 0 \end{bmatrix}. \tag{5.85}$$

(ii) *Eigenvalues of the iteration matrix of the Jacobi method.*

The iteration matrix of the Jacobi method is:

$$H_J = -D^{-1}(E + F) = = \begin{bmatrix} 0 & -a & -a \\ -a & 0 & 0 \\ -a & 0 & 0 \end{bmatrix}, \tag{5.86}$$

and its eigenvalues are the roots of the equation:

$$P(\lambda) = \det(H_J - \lambda I) = \begin{bmatrix} -\lambda & -a & -a \\ -a & -\lambda & 0 \\ -a & 0 & -\lambda \end{bmatrix} = -(1)^3 \begin{bmatrix} \lambda & a & a \\ a & \lambda & 0 \\ a & 0 & \lambda \end{bmatrix} = 0. \tag{5.87}$$

From Eq. (5.87) one obtains:

$$\lambda(\lambda^2 - 2a^2) = 0, \tag{5.88}$$

which has solutions:

$$\lambda_1 = 0, \tag{5.89a}$$

$$\lambda_2 = a\sqrt{2}, \tag{5.89b}$$

$$\lambda_3 = -a\sqrt{2}. \tag{5.89c}$$

(iii) *Eigenvalues of the iteration matrix of the Gauss-Seidel method.*

We define the auxiliary matrix:

$$C = (E + D) = \begin{bmatrix} 1 & 0 & 0 \\ a & 1 & 0 \\ a & 0 & 1 \end{bmatrix}, \tag{5.90}$$

which has determinant $|C| = 1$, and inverse:

$$C^{-1} = (E + D)^{-1} = \begin{bmatrix} 1 & 0 & 0 \\ -a & 1 & 0 \\ -a & 0 & 1 \end{bmatrix}. \tag{5.91}$$

The iteration matrix of the Gauss-Seidel method is [see Eq. (5.29)]:

$$H_{GS} = -(E + D)^{-1}F = \begin{bmatrix} -1 & 0 & 0 \\ a & -1 & 0 \\ a & 0 & -1 \end{bmatrix} \begin{bmatrix} 0 & a & a \\ 0 & 0 & 0 \\ 0 & 0 & 0 \end{bmatrix} = \begin{bmatrix} 0 & -a & -a \\ 0 & a^2 & a^2 \\ 0 & a^2 & a^2 \end{bmatrix}. \tag{5.92}$$

The eigenvalues of the iteration matrix H_{GS} are the solutions of the algebraic equation:

$$Q(\eta) = \det(H_{GS} - \eta I) = \begin{bmatrix} -\eta & -a & -a \\ 0 & a^2 - \eta & a^2 \\ 0 & a^2 & a^2 - \eta \end{bmatrix} = \eta^2 \left[\eta - 2a^2\right] = 0. \tag{5.93}$$

Equation (5.93) has solutions:

$$\eta_1 = 0, \tag{5.94a}$$

$$\eta_2 = 0, \tag{5.94b}$$

$$\eta_3 = 2a^2. \tag{5.94c}$$

To the solution $\lambda_1 = 0$ (5.89a) corresponds $\eta_1 = 0$; to the couple $\lambda_2 = a\sqrt{2}$, $\lambda_3 = -a\sqrt{2}$, (5.89b) and (5.89c), corresponds the couple $\eta_2 = 0$ and $\eta_3 = 2a^2$.

Example 5.6

(i) *Statement of the problem.*

Given the matrix:

$$A = \begin{bmatrix} 1 & a & a \\ a & 1 & a \\ a & a & 1 \end{bmatrix} \tag{5.95}$$

which has not "Property A," show that the theoretical relationships given in Table 5.1 *are not satisfied. In other words, show that "Property A" is a necessary condition for the theory of SOR to be applicable.*

Matrix A may represent a physical problem concerning three bodies. Each one of these bodies is coupled to the other two by means of the link represented by the number <a> in the matrix.

According to Eq. (5.12), we define:

$$D = D^{-1} = \begin{bmatrix} 1 & 0 & 0 \\ 0 & 1 & 0 \\ 0 & 0 & 1 \end{bmatrix} = I, \tag{5.96}$$

$$E = \begin{bmatrix} 0 & 0 & 0 \\ a & 0 & 0 \\ a & a & 0 \end{bmatrix}, \tag{5.97}$$

$$F = \begin{bmatrix} 0 & a & a \\ 0 & 0 & a \\ 0 & 0 & 0 \end{bmatrix}, \tag{5.98}$$

$$E + F = \begin{bmatrix} 0 & a & a \\ a & 0 & a \\ a & a & 0 \end{bmatrix}. \tag{5.99}$$

(ii) *Eigenvalues of the iteration matrix of the Jacobi method.*

The iteration matrix of the Jacobi method is:

$$H_J = -D^{-1}(E+F) = \begin{bmatrix} 0 & -a & -a \\ -a & 0 & -a \\ -a & -a & 0 \end{bmatrix}, \quad (5.100)$$

and its eigenvalues are the roots of the equation:

$$P(\lambda) = \det(H_J - \lambda I) = \begin{bmatrix} -\lambda & -a & -a \\ -a & -\lambda & -a \\ -a & -a & -\lambda \end{bmatrix} = -(1)^3 \begin{bmatrix} \lambda & a & a \\ a & \lambda & a \\ a & a & \lambda \end{bmatrix} = 0. \quad (5.101)$$

Developing the determinant in Eq. (5.101), one derives:

$$P(\lambda) = \lambda^3 - 3a^2\lambda + 2a^3 = 0. \quad (5.102)$$

Equation (5.102) has root $\lambda = a$ with order of multiplicity equal to 2. In fact $\lambda = a$ is also a root of the derivative $\frac{dP(\lambda)}{d\lambda} = 3\lambda^2 - 3a^2$. Therefore the polynomial $P(\lambda) = \lambda^3 - 3a^2\lambda + 2a^3$ can be divided by$(\lambda - a)^2$, which yields the factorization

$$P(\lambda) = (\lambda - a)^2(\lambda + 2a) = 0. \quad (5.103)$$

Therefore the roots of Eq. (5.102) are:

$$\lambda_1 = \lambda_2 = a, \quad (5.104a)$$

$$\lambda_3 = -2a. \quad (5.104b)$$

(iii) *Eigenvalues of the iteration matrix of the Gauss-Seidel method.*

We define the auxiliary matrix:

$$C = (E+D) = \begin{bmatrix} 1 & 0 & 0 \\ a & 1 & 0 \\ a & a & 1 \end{bmatrix}, \quad (5.105)$$

which has determinant $|C| = 1$, and inverse:

$$C^{-1} = (E+D)^{-1} = \begin{bmatrix} 1 & 0 & 0 \\ -a & 1 & 0 \\ (a^2 - a) & -a & 1 \end{bmatrix}. \quad (5.106)$$

The iteration matrix of the Gauss-Seidel method is [see Eq. (5.29)]:

$$\begin{aligned} H_{GS} &= -(E+D)^{-1}F = -\begin{bmatrix} 1 & 0 & 0 \\ -a & 1 & 0 \\ (a^2-a) & -a & 1 \end{bmatrix}\begin{bmatrix} 0 & a & a \\ 0 & 0 & a \\ 0 & 0 & 0 \end{bmatrix} \\ &= \begin{bmatrix} 0 & -a & -a \\ 0 & a^2 & (a^2-a) \\ 0 & (a^2-a^3) & (2a^2-a^3) \end{bmatrix}. \end{aligned} \tag{5.107}$$

The eigenvalues of the iteration matrix H_{GS} are the solutions of the algebraic equation:

$$\begin{aligned} Q(\eta) &= \det(H_{GS}-\eta I) = \begin{bmatrix} -\eta & -a & -a \\ 0 & a^2-\eta & (a^2-a) \\ 0 & (a^2-a^3) & (2a^2-a^3)-\eta \end{bmatrix} \\ &= \eta\left[\eta^2+\eta\left(a^3-3a^2\right)+a^3\right] = 0. \end{aligned} \tag{5.108}$$

Equation (5.108) has solutions:

$$\eta_1 = 0, \tag{5.109a}$$

$$\eta_2 = \frac{1}{2}\left[-\left(a^3-3a^2\right)+\sqrt{(a^3-3a^2)^2-4a^3}\right], \tag{5.109b}$$

$$\eta_3 = \frac{1}{2}\left[-\left(a^3-3a^2\right)-\sqrt{(a^3-3a^2)^2-4a^3}\right]. \tag{5.109c}$$

Because $\eta_i \neq \lambda_i^2$ $(i = 2, 3)$ the Young-Frankel theory of SOR does not apply; an optimum relaxation parameter cannot be computed nor is guaranteed that the above explained iteration algorithms converge.

Acknowledgments Results of this example are given without proof in Forsythe and Wasow (1960).

5.3 Relaxation Methods by Minimization of a Functional (Variational Methods)

Variational methods consist in the minimization of a functional which leads to the solution of a system of linear algebraic equations. In this section we consider a variant of variational methods which leads to a general formulation of relaxation methods, while in the next section we will consider variants yielding several classes of gradient methods.

5.3.1 N-Dimensional Case

Let X and $\overline{X}$ be an approximate solution and the exact solution, respectively, of the matrix equation:

$$AX = B, \tag{5.110}$$

where A is matrix of order N, assumed to be positive definite and symmetric. Let:

$$Y = \overline{X} - X, \tag{5.111}$$

be defined as error vector and:

$$R = \sum\nolimits_i r_i = AX - B = AX - A\overline{X} = A\left(X - \overline{X}\right) = -AY, \tag{5.112}$$

be the residual vector. Let $(u, v) = \sum_i u_i v_i$ denote the scalar product of any two vectors u, v.

We recall that, denoting with A^H the Hermitian (transpose conjugate) matrix of A, it always holds:

$$(Au, v) = \left(A^H v, u\right), \tag{5.113}$$

for any vectors u, v. If A is real $A^H = A^T$ (transpose). If A is real and symmetric $A^H = A^T = A$; hence

$$(Au, v) = (Av, u). \tag{5.114}$$

We define $f(X) = (AY, Y)$ as error function. This functional can be computed by:

$$\begin{aligned} f(X) &= (AY, Y) = \left(A\left(\overline{X} - X\right), \overline{X} - X\right) \\ &= \left(A\overline{X}, \overline{X}\right) - \left(A\overline{X}, X\right) - \left(AX, \overline{X}\right) + (AX, X) \\ &= \left(A\overline{X}, \overline{X}\right) - 2\left(A\overline{X}, X\right) + (AX, X), \\ &= (AX, X) - 2(B, X) + \left(B, \overline{X}\right), \end{aligned} \tag{5.115}$$

having used Eq. (5.114) and $A\overline{X} = B$.

The functional $f(X)$, by the assumptions that the matrix A is positive definite, is always positive, but for $X = \overline{X}$ the functional reaches its minimum value $f\left(\overline{X}\right) = 0$. To find the solution of Eq. (5.110) therefore implies to minimize the functional $f(X)$. Because the last term at the right-hand side of Eq. (5.115), $\left(B, \overline{X}\right)$, is a constant, we can consider as well the functional:

$$\phi(X) = (AX, X) - 2(B, X) \quad \left[= f(X) - \left(B, \overline{X}\right)\right] \text{ [From Eq. (5.115)]}$$

$$= \sum_{ij} a_{ij} x_j x_i - 2\sum_i b_i x_i, \tag{5.116}$$

which, for $X = \overline{X}$, has the minimum value $-(B, \overline{X})$.

Let us define the unit vector $\xi_i = |0, 0, \ldots 1, 0, \ldots 0|^T$ of order N which has all components equal to zero, but the i-th component which is one. Starting from an approximate solution X, we try to obtain a better solution X' just varying the i-th component x_i of X, by letting:

$$X' = X + \alpha\xi_i, \tag{5.117}$$

where α is a constant to be determined, such that $f(X') < f(X)$.

The error vector, corresponding to $\overline{X}'$, is:

$$Y' = \overline{X} - X' = \overline{X} - X - \alpha\xi_i = Y - \alpha\xi_i, \tag{5.118}$$

and the function error $f(X')$ is:

$$\begin{aligned} f(X') &= f(X + \alpha\xi_i) = (AY', Y') = (A(Y - \alpha\xi_i), (Y - \alpha\xi_i)) \\ &= (AY, Y) - \alpha(AY, \xi_i) - \alpha(A\xi_i, Y) + \alpha^2(A\xi_i, \xi_i) \\ &= (AY, Y) - 2\alpha(AY, \xi_i) + \alpha^2(A\xi_i, \xi_i) \\ &= (AY, Y) + 2\alpha(R, \xi_i) + \alpha^2(A\xi_i, \xi_i) \\ &= f(X) + 2\alpha r_i + \alpha^2 a_{ii}. \end{aligned} \tag{5.119}$$

The value of the parameter α which minimizes $f(X')$ is obtained by:

$$\frac{df(X')}{d\alpha} = 2r_i + 2\alpha a_{ii} = 0, \tag{5.120}$$

which yields:

$$\alpha = -\frac{r_i}{a_{ii}}. \tag{5.121}$$

A relaxation parameter ω can also be used by letting:

$$\alpha = -\omega\frac{r_i}{a_{ii}}. \qquad (0 < \omega < 2) \tag{5.122}$$

The condition $\omega < 2$ arises from the condition that $f(X')$ must be smaller than $f(X)$ if the solution has to be improved; hence it must be $f(X') - f(X) < 0$.

From Eq. (5.119), using Eq. (5.122), the condition $f(X') - f(X) < 0$ becomes

Table 5.2 Computational scheme to relax the i-th component ($i = 1, 2, \ldots N$) of the approximate solution X

Old residuals	Subsequent approximations	Parameters α_i	New residuals
	$X^0(=X)$		
$R^1 = AX^0 - B$	$X^1 = X^0 + \alpha_1\xi_1$	$\alpha_1 = -r_1/a_{11}$	$R'^1 = AX^1 - B$
$R^2 = AX^1 - B$	$X^2 = X^1 + \alpha_2\xi_2$	$\alpha_2 = -r_2/a_{22}$	$R'^2 = AX^2 - B$
$R^i = AX^{i-1} - B$	$X^i = X^{i-1} + \alpha_i\xi_i$	$\alpha_i = -r_i/a_{ii}$	$R'^i = AX^i - B$
$R^N = AX^{N-1} - B$	$X^N = X^{N-1} + \alpha_N\xi_N(=X')$	$\alpha_N = -r_N/$ $a_{NN}R'^N = AX^N - B$	

$$f(X') - f(X) = 2\alpha r_i + \alpha^2 a_{ii} = -2\frac{\omega r_i}{a_{ii}} r_i + \frac{\omega^2 r_i^2}{a_{ii}^2} a_{ii} = \frac{\omega r_i^2}{a_{ii}}(\omega - 2) < 0, \tag{5.123}$$

hence $\omega < 2$.

Let us assume $\omega = 1$. Equation (5.117), with α given by Eq. (5.121), can be used for every component of the approximate solution X to obtain better solution X', with the scheme of Table 5.2, starting from $X^0 = X$ to obtain$X^N = X'$:

At every step the corresponding component of the new residual vector is zeroed (or "relaxed"). For instance, at the i-th step the component $r'_i = \left(R'^i, \xi_i\right)$ of R' is zeroed:

$$\begin{aligned} r'_i &= \left(R'^i, \xi_i\right) = \left(AX'^i - B, \xi_i\right) = \left(A\left(X^{i-1} + \alpha_i\xi_i\right) - B, \xi_i\right) \\ &= A\left(X^{i-1}, \xi_i\right) + \alpha_i(A\xi_i, \xi_i) - (B, \xi_i) \\ &= \left(AX^{i-1} - B, \xi_i\right) + \alpha_i(A\xi_i, \xi_i) \\ &= \left(R^i, \xi_i\right) + \alpha_i A(\xi_i, \xi_i) \\ &= r_i + \alpha_i a_{ii} = 0, \end{aligned} \tag{5.124}$$

because of the choice of α_i [Eq. (5.121)].

5.3.2 *Geometrical Representation of the Two-Dimensional Case*

Let us consider a two-dimensional case for the matrix equation $AX = B$, in the space (x_1, x_2), with:

$$A = \begin{vmatrix} a_{11} & a_{12} \\ a_{21} & a_{22} \end{vmatrix}, \qquad (a_{12} = a_{21}) \tag{5.125}$$

$$B = \begin{vmatrix} b_1 \\ b_2 \end{vmatrix}, \tag{5.126}$$

$$\overline{X} = \begin{vmatrix} \overline{x}_1 \\ \overline{x}_2 \end{vmatrix}, \qquad \text{(exact solution)} \tag{5.127}$$

$$X = \begin{vmatrix} x_1 \\ x_2 \end{vmatrix}, \qquad \text{(approximate solution)} \tag{5.128}$$

$$Y = \overline{X} - X = \begin{vmatrix} y_1 \\ y_2 \end{vmatrix} = \begin{vmatrix} \overline{x}_1 - x_1 \\ \overline{x}_2 - x_2 \end{vmatrix}, \qquad \text{(error vector)} \tag{5.129}$$

$$R = AX - B = \begin{vmatrix} r_1 \\ r_2 \end{vmatrix} = \begin{vmatrix} a_{11}x_1 + a_{12}x_2 - b_1 \\ a_{21}x_1 + a_{22}x_2 - b_2 \end{vmatrix}. \qquad \text{(residual vector)} \tag{5.130}$$

Hence one computes:

$$(AX, X) = \left(\begin{vmatrix} a_{11} & a_{12} \\ a_{21} & a_{22} \end{vmatrix} \cdot \begin{vmatrix} x_1 \\ x_2 \end{vmatrix}, \begin{vmatrix} x_1 \\ x_2 \end{vmatrix} \right), \tag{5.131}$$

$$= (a_{11}x_1 + a_{12}x_2)x_1 + (a_{21}x_1 + a_{22}x_2)x_2, \tag{5.132}$$

$$(B, \overline{X}) = b_1\overline{x}_1 + b_2\overline{x}_2, \tag{5.133}$$

$$(B, X) = b_1x_1 + b_2x_2, \tag{5.134}$$

and the functional $f(X)$ becomes [Cfr. Eq. (5.115)]:

$$\begin{aligned} f(X) &= (AY, Y) = (AX, X) - 2(B, X) + (B, \overline{X}) \\ &= (a_{11}x_1 + a_{12}x_2)x_1 + (a_{21}x_1 + a_{22}x_2)x_2 - 2(b_1x_1 + b_2x_2) + b_1\overline{x}_1 + b_2\overline{x}_2 \\ &= a_{11}x_1^2 + 2a_{12}x_1x_2 + a_{22}x_2^2 - 2b_1x_1 - 2b_2x_2 + b_1\overline{x}_1 + b_2\overline{x}_2 \end{aligned} \tag{5.135}$$

The functional $\phi(X)$ is given by [Cfr. Eq. (5.116)]:

$$\begin{aligned} \phi(X) &= f(X) - (B, \overline{X}) = (AX, X) - 2(B, X) \\ &= a_{11}x_1^2 + 2a_{12}x_1x_2 + a_{22}x_2^2 - 2b_1x_1 - 2b_2x_2. \end{aligned} \tag{5.136}$$

The equation:

$$\phi(X) = a_{11}x_1^2 + 2a_{12}x_1x_2 + a_{22}x_2^2 - 2b_1x_1 - 2b_2x_2 = m \tag{5.137}$$

represents a family of ellipses because detA $= a_{11}a_{22} - a_{12}^2 > 0$ for the assumption made that the matrix A is positive definite. Varying the parameter m, the ellipse of

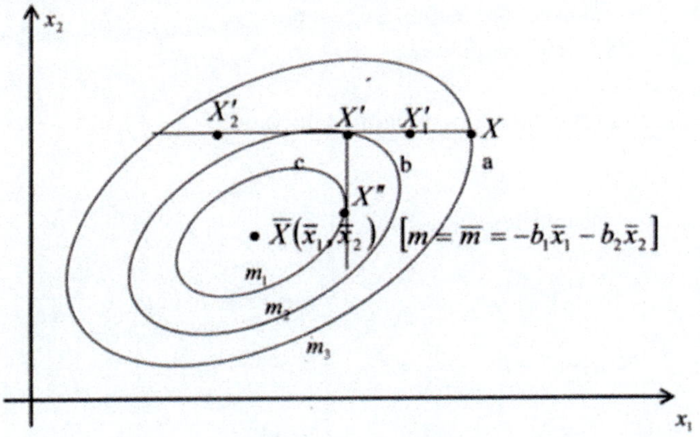

Fig. 5.6 Family of ellipses of equation $\phi(X) = m$ [Cfr. Eq. (5.137)]

the family varies. For $x_1 = \bar{x}_1$, $x_2 = \bar{x}_2$ (real solution), one has $f(\bar{X}) = 0$ and $\phi(\bar{X}) = \bar{m} = -b_1\bar{x}_1 - b_2\bar{x}_2$ which is the minimum value of m for which the ellipses of the family (5.137) are real. Actually for $m = \bar{m} = -b_1\bar{x}_1 - b_2\bar{x}_2$ the ellipse $\phi(X) = \bar{m}$ collapses to the point $\bar{X}(\bar{x}_1, \bar{x}_2)$; for $m < \bar{m}$ the ellipses become imaginary. This situation is shown in Fig. 5.6.

If, in Fig. 5.6, the point X on the ellipse represents a first approximate solution, an improved solution X', on the ellipse b, is obtained by relaxing the x_1 component of X (x_1, x_2) by letting:

$$\begin{aligned} x_1' &= x_1 + \alpha_1 \qquad (\alpha_1 = -r_1/a_{11}) \\ x_2' &= x_2. \end{aligned} \tag{5.138}$$

The new residuals become:

$$R' = AX' - B = \begin{vmatrix} a_{11}x_1' + a_{12}x_2' - b_1 \\ a_{21}x_1' + a_{22}x_2' - b_2 \end{vmatrix} = \begin{vmatrix} a_{11}\left(x_1 - \dfrac{r_1}{a_{11}}\right) + a_{12}x_2 - b_1 \\ a_{21}\left(x_1 - \dfrac{r_1}{a_{11}}\right) + a_{22}x_2 - b_2 \end{vmatrix}$$
$$= \begin{vmatrix} a_{11}x_1 + a_{12}x_2 - b_1 - r_1 \\ a_{21}x_1 + a_{22}x_2 - b_2 - \dfrac{a_{21}}{a_{11}}r_1 \end{vmatrix} = \begin{vmatrix} r_1 - r_1 \\ r_2 - \dfrac{a_{21}}{a_{11}}r_1 \end{vmatrix} = \begin{vmatrix} 0 \\ r_2 - \dfrac{a_{21}}{a_{11}}r_1 \end{vmatrix} = \begin{vmatrix} r_1' \\ r_2' \end{vmatrix}. \tag{5.139}$$

If one chooses $\alpha = -\omega r_1/a_{11}$, with $\omega < 1$, one goes from point X to point X_1' (cfr. Fig. 5.6) because an under-relaxation is used; if $1 < \omega < 2$ one goes from point X to point X_2' with over-relaxation.

The approximate solution X' is now used to relax the second component x_2, thus going from point X' to X'', by letting:

$$\begin{aligned} x_1'' &= x_1', \\ x_2'' &= x_2' + \alpha_2. \quad \left(\alpha_2 = -r_2'/a_{22}\right) \end{aligned} \tag{5.140}$$

After this second relaxation, the residuals become:

$$\begin{aligned} R'' &= AX'' - B = \begin{vmatrix} a_{11}x_1'' + a_{12}x_2'' - b_1 \\ a_{21}x_1'' + a_{22}x_2'' - b_2 \end{vmatrix} = \begin{vmatrix} a_{11}x_1' + a_{12}\left(x_2' - \dfrac{r_2'}{a_{22}}\right) - b_1 \\ a_{21}x_1' + a_{22}\left(x_2' - \dfrac{r_2'}{a_{22}}\right) - b_2 \end{vmatrix} \\ &= \begin{vmatrix} a_{11}x_1' + a_{12}x_2' - b_1 - \dfrac{a_{12}}{a_{22}}r_2' \\ a_{21}x_1' + a_{22}x_2' - b_2 - r_2' \end{vmatrix} = \begin{vmatrix} r_1' - \dfrac{a_{12}}{a_{22}}r_2' \\ r_2' - r_2' \end{vmatrix} = \begin{vmatrix} 0 - \dfrac{a_{12}}{a_{22}}r_2' \\ 0 \end{vmatrix} = \begin{vmatrix} r_1'' \\ r_2'' \end{vmatrix}. \end{aligned} \tag{5.141}$$

One continues relaxing one component x_1 or x_2 at the time and the solution approaches point $\overline{X}(\overline{x}_1, \overline{x}_2)$.

Example 5.7

Given the system $\begin{cases} x_1 - x_2 = 1 \\ -x_1 + 2x_2 = 2 \end{cases}$, one has $A = \begin{vmatrix} 1 & -1 \\ -1 & 2 \end{vmatrix}$, $B = \begin{vmatrix} 1 \\ 2 \end{vmatrix}$. The exact solution is $\overline{X} = |4, 3|^T$. Starting from $X^0 = |0, 0|^T$, one finds the subsequent approximate solutions and residuals:

$$X^0 = \begin{vmatrix} 0 \\ 0 \end{vmatrix}, \quad R^0 = \begin{vmatrix} -1 \\ -2 \end{vmatrix},$$

$$X^1 = \begin{vmatrix} 1 \\ 0 \end{vmatrix}, \quad R^1 = \begin{vmatrix} 0 \\ -3 \end{vmatrix},$$

$$X^2 = \begin{vmatrix} 1 \\ 1.5 \end{vmatrix}, \quad R^2 = \begin{vmatrix} -1.5 \\ 0 \end{vmatrix},$$

$$X^3 = \begin{vmatrix} 2.5 \\ 1.5 \end{vmatrix}, \quad R^3 = \begin{vmatrix} 0 \\ -1.5 \end{vmatrix},$$

$$X^4 = \begin{vmatrix} 2.5 \\ 9/4 \end{vmatrix}, \quad R^4 = \begin{vmatrix} -3/4 \\ 0 \end{vmatrix},$$

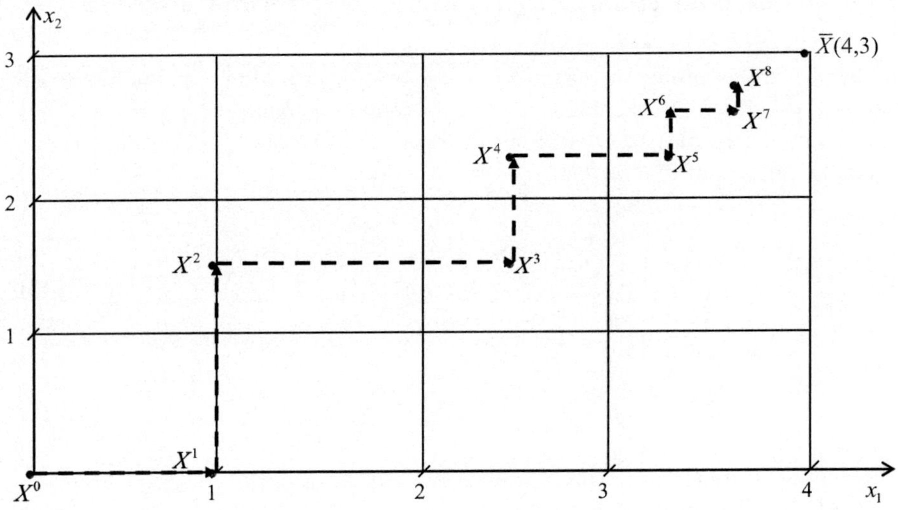

Fig. 5.7 Subsequent approximate solutions for Example 5.7

$$X^5 = \begin{vmatrix} 13/4 \\ 9/4 \end{vmatrix}, \quad R^5 = \begin{vmatrix} 0 \\ -3/4 \end{vmatrix},$$

$$X^6 = \begin{vmatrix} 13/4 \\ 21/8 \end{vmatrix}, \quad R^6 = \begin{vmatrix} -3/8 \\ 0 \end{vmatrix},$$

$$X^7 = \begin{vmatrix} 29/8 \\ 21/8 \end{vmatrix}, \quad R^7 = \begin{vmatrix} 0 \\ -3/8 \end{vmatrix},$$

$$X^8 = \begin{vmatrix} 29/8 \\ 45/16 \end{vmatrix}, \quad R^8 = \begin{vmatrix} -3/16 \\ 0 \end{vmatrix}.$$

These results are plotted in Fig. 5.7.

5.4 Gradient Methods as Subclass of Variational Methods

Gradient methods can also be considered as a class of variational methods because they are obtained by minimization of a functional. Depending on the choice of the functional one has different variants of the gradient methods.

5.4.1 Gradient Methods Proper (Methods of Fastest Descent)

Gradient methods proper consists in moving from a provisional solution X to an improved solution X' proceeding along the gradient of the family of hyper- ellipsoids (in the N-dimensional space) defined by the equation $\phi(X) = m$. We consider the two following variants of the method.

Variant 1 Set:

$$X' = X - \alpha R, \tag{5.142}$$

and minimize the functional:

$$\phi(X') = (AX', X') - 2(B, X'). \tag{5.143}$$

Recalling that $\phi(X) = (AX, X) - 2(B, X)$ [Crf. Eq. (5.116)] and that$(AX, R) = (AR, X)$, one obtains:

$$\begin{aligned}
\phi(X') &= (AX', X') - 2(B, X') \\
&= (A(X - \alpha R), X - \alpha R) - 2(B, X - \alpha R) \\
&= (AX, X) - \alpha(AX, R) - \alpha(AR, X) + \alpha^2(AR, R) - 2(B, X) + 2\alpha(B, R) \\
&= \phi(X) - 2\alpha(AX, R) + \alpha^2(AR, R) + 2\alpha(B, R) \\
&= \phi(X) - 2\alpha(AX - B, R) + \alpha^2(AR, R) \\
&= \phi(X) - 2\alpha(R, R) + \alpha^2(AR, R).
\end{aligned} \tag{5.144}$$

The optimum value of the parameter α, which minimizes the functional $\phi(X')$, is obtained by:

$$\frac{d\phi(X')}{d\alpha} = -2(R, R) + 2\alpha(AR, R) = 0, \tag{5.145}$$

and is:

$$\alpha = \frac{(R, R)}{(AR, R)}. \tag{5.146}$$

A sequence of solutions converging to the real solution $\overline{X}$ is obtained by the scheme:

$$X^{r+1} = X^r - \alpha^r R^r, \tag{5.147}$$

with:

$$\alpha^r = \frac{(R^r, R^r)}{(AR^r, R^r)}. \tag{5.148}$$

Variant 2 Set:

$$X' = X - \beta R, \tag{5.149}$$

and minimize the scalar product (R', R') of the residuals:

$$\begin{aligned} R' &= AX' - B \\ &= A(X - \beta R) - B = AX - B - \beta AR = R - \beta AR. \end{aligned} \tag{5.150}$$

One obtains:

$$\begin{aligned} (R', R') &= (R - \beta AR, R - \beta AR) \\ &= (R, R) - \beta(R, AR) - \beta(AR, R) + \beta^2(AR, AR) \\ &= (R, R) - 2\beta(AR, R) + \beta^2(AR, AR). \end{aligned} \tag{5.151}$$

The optimum value of the parameter β is obtained by deriving the scalar product (R', R') with respect to β:

$$\frac{d(R', R')}{d\beta} = -2(AR, R) + 2\beta(AR, AR) = 0, \tag{5.152}$$

and is:

$$\beta = \frac{(AR, R)}{(AR, AR)}. \tag{5.153}$$

A sequence of subsequent improved solution is obtained by the scheme:

$$X^{r+1} = X^r - \beta^r R^r, \tag{5.154}$$

with:

$$\beta^r = \frac{(AR^r, R^r)}{(AR^r, AR^r)}. \tag{5.155}$$

Example 5.8
Let us consider again the system:

$$\begin{cases} x_1 - x_2 = 1 \\ -x_1 + 2x_2 = 2 \end{cases},$$

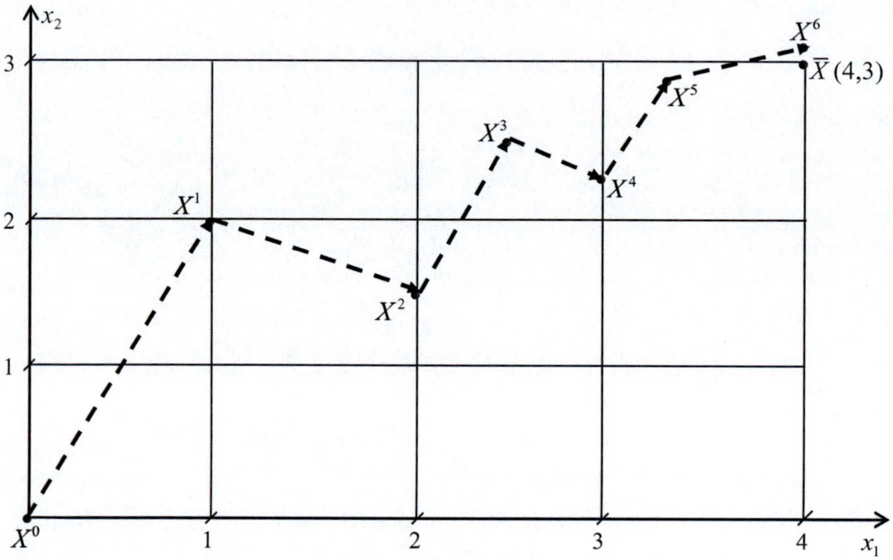

Fig. 5.8 Subsequent approximate solutions for Example 5.8

which has exact solution $\overline{X} = (4, 3)^T$. Using $X^{r+1} = X^r - \alpha^r R^r$ and $\alpha^r = \frac{(R^r, R^r)}{(AR^r, R^r)}$, and starting from $X^0 = (0, 0)^T$, one finds the subsequent approximate solutions:

$$X^0 = (0, 0)^T,$$

$$X^1 = (1, 2)^T,$$

$$X^2 = (2, 1.5)^T,$$

$$X^3 = (2.5, 2.5)^T,$$

$$X^4 = (3, 2.25)^T,$$

$$X^5 = (53/16, 23/8)^T \approx (3.31, 2.88)^T,$$

$$X^6 = (1972/488, 1519/488)^T \approx (4.05, 3.11)^T.$$

These results are plotted in Fig. 5.8.

References

Forsythe, G. E., & Wasow, W. R. (1960). *Finite-difference methods for partial differential equations*. John Wiley & Sons.

Young, D. M. (1971). *Iterative solution of large linear systems*. Academic Press.

Chapter 6
Numerical Solution of Poisson Equation

In this chapter, we consider a model Poisson equation for which we apply several classical methods yielding an almost exact solution. These methods are applicable in case of a “real” Poisson equation in which the coefficients are constant. In the next chapter, we will deal with several problems in which a Poisson-like equation is obtained. We call “Poisson like” an equation which, in its discretized form, is similar to a discretized Poisson equation, apart from the fact that both the coefficients and the right-hand side (function f) can be space and time dependent.

6.1 Model Equation

Let

$$\nabla^2 u = f, \tag{6.1}$$

with $u = u(x, y), f = f(x,y)$, be a Poisson equation defined in a domain $\boldsymbol{\Omega}$ of the (x, y) plane, with given boundary conditions on the boundary $\boldsymbol{\Gamma}$ of $\boldsymbol{\Omega}$. In the following, we assume that the definition domain is the square $\boldsymbol{\Omega} = [0 \leq x \leq 1; 0 \leq y \leq 1]$ and that $u = 0$ on $\boldsymbol{\Gamma}$.

If:

$$f(x, y) = 6xy(xy + x + y - 3)e^{x+y}, \tag{6.2}$$

Equation (6.1) with the boundary condition $u = 0$ on $\boldsymbol{\Gamma}$ has the analytical solution (Houstis et al., 1978):

M. Bottoni, *Physical Modeling and Computational Techniques for Thermal and Fluid-dynamics*, Mechanical Engineering Series,
https://doi.org/10.1007/978-3-030-79717-1_6

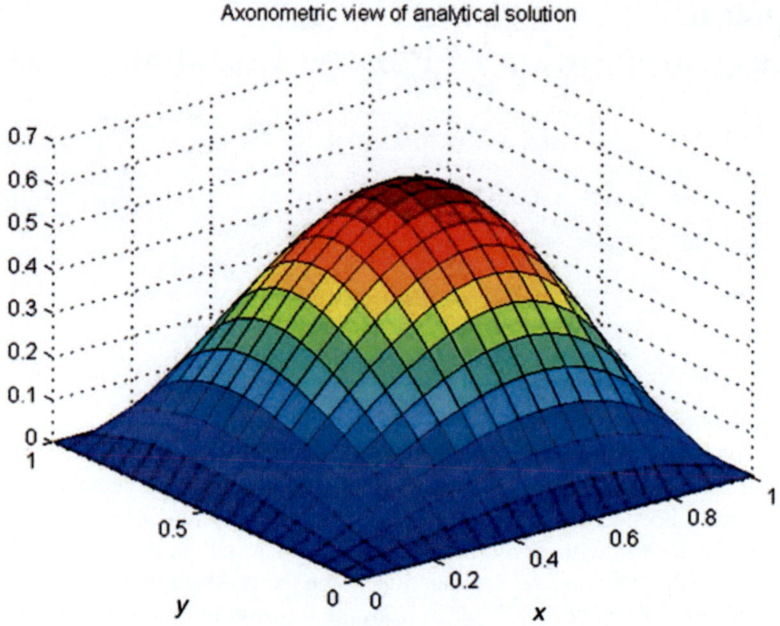

Fig. 6.1 Axonometric view of the analytical solution (6.3) of model Eqs. (6.1) and (6.2)

$$u(x,y) = 3\left(x - x^2\right)\left(y - y^2\right)e^{x+y}. \tag{6.3}$$

The maximum of this function is $u_{\max} = 0.575$, for $x_m = y_m = 0.618$. Figure 6.1 shows an axonometric view of the analytical solution (6.3).

We examine the following numerical methods to solve Poisson Eq. (6.1) with the known function $f(x,y)$ given by Eq. (6.2):

(i) 5-point formula
(ii) 9-point formula
(iii) Classical Ritz method.
(iv) Residual Method of Orthogonal Collocations (RMOC)—Galerkin variant.
(v) Residual Method of Orthogonal Collocations (RMOC)—subdomain collocations.

For the numerical treatment of methods (iii) through (v), we follow the approach outlined in (Norrie & de Vries, 1973).

6.2 5-Point Formula

Discretization of Eq. (6.1) with central differences yields:

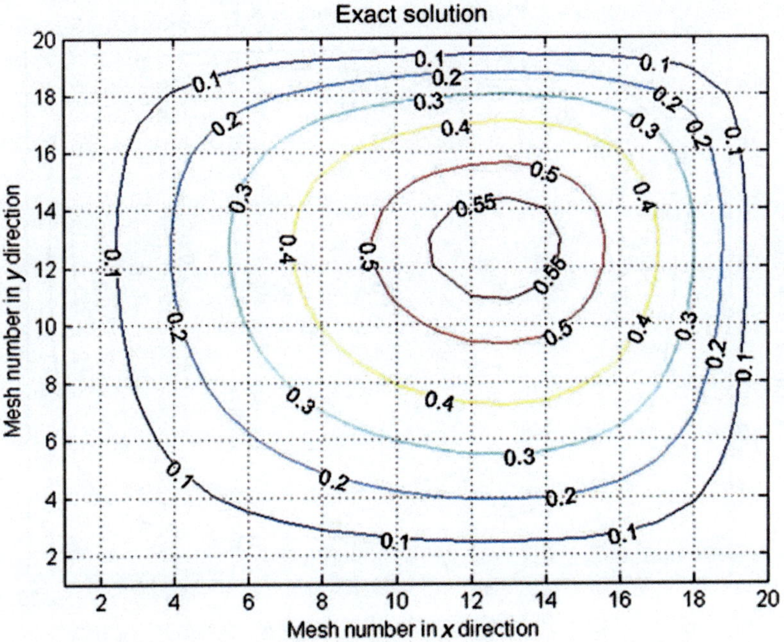

Fig. 6.2 Map of the analytical solution for the Poisson equation of example (6.1)–(6.2)

$$\frac{1}{\Delta x^2}\left(u_{i-1,j} - 2u_{i,j} + u_{i+1,j}\right) + \frac{1}{\Delta y^2}\left(u_{i,j+1} - 2u_{i,j} + u_{i,j-1}\right) = f_{ij}. \tag{6.4}$$

In case of uniform meshes ($\Delta x = \Delta y = h$), one obtains:

$$u_{i+1,j} + u_{i-1,j} + u_{i,j+1} + u_{i,j-1} - 4u_{i,j} = h^2 f_{i,j}. \tag{6.5}$$

This 5-point formula, based on central differences for the approximation of the second-order derivatives, has an accuracy of order h^2.

Numerical solution of the system of Eq. (6.5), written for all nodes in the domain $\boldsymbol{\Omega}$, is made for the model problem in the subroutine FIVEPF of the program package **PROG_VI_1** [POISSON].

Figure 6.2 shows the distribution of the analytical solution of the model problem in the domain $\boldsymbol{\Omega}$. Figure 6.3 shows the results obtained with the 5-point formula, which are practically identical with the analytical ones. In fact, the maximum residuals, as shown in Fig. 6.4, are of the order of 10^{-13}. Discretization of the domain $\boldsymbol{\Omega}$ was made with 20 × 20 cells.

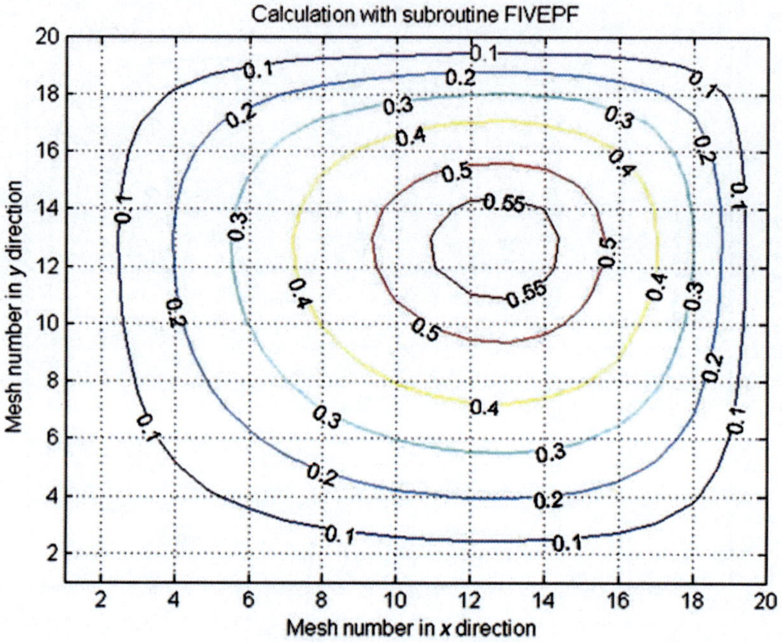

Fig. 6.3 Solution with the 5-point formula of the Poisson equation of example (6.1)–(6.2)

6.3 9-Point Formula

The 9-point formula, which has an accuracy of order h^4, can be derived as follows (Smith, 1969). Let us define the differential operators:

$$\xi = h\frac{\partial}{\partial x}, \tag{6.6}$$

$$\eta = h\frac{\partial}{\partial y}, \tag{6.7}$$

$$D^2 = \frac{\partial^2}{\partial x \partial y}. \tag{6.8}$$

From these definitions, one derives:

$$\xi\eta = h^2 D^2, \tag{6.9}$$

$$\xi^2 + \eta^2 = h^2\frac{\partial^2}{\partial x^2} + h^2\frac{\partial^2}{\partial y^2} = h^2\nabla^2, \tag{6.10}$$

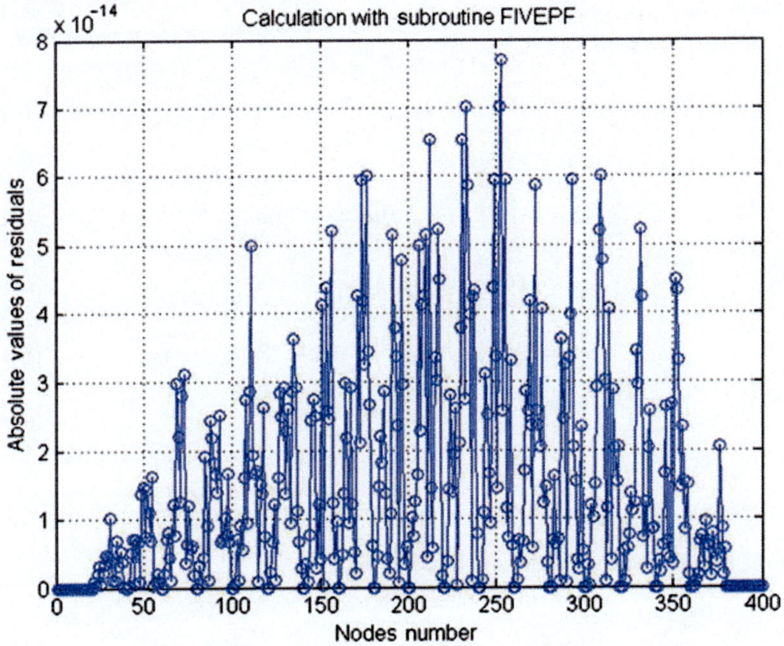

Fig. 6.4 Residuals of the solution with the 5-point formula of the Poisson equation of example (6.1)–(6.2)

$$\xi^4 + \eta^4 = \left(\xi^2 + \eta^2\right)^2 - 2\xi^2\eta^2 = h^4\left(\nabla^4 - 2D^4\right), \tag{6.11}$$

$$\xi^6 + \eta^6 = \left(\xi^2 + \eta^2\right)^3 - 3\xi^2\eta^2\left(\xi^2 + \eta^2\right) = h^6\left(\nabla^6 - 3D^4\nabla^2\right), \tag{6.12}$$

$$\xi^4\eta^2 + \xi^2\eta^4 = \xi^2\eta^2\left(\xi^2 + \eta^2\right) = h^6 D^4\nabla^2. \tag{6.13}$$

Recalling the exponential series expansion $e^n = 1 + n + n^2/2\,! + \cdots + n^n/n\,! + \cdots$, one can represent the Taylor expansion of $u = (x + h)$ around point x as follows:

$$\begin{aligned} u(x+h) =& u(x) + hu^{I}(x) + \frac{h^2}{2!}u^{II}(x) + \cdots + \frac{h^n}{n!}u^{(n)}(x) + \cdots \\ =& \left[1 + h\frac{d}{dx} + \frac{h^2}{2!}\frac{d^2}{dx^2} + \cdots + \frac{h^n}{n!}\frac{d^n}{dx^n} + \cdots\right]u(x) \\ =& \left[e^{h\frac{d}{dx}}\right]u(x) = e^{\xi}u(x) \end{aligned} \tag{6.14}$$

Let us refer to nine nodes, labelled 0–8, at the following locations on a regular grid of spacing h: $0 = (i, j)$, $1 = (i - 1, j)$, $2 = (i + 1, j)$, $3 = (i, j - 1)$, $4 = (i, j + 1)$, $5 = (i + 1, j + 1)$, $6 = (i - 1, j + 1)$, $7 = (i - 1, j - 1)$, $8 = (i + 1, j - 1)$.

With reference to these nodes, we can write:

$$u_2 = u(x+h,y) = e^{\xi}u_0, \tag{6.15a}$$

$$u_1 = u(x-h,y) = e^{-\xi}u_0, \tag{6.15b}$$

$$u_4 = u(x,y+h) = e^{\eta}u_0, \tag{6.15c}$$

$$u_3 = u(x,y-h) = e^{-\eta}u_0, \tag{6.15d}$$

$$u_5 = \left[e^{h\frac{d}{dy}}\right]u_2 = e^{\eta}u_2 = e^{\xi+\eta}u_0, \tag{6.16a}$$

$$u_6 = \left[e^{h\frac{d}{dy}}\right]u_1 = e^{\eta}u_1 = e^{-\xi+\eta}u_0, \tag{6.16b}$$

$$u_7 = \left[e^{h\frac{d}{dy}}\right]u_2 = e^{-\eta}u_1 = e^{-\xi-\eta}u_0, \tag{6.16c}$$

$$u_8 = \left[e^{h\frac{d}{dy}}\right]u_2 = e^{-\eta}u_2 = e^{\xi-\eta}u_0. \tag{6.16d}$$

We define $S_{1-4} = \sum_{i=1}^{4} u_i$ and $S_{5-8} = \sum_{i=5}^{8} u_i$ and then find the values of these sums in terms of the differential factors defined above:

$$S_{1-4} = \sum_{i=1}^{4} u_i = \left(e^{\xi} + e^{-\xi} + e^{\eta} + e^{-\eta}\right)u_0. \tag{6.17}$$

Expanding the exponentials up to the sixth-order terms inclusive, and cancelling terms of opposite sign are obtained:

$$\begin{aligned} S_{1-4} &= \left[4 + \xi^2 + \eta^2 + \frac{\xi^4+\eta^4}{12} + \frac{\xi^6+\eta^6}{360}\right]u_0 \\ &= 4u_0 + h^2\nabla^2 u_0 + \frac{1}{12}h^4\left(\nabla^4 - 2D^4\right)u_0 + \frac{1}{360}h^6\left(\nabla^6 - 3D^4\nabla^2\right)u_0. \end{aligned} \tag{6.18}$$

Likewise, with expansion of the exponential up to the sixth-order terms, one obtains:

$$\begin{aligned} S_{5-8} &= u_5 + u_7 + u_8 + u_6 = \left(e^{\xi+\eta} + e^{-\xi+\eta} + e^{\xi-\eta} + e^{-(\xi-\eta)}\right)u_0 \\ &= \Big\{\underbrace{4 + (\xi+\eta)^2 + (\xi-\eta)^2}_{\text{Term (A)}} + \underbrace{\tfrac{1}{12}\left[(\xi+\eta)^4 + (\xi-\eta)^4\right]}_{\text{Term (B)}} + \underbrace{\tfrac{1}{360}\left[(\xi+\eta)^6 + (\xi-\eta)^6\right]}_{\text{Term (C)}}\Big\}u_0. \end{aligned} \tag{6.19}$$

Terms A, B, and C in Formula (6.19) are computed as follows:

$$\text{Term A} = (\xi+\eta)^2 + (\xi-\eta)^2 = 2(\xi^2+\eta^2) = 2h^2\nabla^2, \tag{6.20}$$

$$\begin{aligned}\text{Term B} &= \frac{1}{12}\left[(\xi+\eta)^4 + (\xi-\eta)^4\right] = \frac{1}{12}\left[2(\xi^4+\eta^4) + 12\xi^2\eta^2\right] \\ &= \frac{1}{6}h^4(\nabla^4 - 2D^4) + h^4D^4 = \frac{1}{6}h^4(\nabla^4 + 4D^4),\end{aligned} \tag{6.21}$$

$$\begin{aligned}\text{Term C} &= \frac{1}{360}\left[(\xi+\eta)^6 + (\xi-\eta)^6\right] = \frac{1}{360}\left[2(\xi^6+\eta^6) + 30\xi^2\eta^2(\xi^2+\eta^2)\right] \\ &= \frac{1}{180}h^6(\nabla^6 - 3D^4\nabla^2) + \frac{1}{12}h^6D^4\nabla^2 = \frac{1}{180}h^6(\nabla^6 + 12D^4\nabla^2).\end{aligned} \tag{6.22}$$

Inserting Eqs. (6.20)–(6.22) into Eq. (6.19) yields:

$$\begin{aligned}S_{5-8} &= 4u_0 + 2h^2\nabla^2u_0 + \frac{1}{6}h^4(\nabla^4 + 4D^4)u_0 \\ &\quad + \frac{1}{180}h^6(\nabla^6 + 12D^4\nabla^2)u_0.\end{aligned} \tag{6.23}$$

Solving both Eqs. (6.18) and (6.23) with respect to D^4u_0 yields:

$$\begin{aligned}2h^4D^4u_0 &= -12S_{1-4} + 48u_0 + 12h^2\nabla^2u_0 + h^4\nabla^4u_0 + \frac{h^6}{30} \\ &\quad \times (\nabla^6 - 3D^4\nabla^2)u_0.\end{aligned} \tag{6.24}$$

$$\begin{aligned}4h^4D^4u_0 &= 6S_{5-8} - 24u_0 - 12h^2\nabla^2u_0 - h^4\nabla^4u_0 - \frac{h^6}{30} \\ &\quad \times (\nabla^6 + 12D^4\nabla^2)u_0.\end{aligned} \tag{6.25}$$

Multiplying Eq. (6.24) by 2 and subtracting Eq. (6.25) yields:

$$-24S_{1-4} - 6S_{5-8} + 120u_0 + 36h^2\nabla^2u_0 + 3h^4\nabla^4u_0 + \frac{h^6}{30}(3\nabla^6 + 6D^4\nabla^2)u_0 = 0. \tag{6.26}$$

Hence, solving for $\nabla^2 u_0$, one obtains:

$$\nabla^2u_0 = \frac{1}{6h^2}(4S_{1-4} + S_{5-8} - 20u_0) - \frac{h^2}{12}\nabla^4u_0 + 0(h^4). \tag{6.27}$$

Because $\nabla^2u_0 = f_0$ and $\nabla^4u_0 = \nabla^2(\nabla^2u_0) = \nabla^2f_0$, Eq. (6.27) yields:

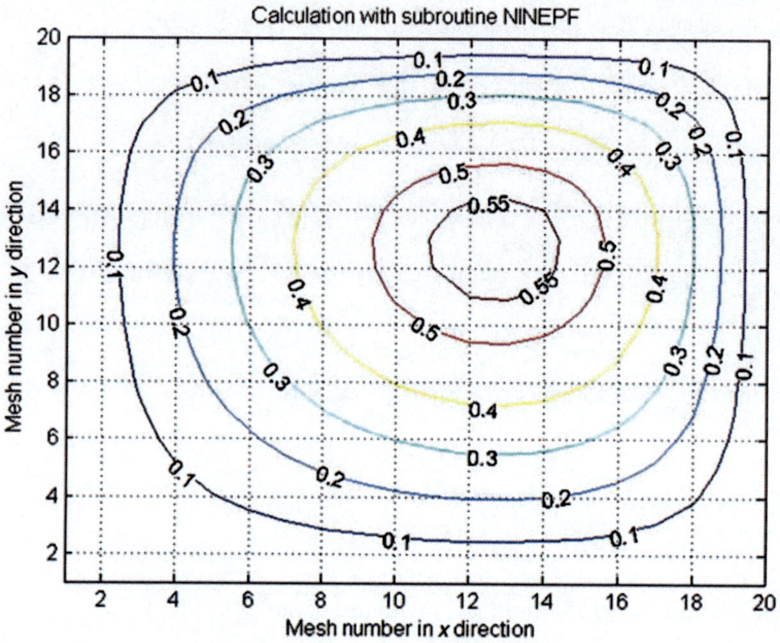

Fig. 6.5 Solution with the 9-point formula of the Poisson equation of example (6.1)–(6.2)

$$-20u_0 + 4(u_1 + u_2 + u_3 + u_4) + u_5 + u_6 + u_7 + u_8 = 6h^2 f_0 + \frac{h^4}{2}\nabla^2 f_0 \quad (6.28)$$

The 9-point formula for $\nabla^2 u = f$ can therefore be represented by:

$$\begin{bmatrix} 1 & 4 & 1 \\ 4 & -20 & 4 \\ 1 & 4 & 1 \end{bmatrix} u = 6h^2 f + \frac{h^4}{2}\nabla^2 f, \quad (6.29)$$

with

$$\nabla^2 f = 2f\left(1 + \frac{1}{x} + \frac{1}{y}\right) + 12e^x e^y (1 + x)(1 + y)(x + y). \quad (6.30)$$

Results obtained with the 9-point formula are shown in Figs. 6.5 and 6.6. Notice that the residuals are almost two orders of magnitude smaller than those obtained with the 5-point formula.

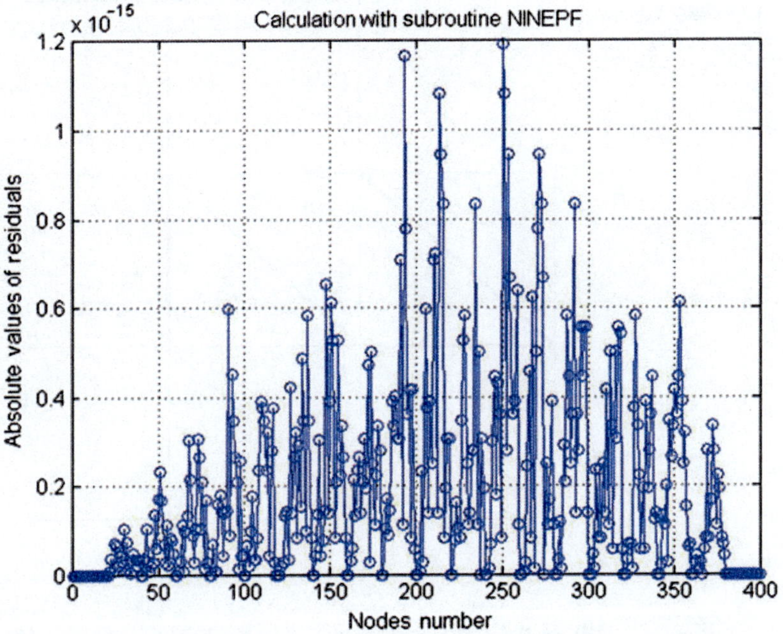

Fig. 6.6 Residuals of the solution with the 9-point formula of the Poisson equation of example (6.1)–(6.2)

6.4 Solution of Model Equation with the Classical Ritz Method

Let us partition the definition domain Ω in the arbitrary number of subdomains $\Omega_e(e = 1, 2, \ldots, NE)$ consisting of triangles with two sides parallel to the coordinate axes (this condition is not compulsory and could be relaxed) as shown in Fig. 6.7.

Let us denote with indices i, j, and m the vertices of one triangular element, with known coordinates (x_i, y_i), (x_j, y_j), and (x_m, y_m), respectively. The area Δ of the triangle is given by (Zienkiewicz, 1971):

$$\Delta = \frac{1}{2}\begin{vmatrix} 1 & x_i & y_i \\ 1 & x_j & y_j \\ 1 & x_m & y_m \end{vmatrix}. \tag{6.31}$$

We assume that the unknown function $u(x,y)$ can be represented over the element Ω_e by a bilinear function, i.e., by the equation of a plane in the space (x, y, Φ):

$$\Phi(x, y) = \alpha_1 + \alpha_2 x + \alpha_3 y. \tag{6.32}$$

When applied to the nodal points i, j, and m, Eq. (6.32) yields:

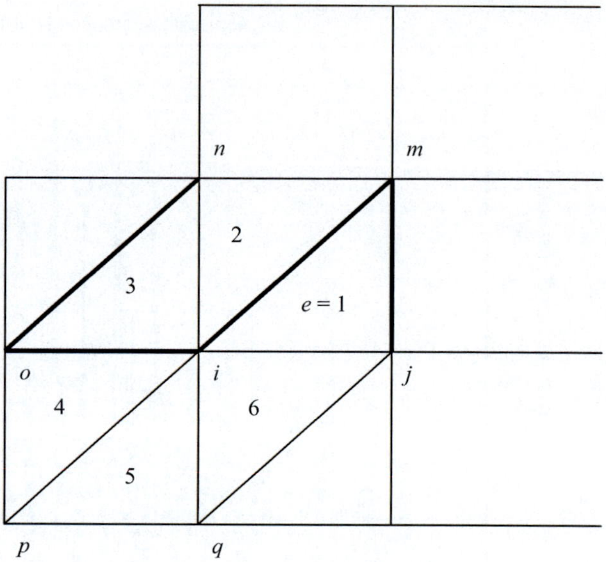

Fig. 6.7 Labelling of nodes for the solution of the Poisson equation of example (6.1)–(6.2)

$$\Phi_i = \Phi(x_i, y_i) = \alpha_1 + \alpha_2 x_i + \alpha_3 y_i, \tag{6.33a}$$

$$\Phi_j = \Phi(x_j, y_j) = \alpha_1 + \alpha_2 x_j + \alpha_3 y_j, \tag{6.33b}$$

$$\Phi_m = \Phi(x_m, y_m) = \alpha_1 + \alpha_2 x_m + \alpha_3 y_m. \tag{6.33c}$$

Coefficients α_1, α_2, α_3 in Eq. (6.33) are uniquely determined in dependence on the nodal values Φ_i, Φ_j, Φ_m because the determinant of the matrix of the coefficients of the system is 2Δ (see Eq. (6.31)) and therefore different from zero. Solving Eq. (6.33) with respect to α_1, α_2, α_3 with Cramer's rule yields:

$$\alpha_1 = \frac{1}{2\Delta}\begin{vmatrix} \Phi_i & x_i & y_i \\ \Phi_j & x_j & y_j \\ \Phi_m & x_m & y_m \end{vmatrix}$$

$$= \frac{1}{2\Delta}\left[\Phi_i(x_j y_m - x_m y_j) + \Phi_j(x_m y_i - x_i y_m) + \Phi_m(x_i y_j - x_j y_i)\right] \tag{6.34a}$$

$$\alpha_2 = \frac{1}{2\Delta}\begin{vmatrix} 1 & \Phi_i & y_i \\ 1 & \Phi_j & y_j \\ 1 & \Phi_m & y_m \end{vmatrix}$$

$$= \frac{1}{2\Delta}\left[\Phi_i(y_j - y_m) + \Phi_j(y_m - y_i) + \Phi_m(y_i - y_j)\right], \tag{6.34b}$$

$$\alpha_3 = \frac{1}{2\Delta}\begin{vmatrix} 1 & x_i & \Phi_i \\ 1 & x_j & \Phi_j \\ 1 & x_m & \Phi_m \end{vmatrix}$$

$$= \frac{1}{2\Delta}\left[\Phi_i(x_m - x_j) + \Phi_j(x_i - x_m) + \Phi_m(x_j - x_i)\right]. \tag{6.34c}$$

Introducing Eq. (6.34) into Eq. (6.32) obtains:

$$\Phi(x, y) = \frac{1}{2\Delta} \times \left[(a_i + b_i x + c_i y)\Phi_i + (a_j + b_j x + c_j y)\Phi_j + (a_m + b_m x + c_m y)\Phi_m\right] \tag{6.35}$$

with the following definitions:

$$a_i = x_j y_m - x_m y_j, \tag{6.36a}$$

$$a_j = x_m y_i - x_i y_m, \tag{6.36b}$$

$$a_m = x_i y_j - x_j y_i, \tag{6.36c}$$

$$b_i = y_j - y_m, \tag{6.36d}$$

$$b_j = y_m - y_i, \tag{6.36e}$$

$$b_m = y_i - y_j, \tag{6.36f}$$

$$c_i = x_m - x_j, \tag{6.36g}$$

$$c_j = x_i - x_m, \tag{6.36h}$$

$$c_m = x_j - x_i. \tag{6.36i}$$

Equation (6.35) can be written as:

$$\Phi(x, y) = N_i(x, y)\Phi_i + N_j(x, y)\Phi_j + N_m(x, y)\Phi_m, \tag{6.37}$$

with the following definitions of shape functions:

$$N_i(x, y) = \frac{1}{2\Delta}(a_i + b_i x + c_i y), \tag{6.38a}$$

$$N_j(x, y) = \frac{1}{2\Delta}(a_j + b_j x + c_j y), \tag{6.38b}$$

$$N_m(x,y) = \frac{1}{2\Delta}(a_m + b_m x + c_m y). \tag{6.38c}$$

Denoting $\boldsymbol{\Phi}^e$ the unknown function over the element e and using the matrix notation, Eq. (6.37) becomes:

$$\boldsymbol{\Phi}^e = \left| N_i \; N_j \; N_m \right| \begin{vmatrix} \boldsymbol{\Phi}_i \\ \boldsymbol{\Phi}_j \\ \boldsymbol{\Phi}_m \end{vmatrix}; \tag{6.39}$$

hence:

$$\boldsymbol{\Phi}^e = \{N^e\}\{\boldsymbol{\Phi}^e\}, \tag{6.40}$$

with:

$$\{N^e\} = |N_i \quad N_j \quad N_m|, \tag{6.41}$$

$$\{\boldsymbol{\Phi}^e\} = \begin{vmatrix} \boldsymbol{\Phi}_i \\ \boldsymbol{\Phi}_j \\ \boldsymbol{\Phi}_m \end{vmatrix}. \tag{6.42}$$

It is well-known from the theory of functionals (Strang & Fix, 1973) that the solution $\boldsymbol{\Phi}$ which satisfies Eq. (6.1) is given by a function ξ which minimizes the functional:

$$\mathrm{X}(\xi) = (A\xi, \xi) - 2(f, \xi) \tag{6.43}$$

when we define the operator $A = -\nabla^2$ and the scalar product of any two functions f_1, f_2 is defined on the domain $\boldsymbol{\Omega}$ by:

$$(f_1, f_2) = \iint_{\Omega} f_1 f_2 d\boldsymbol{\Omega}. \tag{6.44}$$

Using this definition of scalar product, Eq. (6.43) becomes:

$$\mathrm{X}(\xi) = \iint_{\Omega} (-\nabla^2 \xi)\xi d\boldsymbol{\Omega} - 2\iint_{\Omega} (f, \xi) d\boldsymbol{\Omega}. \tag{6.45}$$

Green's identity applied to two functions u, ξ is:

$$\iint_{\Omega}(-\nabla^2\xi)ud\Omega = \int_{\Omega}\nabla u\cdot\nabla\xi d\Omega - \int_{\Gamma}u\frac{\partial\xi}{\partial n}d\Gamma, \tag{6.46}$$

where $\vec{n}$ is in the normal to the contour Γ. In our case the contour integral vanishes because $u = 0$ on Γ. Using the Green's identity for $u = \xi$ yields:

$$\iint_{\Omega}(-\nabla^2\xi)\xi d\Omega = \iint_{\Omega}\nabla\xi\cdot\nabla\xi d\Omega = \iint_{\Omega}(\nabla\xi)^2 d\Omega. \tag{6.47}$$

Using the identity (6.47) in Eq. (6.45) yields:

$$\mathbf{X}(\xi) = \iint_{\Omega}\left[(\nabla\xi)^2 - 2f\xi\right]d\Omega = \iint_{\Omega}\left[\left(\frac{\partial\xi}{\partial x}\right)^2 + \left(\frac{\partial\xi}{\partial y}\right)^2 - 2f\xi\right]dxdy. \tag{6.48}$$

In our particular case, with $\xi = \Phi(x, y)$, given by Eq. (6.37), the functional for one element e becomes:

$$\mathbf{X}_e = \iint_{\Omega_e}\left[\left(\frac{\partial\Phi(x,y)}{\partial x}\right)^2 + \left(\frac{\partial\Phi(x,y)}{\partial y}\right)^2 - 2\Phi(x,y)f(x,y)\right]dxdy. \tag{6.49}$$

Differentiating Eq. (6.37) with the definitions (6.38) yields:

$$\frac{\partial\Phi(x,y)}{\partial x} = \frac{1}{2\Delta}\left[b_i\Phi_i + b_j\Phi_j + b_m\Phi_m\right], \tag{6.50a}$$

$$\frac{\partial\Phi(x,y)}{\partial y} = \frac{1}{2\Delta}\left[c_i\Phi_i + c_j\Phi_j + c_m\Phi_m\right]. \tag{6.50b}$$

Introducing Eq. (6.50) into Eq. (6.49) and rearranging, one obtains:

$$\begin{aligned}\mathbf{X}_e = \iint_{\Omega_e}\Bigg\{&\frac{b_i^2 + c_i^2}{4\Delta^2}\Phi_i^2 + \frac{b_j^2 + c_j^2}{4\Delta^2}\Phi_j^2 + \frac{b_m^2 + c_m^2}{4\Delta^2}\Phi_m^2 \\ &+ \frac{b_ib_j + c_ic_j}{2\Delta^2}\Phi_i\Phi_j + \frac{b_ib_m + c_ic_m}{2\Delta^2}\Phi_i\Phi_m + \frac{b_jb_m + c_jc_m}{2\Delta^2}\Phi_j\Phi_m \\ &- \frac{1}{\Delta}\left[(a_i + b_ix + c_iy)f(x,y)\right]\Phi_i - \frac{1}{\Delta}\left[(a_j + b_jx + c_jy)f(x,y)\right]\Phi_j \\ &- \frac{1}{\Delta}\left[(a_m + b_mx + c_my)f(x,y)\right]\Phi_m\Bigg\}dxdy\end{aligned} \tag{6.51}$$

The first six terms in the integrand of Eq. (6.51) are computed directly. For instance, the first term, being $\iint_{\Omega_e}dxdy = \Delta$, yields:

$$\iint_{\Omega_e} \frac{b_i^2 + c_i^2}{4\Delta^2} \Phi_i^2 dxdy = \frac{b_i^2 + c_i^2}{4\Delta^2} \Phi_i^2 \iint_{\Omega_e} dxdy = \frac{b_i^2 + c_i^2}{4\Delta} \Phi_i^2. \tag{6.52}$$

For the last three terms in the integrand of Eq. (6.51), we define:

$$F_i^e = \iint_{\Omega_e} (a_i + b_i x + c_i y) f(x, y) dxdy, \tag{6.53a}$$

$$F_j^e = \iint_{\Omega_e} (a_j + b_j x + c_j y) f(x, y) dxdy, \tag{6.53b}$$

$$F_m^e = \iint_{\Omega_e} (a_m + b_m x + c_m y) f(x, y) dxdy. \tag{6.53c}$$

These integrals F_i^e, F_j^e, F_m^e can be computed numerically when the function $f(x, y)$ is given for a particular problem.

Hence, the functional (6.51) becomes:

$$\begin{aligned} X_e = & \frac{b_i^2 + c_i^2}{4\Delta} \Phi_i^2 + \frac{b_j^2 + c_j^2}{4\Delta} \Phi_j^2 + \frac{b_m^2 + c_m^2}{4\Delta} \Phi_m^2 \\ & + \frac{1}{2\Delta} (b_i b_j + c_i c_j) \Phi_i \Phi_j + \frac{1}{2\Delta} (b_i b_m + c_i c_m) \Phi_i \Phi_m + \frac{1}{2\Delta} (b_j b_m + c_j c_m) \Phi_j \Phi_m \\ & - \frac{F_i^e}{\Delta} \Phi_i - \frac{F_j^e}{\Delta} \Phi_j - \frac{F_m^e}{\Delta} \Phi_m. \end{aligned} \tag{6.54}$$

The derivatives of the functional X_e with respect to the nodal values Φ_i, Φ_j, Φ_m are given by:

$$\frac{\partial X_e}{\partial \Phi_i} = \frac{1}{2\Delta} (b_i^2 + c_i^2) \Phi_i + \frac{1}{2\Delta} (b_i b_j + c_i c_j) \Phi_j + \frac{1}{2\Delta} (b_i b_m + c_i c_m) \Phi_m - \frac{F_i^e}{\Delta}, \tag{6.55a}$$

$$\frac{\partial X_e}{\partial \Phi_j} = \frac{1}{2\Delta} (b_j^2 + c_j^2) \Phi_j + \frac{1}{2\Delta} (b_i b_j + c_i c_j) \Phi_i + \frac{1}{2\Delta} (b_j b_m + c_j c_m) \Phi_m - \frac{F_j^e}{\Delta}, \tag{6.55b}$$

$$\frac{\partial X_e}{\partial \Phi_m} = \frac{1}{2\Delta} (b_m^2 + c_m^2) \Phi_m + \frac{1}{2\Delta} (b_i b_m + c_i c_m) \Phi_i + \frac{1}{2\Delta} \times (b_j b_m + c_j c_m) \Phi_j - \frac{F_m^e}{\Delta}. \tag{6.55c}$$

The condition of minimizing the functional $X_e = X_e\left(\Phi_p^e\right)(p = i, j, m)$ is given by: $\frac{\partial X_e}{\partial \Phi_p^e} = 0(p = i, j, m)$; hence:

$$\begin{vmatrix} \frac{\partial X_e}{\partial \Phi_i} \\ \frac{\partial X_e}{\partial \Phi_j} \\ \frac{\partial X_e}{\partial \Phi_m} \end{vmatrix} = \frac{1}{2\Delta}\begin{vmatrix} b_i^2 + c_i^2 & b_ib_j + c_ic_j & b_ib_m + c_ic_m \\ b_ib_j + c_ic_j & b_j^2 + c_j^2 & b_jb_m + c_jc_m \\ b_ib_m + c_ic_m & b_jb_m + c_jc_m & b_m^2 + c_m^2 \end{vmatrix} \cdot \begin{vmatrix} \Phi_i \\ \Phi_j \\ \Phi_m \end{vmatrix} - \frac{1}{\Delta}\begin{vmatrix} F_i^e \\ F_j^e \\ F_m^e \end{vmatrix} = 0. \tag{6.56}$$

Equation (6.56) can be written in the compact form:

$$\frac{\partial X_e}{\partial \Phi_p^e} = M^e\{\Phi^e\} - \{F^e\} = 0, \tag{6.57}$$

where M^e is the element matrix:

$$M^e = \frac{1}{2\Delta}\begin{vmatrix} b_i^2 + c_i^2 & b_ib_j + c_ic_j & b_ib_m + c_ic_m \\ b_ib_j + c_ic_j & b_j^2 + c_j^2 & b_jb_m + c_jc_m \\ b_ib_m + c_ic_m & b_jb_m + c_jc_m & b_m^2 + c_m^2 \end{vmatrix} \tag{6.58}$$

and:

$$\{F^e\} = \frac{1}{\Delta}\begin{vmatrix} F_i^e \\ F_j^e \\ F_m^e \end{vmatrix}. \tag{6.59}$$

The element in Eq. (6.57) must be assembled for every internal node, taking into account the contributions of all nodes surrounding the one considered. For instance, referring to Fig. 6.7, node i is surrounded by six other nodes (j, m, n, o, p, q) belonging to six elements labelled from 1 to 6 counterclockwise. The contribution to node i from element $e = 1$, hence from nodes j and m, is the one already computed, namely:

$$\frac{\partial X_1}{\partial \Phi_i} = \frac{1}{2\Delta_1} \left| b_i^2 + c_i^2 \quad b_i b_j + c_i c_j \quad b_i b_m + c_i c_m \right| \cdot \begin{vmatrix} \Phi_i \\ \Phi_j \\ \Phi_m \end{vmatrix} - \frac{1}{\Delta_1} \begin{vmatrix} F_i^1 \\ F_j^1 \\ F_m^1 \end{vmatrix} = 0. \tag{6.60a}$$

The contribution to node i from the element $e = 2$, hence from nodes m and n, is:

$$\frac{\partial X_2}{\partial \Phi_i} = \frac{1}{2\Delta_2} \left| b_i^2 + c_i^2 \quad b_i b_m + c_i c_m \quad b_i b_n + c_i c_n \right| \cdot \begin{vmatrix} \Phi_i \\ \Phi_m \\ \Phi_n \end{vmatrix} - \frac{1}{\Delta_2} \begin{vmatrix} F_i^2 \\ F_m^2 \\ F_n^2 \end{vmatrix} = 0. \tag{6.60b}$$

The contribution to node i from the element $e = 3$, hence from nodes n and o, is:

$$\frac{\partial X_3}{\partial \Phi_i} = \frac{1}{2\Delta_3} \left| b_i^2 + c_i^2 \quad b_i b_n + c_i c_n \quad b_i b_o + c_i c_o \right| \cdot \begin{vmatrix} \Phi_i \\ \Phi_n \\ \Phi_o \end{vmatrix} - \frac{1}{\Delta_2} \begin{vmatrix} F_i^3 \\ F_n^3 \\ F_0^3 \end{vmatrix} = 0. \tag{6.60c}$$

The contribution to node i from the element $e = 4$, hence from nodes o and p, is:

$$\frac{\partial X_4}{\partial \Phi_i} = \frac{1}{2\Delta_4} \left| b_i^2 + c_i^2 \quad b_i b_o + c_i c_o \quad b_i b_p + c_i c_p \right| \cdot \begin{vmatrix} \Phi_i \\ \Phi_o \\ \Phi_p \end{vmatrix} - \frac{1}{\Delta_2} \begin{vmatrix} F_i^4 \\ F_o^4 \\ F_p^4 \end{vmatrix} = 0. \tag{6.60d}$$

The contribution to node i from the element $e = 5$, hence from nodes p and p, is:

$$\frac{\partial X_5}{\partial \Phi_i} = \frac{1}{2\Delta_5} \left| b_i^2 + c_i^2 \quad b_i b_p + c_i c_p \quad b_i b_q + c_i c_q \right| \cdot \begin{vmatrix} \Phi_i \\ \Phi_p \\ \Phi_q \end{vmatrix} - \frac{1}{\Delta_2} \begin{vmatrix} F_i^5 \\ F_p^5 \\ F_q^5 \end{vmatrix} = 0. \tag{6.60e}$$

The contribution to node i given by element $e = 6$, hence from nodes q and j, is:

Table 6.1 Coefficients linking node i with neighboring nodes in the numerical solution of the Poisson equation with the Ritz method

$e = 1$	$e = 2$	$e = 3$	$e = 4$	$e = 5$	$e = 6$	
c_{1i}	c_{2i}	c_{3i}	c_{4i}	c_{5i}	c_{6i}	Φ_i
c_{1j}					c_{6j}	Φ_j
c_{1m}	c_{2m}					Φ_m
	c_{2n}	c_{3n}				Φ_n
		c_{3o}	c_{4o}			Φ_o
			c_{4p}	c_{5p}		Φ_p
				c_{5q}	c_{6q}	Φ_q

$$\frac{\partial \mathrm{X}_6}{\partial \Phi_i} = \frac{1}{2\Delta_6} \left| b_i^2 + c_i^2 \quad b_i b_q + c_i c_q \quad b_i b_j + c_i c_j \right| \cdot \left| \begin{matrix} \Phi_i \\ \Phi_q \\ \Phi_j \end{matrix} \right| - \frac{1}{\Delta_6} \left| \begin{matrix} F_i^6 \\ F_q^6 \\ F_j^6 \end{matrix} \right|$$

$$= 0. \tag{6.60f}$$

Therefore, the total contribution to node i given from six neighboring nodes is:

$$\frac{\partial \mathrm{X}_{1-6}}{\partial \Phi_i} = \frac{\partial \mathrm{X}_1}{\partial \Phi_i} + \frac{\partial \mathrm{X}_2}{\partial \Phi_i} + \frac{\partial \mathrm{X}_3}{\partial \Phi_i} + \frac{\partial \mathrm{X}_4}{\partial \Phi_i} + \frac{\partial \mathrm{X}_5}{\partial \Phi_i} + \frac{\partial \mathrm{X}_6}{\partial \Phi_i}. \tag{6.61}$$

The contributions to node i arising from the first term at the right-hand side of Eqs. (6.60) are visualized by Table 6.1. A table with similar structure would arises from the $\left\{ F_q^e \right\}$ ($q = i, j, m$), ($e = 1, 2, \ldots, 6$), (Eq. 6.59), in the second terms at the right-hand side of Eq. (6.60).

The first row of this table implies that all six triangles give a contribution to node i. The second row implies that the contribution of the node j to node i comes from triangles 1 and 6. The third row implies that contribution from node m to node i comes from triangles 1 and 2 and so on.

The coefficients in the table are given by:

$$c_{1i} = \frac{1}{2\Delta_1} \left(b_i^2 + c_i^2 \right) \tag{6.62a}$$

$$c_{2i} = \frac{1}{2\Delta_2} \left(b_i^2 + c_i^2 \right) \tag{6.62b}$$

$$c_{3i} = \frac{1}{2\Delta_3} \left(b_i^2 + c_i^2 \right) \tag{6.62c}$$

$$c_{4i} = \frac{1}{2\Delta_4} \left(b_i^2 + c_i^2 \right) \tag{6.62d}$$

$$c_{5i} = \frac{1}{2\Delta_5} \left(b_i^2 + c_i^2 \right) \tag{6.62e}$$

$$c_{6i} = \frac{1}{2\Delta_6}\left(b_i^2 + c_i^2\right) \tag{6.62f}$$

$$c_{1j} = \frac{1}{2\Delta_1}\left(b_i b_j + c_i c_j\right) \tag{6.62g}$$

$$c_{6j} = \frac{1}{2\Delta_6}\left(b_i b_j + c_i c_j\right) \tag{6.62h}$$

$$c_{1m} = \frac{1}{2\Delta_1}\left(b_i b_m + c_i c_m\right) \tag{6.62i}$$

$$c_{2m} = \frac{1}{2\Delta_2}\left(b_i b_m + c_i c_m\right) \tag{6.62j}$$

$$c_{2n} = \frac{1}{2\Delta_2}\left(b_i b_n + c_i c_n\right) \tag{6.62k}$$

$$c_{3n} = \frac{1}{2\Delta_3}\left(b_i b_n + c_i c_n\right) \tag{6.62l}$$

$$c_{3o} = \frac{1}{2\Delta_3}\left(b_i b_o + c_i c_o\right) \tag{6.62m}$$

$$c_{4o} = \frac{1}{2\Delta_4}\left(b_i b_o + c_i c_o\right) \tag{6.62n}$$

$$c_{4p} = \frac{1}{2\Delta_4}\left(b_i b_p + c_i c_p\right) \tag{6.62o}$$

$$c_{5p} = \frac{1}{2\Delta_5}\left(b_i b_p + c_i c_p\right) \tag{6.62p}$$

$$c_{5q} = \frac{1}{2\Delta_5}\left(b_i b_q + c_i c_q\right) \tag{6.62q}$$

$$c_{6q} = \frac{1}{2\Delta_6}\left(b_i b_q + c_i c_q\right) \tag{6.62r}$$

Taking into account all contributions to node i, Eq. (6.61) can be written as:

$$\frac{\partial \mathrm{X}_{1-6}}{\partial \Phi_i} = \begin{vmatrix} c_i & c_j & c_m & c_n & c_o & c_p & c_q \end{vmatrix} \cdot \begin{vmatrix} \Phi_i \\ \Phi_j \\ \Phi_m \\ \Phi_n \\ \Phi_o \\ \Phi_p \\ \Phi_q \end{vmatrix} - \begin{vmatrix} B_i \\ B_j \\ B_m \\ B_n \\ B_o \\ B_p \\ B_q \end{vmatrix}, \tag{6.63}$$

with:

$$c_i = c_{1i} + c_{2i} + c_{3i} + c_{4i} + c_{5i} + c_{6i}, \tag{6.64a}$$

$$c_j = c_{1j} + c_{6j}, \tag{6.64b}$$

$$c_m = c_{1m} + c_{2m}, \tag{6.64c}$$

$$c_n = c_{2n} + c_{3n}, \tag{6.64d}$$

$$c_o = c_{3o} + c_{4o}, \tag{6.64e}$$

$$c_p = c_{4p} + c_{5p}, \tag{6.64f}$$

$$c_q = c_{5q} + c_{6q}, \tag{6.64g}$$

$$B_i = F_i^1/\Delta_1 + F_i^2/\Delta_2 + F_i^3/\Delta_3 + F_i^4/\Delta_4 + F_i^5/\Delta_5 + F_i^6/\Delta_6, \tag{6.65a}$$

$$B_j = F_j^1/\Delta_1 + F_j^6/\Delta_6, \tag{6.65b}$$

$$B_m = F_m^1/\Delta_1 + F_m^2/\Delta_2, \tag{6.65c}$$

$$B_n = F_n^2/\Delta_2 + F_n^3/\Delta_3, \tag{6.65d}$$

$$B_o = F_o^3/\Delta_3 + F_o^4/\Delta_4, \tag{6.65e}$$

$$B_p = F_p^4/\Delta_4 + F_p^5/\Delta_5, \tag{6.65f}$$

$$B_q = F_q^5/\Delta_5 + F_q^6/\Delta_6. \tag{6.65g}$$

Formulas (6.62) through (6.65) can be simplified, and computations reduced, taking into account that, with the adopted discretization of the domain, the surfaces of the triangles are equal.

Writing the element equation (6.63) for every internal node and assembling the resulting block matrix $\{C^e\}$, for the element e, into a global matrix $\{C\}$ for the full domain, together with the boundary conditions for the boundary nodes, one obtains a matrix equation:

$$\{C\}\{\Phi\} = \{B\} \tag{6.66}$$

which is solved for the vector $\{\Phi\}$ of the nodal values.

6.5 Alternative Derivation of the Element Equation Based on the Concept of Hilbert Space

In this section, we explain an alternative derivation of the element equations for the model problem (6.1) obtained directly from the functional, defined in a Hilbert space, of the functions ξ and using the fact that the differential operator $A = \nabla^2$ is self-adjoint. Information about definitions and properties of a Hilbert space is given in the formula synopsis.

The problem consists in minimizing the functional:

$$\mathrm{X}(\xi) = (A\xi, \xi) - 2(f, \xi), \tag{6.67}$$

using the approximation:

$$\Phi(x, y) = \sum_{p=1}^{N_0} N_p(x, y)\Phi_p, \tag{6.68}$$

in terms of the nodal values $\boldsymbol{\Phi}_p$ and of the shape functions:

$$N_p(x, y) = \frac{1}{2\Delta}\left(a_p + b_p x + c_p y\right). \qquad (p = 1, 2, \ldots, N_0) \tag{6.69}$$

N_0 is the total number of inner nodes in the definition domain.

Inserting Eq. (6.68) into the functional (6.67), one obtains:

$$\begin{aligned} \mathrm{X} &= \left(\sum_{p=1}^{N_0} AN_p\Phi_p, \sum_{q=1}^{N_0} N_q\Phi_q\right) - 2\left(f, \sum_{p=1}^{N_0} N_p\Phi_p\right) \\ &= \sum_{p=1}^{N_0}\sum_{q=1}^{N_0}\left(AN_p, N_q\right)\Phi_p\Phi_q - 2\sum_{p=1}^{N_0}(f, N_p)\Phi_p. \end{aligned} \tag{6.70}$$

In the treatment of Eq. (6.70), we follow the approach by Norrie and de Vries (1973). In the sequence of integers ($p = 1, 2, \ldots, N_0$) and ($q = 1, 2, \ldots, N_0$), let us consider one integer r. The functional (6.70) can be written in terms of $\boldsymbol{\Phi}_r$ as follows:

$$\begin{aligned} \mathrm{X} = {} & (AN_r, N_r)\Phi_r^2 + \sum_{p(\neq r)=1}^{N_0}\left(AN_r, N_p\right)\Phi_r\Phi_p + \sum_{q(\neq r)=1}^{N_0}\left(AN_q, N_r\right)\Phi_q\Phi_r \\ & - 2(f, N_r)\Phi_r + \mathrm{T}exr, \end{aligned} \tag{6.71}$$

where Texr denotes all terms of the functional which do not contains $\boldsymbol{\Phi}_r$.

Because the operator $A = \nabla^2$ is self-adjoint, it holds:

$$\left(AN_r, N_q\right) = \left(N_r, AN_q\right), \tag{6.72}$$

and the two sums in the functional (6.71) are equal. Thus, the functional (6.71) becomes:

$$\mathrm{X} = (AN_r, N_r)\Phi_r^2 + 2\sum_{1=1,\, q\neq r}^{N_0} (AN_r, N_q)\Phi_r\Phi_q - 2(f, N_r)\Phi_r + \mathrm{T}exr. \tag{6.73}$$

The condition for the functional X to be stationary is:

$$\frac{\partial \mathrm{X}}{\partial \Phi_r} = 0. \qquad (r = 1, 2, \ldots, N_0) \tag{6.74}$$

Differentiating Eq. (6.73) and dividing by the constant factor 2, one obtains:

$$(AN_r, N_r)\Phi_r + \sum_{q(\neq r)=1}^{N_0} (AN_r, N_q)\,\Phi_q - (f, N_r) = 0, \tag{6.75}$$

or:

$$\sum_{1=1,\, q\neq 1}^{N_0} (AN_r, N_q)\,\Phi_q = (f, N_r). \tag{6.76}$$

Equation (6.76) can be written in matrix form as:

$$\begin{vmatrix} (AN_1, N_1) & (AN_1, N_2) & \cdots & (AN_1, N_{N_0}) \\ (AN_2, N_1) & (AN_2, N_2) & \cdots & (AN_2, N_{N_0}) \\ \cdots & \cdots & \cdots & \cdots \\ (AN_{N_0}, N_1) & (AN_{N_0}, N_2) & \cdots & (AN_{N_0}, N_{N_0}) \end{vmatrix} \cdot \begin{vmatrix} \Phi_1 \\ \Phi_2 \\ \cdots \\ \Phi_{N_0} \end{vmatrix} = \begin{vmatrix} (f, N_1) \\ (f, N_2) \\ \cdots \\ (f, N_{N_0}) \end{vmatrix}. \tag{6.77}$$

Recalling that the definition domain $\boldsymbol{\Omega}$ has been partitioned in NE domains $\boldsymbol{\Omega}_e$, $(e = 1, 2, \ldots, NE)$, with $\boldsymbol{\Omega} = U\boldsymbol{\Omega}_e$, with area $\boldsymbol{\Delta}_e$, the generic term of the matrix at the left side of Eq. (6.77) can be written as:

$$C_{\gamma\xi} = (AN_\gamma, N_\xi) = \iint_{\Omega} (AN_\gamma) N_\xi d\Omega = \sum_{e=1}^{NE} \iint_{\Omega_e} (\nabla^2 N_\gamma) N_\xi dxdy. \tag{6.78}$$

Using the Green's identity, one obtains:

$$\begin{aligned} C_{\gamma\xi} = (AN_\gamma, N_\xi) &= \sum_{e=1} e \iint_{\Omega_e} (\nabla N_\gamma) \cdot \nabla N_\xi dxdy \\ &= \sum_{e=1}^{NE} \iint_{\Omega_e} \left(\frac{\partial N_\gamma}{\partial x} \frac{\partial N_\xi}{\partial x} + \frac{\partial N_\gamma}{\partial y} \frac{\partial N_\xi}{\partial y} \right) dxdy. \end{aligned} \tag{6.79}$$

From Eq. (6.69), one derives:

$$\frac{\partial N_\gamma}{\partial x} = \frac{b_\gamma}{2\Delta_e}, \tag{6.80a}$$

$$\frac{\partial N_\gamma}{\partial y} = \frac{c_\gamma}{2\Delta_e}, \tag{6.80b}$$

$$\frac{\partial N_\xi}{\partial x} = \frac{b_\xi}{2\Delta_e}, \tag{6.81a}$$

$$\frac{\partial N_\xi}{\partial y} = \frac{c_\xi}{2\Delta_e}. \tag{6.81b}$$

Inserting (6.80) and (6.81) into the functional (6.79) and using $\iint_{\Omega_e} dxdy = \Delta_e$, one obtains:

$$C_{\gamma\xi} = \sum_{e=1}^{NE} \iint_{\Omega_e} \left[\frac{1}{4\Delta_e{}^2}\left(b_\gamma b_\xi + c_\gamma c_\xi\right)\right] dxdy = \sum_{e=1}^{NE} \frac{b_\gamma b_\xi + c_\gamma c_\xi}{4\Delta_e}. \tag{6.82}$$

For the right-hand side of Eq. (6.77), one obtains:

$$\begin{aligned} B_\gamma =& \left(f, N_\gamma\right) = \iint_\Omega \left[f(x,y)N_\gamma\right] d\Omega = \sum_{e=1}^{NE} \iint_{\Omega_e} \left(f(x,y)N_\gamma\right) dxdy \\ =& \sum_{e=1}^{NE} \iint_\Omega \frac{1}{2\Delta_e}\left(a_\gamma + b_\gamma x + c_\gamma y\right) f(x,y) dxdy = \sum_1^{NE} e \frac{F_\gamma^e}{2\Delta_e}, \end{aligned} \tag{6.83}$$

with the definition:

$$F_\gamma^e = \iint_{\Omega_e} \left(a_\gamma + b_\gamma x + c_\gamma y\right) f(x,y) dxdy. \tag{6.84}$$

Apart from the factor (1/2), which is eliminated, coefficients $C_{\gamma\xi}$ (see Eq. 6.56) and B_γ (see Eqs. (6.53) and (6.65)) coincide with those obtained in the previous section.

6.6 Residual Method of Orthogonal Collocations (RMOC): Galerkin Variant

Let the model equation (6.1) be represented in the domain Ω by:

$$A\boldsymbol{\Phi} - f = 0, \tag{6.85}$$

with the definition of the operator $A = \nabla^2$ and with $\boldsymbol{\Phi} = 0$ on the boundary of $\boldsymbol{\Omega}$.

We look for an approximation to the solution of Eq. (6.85) in the form:

$$\boldsymbol{\Phi} \cong \boldsymbol{\Phi}_N = \sum_{k=1}^{N} a_K N_K, \tag{6.86}$$

where a_K are nodal values to be specified and N_K are shape functions defined hereafter.

Introducing the approximation given by Eq. (6.86) into Eq. (6.85) does not satisfy exactly Eq. (6.85) but leads to residuals $\mathfrak{R}$ given by:

$$\mathfrak{R} = A\sum_{k=1}^{N} a_K N_K - f. \tag{6.87}$$

We want to minimize the residuals by imposing the orthogonality condition to a set of weight functions $W = \{W_j\}$, $(j = 1, 2, \ldots, M)$:

$$(\mathfrak{R}, W_j) = \left(A\sum_{k=1}^{N} a_K N_K - f, W_j\right) = 0, \qquad (j = 1, 2, \ldots, M) \tag{6.88}$$

where the scalar product is defined, for any two functions f_1 and f_2, by Eq. (6.44). Equation (6.88) can therefore be written as:

$$\mathfrak{I}_j = \iint_{\Omega} W_j\left[A\sum_{k=1}^{N} a_K N_K - f\right] d\Omega = 0. \qquad (j = 1, 2, \ldots, M) \tag{6.89}$$

In the Galerkin variant of the Residual Method of Orthogonal Collocations (RMOC), the weight functions are chosen to be the same as the shape functions; hence, $W_j = N_j$, and $M = N$. Let us partition the domain $\boldsymbol{\Omega}$ into NE disjoint (not overlapping) but adjacent elements $\boldsymbol{\Omega}_e$, with $\boldsymbol{\Omega} = U\boldsymbol{\Omega}_e$. Summing the integrals of Eq. (6.89) for every subdomain $\boldsymbol{\Omega}_e$ yields:

$$\begin{aligned}\mathfrak{I} = \sum_{e=1}^{NE} \mathfrak{I}^e &= \sum_{e=1}^{NE} \iint_{\Omega_e} N_j\left[A\sum_{k=1}^{N} K a_K N_K - f\right] d\Omega_e \\ &= \sum_{e=1}^{NE} \iint_{\Omega_e} N_j\left[\nabla^2 \Phi_N - f\right] d\Omega_e = 0, \qquad (j = 1, 2, \ldots, N)\end{aligned} \tag{6.90}$$

where:

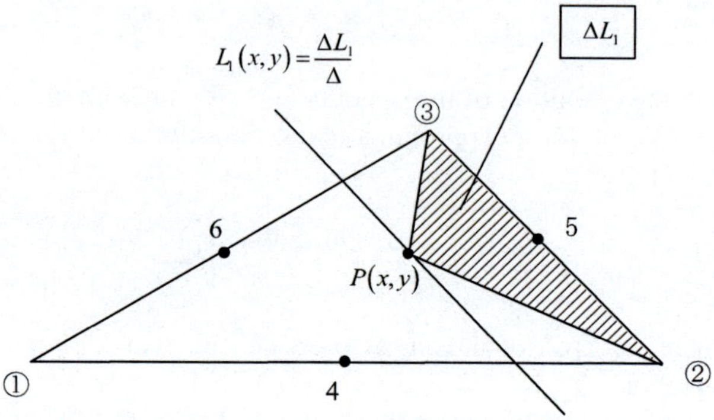

Fig. 6.8 Nodal points and definition of area coordinates

$$\Im_j^e = \iint_{\Omega_e} N_j(\nabla^2\Phi_N - f)d\Omega_e = 0. \qquad (j = 1, 2, \ldots, N) \tag{6.91}$$

Application of the Green's identity to the first term at the right-hand side of Eq. (6.91) yields:

$$\iint_{\Omega_e} N_j\nabla^2\Phi_N d\Omega_e = -\iint_{\Omega_e} (\nabla N_j \cdot \nabla\Phi_N)d\Omega_e + \int_\Gamma N_j\frac{\partial\Phi_N}{\partial n}d\Gamma. \tag{6.92}$$

The last integral occurs only for elements on the boundary of Ω but can be dropped because $\Phi = 0$ on Γ. Using the identity (6.92), Eq. (6.91) becomes:

$$\begin{aligned}\Im_j^e &= -\iint_{\Omega_e} (\nabla N_j \cdot \nabla\Phi_N)d\Omega_e - \iint_{\Omega_e} N_j f d\Omega_e \\ &= 0. \quad (e = 1, 2, \ldots, NE)(\, j = 1, 2, \ldots, N)\end{aligned} \tag{6.93}$$

We refer to Eq. (6.93) as the "element equation." In the following, we consider a partition of the domain Ω in triangular elements Ω_e.

With reference to Fig. 6.8, we choose six nodal points (hence $N = 6$) coincident with the three vertices of a triangular elements ($j = 1, 2, 3$) and with the midpoints of the sides ($j = 4, 5, 6$), ordered in a counterclockwise sequence. A point $P(x, y)$ belonging to the triangle Ω_e of area Δ is identified by its area coordinates, defined by:

$$L_1(x, y) = \frac{\Delta L_1}{\Delta} = \frac{1}{2\Delta}\begin{vmatrix} 1 & x & y \\ 1 & x_2 & y_2 \\ 1 & x_3 & y_3 \end{vmatrix} = \frac{1}{2\Delta}(a_1 + b_1 x + c_1 y) \tag{6.94a}$$

$$L_2(x,y) = \frac{\Delta L_2}{\Delta} = \frac{1}{2\Delta}\begin{vmatrix} 1 & x & y \\ 1 & x_3 & y_3 \\ 1 & x_1 & y_1 \end{vmatrix} = \frac{1}{2\Delta}(a_2 + b_2x + c_2y) \tag{6.94b}$$

$$L_3(x,y) = \frac{\Delta L_3}{\Delta} = \frac{1}{2\Delta}\begin{vmatrix} 1 & x & y \\ 1 & x_1 & y_1 \\ 1 & x_2 & y_2 \end{vmatrix} = \frac{1}{2\Delta}(a_3 + b_3x + c_3y) \tag{6.94c}$$

where $\boldsymbol{\Delta}$ is the area of the triangle given, using the coordinates of the vertices, by Eq. (6.31), and the coefficients, obtained by expanding the determinants, are:

$$a_1 = x_2y_3 - x_3y_2, \tag{6.95a}$$

$$b_1 = y_2 - y_3, \tag{6.95b}$$

$$c_1 = x_3 - x_2, \tag{6.95c}$$

$$a_2 = x_3y_1 - x_1y_3, \tag{6.95d}$$

$$b_2 = y_3 - y_1, \tag{6.95e}$$

$$c_2 = x_1 - x_3, \tag{6.95f}$$

$$a_3 = x_1y_2 - x_2y_1, \tag{6.95g}$$

$$b_3 = y_1 - y_2, \tag{6.95h}$$

$$c_3 = x_2 - x_1. \tag{6.95i}$$

A relation between the Cartesian coordinates (x, y) and the area coordinates $L_1(x, y)$, $L_2(x, y)$, and $L_3(x, y)$ is given by:

$$\begin{vmatrix} x \\ y \\ 1 \end{vmatrix} = \begin{bmatrix} x_1 & x_2 & x_3 \\ y_1 & y_2 & y_3 \\ 1 & 1 & 1 \end{bmatrix}\begin{vmatrix} L_1 \\ L_2 \\ L_3 \end{vmatrix}, \tag{6.96}$$

which also implies (third relation):

$$1 = L_1 + L_2 + L_3. \tag{6.97}$$

Parabolic shape functions N_j, $(j = 1, 2, \ldots, 6)$ are defined on the triangle $\boldsymbol{\Omega}_e$ (Norrie & de Vries, 1973) by:

$$N_1 = (2L_1 - 1)L_1 = \frac{1}{2\Delta^2}(a_1 + b_1x + c_1y)^2 - \frac{1}{2\Delta}(a_1 + b_1x + c_1y), \tag{6.98a}$$

$$N_2 = (2L_2 - 1)L_2 = \frac{1}{2\Delta^2}(a_2 + b_2x + c_2y)^2 - \frac{1}{2\Delta}(a_2 + b_2x + c_2y), \tag{6.98b}$$

$$N_3 = (2L_3 - 1)L_3 = \frac{1}{2\Delta^2}(a_3 + b_3x + c_3y)^2 - \frac{1}{2\Delta}(a_3 + b_3x + c_3y), \tag{6.98c}$$

$$N_4 = 4L_1L_2 = \frac{1}{\Delta^2}(a_1 + b_1x + c_1y)(a_2 + b_2x + c_2y), \tag{6.98d}$$

$$N_5 = 4L_2L_3 = \frac{1}{\Delta^2}(a_2 + b_2x + c_2y)(a_3 + b_3x + c_3y), \tag{6.98e}$$

$$N_6 = 4L_1L_3 = \frac{1}{\Delta^2}(a_1 + b_1x + c_1y)(a_3 + b_3x + c_3y). \tag{6.98f}$$

A useful formula to compute integrals of the shape functions, in terms of the area coordinates, over the triangular domain $\boldsymbol{\Omega_e}$ of area $\boldsymbol{\Delta}$ is:

$$\iint_{\Omega_e} {L_1}^a {L_2}^b {L_3}^c dxdy = \frac{a!b!c!}{(a+b+c+2)!} 2\Delta, \tag{6.99}$$

where a, b, and c are integers (0 included, with $0! = 1$).

With these shape functions, the gradients in the first integral of the element equation (6.93) become:

$$\begin{aligned} \nabla N_j \cdot \nabla \Phi_N &= \nabla N_j \cdot \nabla \sum_{k=1}^{6} a_K N_K = \left(\bar{i}\frac{\partial N_j}{\partial x} + \bar{j}\frac{\partial N_j}{\partial y}\right) \cdot \sum_{k=1}^{6} a_K \left(\frac{\partial N_K}{\partial x}\bar{i} + \frac{\partial N_K}{\partial y}\bar{j}\right) \\ &= \frac{\partial N_j}{\partial x} \sum_{k=1}^{6} a_K \frac{\partial N_K}{\partial x} + \frac{\partial N_j}{\partial y} \sum_{k=1}^{6} a_K \frac{\partial N_K}{\partial y}. \end{aligned} \tag{6.100}$$

The element equation (6.93) becomes:

$$\begin{aligned} -\mathfrak{I}_j^e = &\iint_{\Omega_e} \left(\frac{\partial N_j}{\partial x} \sum_{k=1}^{6} a_K \frac{\partial N_K}{\partial x} + \frac{\partial N_j}{\partial y} \sum_{k=1}^{6} a_K \frac{\partial N_K}{\partial y}\right) dxdy \\ &+ \iint_{\Omega_e} N_j f dxdy = 0 \qquad (j = 1, 2, \ldots, N = 6). \end{aligned} \tag{6.101}$$

The derivatives of the shape functions (6.100), needed to compute the integrals in Eq. (6.101), are given in terms of the area coordinates by:

$$\frac{\partial N_1}{\partial x} = \frac{\partial}{\partial x}\left(2L_1{}^2 - L_1\right) = 4L_1\frac{\partial L_1}{\partial x} - \frac{\partial L_1}{\partial x} = (4L_1 - 1)\frac{\partial L_1}{\partial x}$$
$$= (4L_1 - 1)\frac{b_1}{2\Delta}, \tag{6.102a}$$

$$\frac{\partial N_1}{\partial y} = (4L_1 - 1)\frac{\partial L_1}{\partial y} = (4L_1 - 1)\frac{c_1}{2\Delta}, \tag{6.102b}$$

$$\frac{\partial N_2}{\partial x} = (4L_2 - 1)\frac{b_2}{2\Delta}, \tag{6.102c}$$

$$\frac{\partial N_2}{\partial y} = (4L_2 - 1)\frac{c_2}{2\Delta}, \tag{6.102d}$$

$$\frac{\partial N_3}{\partial x} = (4L_3 - 1)\frac{b_3}{2\Delta}, \tag{6.102e}$$

$$\frac{\partial N_3}{\partial y} = (4L_3 - 1)\frac{c_3}{2\Delta}, \tag{6.102f}$$

$$\frac{\partial N_4}{\partial x} = 4\left(L_1\frac{\partial L_2}{\partial x} + L_2\frac{\partial L_1}{\partial x}\right) = 4\left(L_1\frac{b_2}{2\Delta} + L_2\frac{b_1}{2\Delta}\right)$$
$$= \frac{2}{\Delta}(L_1b_2 + L_2b_1), \tag{6.102g}$$

$$\frac{\partial N_4}{\partial y} = 4\left(L_1\frac{\partial L_2}{\partial y} + L_2\frac{\partial L_1}{\partial y}\right) = \frac{2}{\Delta}(L_1c_2 + L_2c_1), \tag{6.102h}$$

$$\frac{\partial N_5}{\partial x} = \frac{2}{\Delta}(L_2b_3 + L_3b_2), \tag{6.102i}$$

$$\frac{\partial N_5}{\partial y} = \frac{2}{\Delta}(L_2c_3 + L_3c_2), \tag{6.102j}$$

$$\frac{\partial N_6}{\partial x} = \frac{2}{\Delta}(L_1b_3 + L_3b_1), \tag{6.102k}$$

$$\frac{\partial N_6}{\partial y} = \frac{2}{\Delta}(L_1c_3 + L_3c_1). \tag{6.102l}$$

The element equation (6.101) is equivalent to the matrix equation:

$$[M^e] \cdot \begin{vmatrix} a_1 \\ a_2 \\ \vdots \\ a_6 \end{vmatrix} = -\begin{vmatrix} B_1 \\ B_2 \\ \vdots \\ B_6 \end{vmatrix}. \tag{6.103}$$

The matrix M^e is symmetric, and its elements are computed as follows: for example, for $j = 1$, $k = 1$ one has:

$$\begin{aligned} M_{11} &= \iint_{\Omega_e} \left[\left(\frac{\partial N_1}{\partial x} \right)^2 + \left(\frac{\partial N_1}{\partial y} \right)^2 \right] dxdy \\ &= \iint_{\Omega_e} \left[(4L_1 - 1)^2 \left(\frac{b_1}{2\Delta} \right)^2 + (4L_1 - 1)^2 \left(\frac{c_1}{2\Delta} \right)^2 \right] dxdy \\ &= \left[\left(\frac{b_1}{2\Delta} \right)^2 + \left(\frac{c_1}{2\Delta} \right)^2 \right] \iint_{\Omega_e} (16L_1{}^2 - 8L_1 + 1) dxdy = \frac{b_1{}^2 + c_1{}^2}{4\Delta}. \end{aligned} \tag{6.104a}$$

Similarly the following are obtained:

$$M_{12} = \iint_{\Omega_e} \left(\frac{\partial N_1}{\partial x} \frac{\partial N_2}{\partial x} + \frac{\partial N_1}{\partial y} \frac{\partial N_2}{\partial y} \right) dxdy = -\frac{1}{12\Delta} (b_1 b_2 + c_1 c_2), \tag{6.104b}$$

$$M_{13} = -\frac{1}{12\Delta} (b_1 b_3 + c_1 c_3), \tag{6.104c}$$

$$M_{14} = \frac{1}{3\Delta} (b_1 b_2 + c_1 c_2), \tag{6.104d}$$

$$M_{15} = 0, \tag{6.104e}$$

$$M_{16} = \frac{1}{3\Delta} (b_1 b_3 + c_1 c_3), \tag{6.104f}$$

$$M_{22} = \frac{1}{4\Delta} (b_2{}^2 + c_2{}^2), \tag{6.104g}$$

$$M_{23} = -\frac{1}{12\Delta} (b_2 b_3 + c_2 c_3), \tag{6.104h}$$

$$M_{24} = \frac{1}{3\Delta} (b_1 b_2 + c_1 c_2) = M_{14}, \tag{6.104i}$$

$$M_{25} = \frac{1}{3\Delta} (b_2 b_3 + c_2 c_3), \tag{6.104j}$$

$$M_{26} = 0, \tag{6.104k}$$

$$M_{33} = \frac{1}{4\Delta} (b_3{}^2 + c_3{}^2), \tag{6.104l}$$

$$M_{34} = 0, \tag{6.104m}$$

$$M_{35} = \frac{1}{3\Delta} (b_2 b_3 + c_2 c_3) = M_{25}, \tag{6.104n}$$

$$M_{36} = \frac{1}{3\Delta} (b_1 b_3 + c_1 c_3) = M_{16}, \tag{6.104o}$$

$$M_{44} = \frac{2}{3\Delta} (b_1{}^2 + b_2{}^2 + b_1 b_2 + c_1{}^2 + c_2{}^2 + c_1 c_2), \tag{6.104p}$$

$$M_{45} = \frac{1}{3\Delta}\left(b_1b_2 + b_2{}^2 + b_2b_3 + 2b_1b_3 + c_1c_2 + c_2{}^2 + c_2c_3 + 2c_1c_3\right), \quad (6.104q)$$

$$M_{46} = \frac{1}{3\Delta}\left(b_1b_2 + b_1b_3 + b_1{}^2 + 2b_2b_3 + c_1c_2 + c_1c_3 + c_1{}^2 + 2c_2c_3\right), \quad (6.104r)$$

$$M_{55} = \frac{2}{3\Delta}\left(b_2{}^2 + b_3{}^2 + 2b_2b_3 + c_2{}^2 + c_3{}^2 + 2c_2c_3\right), \quad (6.104s)$$

$$M_{56} = \frac{1}{3\Delta}\left(2b_1b_2 + b_1b_3 + b_2b_3 + b_3{}^2 + 2c_1c_2 + c_1c_3 + c_2c_3 + c_3{}^2\right), \quad (6.104t)$$

$$M_{66} = \frac{2}{3\Delta}\left(b_1{}^2 + b_3{}^2 + b_1b_3 + c_1{}^2 + c_3{}^2 + c_1c_3\right). \quad (6.104u)$$

At the right-hand side of Eq. (6.103), we have:

$$B_j = \iint_{\Omega_e} N_j(x,y) f(x,y)\, dxdy, \qquad (j = 1, 2, \ldots, N = 6) \quad (6.105)$$

which can be computed numerically with Gaussian quadratures.

Equation (6.103) is built for every element and assembled in a global equation $[M] \cdot [a] = [B]$.

Solution with respect to $[a] = [M^{-1}][B]$ yields the coefficients of (6.86), hence the numerical solution Φ_N.

6.7 Residual Method of Orthogonal Collocations (RMOC): Subdomain Collocations

For the model problem (6.1), we search again an approximate solution

$$u \simeq u_N \simeq \sum_{k=1}^{N} a_k \Phi_k, \quad (6.106)$$

where $\Phi_K(x, y)$ are shape functions to be chosen and the constants a_K are to be determined. The equation for the residuals (6.88) can be written as:

$$\sum_{k=1}^{N} a_K \left[\left(\frac{\partial^2 \Phi_K}{\partial x^2}, W_j\right) + \left(\frac{\partial^2 \Phi_K}{\partial y^2}, W_j\right)\right] = (f, W_j). \qquad (j = 1, 2, \ldots, N) \quad (6.107)$$

The subdomain collocation variant of the method of orthogonal collocations is obtained by choosing the weigh functions:

$$\begin{aligned} W_{j,e} &= 1 \qquad x \in \Omega_e, \qquad (j = 1, 2, \ldots, N) \qquad (e = 1, 2, \ldots, NE) \\ W_{j,e} &= 0 \qquad x \notin \Omega_e. \qquad (j = 1, 2, \ldots, N) \qquad (e = 1, 2, \ldots, NE) \end{aligned} \tag{6.108}$$

With this choice, from Eq. (6.107), one obtains the element equations:

$$\sum_{k=1}^{N} \int_{\Omega_e} a_K \left(\frac{\partial^2 \Phi_K}{\partial x^2} + \frac{\partial^2 \Phi_K}{\partial y^2} \right) dxdy = \int_{\Omega_e} fdxdy. \qquad (e = 1, 2, \ldots, NE) \tag{6.109}$$

Let us consider a square element Ω_e defined by local coordinates x, y with $(0 \leq x \leq \Delta x,\ 0 \leq y \leq \Delta y)$ with $\Delta x = \Delta y = h$. Let us denote nodes $x = 0$, $x = \Delta x/2$, and $x = \Delta x$ with subscripts 0, 1/2, and 1, respectively, and likewise for the y direction. A parabolic approximation between nodal values a_0, $a_{1/2}$, a_1 is given by:

$$\begin{aligned} \mathbf{X}(x) &= a_0 + \frac{x}{\Delta x}\left(-3a_0 + 4a_{1/2} - a_1\right) + \left(\frac{x}{\Delta x}\right)^2 \left(2a_0 - 4a_{1/2} + 2a_1\right) \\ &= a_0\Phi_0 + a_{1/2}\Phi_{1/2} + a_1\Phi_1, \end{aligned} \tag{6.110}$$

with the shape functions already introduced with Eqs. (3.19):

$$\Phi_0 = 1 - \frac{3x}{\Delta x} + \frac{2x^2}{\Delta x^2}, \tag{6.111a}$$

$$\Phi_{1/2} = \frac{4x}{\Delta x} - \frac{4x^2}{\Delta x^2}, \tag{6.111b}$$

$$\Phi_1 = -\frac{x}{\Delta x} + \frac{2x^2}{\Delta x^2}. \tag{6.111c}$$

With reference to a local normalized coordinate:

$$\xi = \frac{x - h/2}{h}, \qquad (-½ \leq \xi \leq ½) \tag{6.112}$$

using $x/\Delta x = x/h = \xi + 1/2$, Eq. (6.110) becomes:

$$\begin{aligned} \mathbf{X}(\xi) =& a_0 + (\xi + 1/2)\left(-3a_0 + 4a_{1/2} - a_1\right) + (\xi + 1/2)^2\left(2a_0 - 4a_{1/2} + 2a_1\right) \\ =& a_{1/2} + \xi(a_{1x} - a_{0x}) + \xi^2\left(2a_{0x} - 4a_{1/2} + 2a_{1x}\right) \end{aligned} \tag{6.113}$$

The subscripts x denotes that the relative nodes are along the $x(\xi)$-axis. Likewise, with reference to a local normalized coordinate along the y-axis:

$$\eta = \frac{y - h/2}{h}, \qquad (-1/2 \leq \eta \leq 1/2) \tag{6.114}$$

one obtains:

$$Y(\eta) = a_{1/2} + \eta(a_{1y} - a_{0y}) + \eta^2(2a_{0y} - 4a_{1/2} + 2a_{1y}). \tag{6.115}$$

The relationship between indices of nodes in the (x, y) coordinate system and in the (ξ, η) system is as follows:

$$\begin{aligned} 1/2 &\rightarrow (i,j), \\ a_{0x} &\rightarrow (i - 1/2, j), \\ a_{1x} &\rightarrow (i + 1/2, j), \\ a_{0y} &\rightarrow (i, j - 1/2), \\ a_{1y} &\rightarrow (i, j + 1/2). \end{aligned}$$

In the plane (ξ, η), we define the function:

$$\begin{aligned} P(\xi,\eta) &= -a_{ij} + X(\xi) + Y(\eta) \\ &= -a_{ij} + a_{ij} + \xi(a_{i+1/2,j} - a_{i-1/2,j}) + \xi^2(2a_{i-1/2,j} - 4a_{ij} + 2a_{i+1/2,j}) \\ &\quad + a_{ij} + \eta(a_{i,j+1/2} - a_{i,j-1/2}) + \eta^2(2a_{i,j-1/2} - 4a_{ij} + 2a_{i,j+1/2}) \\ &= a_{ij}(1 - 4\xi^2 - 4\eta^2) + a_{i-1/2,j}(-\xi + 2\xi^2) + a_{i+1/2,j}(\xi + 2\xi^2) \\ &\quad + a_{i,j-1/2}(-\eta + 2\eta^2) + a_{i,j+1/2}(\eta + 2\eta^2). \end{aligned} \tag{6.116}$$

Derivatives of $P(\xi, \eta)$ are given by:

$$\frac{\partial P(\xi,\eta)}{\partial \xi} = -8a_{ij}\xi + a_{i-1/2,j}(-1 + 4\xi) + a_{i+1/2,j}(1 + 4\xi), \tag{6.117a}$$

$$\frac{\partial P(\xi,\eta)}{\partial \eta} = -8a_{ij}\eta + a_{i,j-1/2}(-1 + 4\eta) + a_{i,j+1/2}(1 + 4\eta), \tag{6.117b}$$

$$\frac{\partial^2 P(\xi,\eta)}{\partial \xi^2} = -8a_{ij} + 4a_{i-1/2,j} + 4a_{i+1/2,j}, \tag{6.118a}$$

$$\frac{\partial^2 P(\xi,\eta)}{\partial \eta^2} = -8a_{ij} + 4a_{i,j-1/2} + 4a_{i,j+1/2}, \tag{6.118b}$$

hence:

$$\left[\frac{\partial P(\xi,\eta)}{\partial \xi}\right]_{\xi=-1/2} = 4a_{ij} - 3a_{i-1/2,j} - a_{i+1/2,j}, \tag{6.119a}$$

$$\left[\frac{\partial P(\xi,\eta)}{\partial \xi}\right]_{\xi=0} = a_{i+1/2,j} - a_{i-1/2,j}, \tag{6.119b}$$

$$\left[\frac{\partial P(\xi,\eta)}{\partial \xi}\right]_{\xi=1/2} = -4a_{ij} + a_{i-1/2,j} + 3a_{i+1/2,j}, \tag{6.119c}$$

$$\left[\frac{\partial P(\xi,\eta)}{\partial \eta}\right]_{\eta=-1/2} = 4a_{ij} - 3a_{i,j-1/2} - a_{i,j+1/2}, \tag{6.120a}$$

$$\left[\frac{\partial P(\xi,\eta)}{\partial \eta}\right]_{\eta=0} = a_{i,j+1/2} - a_{i,j-1/2}, \tag{6.120b}$$

$$\left[\frac{\partial P(\xi,\eta)}{\partial \eta}\right]_{\eta=1/2} = -4a_{ij} + a_{i,j-1/2} + 3a_{i,j+1/2}. \tag{6.120c}$$

Using Eqs. (6.118a) and (6.118b) and $\int_{\Omega_e} d\xi d\eta = 1$, the left-hand side of the element equation (6.109) becomes:

$$\begin{aligned}\int_{\Omega_e}\sum_{k=1}^{N} a_K\left[\frac{\partial^2\Phi_K}{\partial x^2}+\frac{\partial^2\Phi_K}{\partial y^2}\right]dxdy &= \int_{\Omega_e}\left[\frac{\partial^2 P(\xi,\eta)}{\partial \xi^2}+\frac{\partial^2 P(\xi,\eta)}{\partial \eta^2}\right]d\xi d\eta\\ &= 4\left(a_{i-1/2,j}+a_{i+1/2,j}+a_{i,j-1/2}+a_{i,j+1/2}-4a_{ij}\right)\end{aligned} \tag{6.121}$$

The right-hand side of the element equation (6.109) can be computed numerically for the given function $f(x, y)$ and can be written as:

$$\int_{\Omega_e} f(x,y)dxdy \cong f_{ij}h^2. \tag{6.122}$$

The discrete form of the element equation (6.109) becomes:

$$4\left(a_{i-1/2,j}+a_{i+1/2,j}+a_{i,j-1/2}+a_{i,j+1/2}-4a_{ij}\right) = f_{ij}h^2. \tag{6.123}$$

Equation (6.123) is an improved 5-point formula, as an alternative to Eq. (6.5). The domain collocation method allows however to obtain higher precision, comparable to that attainable with the 9-point formula, by imposing continuity of the derivatives of the solution at the boundaries between meshes. At the boundary $i + 1/2, j$ and $i, j + 1/2$, we can impose the conditions:

Table 6.2 Comparison of method for the numerical solution of the model Poisson equation

Subroutine and method	Number of cells	Number of nodes	Maximum residuum
FIVEPF (5-point formula)	19 × 19	400	7.01×10^{-14}
NINEPF (9-point formula)	19 × 19	400	1.08×10^{-15}
CLRITZ (classical Ritz)	19 × 19	400	1.691×10^{-18}
GALERK (Galerkin method)	9 × 9	361	2.86×10^{-16}
PORTCO (subdomain collocation)	11 × 11	385	2.77×10^{-16}

$$\left[\frac{\partial P(\xi,\eta)}{\partial \xi}\right]_{(i+1/2,j)} = \left[\frac{\partial P(\xi,\eta)}{\partial \xi}\right]_{(i+1-1/2,j)}, \tag{6.124}$$

$$\left[\frac{\partial P(\xi,\eta)}{\partial \eta}\right]_{(i,j+1/2)} = \left[\frac{\partial P(\xi,\eta)}{\partial \eta}\right]_{(i,j+1-1/2)}. \tag{6.125}$$

The derivatives in Eq. (6.124) are replaced by (6.119c) for node $i + 1/2, j$ and by (6.119a) for node $(i + 1 - 1/2, j)$. Likewise the derivatives in Eq. (6.125) are replaced by (6.120c) for node $i, j + 1/2$ and by (6.120a) for node $(i, j + 1 - 1/2)$. This yields two continuity conditions:

$$a_{i-1/2,j} - 4a_{ij} + 6a_{i+1/2,j} - 4a_{i+1,j} + a_{i+3/2,j} = 0, \tag{6.126}$$

$$a_{i,j-1/2} - 4a_{ij} + 6a_{i,j+1/2} - 4a_{i,j+1} + a_{i,j+3/2} = 0. \tag{6.127}$$

The element equation (6.123) and the two continuity equations (6.126), (6.127), written for every subdomain $\boldsymbol{\Omega}_e$ $(e = 1, 2, \ldots, NE)$ in $\boldsymbol{\Omega}$, are assembled in a matrix equation $[M][A] = [F]$ which is solved for an unknown vector $[A]$. Then Eq. (6.106) yields the numerical solution for all nodes of the domain.

The computer program **PROG_VI_1_POISSON**, given in Appendix A.VI, has been used to solve the model Poisson equation with five of the above-explained method. The program is organized to compute all five test cases in a single run. Table 6.2 summarizes some results of the numerical solutions of the Poisson model equation obtained with the methods explained above. For the Galerkin method and the subdomain collocation method, the definition domain has been divided in the maximum possible number of cells under the constraint that the resulting number of nodes does not exceed 400, which is the number of nodes (including boundary nodes) used on the other methods.

Computational times were for all cases less than 0.25 s. They were not minimized in the sense that for matrix inversion, a subroutine (MINV) was used which does not exploit the band structure of the matrix to be inverted. In all cases, the residuals were smaller than 10^{-12}. The best performance, from the viewpoint of the residuals, is obtained with the classical Ritz method.

References

Houstis, E. N., Lynch, R. E., Rice, J. R., & Papatheodorou, T. S. (1978). Evaluation of numerical methods for elliptic partial differential equations. *Journal of Computational Physics, 27*, 323–350.

Norrie, D. H., & de Vries, G. (1973). *The finite element method*. Academic Press.

Smith, G. D. (1969). *Numerical Solution of Partial Differential Equations*. Oxford University Press.

Strang, G., & Fix, G. J. (1973). *An analysis of the finite element method*. Prentice-Hall.

Zienkiewicz, O. C. (1971). *The finite element method in engineering science*. McGraw Hill.

Chapter 7
Derivation and Numerical Solutions of Poisson-Like Equations

As mentioned earlier, we call "Poisson-like" equation a linear algebraic equation which is formally similar to the equation which would be obtained by discretizing a Poisson equation, apart from the fact that the coefficients are not equal and not necessarily constant. When the solution of the Poisson equation is inserted in a time-dependent computer program, the coefficients may change with time in the course of the calculation. In this case, the standard algorithms, based on the 5-point or 9-points formula, are not applicable, but more general algorithms are required. These more general algorithms are applicable also to the true Poisson equation with constant coefficients, although less efficiently than the algorithms which were specially devised for the Poisson equation.

These most general algorithms can be divided in direct and iterative. Among the direct algorithms, the most commonly used are Gaussian elimination and Cholesky decomposition of the matrix of coefficients. Among the iterative algorithms, the Alternating Direction Implicit (ADI) technique has been more frequently applied in the past years. In the next sections, we examine these methods in some detail.

One variant of the Cholesky LU (lower, upper) matrix decomposition is the algorithm by Doolittle (Fox, 1964) which is in practice applicable with today's computers, for reasons of computing time and of core storage, to problems with up to about 50,000 nodes. This numerical technique is outlined in detail in Sect. 7.4.

The need for the simulation of ever larger problems made it necessary to provide alternative computational methods to those mentioned above. Classical ADI methods, both two-dimensional (Peaceman & Rachford, 1955) and three-dimensional (Douglas, 1962), are vectorizable only to a limited extent. In subsequent years, more advanced variants of the ADI scheme have been proposed (Press et al., 1986; Messina & Londrillo, 1988), which use convergence accelerating parameters and are based on algorithms vectorizable to a high degree. We refer to the last mentioned technique as ML-ADI variant.

When a convergence parameter is used, and not kept constant, it must be varied cyclically within an optimum range. A heuristic way, in the original meaning of a

M. Bottoni, *Physical Modeling and Computational Techniques for Thermal and Fluid-dynamics*, Mechanical Engineering Series,
https://doi.org/10.1007/978-3-030-79717-1_7

method helping to discover or learn, can be devised for approximating the optimum range of variation. This approach is first explained in the following for a model 2D problem, and then it is extrapolated to the 3D case. Solutions obtained with the ML-ADI technique are compared with those obtained with a vectorized Gaussian elimination technique. In a final section, the ADI algorithm is implemented for the treatment of two-phase flow problems. An example of application is given, taken from real problems concerning safety analysis of fuel bundles in cores of fast nuclear reactors.

7.1 The Pressure Method: Poisson Equation for a Divergence-Free Velocity Field

An alternative to the solution of the continuity and momentum equations of fluid dynamics in terms of the primitive variables (density and velocity components) is offered by the pressure method. This method yields a Poisson-like equations for the pressure field in the domain of interest, which can be solved with standard algorithms for elliptic operators. Solution of the Poisson equation yields the pressure distribution from which then the velocity components are obtained.

We present first the pressure method from the theoretical viewpoint assuming a divergence-free velocity field and then the derivation of the Poisson equation from the practical viewpoint, taking into account the compressibility of the fluid.

Let us assume a velocity field $\vec{V} = (u, v, w)$, described by the continuity and momentum equations:

$$\frac{\partial \rho}{\partial t} + \nabla \cdot \left(\rho \vec{V}\right) = 0, \tag{7.1}$$

$$\frac{\partial}{\partial t}\left(\rho \vec{V}\right) + \nabla \cdot \left(\rho \vec{V}\vec{V}\right) = -\nabla p + \nabla \cdot \left(\mu \nabla \vec{V}\right) + \rho \vec{g} - \vec{R}. \tag{7.2}$$

In these equations, ρ and μ are the fluid density and dynamical viscosity, p is the fluid pressure, $\vec{g}$ is the gravity acceleration, $\vec{R}$ is a friction force, and t is the time.

In the following, we will neglect the acceleration of gravity and the friction force because they are irrelevant for the explanation of the pressure method.

Next, we assume ρ and μ constant; hence $\nu = \mu/\rho$, the kinematic viscosity, is also constant.

Under this assumption, Eq. (7.1) yields:

$$\nabla \cdot \vec{V} = 0. \tag{7.3}$$

Dividing Eq. (7.2) by ρ, one obtains:

$$\frac{\partial \vec{V}}{\partial t} + \nabla \cdot \vec{V}\vec{V} = -\frac{1}{\rho}\nabla p + \nu \nabla^2 \vec{V}. \tag{7.4}$$

Taking into account Eq. (7.3), the convective term can be written as:

$$\nabla \cdot \overline{VV} = \overline{V} \cdot \nabla \overline{V} + \vec{V}(\nabla \cdot \overline{V}) = \overline{V} \cdot \nabla \overline{V}. \tag{7.5}$$

The pressure method consists essentially taking the divergence of Eq. (7.4) which yields:

$$\frac{\partial}{\partial t}\left(\nabla \cdot \vec{V}\right) + \nabla \cdot \left[\overline{V} \cdot \nabla \overline{V}\right] = -\frac{1}{\rho}\nabla \cdot (\nabla p) + \nu \nabla \cdot \nabla^2 \overline{V}. \tag{7.6}$$

The first term of the left-hand side is zero, because of Eq. (7.3). The second term can be written, identically, as:

$$\nabla \cdot \left[\overline{V} \cdot \nabla \overline{V}\right] = \nabla \overline{V} : \nabla \overline{V}. \tag{7.7}$$

Solving Eq. (7.6) for $\nabla \cdot \ \nabla p = \nabla^2 p$, the following is obtained:

$$\nabla^2 p = -\rho\left(\nabla \overline{V} : \nabla \overline{V}\right) + \mu \nabla \cdot \nabla^2 \overline{v}, \tag{7.8}$$

which can be written as a Poisson-like equation:

$$\nabla^2 p = f, \tag{7.9}$$

with the formal definition:

$$f = -\rho\left(\nabla \overline{V} : \nabla \overline{V}\right) + \mu \nabla \cdot \nabla^2 \overline{V}. \tag{7.10}$$

7.2 Schematic Derivation of the Discrete Poisson Equation

Let us introduce a discretization index (superscript) n referring to time level t_n and indices (subscripts) i, j, and k to denote discretization with respect to space in the three coordinate directions x, y, and z, respectively. Thus, the point P_{ijk} refers to the coordinates (x_i, y_i, z_k) or $(i\Delta x, j\Delta y, k\Delta z)$ (if the mesh is uniform). In the following, we consider staggered meshes, namely, central meshes taken around nodes (i, j, k) for the continuity equation and displaced meshes taken around nodes $(i \pm 1/2, j, k)$ or $(i, j \pm 1/2, k)$ or $(i, j, k \pm 1/2)$ for the three scalar components of the momentum equation.

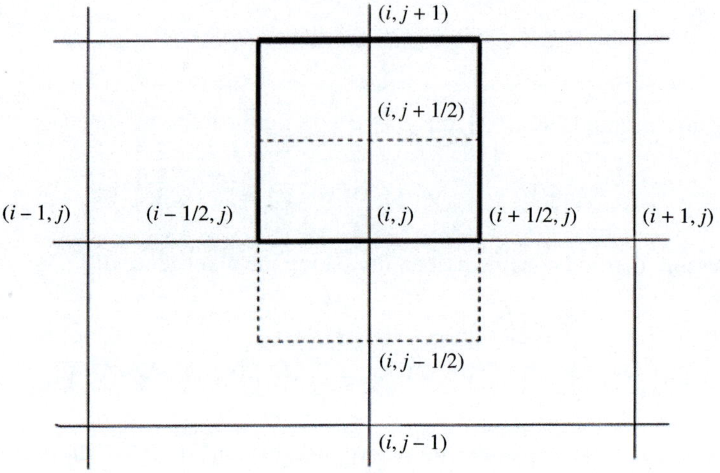

Fig. 7.1 Indexing of nodes used for discretization of the Poisson equation

Figure 7.1 shows in a two-dimensional case, a central mesh (dashed lines) around node (i, j) and a displaced mesh (continuous dark line) around node $(i, j + \frac{1}{2})$.

In staggered meshes, scalar quantities (pressure, density, viscosity, etc.) are usually defined in the center points of the central meshes (e.g., in nodes (i, j, k) or $(i, j + 1, k)$), while the scalar components of vector quantities, for instance, the velocity components u, v, and w, are defined in the center points of the displaced meshes (e.g., in the nodes) $[(i, j + \frac{1}{2}, k)$ or $(i, j, k - \frac{1}{2})]$. Other conventions are however possible.

We start discretizing Eq. (7.2), but differently from Sect. 7.1, we do not make the limiting assumption that the density is constant. Thus [Cfr. Eq. (7.15)], the isentropic sound speed can be defined. Neglecting gravity and friction terms, the component along the x-direction of the momentum equation (7.2) is:

$$\frac{\partial}{\partial t}(\rho u) + \frac{\partial}{\partial x}(\rho uu) + \frac{\partial}{\partial y}(\rho vu) + \frac{\partial}{\partial z}(\rho wu)$$
$$= -\frac{\partial p}{\partial x} + \frac{\partial}{\partial x}\left(\mu \frac{\partial u}{\partial x}\right) + \frac{\partial}{\partial y}\left(\mu \frac{\partial u}{\partial y}\right) + \frac{\partial}{\partial z}\left(\mu \frac{\partial u}{\partial z}\right). \tag{7.11}$$

We discretize Eq. (7.11) with reference to a displaced mesh around node $(i + 1/2, j, k)$ treating the pressure gradient term implicitly (at time level $n + 1$) and all convective and diffusive terms explicitly (at time level n). All these explicit terms are brought to the second member of the equation and denoted collectively with TX. This procedure is followed for the only purpose of emphasizing the main idea of the pressure method, which consist in treating implicitly the pressure terms. In practice, however, also (at least) the convective terms (and better also the diffusive terms) should be treated implicitly.

Discretization of Eq. (7.11) yields:

$$\frac{1}{\Delta t}\left[(\rho u)^{n+1}_{i+½,j,k} - (\rho u)^{n}_{i+½,j,k}\right] = -\frac{p^{n+1}_{i+1,j,k} - p^{n+1}_{i,j,k}}{\Delta x_{i+½}} + \mathrm{TX}^{n}_{i+½}. \tag{7.12}$$

Multiplying by Δt and collecting also the term $\Delta t(\rho u)^{n}_{i+½,j,k}$ in TX, one obtains:

$$(\rho u)^{n+1}_{i+½,j,k} = -\frac{\Delta t}{\Delta x_{i+½}}\left(p^{n+1}_{i+1,j,k} - p^{n+1}_{i,j,k}\right) + \mathrm{TX}^{n}_{i+½}. \tag{7.13a}$$

Similarly, discretizing the x-components of the momentum equation around node $(i - 1/2, j, k)$, one derives:

$$(\rho u)^{n+1}_{i-½,j,k} = -\frac{\Delta t}{\Delta x_{i-½}}\left(p^{n+1}_{i,j,k} - p^{n+1}_{i-1,j,k}\right) + \mathrm{TX}^{n}_{i-½}. \tag{7.13b}$$

We proceed in a similar way by discretizing the y-components of the momentum equation around nodes $(i, j + 1/2, k)$ and $(i, j - 1/2, k)$ and the z-component around nodes $(i, j, k + 1/2)$ and $(i, j, k - 1/2)$. Thus, one obtains:

$$(\rho v)^{n+1}_{i,j+½,k} = -\frac{\Delta t}{\Delta y_{j+½}}\left(p^{n+1}_{i,j+1,k} - p^{n+1}_{i,j,k}\right) + \mathrm{TY}^{n}_{j+½}, \tag{7.13c}$$

$$(\rho v)^{n+1}_{i,j-½,k} = -\frac{\Delta t}{\Delta y_{j-½}}\left(p^{n+1}_{i,j,k} - p^{n+1}_{i,j-1,k}\right) + \mathrm{TY}^{n}_{j-½}, \tag{7.13d}$$

$$(\rho w)^{n+1}_{i,j,k+½} = -\frac{\Delta t}{\Delta z_{z+½}}\left(p^{n+1}_{i,j,k+1} - p^{n+1}_{i,j,k}\right) + \mathrm{TZ}^{n}_{k+½}, \tag{7.13e}$$

$$(\rho w)^{n+1}_{i,j,k-½} = -\frac{\Delta t}{\Delta z_{z-½}}\left(p^{n+1}_{i,j,k} - p^{n+1}_{i,j,k-1}\right) + \mathrm{TZ}^{n}_{k-½}. \tag{7.13f}$$

Assuming an equation of state $\rho = \rho(p)$, the time derivative of density in the continuity equation (7.1) can be written as:

$$\frac{\partial \rho}{\partial t} = \frac{\partial \rho}{\partial p}\frac{\partial p}{\partial t} = \frac{1}{c^2}\frac{\partial p}{\partial t}, \tag{7.14}$$

with the definition of the isentropic speed of sound:

$$c = \left(\frac{\partial p}{\partial \rho}\right)^{½}. \tag{7.15}$$

Using Eq. (7.14), discretization of the continuity Eq. (7.1) in a central mesh around node (i, j, k) yields:

$$\frac{1}{\Delta tc^2}\left[p_{i,j,k}^{n+1} - p_{i,j,k}^{n}\right] + \frac{1}{\Delta x_i}\left[(\rho u)_{i+1/2,j,k}^{n+1} - (\rho u)_{i-1/2,j,k}^{n+1}\right] + \frac{1}{\Delta y_j}\left[(\rho v)_{i,j+1/2,k}^{n+1} - (\rho v)_{i,j-1/2,k}^{n+1}\right]$$
$$+\frac{1}{\Delta z_k}\left[(\rho w)_{i,j,k+1/2}^{n+1} - (\rho u)_{i,j,k-1/2}^{n+1}\right]. \tag{7.16}$$

Introducing the discretized momentum Eq. (7.13) into Eq. (7.16) yields:

$$\frac{1}{\Delta tc^2}\left[p_{i,j,k}^{n+1} - p_{i,j,k}^{n}\right] + \frac{1}{\Delta x_i}\left[\frac{\Delta t}{\Delta x_{i+1/2}}\left(-p_{i+1,j,k}^{n+1} + p_{i,j,k}^{n+1}\right) + \frac{\Delta t}{\Delta x_{i-1/2}}\left(p_{i,j,k}^{n+1} - p_{i-1,j,k}^{n+1}\right)\right]$$
$$+\frac{1}{\Delta y_j}\left[\frac{\Delta t}{\Delta y_{j+1/2}}\left(-p_{i,j+1,k}^{n+1} + p_{i,j,k}^{n+1}\right) + \frac{\Delta t}{\Delta y_{j-1/2}}\left(p_{i,j,k}^{n+1} - p_{i,j-1,k}^{n+1}\right)\right]$$
$$+\frac{1}{\Delta z_k}\left[\frac{\Delta t}{\Delta z_{k+1/2}}\left(-p_{i,j,k+1}^{n+1} + p_{i,j,k}^{n+1}\right) + \frac{\Delta t}{\Delta z_{k-1/2}}\left(p_{i,j,k}^{n+1} - p_{i,j,k-1}^{n+1}\right)\right] = G_{i.j.k}^{n}. \tag{7.17}$$

All contributions at time level n have been collected in the term $G_{i,j,k}^n$ at the right-hand side.

Equation (7.17) is an equation for the seven unknowns $p_{i,j,k}^{n+1}, p_{i\pm1,j,k}^{n+1}, p_{i,j\pm1,k}^{n+1}, p_{i,j,k\pm1}^{n+1}$. Collecting terms it can be written as:

$$p_{i,j,k}^{n+1}\left[\frac{1}{\Delta tc^2} + \frac{\Delta t}{\Delta x_i \Delta x_{i+1/2}} + \frac{\Delta t}{\Delta x_i \Delta x_{i-1/2}} + \frac{\Delta t}{\Delta y_j \Delta y_{j+1/2}} + \frac{\Delta t}{\Delta y_j \Delta y_{j-1/2}} + \frac{\Delta t}{\Delta z_k \Delta z_{k+1/2}} + \frac{\Delta t}{\Delta z_k \Delta z_{k-1/2}}\right]$$
$$+p_{i+1,j,k}^{n+1}\left(-\frac{\Delta t}{\Delta x_i \Delta x_{i+1/2}}\right) + p_{i-1,j,k}^{n+1}\left(-\frac{\Delta t}{\Delta x_i \Delta x_{i-1/2}}\right)$$
$$+p_{i,j+1,k}^{n+1}\left(-\frac{\Delta t}{\Delta y_j \Delta y_{j+1/2}}\right) + p_{i,j-1,k}^{n+1}\left(-\frac{\Delta t}{\Delta y_j \Delta y_{j-1/2}}\right)$$
$$+p_{i,j,k+1}^{n+1}\left(-\frac{\Delta t}{\Delta z_k \Delta z_{k+1/2}}\right) + p_{i,j,k-1}^{n+1}\left(-\frac{\Delta t}{\Delta z_k \Delta z_{k-1/2}}\right) = G_{i,j,k}^{n}. \tag{7.18}$$

Next, we use the following subscripts:

$$\begin{aligned}
&0 \quad \text{for } i,j,k; \quad 1 \quad \text{for } i-1,j,k; \quad 2 \quad \text{for } i+1,j;\\
&3 \quad \text{for } i,j-1,k; \quad 4 \quad \text{for } i,j+1,k;\\
&5 \quad \text{for } i,j,k-1; \quad 6 \quad \text{for } i,j,k+1.
\end{aligned}$$

With these conventions, Eq. (7.18) may be written as:

$$a_0 p_0^{n+1} + a_1 p_1^{n+1} + a_2 p_2^{n+1} + a_3 p_3^{n+1} + a_4 p_4^{n+1} + a_5 p_5^{n+1} + a_6 p_6^{n+1} = G_0^n \quad (7.19)$$

with the definition of the coefficients:

$$a_0 = \frac{1}{\Delta t c^2} + \frac{\Delta t}{\Delta x_i \Delta x_{i+1/2}} + \frac{\Delta t}{\Delta x_i \Delta x_{i-1/2}} + \frac{\Delta t}{\Delta y_j \Delta y_{j+1/2}} + \frac{\Delta t}{\Delta y_j \Delta y_{j-1/2}} + \frac{\Delta t}{\Delta z_k \Delta z_{k+1/2}} + \frac{\Delta t}{\Delta z_k \Delta z_{k-1/2}}, \quad (7.20a)$$

$$a_1 = -\frac{\Delta t}{\Delta x_i \Delta x_{i-1/2}}, \quad (7.20b)$$

$$a_2 = -\frac{\Delta t}{\Delta x_i \Delta x_{i+1/2}}, \quad (7.20c)$$

$$a_3 = -\frac{\Delta t}{\Delta y_j \Delta y_{j-1/2}}, \quad (7.20d)$$

$$a_4 = -\frac{\Delta t}{\Delta y_j \Delta y_{j+1/2}}, \quad (7.20e)$$

$$a_5 = -\frac{\Delta t}{\Delta z_k \Delta z_{k-1/2}}, \quad (7.20f)$$

$$a_6 = -\frac{\Delta t}{\Delta z_k \Delta z_{k+1/2}}. \quad (7.20g)$$

Equation (7.19) is formally identical (apart from the fact that the coefficients are not constant) with the equation which would be obtained by discretization of a Poisson equation $\nabla^2 p = G$. It can be written in the following compact form:

$$a_0 p_0^{n+1} + \sum_{\alpha=1}^{6} a_\alpha p_\alpha^{n+1} = G_0^n. \quad (7.21)$$

and we refer to it as “Poisson-like” equation.

7.2.1 Stability Analysis of Poisson Equation with the Fourier Method

Let us refer again to a discretized Poisson-like equation written in the form (7.21) and assume it must be solved in the domain of a parallelepiped with edges of length L_x, L_y, and L_z, respectively, each one partitioned in N meshes of lengths $\Delta x = L_x/N$, $\Delta y = L_y/N$, and $\Delta z = L_z/N$. Let us further assume we are applying an iterative method so that at two subsequent iteration steps, r and $r + 1$, the current solutions for the unknown pressure approximately satisfy the equations:

$$\sum_{\alpha=0}^{6} a_\alpha p_\alpha^r = G_0^n, \tag{7.22}$$

$$\sum_{\alpha=0}^{6} a_\alpha p_\alpha^{r+1} = G_0^n. \tag{7.23}$$

Subtraction of these two equations yields

$$\sum_{\alpha=0}^{6} a_\alpha p_\alpha^{r+1} = \sum_{\alpha=0}^{6} a_\alpha p_\alpha^r \tag{7.24}$$

We can assume that the numerical solution p, defined in all grid points, differs from the unknown correct solution $\overline{p}$ by an error $\widetilde{p}$, so that:

$$p = \overline{p} - \widetilde{p}. \tag{7.25}$$

Substituting Eq. (7.25) into Eq. (7.24) and keeping in mind that the correct, though unknown, solution $\overline{p}$ does not depend on the iteration index, one derives:

$$\sum_{\alpha=0}^{6} a_\alpha \left(\overline{p}_\alpha - \widetilde{p}_\alpha^{r+1}\right) = \sum_{\alpha=0}^{6} a_\alpha \left(\overline{p}_\alpha - \widetilde{p}_\alpha^r\right), \tag{7.26}$$

hence:

$$\sum_{\alpha=0}^{6} a_\alpha \widetilde{p}_\alpha^{r+1} = \sum_{\alpha=0}^{6} a_\alpha \widetilde{p}_\alpha^r. \tag{7.27}$$

In this equation, with the same conventions used in the previous section, $\alpha = 0$ refers to the central node (i, j, k) of a discretization cell; $\alpha = 1$ and $\alpha = 2$ refer to the neighboring nodes $(i - 1, j, k)$ and $(i + 1, j, k)$ in the x direction; $\alpha = 3$ and $\alpha = 4$ refer to the neighboring nodes $(i, j-1, k)$ and $(i, j+1, k)$ in the y direction; and $\alpha = 5$ and $\alpha = 6$ refer to the neighboring nodes $(i, j, k-1)$ and $(i, j, k+1)$ in the z direction.

Equation (7.27) states simply that the distribution over the full definition domain of the error in the pressure solution satisfies the same Eq. (7.24) for the unknown pressure.

Following the method explained for a one-dimensional domain in Sect. 4.4.2, we assume that the distribution of errors can be represented by a Fourier series with M terms. In analogy with formula (Sect. 4.4.2.1), we can write for the three-dimensional cases:

$$e\left(x_i, y_j, z_k\right) = \sum_{p=1}^{M} A_p^r \exp\left(\lambda_p x_i + \mu_p y_j + \nu_p z_k\right), \tag{7.28}$$

where A_p^r is the amplitude of the error component p:

$$\lambda_p = \sqrt{-1} p\pi / L_x, \tag{7.29a}$$

$$\mu_p = \sqrt{-1} p\pi / L_y, \tag{7.29b}$$

$$\nu_p = \sqrt{-1} p\pi / L_z, \tag{7.29c}$$

and we used the short notations $x_i = i\Delta x$, $y_j = j\Delta y$, and $z_k = k\Delta z$.

Because the equations are linear, and every component of a linear combinations of solutions is a solution, we need not consider every one of the M terms in the above expansions but can restrict the analysis to only one component, say:

$$e(i, j, k) = A^r \exp J(i\lambda + \mu j + \nu k), \tag{7.30}$$

where $\lambda = p\pi/L_x$, $\mu = p\pi/L_y$, $\nu = p\pi/L_z$, and the imaginary unit is denoted $J = \sqrt{-1}$.

Carrying on the numerical solution of the Poisson equation with the iterative scheme, the solution will be attained if the numerical errors vanish in the course of the iterations. This implies that the amplification factor of the errors, defined as the ratio $G(\lambda, \mu, \nu) = A^{r+1}/A^r$ of two successive amplitudes of the error component considered, must vanish with increasing iteration number. This implies that the modulus of $G(\lambda, \mu, \nu)$ must be smaller than 1. We may investigate the behavior of $|G|$ assigning to the members of Eq. (7.27) weights Φ and $1 - \Phi$, respectively. Inserting Eq. (7.29) in Eq. (7.27), and multiplying each member by the respective weight, one obtains:

$$
\begin{aligned}
&\Phi[a_o A^{r+1} \exp J[i\lambda + \mu j + \nu k] + a_1 A^{r+1} \exp J[\lambda(i-1) + \mu j + \nu k] \\
&\quad + a_2 A^{r+1} \exp J[\lambda(i+1) + \mu j + \nu k] \\
&+ a_3 A^{r+1} \exp J[\lambda + \mu(j-1) + \nu k] + a_4 A^{r+1} \exp J[\lambda + \mu(j+1) + \nu k] \\
&+ a_5 A^{r+1} \exp J[\lambda + \mu + \nu(k-1)] + a_6 A^{r+1} \exp J[\lambda + \mu + \nu(k+1)]] \\
&= (1-\Phi)[a_o A^r \exp J[i\lambda + \mu j + \nu k] + a_1 A^r \exp J[\lambda(i-1) + \mu j + \nu k] \\
&\quad + a_2 A^r \exp J[\lambda(i+1) + \mu j + \nu k] \\
&+ a_3 A^r \exp J[\lambda + \mu(j-1) + \nu k] + a_4 A^r \exp J[\lambda + \mu(j+1) + \nu k] \\
&+ a_5 A^r \exp J[\lambda + \mu + \nu(k-1)] + a_6 A^r \exp J[\lambda + \mu + \nu(k+1)]]
\end{aligned}
\quad . \quad (7.31)
$$

Dividing each term of Eq. (7.31) by exp$J(i\lambda + \mu j + \nu k)$ yields:

$$
\begin{aligned}
&\Phi A^{r+1}\left(a_0 + a_1 e^{-J\lambda} + a_2 e^{J\lambda} + a_3 e^{-J\mu} + a_4 e^{J\mu} + a_5 e^{-J\nu} + a_6 e^{J\nu}\right) \\
&= (1-\Phi) A^r \left(a_0 + a_1 e^{-J\lambda} + a_2 e^{J\lambda} + a_3 e^{-J\mu} + a_4 e^{J\mu} + a_5 e^{-J\nu} + a_6 e^{J\nu}\right),
\end{aligned}
\quad (7.32)
$$

from which the amplification factor is derived as:

$$
G = \frac{A^{r+1}}{A^r} = \frac{1-\Phi}{\Phi}. \quad (7.33)
$$

Because $G(\Phi = 1) = 0$ and $G(\Phi = 1/2) = 1$, while to $\Phi < 1/2$ corresponds $G > 1$, one deduces that the range of stability of the numerical solution of the Poisson equation, for which $|G| < 1$, is given by $(1/2 < \Phi < 1)$. This implies that in the iterative numerical method used for approaching the solution, the preliminary pressure values at iteration step $r + 1$ must be given a weight larger than that given to the terms at iteration step r.

7.3 Derivation of Poisson-Like Equations for Pressure and Enthalpy with Fully Implicit Treatment

In the following sections, reference is made to real problems treated at the Nuclear Research Centre of Karlsruhe (Kernforschungszentrum Karlsruhe, KfK), Germany, in the framework of safety analysis of liquid metal fast breeder reactors (LMFBRs). In this framework, KfK has sponsored the development of a computer program describing the multidimensional thermal-hydraulic behavior of a fuel rod bundle under nominal and accident conditions. The simulation of the coolant behavior in the reactor core, including incoherencies (time shifts) of boiling initiation of the coolant

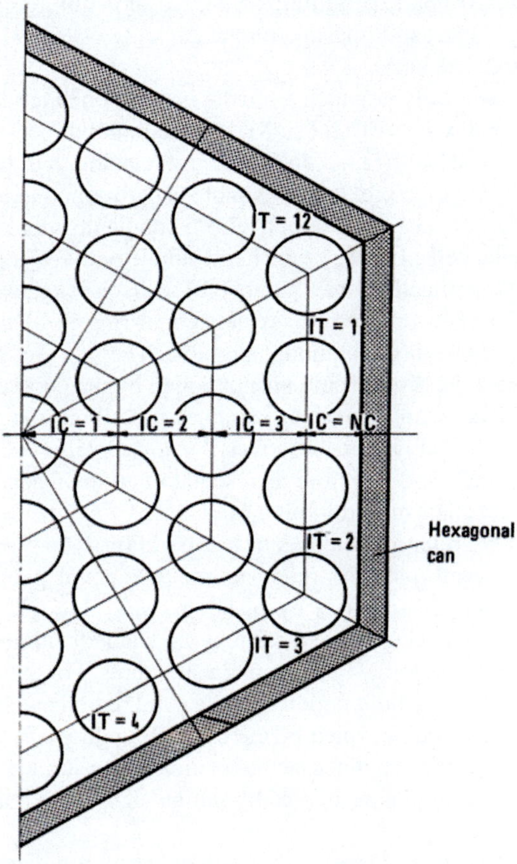

Fig. 7.2 Discretization of a 37-pin bundle with indexing of control cells in radial and azimuthal directions

in case of hypothetical accidents, requires the correct representation of the coolant channels and of the geometrical configuration of the fuel rods. Therefore, a correct three-dimensional representation of the fuel bundle is mandatory. Computer programs dedicated to these analyses must be able to describe not only single-phase flow conditions but, in case of core accidents, also two-phase flow conditions (liquid and vapor) in which the coolant of the reactor (liquid sodium in case of fast reactors) boils. At KfK, the computer program BACCHUS-3D/TP (three-dimensional, two phases) (Bottoni et al., 1992) has been developed over several years. In the following, a simplified description is given of the geometrical model used in this code to represent correctly a typical fuel bundle mainly consisting of rods of fuel, coolant flowing along the rods, and an outer hexagonal canning.

The hexagonal arrangement of fuel rods in bundles typical of LMFBRs is shown in Fig. 7.2 for the half of a 37-pin bundle, although in the reactors, much larger bundles with up to 397 fuel rods can be used. Control cells for mass and enthalpy balances in the coolant are bounded in axial direction by planes perpendicular to the z-axis of the bundle and Δz apart; in radial directions by planes through the pin axes and Δr apart; and in the azimuthal direction by planes passing through the bundle axis with azimuthal aperture of $\Delta\theta = 2\pi/\text{NTH}$ (NTH is the number of azimuthal cells used; see later).

The cells in which a bundle is partitioned are labelled in the axial direction with an index JC = 1, 2, . . ., MC; in the radial direction (as shown in the figure) with an index IC = 1, 2, . . ., NC; and in the azimuthal direction with index IT = 1, 2, . . ., NTH. Therefore, the full bundle is partitioned in a total number NP = MC × NC × NTH of cells. In case of a 37-pin bundle (as represented in the figure), there are four radial cells. In case of the large bundle with 397-pin rods, there are 12 radial cells. In radial direction, the number of cells is suggested by the number of hexagons (in jargon referred to as "hexagonal rings") of the bundles. A finer discretization would be difficult, though possible. The same holds for the azimuthal discretization. The axial discretization is not bound by the geometrical configuration of the bundle.

The main dependent variables describing the coolant behavior (mass, pressure, and enthalpy) are described by conservation equations integrated over the cell volumes and therefore representative of the mean values in the cells, for this reason also called "control volumes."

We refer to the domain shown in Fig. 7.2 as "pseudo"-cylindrical domain, where the term "pseudo" refers to the hexagonal geometry with cells bounded in the azimuthal directions by planes passing through the z-axis. The radial, axial, and azimuthal coordinate directions are labeled with indices r, z, and s, respectively. The simulated domain may contain structures. For these reasons, computational cells are divided in a fluid region and in a solid region and have a volume porosity, defined as the fraction occupied by the fluid. Bounding surfaces are characterized by a surface permeability, defined as the fraction of the surface open for the flow. The sub-case of a continuum is recovered by setting all volume porosities and surface permeabilities to one.

Centered control cells are denoted with indices (i, j, k) for the radial, axial, and azimuthal directions, respectively. These indices actually refer to the central node of a cell, and bounding surfaces are denoted by $(i \pm 1/2, j \pm 1/2, k \pm 1/2)$. Cells displaced (staggered) in the three coordinate directions have centered point labeled as $(i \pm 1/2, j, k)$, $(i, j \pm 1/2, k)$, and $(i, j, k \pm 1/2)$, respectively.

Space discretization of the conservation equations is done with reference to the staggered meshes. Scalar quantities, like coolant pressure, enthalpy, and other physical properties of the fluid, are defined at the center point (i, j, k) of a control volume. Velocity components of the coolant (u, w, and v for the r, z, and s directions, respectively) are defined at the midpoints of the boundary faces.

All cells are characterized by a total volume V, a volume occupied by the fluid V_f, by the areas of the lateral faces S_t, S_b (top, bottom, perpendicular to the z-axis of the bundle); S_i, S_e (internal, external, perpendicular to the radial coordinate r); and S_m, S_p (bounding the cell in the azimuthal direction, where the subscripts m (minus) and p (plus) denote the sequence considered in the positive clockwise direction). These geometrical elements are used to define volumetric porosities and surface permeabilities for every cell.

Let $S_{f_t}, S_{f_b}, S_{f_i}, S_{f_e}, S_{f_m}, S_{f_p}$ be the cross-flow areas of the bounding faces. For every cell, we define surface permeabilities ζ, Ψ, ξ for the axial, radial, and azimuthal directions, respectively, as the ratio of the respective cross-flow area to total area, for instance, $\Psi_i = S_{f_i}/S_i$, $\zeta_b = S_{f_b}/S_b$, etc. The volume porosity of a cell is defined as the ratio of the volume occupied by the fluid to the total cell volume, i.e., $\varepsilon = V_f/V$. In an undisturbed geometry, the volume porosity is equal to the surface permeabilities in the axial direction.

7.3.1 Governing Equations

In this section, we consider the conservation equations for mass, momentum, and enthalpy for a single-phase flow or for a two-phase flow (for instance, water and water vapor) treated with the slip model (SM) in which a difference of phase velocities is allowed. In case there is not a slip between the phase velocities, the so-called homogeneous equilibrium model (HEM) is simulated. The homogeneous equilibrium model can be obtained as a sub-case of the slip model dropping the terms which depend on the slip velocity.

The governing equations for the conservation of mass, momentum, and energy of the coolant for the slip model have been derived in reference (Bottoni & Sengpiel, 1992), from the conservation equations for single-phase flow (Bird et al., 1960). Neglecting some terms (like viscous dissipation in the fluid) which are small compared with the dominant ones (like power source), these equations can be written as follows:

Continuity equation:

$$\frac{\partial \rho_m}{\partial t} + \nabla \cdot (\rho_m \overline{u}_m) = 0; \tag{7.34}$$

Momentum equation:

$$\begin{aligned} &\frac{\partial}{\partial t}(\rho_m \overline{u}_m) + \nabla \cdot (\rho_m \overline{u}_m \overline{u}_m) + \nabla \cdot \left| x(1-x)\rho_m \overline{V}_{Sl} \overline{V}_{Sl} \right| \\ &= -\nabla p + \nabla \cdot (\mu \nabla \overline{u}_m) + \rho_m \overline{g} - \overline{R}; \end{aligned} \tag{7.35}$$

Energy equation:

$$\begin{aligned} &\frac{\partial}{\partial t}(\rho_m h_m) + \nabla \cdot (\rho_m \overline{u}_m h_m) + \nabla \cdot \left[x(1-x)\rho_m (h_v - h_l) \overline{V}_{Sl} \right] \\ &= \frac{\partial p}{\partial t} + \overline{u}_m \cdot \nabla p + \left(\frac{1}{\rho_m} \alpha_l \alpha_v (\rho_l - \rho_v) \overline{V}_{Sl} \right) \cdot \nabla p - \nabla \cdot \overline{q} + Q. \end{aligned} \tag{7.36}$$

Subscripts v, l, and m refer to vapor, liquid, and coolant mixture, respectively. The term $\overline{R}$ in the momentum equation represents the drag force per unit volume, and $\overline{V}_{\mathrm{Sl}}$ is the slip velocity. The dynamic viscosity μ is the effective one, defined as the sum of molecular and turbulent contributions. The thermodynamic quality of the flow, x, is defined, in terms of the specific enthalpies of the phases, by:

$$x = (h_{\mathrm{m}} - h_{\mathrm{l}})/(h_{\mathrm{v}} - h_{\mathrm{l}}). \tag{7.37}$$

The density and specific enthalpy of the mixture are given by:

$$\rho_{\mathrm{m}} = \alpha_{\ell}\rho_{\ell} + \alpha_{\mathrm{v}}\rho_{\mathrm{v}}, \tag{7.38}$$

$$h_{\mathrm{m}} = \frac{1}{\rho_{\mathrm{m}}}(\alpha_{\ell}\rho_{\ell}h_{\ell} + \alpha_{\mathrm{v}}\rho_{\mathrm{v}}h_{\mathrm{v}}), \tag{7.39}$$

where α_{ℓ} and α_{v} are the liquid and vapor volume fractions, with $\alpha_{\ell} + \alpha_{\mathrm{v}} = 1$.

The continuity equation, Eq. (7.34), is formally identical to the continuity equation for single-phase flow and reduces to it if the void fraction is zero. The momentum and energy equations, Eqs. (7.35) and (7.36), are similar to those for single-phase flow, but contain additional terms, which depend on the slip velocity V_{Sl}, and are referred to as "momentum slip" and "energy slip," respectively. In case the slip velocity is zero (i.e., the phases have the same velocity), the equations obtained are those of the homogeneous equilibrium model (HEM), as a subcase of the slip model.

7.3.2 *Calculation of Liquid and Vapor Velocity Components from Given Slip Velocity or Slip Ratio*

For every coordinate direction, we define the slip velocity as the difference between the velocity of the vapor and liquid phase. With reference to the z coordinate direction, we have:

$$W_{\mathrm{Sl}} = w_{\mathrm{v}} - w_{\mathrm{l}}. \tag{7.40}$$

The normalized slip velocity is defined as the ratio of the slip velocity to the mixture velocity component in the same coordinate direction:

$$W_{\mathrm{Sl}}^{N} = \frac{W_{\mathrm{Sl}}}{w_{\mathrm{m}}} = \frac{w_{\mathrm{v}} - w_{\mathrm{l}}}{w_{\mathrm{m}}}. \tag{7.41}$$

Similar definitions apply for the r and s coordinate directions. The mixture velocity is given by:

$$w_m = \frac{1}{\rho_m}(\alpha_l \rho_l w_l + \alpha_v \rho_v w_v). \tag{7.42}$$

Relationships between the slip velocity, mixture velocity, and the velocity components of either phase can be obtained under the assumption of thermodynamic equilibrium between the phases ($T_v = T_l$). In this case, one has (Bottoni et al., 1992):

$$x = \frac{\alpha_v \rho_v}{\rho_m}, \tag{7.43}$$

$$1 - x = \frac{\alpha_l \rho_l}{\rho_m}. \tag{7.44}$$

From Eqs. (7.42) through (7.44), the velocity components of the two phases are derived, as function of the mixture velocity and of the slip velocity, by:

$$w_l = w_m - xW_{Sl}, \tag{7.45}$$

$$w_v = w_m + (1 - x)W_{Sl}. \tag{7.46}$$

Combining the two previous equations yields the following relationship between the slip ratio $H_z = w_v/w_l$ and the slip velocity W_{Sl}:

$$H_z = \frac{w_m + (1 - x)W_{Sl}}{w_m - xW_{Sl}}, \tag{7.47}$$

or its inverse:

$$W_{Sl} = w_m \frac{H_z - 1}{1 + x(H_z - 1)}. \tag{7.48}$$

In the case of $H = 1$ (the homogeneous equilibrium model), the slip velocity vanishes.

The slip velocity can be eliminated from Eqs. (7.45) and (7.46), thus obtaining the velocity components of either phase as a function of the mixture velocity w_m and of the slip ratio H_z:

$$w_l = \frac{w_m}{1 - x + xH_z} \tag{7.49}$$

$$w_v = H_z w_l. \tag{7.50}$$

Two options are commonly used for imposing either the slip ratio or the slip velocity:

Option 1:
The slip ratios are prescribed for the three coordinate directions $\alpha \equiv (r, z, s)$, as:

$$H_\alpha = \frac{u_{\mathrm{v}\alpha}}{u_{\mathrm{l}\alpha}}. \tag{7.51}$$

In this case, Eqs. (7.49) and (7.50) and similar ones in the r and s coordinate directions give the velocity components of either phase (to be used for calculating the "momentum slip" and "energy slip," i.e., the terms containing the slip velocity, in the momentum and energy equations):

$$u_{\mathrm{l}\alpha} = \frac{u_{\mathrm{m}\alpha}}{1 - x + xH_\alpha} \tag{7.52}$$

$$u_{\mathrm{v}\alpha} = H_\alpha u_{\mathrm{l}\alpha}. \tag{7.53}$$

The homogeneous equilibrium model (HEM) is obtained as a sub-case of the slip model (SM), setting $H_r = H_s = H_z = 1$.

Option 2:
In this case, we define the normalized slip velocities by:

$$U^N_{\mathrm{Sl}.\alpha} = \frac{u_{\mathrm{v}\alpha} - u_{\mathrm{l}\alpha}}{u_{\mathrm{m}\alpha}}. \tag{7.54}$$

In this option, the velocity components of the two phases are given by:

$$u_{\mathrm{l}\alpha} = u_{\mathrm{m}\alpha}\left(1 - xU^N_{\mathrm{Sl},\alpha}\right), \tag{7.55}$$

$$u_{\mathrm{v}\alpha} = u_{\mathrm{l}\alpha}\left(1 + \frac{U^N_{\mathrm{Sl},\alpha}}{1 - xU^N_{\mathrm{Sl},\alpha}}\right) = u_{\mathrm{l}\alpha} + u_{\mathrm{m}\alpha}U^N_{\mathrm{Sl},\alpha}. \tag{7.56}$$

The HEM is obtained as a sub-case in this second option setting $U^N_{\mathrm{Sl}} = V^N_{\mathrm{Sl}} = W^N_{\mathrm{Sl}} = 0$.

7.3.3 *Finite Difference Form of the Continuity Equation*

We treat the convective terms in Eq. (7.1) fully implicitly. Integrating this equation over a centered cell and replacing the volume integral of the divergence term by means of surface fluxes, one derives the following finite difference form:

$$\begin{aligned}
&\varepsilon_{ijk}\left(\frac{\partial\rho_{\mathrm{m}}}{\partial t}\right)_{ijk}^{n+1}+\frac{1}{\Delta z_j}\left[\zeta_{\,j+1/2}(\rho_{\mathrm{m}}w_{\mathrm{m}})_{i,j+1/2,k}^{n+1}-\zeta_{\,j-1/2}(\rho_{\mathrm{m}}w_{\mathrm{m}})_{i,j-1/2,k}^{n+1}\right]+\\
&+\frac{1}{\Delta r_i}\left[(\Psi F_{\mathrm{c}}\rho_{\mathrm{m}}u_{\mathrm{m}})_{i+1/2,j,k}^{n+1}-(\Psi F_{\mathrm{c}}\rho_{\mathrm{m}}u_{\mathrm{m}})_{i-1/2,j,k}^{n+1}\right]+\\
&+\frac{1}{\Delta s_k}\left[\left(\frac{\xi\rho_{\mathrm{m}}v_{\mathrm{m}}}{\cos\beta}\right)_{i,j,k+1/2}^{n+1}-\left(\frac{\xi\rho_{\mathrm{m}}v_{\mathrm{m}}}{\cos\beta}\right)_{i,j,k-1/2}^{n+1}\right]=0.
\end{aligned} \tag{7.57}$$

Equation (7.57) can be written in compact form as:

$$\varepsilon_{ijk}\left(\frac{\partial\rho_{\mathrm{m}}}{\partial t}\right)_{ijk}^{n+1}+\sum_{\alpha=1}^{3}\frac{1}{\Delta l_\alpha}\Delta(\gamma_\alpha c_\alpha\langle\rho_{\mathrm{m}}u_{\mathrm{m}\alpha}\rangle)^{n+1}=0 \qquad (\alpha=r,s,z) \tag{7.58}$$

where the parenthesis $\langle\rangle$ denotes mean value over the surfaces of the cells and:

γ_α = surface permeability.
ε = volume porosity.
$c_z = 1$.
$c_r = F_{\mathrm{c}} = S_{i\,-\,1/2}/S_i$ for inner surface.
$c_s = F_{\mathrm{c}} = S_{i\,+\,1/2}/S_i$ for outer surface.
$c_s = 1/\cos\beta$; β = angle between r-axis and azimuthal boundary surface of centered control volume.

Equation (7.57) is used to derive a Poisson-like equation for the pressure distribution, as explained in Sect. 7.3.5. It is also used to derive the updated value of the void fraction. Discretizing the time derivative and solving with respect to ρ_{m}^{n+1}, one has:

$$\rho_{\mathrm{m}}^{n+1}=\rho_{\mathrm{m}}^{n}-\frac{\Delta t}{\varepsilon}\sum_{\alpha=1}^{3}\frac{1}{\Delta l_\alpha}\Delta(\gamma_\alpha c_\alpha\langle\rho_{\mathrm{m}}u_{\mathrm{m}\alpha}\rangle). \tag{7.59}$$

Hence, from the definition $\rho_{\mathrm{m}}=\alpha_{\mathrm{v}}\rho_{\mathrm{v}}+\alpha_{\mathrm{l}}\rho_{\mathrm{l}}$:

$$\begin{aligned}
\alpha_{\mathrm{v}}^{n+1}&=\frac{\rho_{\mathrm{l}}^{n+1}-\rho_{\mathrm{m}}^{n+1}}{\rho_{\mathrm{l}}^{n+1}-\rho_{\mathrm{v}}^{n+1}}\\
&=\frac{1}{\rho_{\mathrm{l}}^{n+1}-\rho_{\mathrm{v}}^{n+1}}\left[\rho_{\mathrm{l}}^{n+1}-\rho_{\mathrm{m}}^{n}-\frac{\Delta t}{\varepsilon}\sum_{\alpha=1}^{3}\frac{1}{\Delta l_\alpha}\Delta(\gamma_\alpha c_\alpha\langle\rho_{\mathrm{m}}u_{\mathrm{m}\alpha}\rangle)^{n+1}\right].
\end{aligned} \tag{7.60}$$

7.3.4 Fully Implicit Treatment of Momentum Conservation Equation

The scalar component of the momentum conservation equation (7.35) in the z-coordinate direction can be written as:

$$\frac{\partial}{\partial t}(\rho_{\rm m} w_{\rm m}) + \frac{\partial}{\partial L_\alpha}\left|(\rho_{\rm m} u_{{\rm m}\alpha}) w_{\rm m} + x(1-x)\rho_{\rm m}(u_{\rm v} - u_{\rm l})_\alpha (w_{\rm v} - w_{\rm l})\right| - \frac{\partial}{\partial L_\alpha}\left(\mu_{\rm m}\frac{\partial w_{\rm m}}{\partial L_\alpha}\right)$$
$$= -\frac{\partial p}{\partial z} - \rho_{\rm m} g_z - R_z. \qquad (\alpha = r, s, z) \tag{7.61}$$

α represents any of the coordinate directions (r, s, z in bundle geometry), and u_α is the respective velocity component.

Letting:

$$G_{\rm Sl} = x(1-x)\rho_{\rm m}, \quad \text{(momentum slip)} \tag{7.62}$$

$$S_z = -\frac{\partial p}{\partial z} - \rho_{\rm m} g_z - R_z, \quad \text{(source term)} \tag{7.63}$$

$$Jw_{{\rm m}_\alpha} = (\rho_{\rm m} u_{{\rm m}_\alpha}) w_{\rm m} + G_{\rm Sl}(u_{\rm v} - u_{\rm l})_\alpha (w_{\rm v} - w_{\rm l}) - \mu_{\rm m}\frac{\partial w_{\rm m}}{\partial L_\alpha}, \quad \text{(convective plus diffusive flux of momentum)} \tag{7.64}$$

Equation (7.61) can be written as:

$$\frac{\partial}{\partial t}(\rho_{\rm m} w_{\rm m}) + \frac{\partial}{\partial L_\alpha} Jw_{{\rm m}\alpha} = S_z. \tag{7.65}$$

Integration over the volume $V_{\rm f}$ of the fluid in a control cell yields:

$$\int_{V_{\rm f}} \frac{\partial}{\partial t}(\rho_{\rm m} w_{\rm m}) dV + \int_{V_{\rm f}} \frac{\partial}{\partial L_\alpha} Jw_{{\rm m}\alpha} dV = \int_{V_{\rm f}} S_z dV. \tag{7.66}$$

The control volume for the z-component of the momentum equation is obtained displacing a centered control cell (i, j, k) by half-mesh in the z direction. Replacing the volume integral of the divergence term by means of the fluxes through the cell bounding surfaces, one derives:

$$\int_{V_f} \frac{\partial}{\partial t}(\rho_m w_m)dV + \int_{A_{f_{i+1/2,j+1/2,k}}} (Jw_m)_r dA - \int_{A_{f_{i-1/2,j+1/2,k}}} (Jw_m)_r dA$$
$$+\int_{A_{f_{i,j+1,k}}} (Jw_m)_z dA - \int_{A_{f_{i,j,k}}} (Jw_m)_z dA + \int_{A_{f_{i,j+1/2,k+1/2}}} (Jw_m)_s dA \tag{7.67}$$
$$-\int_{A_{f_{i,j+1/2,k-1/2}}} (Jw_m)_s dA = \int_{V_f} S_z dV.$$

We discretize this equation with respect to time treating convective, diffusive, and the pressure gradient term fully implicitly and the friction term half-implicitly. Replacing the surface integrals by the mean values over the surfaces, one obtains:

$$\frac{V_{f_{i,j+1/2,k}}}{\Delta t}\left[(\rho_m w_m)^{n+1} - (\rho_m w_m)^n\right]_{i,j+1/2,k} + \left\langle (Jw_m)_r A_f \right\rangle^{n+1}_{i+1/2,j+1/2,k}$$
$$-\left\langle (Jw_m)_r A_f \right\rangle^{n+1}_{i-1/2,j+1/2,k} + \left\langle (Jw_m)_z A_f \right\rangle^{n+1}_{i,j+1,k} - \left\langle (Jw_m)_z A_f \right\rangle^{n+1}_{ijk}$$
$$+\left\langle (Jw_m)_s A_f \right\rangle^{n+1}_{i,j+1/2,k+1/2} - \left\langle (Jw_m)_s A_f \right\rangle^{n+1}_{i,j+1/2,k-1/2}$$
$$= -\left(\frac{V}{\Delta z}\right)_{i,j+1/2,k}\left(p^{n+1}_{i,j+1,k} - p^{n+1}_{ijk}\right) - \left(V_f \rho_m g_z\right)^n_{i,j+1/2,k} - \left[\frac{V_f}{2D_h} f |w_m|^n (\rho_m w_m)^{n+1}\right]_{i,j+1/2,k}. \tag{7.68}$$

The last term in Eq. (7.68) represents the fiction force depending on the hydraulic diameter of the flow duct and on a given function $f(|w_m|)$ of the velocity of the mixture.

The averaged flux of the vector Jw_m in the positive r-direction is evaluated by:

$$J_r + = \int_{A_{f_{i+1/2,j+1/2,k}}} (Jw_m)_r dA = \langle Jw_m \cdot A_f \rangle_{i+1/2,j+1/2,k}$$
$$= \left\{ A_f \rho_m u_m w_m + A_f G_{SI}(u_v - u_l)(w_v - w_l) - \frac{A_f \mu_m}{\rho_m \Delta r}\Delta(\rho_m w_m) \right\}_{i+1/2,j+1/2,k}$$
$$= \{F\rho_m w_m + G(w_v - w_l) - D \cdot \Delta(\rho_m w_m)\}_{i+1/2,j+1/2,k}, \tag{7.69}$$

with the definitions:

$$F_{i+1/2,j+1/2,k} = (A_f u_m)_{i+1/2,j+1/2,k}, \tag{7.70}$$

$$G_{i+1/2,j+1/2,k} = [A_f G_{Sl}(u_v - u_l)]_{i+1/2,j+1/2,k}, \tag{7.71}$$

$$D_{i+1/2,j+1/2,k} = \left(A_f \frac{\mu_m}{\rho_m \Delta r}\right)_{i+1/2,j+1/2,k}, \tag{7.72}$$

$$\Delta(\rho_m w_m)_{i+1/2,j+1/2,k} = (\rho_m w_m)_{i+1,j+1/2,k} - (\rho_m w_m)_{i,j+1/2,k}. \tag{7.73}$$

For both $u_{i+1/2,\, j+1/2,\, k}$ positive and negative, Eq. (7.69) can be written as:

$$\begin{aligned} J_r+ &= \left[0, F_{i+1/2,j+1/2,k}\right](\rho_m w_m)_{i,j+1/2,k} - \left[0, -F_{i+1/2,j+1/2,k}\right](\rho_m w_m)_{i+1,j+1/2,k} \\ &+ \left[0, G_{i+1/2,j+1/2,k}\right](w_v - w_l)_{i,j+1/2,k} - \left[0, -G_{i+1/2,j+1/2,k}\right]\left(w_g - w_l\right)_{i+1,j+1/2,k} \\ &+ D_{i+1/2,j+1/2,k}\left\{(\rho_m w_m)_{i,j+1/2,k} - (\rho_m w_m)_{i+1,j+1/2,k}\right\}. \end{aligned} \tag{7.74}$$

The symbol $[a, b]$ denotes the maximum of the two real numbers a and b. Similar definitions of the fluxes hold for the other surfaces bounding a cell.

Let us introduce the following subscripts to index the central node considered and the six neighboring nodes in the three coordinate directions:

$$\begin{array}{lll} 0 & \text{for} & i, j+1/2, k \\ 1 & \text{for} & i-1, j+1/2, k \\ 2 & \text{for} & i+1, j+1/2, k \\ 3 & \text{for} & i, j+1/2, k-1 \\ 4 & \text{for} & i, j+1/2, k+1 \\ 5 & \text{for} & i, j-1/2, k \\ 6 & \text{for} & i, j+3/2, k \end{array}$$

The flux J_r+ (and, similarly, the other ones in Eq. (7.69)) can be written as:

$$\begin{aligned} J_r+ &= \int_{A_{f_{i+1/2,j+1/2,k}}} (Jw_m)_r dA = \left\langle (Jw_m)_r \cdot A_f \right\rangle_{i+1/2,j+1/2,k} \\ &= \left[0, F_{i+1/2,j+1/2,k}\right](\rho_m w_m)_0 - \left[0, -F_{i+1/2,j+1/2,k}\right](\rho_m w_m)_2 \\ &+ \left[0, G_{i+1/2,j+1/2,k}\right](w_v - w_l)_0 - \left[0, -G_{i+1/2,j+1/2,k}\right](w_v - w_l)_2 \\ &+ D_{i+1/2,j+1/2,k}\left[(\rho_m w_m)_0 - (\rho_m w_m)_2\right]. \end{aligned} \tag{7.75}$$

Using Eq. (7.48), we can write:

$$U_{\mathrm{Sl},\alpha} = u_{\mathrm{v}\alpha} - u_{\mathrm{l}\alpha} = u_{\mathrm{m}\alpha}\frac{H_\alpha - 1}{1 + x(H_\alpha - 1)} = \rho_\mathrm{m} u_{\mathrm{m}\alpha}\frac{\widetilde{H}_\alpha}{\rho_\mathrm{m}}, \quad (\alpha = r, z, s) \tag{7.76}$$

with the definitions:

$$\widetilde{H}_\alpha = \frac{H_\alpha - 1}{1 + x(H_\alpha - 1)}. \quad (\alpha = r, s, z) \tag{7.77}$$

With this artifice, we can calculate the sum of the first and third convective terms in Eq. (7.75) as:

$$\begin{aligned}&\left[0, F_{i+1/2,j+1/2,k}\right](\rho_\mathrm{m} w_\mathrm{m})_0 + \left[0, +G_{i+1/2,j+1/2,k}\right](w_\mathrm{v} - w_\mathrm{l})_0 \\ &= \left[0, F_{i+1/2,j+1/2,k} + G_{i+1/2,j+1/2,k}\left(\frac{\widetilde{H}_z}{\rho_\mathrm{m}}\right)_0\right](\rho_\mathrm{m} w_\mathrm{m})_0\end{aligned} \tag{7.78}$$

and similarly for the other convective fluxes. With this procedure, slip velocities are eliminated, being replaced by the mass fluxes of the mixture.

Thus, Eq. (7.75) (and similarly for the other convective fluxes) can be written in more compact form:

$$\begin{aligned}J_r + =& D_{i+1/2}\left[(\rho_\mathrm{m} w_\mathrm{m})_0 - (\rho_\mathrm{m} w_\mathrm{m})_2\right] + \left[0, F_{i+1/2} + G_{i+1/2}\left(\frac{\widetilde{H}_{z0}}{\rho_{\mathrm{m}0}}\right)\right](\rho_\mathrm{m} w_\mathrm{m})_0 \\ &- \left[0, -\left(F_{i+1/2} + G_{i+1/2}\frac{\widetilde{H}_{z2}}{\rho_{\mathrm{m}2}}\right)\right](\rho_\mathrm{m} w_\mathrm{m})_2.\end{aligned} \tag{7.79}$$

Introducing the convective and diffusive fluxes (Eq. (7.79), and likewise for the fluxes in the other coordinate directions) evaluated at time level $n + 1$, into Eq. (7.68) (which—we recall—arises from the z-component of the momentum equation), and rearranging, one derives:

$$\alpha_0^w (\rho_\mathrm{m} w_\mathrm{m})_0^{n+1} - \sum_{\beta=1}^{6} \alpha_\beta^w (\rho_\mathrm{m} w_\mathrm{m})_\beta^{n+1} = b_0^w - d_0^w\left(p_{i,j+1,k}^{n+1} - p_{ijk}^{n+1}\right) \tag{7.80}$$

with the following definitions of the coefficients:

$$
\begin{aligned}
a_0^w =& \frac{(\varepsilon V)_0}{\Delta t} + \left(\frac{\varepsilon V}{2D_\mathrm{h}} f|w_\mathrm{m}|^n\right)_0 + \left[0, F^{n+1}_{i+1/2,j+1/2,k} + G^{n+1}_{i+1/2,j+1/2,k}\left(\frac{\widetilde{H}_z}{\rho_\mathrm{m}}\right)_0\right] \\
&+ \left[0, -\left(F^{n+1}_{i-1/2,j+1/2,k} + G^{n+1}_{i-1/2,j+1/2,k}\right)\left(\frac{\widetilde{H}_z}{\rho_\mathrm{m}}\right)_0\right] \\
&+ \left[0, F^{n+1}_{i,j+1,k} + G^{n+1}_{i,j+1,k}\left(\frac{\widetilde{H}_z}{\rho_\mathrm{m}}\right)_0\right] + \left[0, -\left(F^{n+1}_{ijk} + G^{n+1}_{ijk}\left(\frac{\widetilde{H}_z}{\rho_\mathrm{m}}\right)_0\right)\right] \\
&+ \left[0, F^{n+1}_{i,j+1/2,k+1/2} + G^{n+1}_{i,j+1/2,k+1/2}\left(\frac{\widetilde{H}_z}{\rho_\mathrm{m}}\right)_0\right] \\
&+ \left[0, -\left(F^{n+1}_{i,j+1/2,k-1/2} + G^{n+1}_{i,j+1/2,k-1/2}\left(\frac{\widetilde{H}_z}{\rho_\mathrm{m}}\right)_0\right)\right] \\
&+ D^{n+1}_{i+1/2,j+1/2,k} + D^{n+1}_{i-1/2,j+1/2,k} + D^{n+1}_{i,j+1,k} + D^{n+1}_{ijk} + D^{n+1}_{i,j+1/2,k+1/2} + D^{n+1}_{i,j+1/2,k-1/2},
\end{aligned}
\tag{7.81a}
$$

$$
a_1^w = \left\{\left[0, F^{n+1}_{i-1/2,j+1/2,k} + G^{n+1}_{i-1/2,j+1/2,k}\left(\frac{\widetilde{H}_z}{\rho_\mathrm{m}}\right)_1\right] + D^{n+1}_{i-1/2,j+1/2,k}\right\}, \tag{7.81b}
$$

$$
a_2^w = \left\{\left[0, -\left(F^{n+1}_{i+1/2,j+1/2,k} + G^{n+1}_{i+1/2,j+1/2,k}\left(\frac{\widetilde{H}_z}{\rho_\mathrm{m}}\right)_2\right)\right] + D^{n+1}_{i+1/2,j+1/2,k}\right\}, \tag{7.81c}
$$

$$
a_3^w = \left\{\left[0, F^{n+1}_{i,j+1/2,k-1/2} + G^{n+1}_{i,j+1/2,k-1/2}\left(\frac{\widetilde{H}_z}{\rho_\mathrm{m}}\right)_3\right] + D^{n+1}_{i,j+1/2,k-1/2}\right\}, \tag{7.81d}
$$

$$
a_4^w = \left\{\left[0, -\left(F^{n+1}_{i,j+1/2,k+1/2} + G^{n+1}_{i,j+1/2,k+1/2}\left(\frac{\widetilde{H}_z}{\rho_\mathrm{m}}\right)_4\right)\right] + D^{n+1}_{i,j+1/2,k+1/2}\right\}, \tag{7.81e}
$$

$$
a_5^w = \left\{\left[0, F^{n+1}_{i,j,k} + G^{n+1}_{i,j,k}\left(\frac{\widetilde{H}_z}{\rho_\mathrm{m}}\right)_5\right] + D^{n+1}_{ijk}\right\}, \tag{7.81f}
$$

$$
a_6^w = \left\{\left[0, -\left(F^{n+1}_{i,j+1,k} + G^{n+1}_{i,j+1,k}\left(\frac{\widetilde{H}_z}{\rho_\mathrm{m}}\right)_6\right)\right] + D^{n+1}_{i,j+1,k}\right\}, \tag{7.81g}
$$

$$
b_0^w = \frac{(\varepsilon V)_0}{\Delta t}(\rho_\mathrm{m} w_\mathrm{m})_0^n - (\varepsilon V \rho_\mathrm{m} g_z)_0^n, \tag{7.81h}
$$

$$
d_0^w = \left(\frac{\varepsilon V}{\Delta z}\right)_0. \tag{7.81i}
$$

Defining:

$$\widehat{w}_6 = \frac{1}{a_0^w}\left\{\sum_{\beta=1}^{6} a_\beta^w (\rho_m w_m)_\beta^{n+1} + b_0^w\right\}, \tag{7.82}$$

$$d_6^w = \frac{d_0^w}{a_0^w} = \frac{1}{a_0^w}\left(\frac{\varepsilon V}{\Delta z}\right)_0, \tag{7.83}$$

Equation (7.80) can be written as:

$$(\rho_m w_m)_{i,j+1/2,k}^{n+1} = \widehat{w}_6 - d_6^w\left(p_{i,j+1,k}^{n+1} - p_{ijk}^{n+1}\right). \tag{7.84a}$$

Similarly, one derives for the radial and azimuthal components of the momentum equation:

$$(\rho_m u_m)_{i+1/2,j,k}^{n+1} = \widehat{u}_2 - d_2^u\left(p_{i+1,j,k}^{n+1} - p_{ijk}^{n+1}\right), \tag{7.84b}$$

$$(\rho_m v_m)_{i,j,k+1/2}^{n+1} = \widehat{v}_4 - d_4^v\left(p_{i,j,k+1}^{n+1} - p_{ijk}^{n+1}\right), \tag{7.84c}$$

with:

$$\widehat{u}_2 = \frac{1}{a_0^u}\left\{\sum_{\beta=1}^{6} a_\beta^u (\rho_m u_m)_\beta^{n+1} + b_0^u\right\}, \tag{7.85}$$

$$\widehat{v}_4 = \frac{1}{a_0^v}\left\{\sum_{\beta=1}^{6} a_\beta^v (\rho_m v_m)_\beta^{n+1} + b_0^v\right\}, \tag{7.86}$$

$$d_2^u = \frac{1}{a_{i+1/2,j,k}^u}\left(\frac{\varepsilon V}{\Delta r}\right)_{i+1/2,j,k}, \tag{7.87}$$

$$d_4^v = \frac{1}{a_{i,j,k+1/2}^v}\left(\frac{\varepsilon V}{\Delta s}\right)_{i,j,k+1/2}. \tag{7.88}$$

For the three components of the momentum equations, written for control volumes displaced by half-cell in the respective backward direction, one has (with similar definitions of symbols):

$$(\rho_m u_m)_{i-1/2,j,k}^{n+1} = \widehat{u}_1 - d_1^u\left(p_{ijk}^{n+1} - p_{i-1,j,k}^{n+1}\right), \tag{7.89a}$$

$$(\rho_m v_m)_{i,j,k-1/2}^{n+1} = \widehat{v}_3 - d_3^v\left(p_{ijk}^{n+1} - p_{i,j,k-1}^{n+1}\right), \tag{7.89b}$$

$$(\rho_m w_m)_{i,j-1/2,k}^{n+1} = \widehat{w}_5 - d_5^w\left(p_{ijk}^{n+1} - p_{i,j-1,k}^{n+1}\right). \tag{7.89c}$$

7.3.5 Poisson Equation Describing the Pressure Distribution

The Poisson equation describing the coolant pressure distribution is obtained by combining the continuity and momentum equations. By introducing Eqs. (7.84) and (7.89) into (7.57), one derives:

$$
\begin{aligned}
&\varepsilon_{ijk}\left(\frac{\partial\rho_\mathrm{m}}{\partial t}\right)_{ijk}^{n+1}+\frac{1}{\Delta z_j}\left\{\begin{array}{l}\zeta_{i+1/2}\left[\widehat{w}_6-d_6^w\left(p_{i,j+1,k}^{n+1}-p_{ijk}^{n+1}\right)\right]\\-\zeta_{j-1/2}\left[\widehat{w}_5-d_5^w\left(p_{ijk}^{n+1}-p_{i,j-1,k}^{n+1}\right)\right]\end{array}\right\}\\
&+\frac{1}{\Delta r_i}\left\{\begin{array}{l}(\Psi F_\mathrm{c})_{i+1/2}\left[\widehat{u}_2-d_2^u\left(p_{i+1,j,k}^{n+1}-p_{ijk}^{n+1}\right)\right]\\-(\Psi F_\mathrm{c})_{i-1/2}\left[\widehat{u}_1-d_1^u\left(p_{ijk}^{n+1}-p_{i-1,j,k}^{n+1}\right)\right]\end{array}\right\}\\
&+\frac{1}{\Delta s_k}\left\{\begin{array}{l}\left(\frac{\xi}{\cos\beta}\right)_{k+1/2}\left[\widehat{v}_4-d_4^v\left(p_{i,j,k+1}^{n+1}-p_{ijk}^{n+1}\right)\right]\\-\left(\frac{\xi}{\cos\beta}\right)_{k-1/2}\left[\widehat{v}_3-d_3^v\left(p_{i,j,k}^{n+1}-p_{i,j,k-1}^{n+1}\right)\right]\end{array}\right\}=0.
\end{aligned}
\tag{7.90}
$$

The time derivative of the coolant mixture density is calculated by using the equation of state $\rho_\mathbf{m}=\rho_\mathbf{m}(p,h)$ as:

$$
\left(\frac{\partial\rho_\mathrm{m}}{\partial t}\right)^{n+1}=\left(\frac{\partial\rho_\mathrm{m}}{\partial p}\right)_{h_\mathrm{m}}^{n}\frac{p_{ijk}^{n+1}-p_{ijk}^{n}}{\Delta t}+\left(\frac{\partial\rho_\mathrm{m}}{\partial h_\mathrm{m}}\right)_{p}^{n}\frac{\left(h_\mathrm{m}^{n+1}-p_\mathrm{m}^{n}\right)_{ijk}}{\Delta t}. \tag{7.91}
$$

Therefore, the first term in Eq. (7.90) is approximated by:

$$
\begin{aligned}
\varepsilon_{ijk}\left(\frac{\partial\rho_\mathrm{m}}{\partial t}\right)_{ijk}^{n+1}&=p_{ijk}^{n+1}\left[\frac{\varepsilon_{ijk}}{\Delta t}\left(\frac{\partial\rho_\mathrm{m}}{\partial p}\right)_{h_\mathrm{m}}^{n}\right]-\frac{\varepsilon_{ijk}}{\Delta t}\\
&\quad\times\left[\left(\frac{\partial\rho_\mathrm{m}}{\partial p}\right)_{h_\mathrm{m}}^{n}p_{ijk}-\left(\frac{\partial\rho_\mathrm{m}}{\partial h_\mathrm{m}}\right)_{p}^{n}\left(h_\mathrm{m}^{n+i}-h_\mathrm{m}^{n}\right)_{ijk}\right].
\end{aligned}
\tag{7.92}
$$

The analytical expressions of the derivatives at the right-hand side of Eq. (7.92) are given by (Bridgman, 1961):

$$
\left(\frac{\partial\rho_\mathrm{m}}{\partial h_\mathrm{m}}\right)_{p}=-\frac{\rho_\mathrm{m}^2}{T_\mathrm{s}\frac{dp}{dT_\mathrm{s}}}, \tag{7.93}
$$

$$\left(\frac{\partial \rho_{\mathrm{m}}}{\partial p}\right)_{h_{\mathrm{m}}} = \frac{x\left(h'_{\mathrm{vs}} + T_{\mathrm{s}}\frac{dp}{dT_{\mathrm{s}}}\frac{\rho'_{\mathrm{vs}}}{\rho_{\mathrm{v}}^2}\right) + (1-x)\left(h'_{\mathrm{ls}} + T_{\mathrm{s}}\frac{dp}{dT_{\mathrm{s}}}\frac{\rho'_{\mathrm{ls}}}{\rho_{\mathrm{l}}^2}\right)}{\frac{T_{\mathrm{s}}}{\rho_{\mathrm{m}}^2}\frac{dp}{dT_{\mathrm{s}}}}, \tag{7.94}$$

where subscript s refers to saturation conditions. The thermodynamic quality x is given by:

$$x = \frac{h_{\mathrm{m}} - h_{\mathrm{ls}}}{h_{\mathrm{vs}} - h_{\mathrm{ls}}}. \tag{7.95}$$

T_{s} is the saturation temperature, and the derivative dp/dT_{s} is given by the Clapeyron's equation:

$$\frac{1}{T'_{\mathrm{s}}} = \frac{dp}{dT_{\mathrm{s}}} = \frac{h_{\mathrm{v}} - h_{\mathrm{l}}}{T_{\mathrm{s}}\left(\frac{1}{\rho_{\mathrm{v}}} - \frac{1}{\rho_{\mathrm{l}}}\right)}. \tag{7.96}$$

The derivatives h'_{ls} and h'_{vs} are obtained from the functional expressions:

$$h_{\mathrm{ls}} = h_{\mathrm{ls}}(p, T_{\mathrm{s}}), \tag{7.97a}$$

$$h_{\mathrm{vs}} = h_{\mathrm{vs}}(p, T_{\mathrm{s}}), \tag{7.97b}$$

$$\rho_{\mathrm{ls}} = \rho_{\mathrm{ls}}(p, h_{\mathrm{ls}}), \tag{7.97c}$$

$$\rho_{\mathrm{vs}} = \rho_{\mathrm{vs}}(p, h_{\mathrm{vs}}). \tag{7.97d}$$

Formal differentiation yields:

$$dh_{\mathrm{ls}} = \left(\frac{\partial h_{\mathrm{ls}}}{\partial p}\right)_{T_{\mathrm{s}}} dp + \left(\frac{\partial h_{\mathrm{ls}}}{\partial T_{\mathrm{s}}}\right)_{T_{\mathrm{s}}} dT_{\mathrm{s}}, \tag{7.98a}$$

$$dh_{\mathrm{vs}} = \left(\frac{\partial h_{\mathrm{vs}}}{\partial p}\right)_{T_{\mathrm{s}}} dp + \left(\frac{\partial h_{\mathrm{vs}}}{\partial T_{\mathrm{s}}}\right)_{T_{\mathrm{s}}} dT_{\mathrm{s}}, \tag{7.98b}$$

$$d\rho_{\mathrm{ls}} = \left(\frac{\partial \rho_{\mathrm{ls}}}{\partial p}\right)_{h_{\mathrm{ls}}} dp + \left(\frac{\partial \rho_{\mathrm{ls}}}{\partial h_{\mathrm{ls}}}\right)_{p} dh_{\mathrm{ls}}, \tag{7.98c}$$

$$d\rho_{\mathrm{vs}} = \left(\frac{\partial \rho_{\mathrm{lv}}}{\partial p}\right)_{h_{\mathrm{ls}}} dp + \left(\frac{\partial \rho_{\mathrm{vs}}}{\partial h_{\mathrm{vs}}}\right)_{p} dh_{\mathrm{vs}}. \tag{7.98d}$$

Using $dT_{\mathrm{s}} = T'_{\mathrm{s}}dp$, $dh_{\mathrm{ls}} = h'_{\mathrm{ls}}dp$, $dh_{\mathrm{vs}} = h'_{\mathrm{vs}}dp$, $d\rho_{\mathrm{ls}} = \rho'_{\mathrm{ls}}dp$, and $d\rho_{\mathrm{vs}} = \rho'_{\mathrm{vs}}dp$, one obtains:

$$h'_{\mathrm{ls}} = \left(\frac{dh_{\mathrm{l}}}{dp}\right)_{\mathrm{s}} = \left(\frac{\partial h_{\mathrm{l}}}{\partial p}\right)_{T_{\mathrm{s}}} + \left(\frac{\partial h_{\mathrm{l}}}{\partial T_{\mathrm{s}}}\right)_{p} T'_{\mathrm{s}}, \tag{7.99a}$$

$$h'_{\mathrm{vs}} = \left(\frac{dh_{\mathrm{v}}}{dp}\right)_{\mathrm{s}} = \left(\frac{\partial h_{\mathrm{v}}}{\partial p}\right)_{T_{\mathrm{s}}} + \left(\frac{\partial h_{\mathrm{v}}}{\partial T_{\mathrm{s}}}\right)_{p} T'_{\mathrm{s}}, \tag{7.99b}$$

$$\rho'_{\mathrm{ls}} = \left(\frac{d\rho_{\mathrm{l}}}{dp}\right)_{\mathrm{s}} = \left(\frac{d\rho_{\mathrm{l}}}{dp}\right)_{h_{\mathrm{ls}}} + \left(\frac{d\rho_{\mathrm{l}}}{dh_{\mathrm{l}}}\right)_{p} h'_{\mathrm{ls}}, \tag{7.99c}$$

$$\rho'_{\mathrm{vs}} = \left(\frac{d\rho_{\mathrm{v}}}{dp}\right)_{\mathrm{s}} = \left(\frac{d\rho_{\mathrm{v}}}{dp}\right)_{h_{\mathrm{vs}}} + \left(\frac{d\rho_{\mathrm{v}}}{dh_{\mathrm{v}}}\right)_{p} h'_{\mathrm{vs}}. \tag{7.99d}$$

Introducing Eq. (7.92) into (7.90) and rearranging, one derives the Poisson-like equation for the pressure distribution:

$$\begin{aligned} &A^{n+1} p_{ijk}^{n+1} - B^{n+1} p_{i,j-1,k}^{n+1} - C^{n+1} p_{i,j+1,k}^{n+1} - D^{n+1} p_{i-1,j,k}^{n+1} - E^{n+1} p_{i+1,j,k}^{n+1} \\ &- J^{n+1} p_{i,j,k-1}^{n+1} - K^{n+1} p_{i,j,k+1}^{n+1} \\ &= R^{n+1}, \end{aligned} \tag{7.100}$$

with the following definitions of the coefficients and right-hand side:

$$B^{n+1} = \frac{\zeta_{\,j-1/2}}{\Delta z_j} d_5^w = \left(\frac{\zeta_{\,j-1/2}}{\Delta z_j} \frac{(\varepsilon V)_{\,j-1/2}}{\Delta z_{\,j-1/2}}\right) \frac{1}{a^w_{\,j-1/2}}, \tag{7.101a}$$

$$C^{n+1} = \frac{\zeta_{\,j+1/2}}{\Delta z_j} d_6^w = \left(\frac{\zeta_{\,j+1/2}}{\Delta z_j} \frac{(\varepsilon V)_{\,j+1/2}}{\Delta z_{\,j+1/2}}\right) \frac{1}{a^w_{\,j+1/2}}, \tag{7.101b}$$

$$D^{n+1} = \frac{(\Psi F_{\mathrm{c}})_{i-1/2}}{\Delta r_i} d_1^u = \frac{(\Psi F_{\mathrm{c}})_{i-1/2}}{\Delta r_i} \frac{(\varepsilon V)_{i-1/2}}{\Delta r_{i-1/2}} \frac{1}{a^u_{i-1/2}}, \tag{7.101c}$$

$$E^{n+1} = \frac{(\Psi F_{\mathrm{c}})_{i+1/2}}{\Delta r_i} d_2^u = \frac{(\Psi F_{\mathrm{c}})_{i+1/2}}{\Delta r_i} \frac{(\varepsilon V)_{i+1/2}}{\Delta r_{i+1/2}} \frac{1}{a^u_{i+1/2}}, \tag{7.101d}$$

$$J^{n+1} = \frac{1}{\Delta s_k} \left(\frac{\xi}{\cos\beta}\right)_{k-1/2} d_3^v = \frac{\left(\frac{\xi}{\cos\beta}\right)_{k-1/2}}{\Delta s_k} \frac{(\varepsilon V)_{k-1/2}}{\Delta s_{k-1/2}} \frac{1}{a^v_{k-1/2}}, \tag{7.101e}$$

$$K^{n+1} = \frac{1}{\Delta s_k} \left(\frac{\xi}{\cos\beta}\right)_{k+1/2} d_4^v = \frac{\left(\frac{\xi}{\cos\beta}\right)_{k+1/2}}{\Delta s_k} \frac{(\varepsilon V)_{k+1/2}}{\Delta s_{k+1/2}} \frac{1}{a^v_{k+1/2}}, \tag{7.101f}$$

$$A^{n+1} = \frac{\varepsilon_{ijk}}{\Delta t}\left(\frac{\partial \rho_{\mathrm{m}}}{\partial p}\right)^{n}_{h_{\mathrm{m}}} + B^{n+1} + C^{n+1} + D^{n+1} + E^{n+1} + J^{n+1} + K^{n+1}, \quad (7.101\mathrm{g})$$

$$R^{n+1} = \frac{\varepsilon_{ijk}}{\Delta t}\left[\left(\frac{\partial \rho_{\mathrm{m}}}{\partial p}\right)^{n}_{h_{\mathrm{m}}} p^{n}_{ijk} - \left(\frac{\partial \rho_{\mathrm{m}}}{\partial h_{\mathrm{m}}}\right)^{n}_{p}\left(h^{n+1}_{\mathrm{m}} - h^{n}_{\mathrm{m}}\right)_{ijk}\right] - \frac{\zeta_{j+1/2}}{\Delta z_j}\widehat{w}_6 + \frac{\zeta_{j-1/2}}{\Delta z_j}\widehat{w}_5$$
$$- \frac{(\Psi F_{\mathrm{c}})_{i+1/2}}{\Delta r_i}\widehat{u}_2 + \frac{(\Psi F_{\mathrm{c}})_{i-1/2}}{\Delta r_i}\widehat{u}_1 - \frac{1}{\Delta s_k}\left(\frac{\xi}{\cos\beta}\right)_{k+1/2}\widehat{v}_4 + \frac{1}{\Delta s_k}\left(\frac{\xi}{\cos\beta}\right)_{k-1/2}\widehat{v}_3. \quad (7.101\mathrm{h})$$

7.3.6 Poisson Equation for Enthalpy Distribution

Using the identities:

$$\nabla\cdot\left(p\vec{u}_{\mathrm{m}}\right) = p\left(\nabla\cdot\vec{u}_{\mathrm{m}}\right) + \vec{u}_{\mathrm{m}}\cdot\nabla p, \quad (7.102)$$

$$\nabla\cdot\left[\frac{p}{\rho_{\mathrm{m}}}\alpha_{\mathrm{l}}\alpha_{\mathrm{v}}(\rho_{\mathrm{l}}-\rho_{\mathrm{v}})\vec{V}_{\mathrm{Sl}}\right] = p\left(\nabla\cdot\frac{\alpha_{\mathrm{l}}\alpha_{\mathrm{v}}}{\rho_{\mathrm{m}}}(\rho_{\mathrm{l}}-\rho_{\mathrm{v}})\vec{V}_{\mathrm{Sl}}\right) + \frac{\alpha_{\mathrm{l}}\alpha_{\mathrm{v}}}{\rho_{\mathrm{m}}}(\rho_{\mathrm{l}}-\rho_{\mathrm{v}})\vec{V}_{\mathrm{Sl}}\cdot\nabla p$$
$$= p\left(\nabla\cdot R_{\mathrm{oSl}}\vec{V}_{\mathrm{Sl}}\right) + R_{\mathrm{oSl}}\vec{V}_{\mathrm{Sl}}\cdot\nabla p \quad (7.103)$$

with the following definitions of "enthalpy-slip" and "density-slip":

$$H_{\mathrm{Sl}} = x(1-x)\rho_{\mathrm{m}}(h_{\mathrm{v}} - h_{\mathrm{l}}), \quad (7.104)$$

$$R_{\mathrm{oSl}} = \frac{\alpha_{\mathrm{v}}\alpha_{\mathrm{l}}}{\rho_{\mathrm{m}}}(\rho_{\mathrm{l}} - \rho_{\mathrm{v}}), \quad (7.105)$$

the energy Eq. (7.36) can be written as:

$$\frac{\partial}{\partial t}(\rho_{\mathrm{m}}h_{\mathrm{m}}) + \nabla\cdot(\rho_{\mathrm{m}}h_{\mathrm{m}}\overline{u}_{\mathrm{m}}) + \nabla\cdot H_{\mathrm{Sl}}\overline{V}_{\mathrm{Sl}} = \frac{\partial p}{\partial t} + \nabla\cdot(p\overline{u}_{\mathrm{m}}) - p(\nabla\cdot\overline{u}_{\mathrm{m}})$$
$$+\nabla\cdot\left(pR_{\mathrm{oSl}}\overline{V}_{\mathrm{Sl}}\right) - p\left(\nabla\cdot R_{\mathrm{oSl}}\overline{V}_{\mathrm{Sl}}\right) - \nabla\cdot\overline{q} + Q. \quad (7.106)$$

We express the heat flux in terms of enthalpy gradients:

$$\overline{q} = -\rho_{\mathrm{m}}\widetilde{\alpha}_{\mathrm{m}}\nabla h_{\mathrm{m}}, \quad (7.107)$$

with:

$$\widetilde{\alpha}_{\mathrm{m}} = \frac{\lambda_{\mathrm{m}}}{\rho_{\mathrm{m}} c_{\mathrm{pm}},}, \tag{7.108}$$

and introduce the following definitions:

$$\overline{J}_{\mathrm{m}}^{E} = \rho_{\mathrm{m}} h_{\mathrm{m}} \overline{u}_{\mathrm{m}} - p\overline{u}_{\mathrm{m}} + H_{\mathrm{Sl}} \overline{U}_{\mathrm{Sl}} - pR_{\mathrm{oSl}} \overline{U}_{\mathrm{Sl}} - \rho_{\mathrm{m}} \widetilde{\alpha}_{\mathrm{m}} \nabla h_{\mathrm{m}}, \tag{7.109}$$

$$\overline{J}_{\mathrm{m}}^{P} = \overline{u}_{\mathrm{m}} + R_{\mathrm{oSl}} \overline{U}_{\mathrm{Sl}}. \tag{7.110}$$

Using these definitions, Eq. (7.106) can be written as:

$$\frac{\partial}{\partial t}(\rho_{\mathrm{m}} h_{\mathrm{m}}) + \nabla \cdot \overline{J}_{\mathrm{m}}^{E} + p\left(\nabla \cdot \overline{J}_{\mathrm{m}}^{P}\right) = \frac{\partial p}{\partial t} + Q. \tag{7.111}$$

Equation (7.111) is formally similar to Eq. (7.65) for the axial component of the momentum equation and is treated likewise, integrating over the volume V_{f} of a cell and applying the Gauss theorem to express the divergence terms by means of fluxes through the bounding surfaces. This procedure leads likewise to a Poisson equation for the discretized energy equation (superscript E used in the coefficients) in the form:

$$\begin{aligned} & A^{n+1} h_{\mathrm{m}}^{n+1} - B^{n+1} h_{\mathrm{m},j-1}^{n+1} - C^{n+1} h_{\mathrm{m},j+1}^{n+1} - D^{n+1} h_{\mathrm{m},i-1}^{n+1} - E^{n+1} h_{\mathrm{m},i+1}^{n+1} - J^{n+1} h_{\mathrm{m},k-1}^{n+1} \\ & - K^{n+1} h_{\mathrm{m},k+1}^{n+1} = R^{n+1}, \end{aligned} \tag{7.112}$$

with:

$$\begin{aligned} A^{n+1} = & \varepsilon \rho_{\mathrm{m}}^{n+1} / \Delta t + \left[0, F_{i+1/2}^{E} + G_{i+1/2}^{E} \cdot \frac{1}{h_{\mathrm{m}}^{n+1}}\right] + \left[0, -\left(F_{i-1/2}^{E} + \overline{G}_{i-1/2}^{E} \frac{1}{h_{\mathrm{m}}^{n+1}}\right)\right] \\ & + \left[0, F_{j+1/2}^{E} + G_{j+1/2}^{E} \cdot \frac{1}{h_{\mathrm{m}}^{n+1}}\right] + \left[0, -\left(F_{j-1/2}^{E} + \overline{G}_{j-1/2}^{E} \frac{1}{h_{\mathrm{m}}^{n+1}}\right)\right] \\ & + \left[0, F_{k+1/2}^{E} + G_{k+1/2}^{E} \cdot \frac{1}{h_{\mathrm{m}}^{n+1}}\right] + \left[0, -\left(F_{k-1/2}^{E} + \overline{G}_{k-1/2}^{E} \frac{1}{h_{\mathrm{m}}^{n+1}}\right)\right] \\ & + D_{i+1/2}^{E,n+1} + D_{i-1/2}^{E,n+1} + D_{j+1/2}^{E,n+1} + D_{j-1/2}^{E,n+1} + D_{k+1/2}^{E,n+1} + D_{k1/2}^{E,n+1}, \end{aligned} \tag{7.113a}$$

$$B^{n+1} = \left[0, F_{j-1/2}^{E} + \frac{\overline{G}_{j-1/2}^{E}}{h_{\mathrm{m},j-1}^{n+1}}\right] + D_{j-1/2}^{n+1}, \tag{7.113b}$$

$$C^{n+1} = \left[0, -\left(F_{j+1/2}^{E} + \frac{\overline{G}_{j+1/2}^{E}}{h_{\mathrm{m},j+1}^{n+1}}\right)\right] + D_{j+1/2}^{n+1}, \tag{7.113c}$$

$$D^{n+1} = \left[0, F^E_{i-1/2} + \frac{\overline{G}^E_{i-1/2}}{h^{n+1}_{\mathrm{m},i-1}}\right] + D^{n+1}_{i-1/2}, \tag{7.113d}$$

$$E^{n+1} = \left[0, -\left(F^E_{i+1/2} + \frac{\overline{G}^E_{i+1/2}}{h^{n+1}_{\mathrm{m},i+1}}\right)\right] + D^{n+1}_{i+1/2}, \tag{7.113e}$$

$$J^{n+1} = \left[0, F^E_{k-1/2} + \frac{\overline{G}^E_{k-1/2}}{h^{n+1}_{\mathrm{m},k-1}}\right] + D^{n+1}_{k-1/2}, \tag{7.113f}$$

$$K^{n+1} = \left[0, -\left(F^E_{k+1/2} + \frac{\overline{G}^E_{k+1/2}}{h^{n+1}_{\mathrm{m},k+1}}\right)\right] + D^{n+1}_{k+1/2}, \tag{7.113g}$$

$$R^{n+1} = \left(\frac{\varepsilon}{\Delta t}\right)\left(p^{n+1} - p^n\right) + \varepsilon\langle Q\rangle^E + \frac{\varepsilon}{\Delta t}(\rho_\mathrm{m} h_\mathrm{m})^n. \tag{7.113h}$$

Alternative solution of the Poisson-like Eq. (7.112): Without discretizing the time-dependent term but expanding it with the chain rule:

$$\frac{\partial(\rho_\mathrm{m} h_\mathrm{m})}{\partial t} = \rho_\mathrm{m}\frac{\partial h_\mathrm{m}}{\partial t} + h_\mathrm{m}\frac{\partial \rho_\mathrm{m}}{\partial t}, \tag{7.114}$$

one derives the following ordinary differential equation of first order:

$$\begin{aligned}\frac{dh_\mathrm{m}}{dt} = &\frac{1}{\varepsilon\rho^{n+1}_\mathrm{m}}\Big\{-\widetilde{A}^{n+1} h^{n+1}_\mathrm{m} + B^{n+1}h^{n+1}_{\mathrm{m},j-1} + C^{n+1}h^{n+1}_{\mathrm{m},j+1} + D^{n+1}h^{n+1}_{\mathrm{m},i-1}\\ &+E^{n+1}h^{n+1}_{\mathrm{m},i+1} + J^{n+1}h^{n+1}_{\mathrm{m},k-1} + K^{n+1}h^{n+1}_{\mathrm{m},k+1}\Big\} + \widetilde{R}^{n+1}.\end{aligned} \tag{7.115}$$

All coefficients coincide with those defined previously (Cfr. Eqs. (7.113b) through (7.113g)), with the exception of:

$$\begin{aligned}\widetilde{A}^{n+1} = &\left[0, F^E_{i+1/2} + G^E_{i+1/2}\cdot\frac{1}{h^{n+1}_\mathrm{m}}\right] + \left[0, -\left(F^E_{i-1/2} + \overline{G}^E_{i-1/2}\frac{1}{h^{n+1}_\mathrm{m}}\right)\right]\\ &+\left[0, F^E_{j+1/2} + G^E_{j+1/2}\cdot\frac{1}{h^{n+1}_\mathrm{m}}\right] + \left[0, -\left(F^E_{j-1/2} + \overline{G}^E_{j-1/2}\frac{1}{h^{n+1}_\mathrm{m}}\right)\right]\\ &+\left[0, F^E_{k+1/2} + G^E_{k+1/2}\cdot\frac{1}{h^{n+1}_\mathrm{m}}\right] + \left[0, -\left(F^E_{k-1/2} + \overline{G}^E_{k-1/2}\frac{1}{h^{n+1}_\mathrm{m}}\right)\right]\\ &+D^{E,n+1}_{i+1/2} + D^{E,n+1}_{i-1/2} + D^{E,n+1}_{j+1/2} + D^{E,n+1}_{j-1/2} + D^{E,n+1}_{k+1/2} + D^{E,n+1}_{k-1/2},\end{aligned} \tag{7.116a}$$

$$\widetilde{R}^{n+1} = \frac{1}{\rho_{\mathrm{m}}^{n+1}} \left\{ \left(p^{n+1} - p^{n}\right)/\Delta t + \langle Q \rangle^{E} - h_{\mathrm{m}}^{n+1} \left(\frac{\partial \rho_{\mathrm{m}}}{\partial t} \right)^{n} \right\}. \tag{7.116b}$$

Equation (7.115), when applied to all cells of the definition domain, yields a system of ordinary first-order differential equations:

$$\left\{ \frac{dh_{\mathrm{m}}}{dt} \right\} = \{ f(h_{\mathrm{m}}) \} \tag{7.117}$$

which can be integrated with usual ODE solvers. A Runge-Kutta algorithm of order four is sufficiently precise for single-phase flow calculations, but may not be appropriate for two-phase flow calculations. Runge-Kutta formulas of fifth to eighth order, given in (Fehlberg, 1968), may be appropriate for two-phase flow calculations.

7.4 Solution of the Matrix Equation $AX = B$ with the Method of Doolittle

Let the matrix A in the equation:

$$AX = B \tag{7.118}$$

be a square matrix of order N. We decompose the matrix A into the product of a lower triangular matrix L and an upper triangular matrix U:

$$A = LU, \tag{7.119}$$

with

$$l_{ij} = 0, \quad (j > i) \tag{7.120}$$

$$u_{ij} = 0. \quad (j < i) \tag{7.121}$$

The elements of the main diagonal of the matrix L (or of the matrix U) can be given an arbitrary value except zero. In the method of Doolittle, one sets:

$$l_{ii} = 1. \tag{7.122}$$

In a variant of the method (Crout algorithm), the condition (7.122) is replaced by $u_{ii} = 1$. Carrying on formally the multiplication (7.119), one obtains the following system of equations:

$$l_{11}u_{1j} = a_{1j}, \qquad (j = 1, 2, \ldots, N) \tag{7.123a}$$

$$\begin{cases} l_{21}u_{11} = a_{21}, \\ l_{21}u_{1j} + l_{22}u_{2j} = a_{2j} \qquad (j = 2, 3, \ldots, N) \end{cases} \tag{7.123b}$$

$$\begin{cases} l_{31}u_{11} = a_{31}, \\ l_{31}u_{12} + l_{32}u_{22} = a_{32}, \\ \sum_{k=1}^{2} l_{3k}u_{kj} + l_{33}u_{3j} = a_{3j} \qquad (j = 3, \ldots, N). \end{cases} \tag{7.123c}$$

etc.

Taking into account Eq. (7.122), Eq. (7.123a) yields the values of the elements of the first row of the matrix U ($u_{1j} = a_{1j}$); the second, Eq. (7.123b), yields l_{21}, u_{22}, u_{23}, ..., u_{2N}; the third, Eq. (7.123c), yields l_{31}, u_{32}, u_{33}, u_{34}, ..., u_{3N}; and so on.

The solution of the system, Eq. (7.123), can be written in the compact form:

$$u_{1j} = a_{1j}, \qquad (j = 1, 2, \ldots, N) \tag{7.124}$$

$$l_{i1} = a_{i1}/u_{11}, \qquad (i = 1, 2, \ldots, N) \tag{7.125}$$

$$l_{ij} = \left(a_{ij} - \sum_{k=1}^{j-1} l_{ik}u_{kj}\right)/u_{jj}, \qquad \begin{matrix} (i = 3, 4, \ldots, N) \\ (j = 2, 3, \ldots, i-1) \end{matrix} \tag{7.126}$$

$$u_{ij} = a_{ij} - \sum_{k=1}^{i-1} l_{ik}u_{kj}. \qquad \begin{matrix} (i = 2, 3, \ldots, N) \\ (j = i, \ldots, N) \end{matrix} \tag{7.127}$$

Combining Eqs. (7.118) and (7.119), one derives:

$$(LU)X = B. \tag{7.128}$$

Defining:

$$UX = Y, \tag{7.129}$$

we can first solve the system:

$$LY = B, \tag{7.130}$$

with respect to Y. From Eq. (7.130), one obtains:

$$l_{11}y_1 = B_1, \tag{7.131a}$$

$$\sum_{k=1}^{i} l_{ik} y_k = B_i, \qquad (i = 2, \ldots, N) \tag{7.131b}$$

hence:

$$y_1 = B_1, \tag{7.132}$$

$$y_i = B_i - \sum_{k=1}^{i-1} l_{ik} y_k. \qquad (i = 2, \ldots, N) \tag{7.133}$$

The solution vector X is then obtained from Eq. (7.129). The last row of this system yields:

$$x_N = \frac{y_N}{u_{NN}}, \tag{7.134}$$

and then, one derives:

$$x_{N-1} = \left(y_{N-1} - \sum_{k=N-i+1}^{n} u_{N-i,k} x_k \right) / u_{N-i,N-i}. \qquad (i = 1, 2, \ldots, N-1) \tag{7.135}$$

Remark on improving the numerical solution. Whatever the numerical method used for solving the matrix equation (7.118), the solution can be improved in the following way.

Le X_1 be a first numerical solution of Eq. (7.118) and compute the residuals:

$$R = AX_1 - B \tag{7.136}$$

Solve now the equation:

$$AX = R = AX_1 - B \tag{7.137}$$

obtained from Eq. (7.118) replacing the right-hand side with the residuals, and compute a second solution X_2. Now it holds:

$$A(X_1 - X_2) = B \tag{7.138}$$

and it turns out that the solution $(X_1 - X_2)$ is more precise than the first solution X_1.

7.5 Numerical Solution of the Poisson Equation with the Alternating Direction Implicit (ADI) Method

7.5.1 *The Classical Variant*

Let us recall the Poisson equation for pressure, Eq. (7.100), which can be written in the form:

$$a_0 p_0{}^{n+1} + a_1 p_1{}^{n+1} + a_2 p_2{}^{n+1} + a_3 p_3{}^{n+1} + a_4 p_4{}^{n+1} + a_5 p_5{}^{n+1} + a_6 p_6{}^{n+1} \\ = G_0^n. \tag{7.139}$$

In Eq. (7.139), coefficients a_1, a_2 represent the coupling between the central node $0 \equiv (i, j, k)$ and the neighboring nodes $1 \equiv (i - 1, j, k)$ and $2 \equiv (i + 1, j, k)$ in the x-direction. Likewise, coefficients a_3, a_4 represent the coupling between the central node and the neighboring nodes in the y-direction; coefficients a_5, a_6 represent the coupling in the z-direction. Let N_x, N_y, N_z be the number of nodes, in the discretized definition domain, in the three coordinate directions. The ADI method consists in integrating Eq. (7.139) successively in the three coordinate directions, taking into account the respective couplings of nodes, within iteration from r to $r + 1$. This iteration proceeds therefore in three steps as follows.

First Step. Solution in the x-coordinate direction.

Let us assume that the pressure field is known at iteration r. Equation (7.139) can be written as:

$$a_0 p_0{}^{(1)} + a_1 p_1{}^{(1)} + a_2 p_2{}^{(1)} = G_0^n - \sum_{\alpha=3}^{6} a_\alpha p_\alpha{}^r \tag{7.140}$$

The initial values $p_\alpha{}^r$ $(r = 1)$ $(\alpha = 3, 4, 5, 6)$ are usually taken as the final values $p_\alpha{}^n$ from the previous time step. Superscript (1) denotes a first provisional value of the updated pressure field. The matrix of coefficients of the system of Eq. (7.140), written for all lines $(i = 1, 2, \ldots, N_x)$, is tri-diagonal, and the system can be solved efficiently with the Thomas algorithm (see Sect. 1.3). After this step, the full pressure field $p^{(1)}{}_{ijk}$ is known.

Second Step. Solution in the y-coordinate direction.

Taking into account that the field $p^{(1)}{}_{ijk}$ is known, Eq. (7.139) can be written as:

$$a_0 p_0{}^{(2)} + a_3 p_3{}^{(2)} + a_4 p_4{}^{(2)} = G_0^n - a_1 p_1{}^{(1)} - a_2 p_2{}^{(1)} - a_5 p_5{}^{(1)} - a_6 p_6{}^{(1)} \tag{7.141}$$

We solve this Eq. (7.141), with the Thomas algorithm, for $p^{(2)}$ along the lines: $(j = 1, 2, \ldots, N_y)$.

Third Step. Solution in the z-coordinate direction

For every line ($k = 1, 2, \ldots, N_z$), we solve with the Thomas algorithm the equation:

$$a_0 {p_0}^{r+1} + a_5 {p_5}^{r+1} + a_6 {p_6}^{r+1} = G_0^n - \sum_{\alpha=1}^{4} a_\alpha {p_\alpha}^{(2)} \tag{7.142}$$

using at the right-hand side the updated pressure field $p^{(2)}$.

The three solution steps are repeated cyclically incrementing the iteration superscript r, till convergence.

The stability of the ADI algorithm can be analyzed in the following way:

- Let X be the matrix formed with the coefficients a_1, a_0, a_2;
- Let Y be the matrix formed with the coefficients a_3, a_0, a_4;
- Let Z be the matrix formed with the coefficients a_5, a_0, a_6;
- Let N_x be the matrix formed with the coefficients a_3, a_4, a_5, a_6;
- Let N_y be the matrix formed with the coefficients a_1, a_2, a_5, a_6;
- Let N_z be the matrix formed with the coefficients a_1, a_2, a_3, a_4.

With these definitions, the algorithm represented by Eqs. (7.140), (7.141), and (7.142) can be written in the following matrix form:

$$XP^{(1)} = G^n - N_x P^r, \tag{7.143a}$$

$$YP^{(2)} = G^n - N_y P^{(1)}, \tag{7.143b}$$

$$ZP^{r+1} = G^n - N_z P^{(2)}. \tag{7.143c}$$

Solving Eqs. (7.143) successively for $P^{(1)}$, $P^{(2)}$, and P^{r+1}, the following are obtained:

$$P^{(1)} = X^{-1}G^n - X^{-1}N_x P^r,$$

$$P^{(2)} = Y^{-1}G^n - Y^{-1}N_y P^{(1)} = Y^{-1}G^n - Y^{-1}N_y X^{-1}G^n + Y^{-1}N_y X^{-1}N_x P^r,$$

$$\begin{aligned} P^{(r+1)} =& Z^{-1}G^n - Z^{-1}N_z P^{(2)} \\ =& Z^{-1}G^n - Z^{-1}N_z Y^{-1}G^n + Z^{-1}N_z Y^{-1}N_y X^{-1}G^n - Z^{-1}N_z Y^{-1}N_y X^{-1}N_x P^r \\ =& \left(-Z^{-1}N_z Y^{-1}N_y X^{-1}N_x\right)P^r + \left(Z^{-1} - Z^{-1}N_z Y^{-1} + Z^{-1}N_z Y^{-1}N_y X^{-1}\right)G^n, \end{aligned}$$

$$P^{(r+1)} = HP^r + KG^n, \tag{7.144}$$

with the definition of the iteration matrix:

$$H = -Z^{-1}N_zY^{-1}N_yX^{-1}N_x, \tag{7.145a}$$

and of:

$$K = Z^{-1} - Z^{-1}N_zY^{-1} + Z^{-1}N_zY^{-1}N_yX^{-1}. \tag{7.145b}$$

The definition of matrix K is however not essential for the stability analysis.

The stability of the algorithm depends on the spectral radius of the iteration matrix H. In algorithm (7.143), there is not the possibility to reduce the spectral radius of the matrix H. For this reason, to be able to reduce the spectral radius, the algorithm is modified as follows, by introducing an optimization parameter ρ:

$$(X + \rho I)P^{(1)} = G^n - (N_x - \rho I)P^r \tag{7.146a}$$

$$(Y + \rho I)P^{(2)} = G^n - (N_y - \rho I)P^{(1)} \tag{7.146b}$$

$$(Z + \rho I)P^{(r+1)} = G^n - (N_z - \rho I)P^{(2)} \tag{7.146c}$$

Successive solution for $P^{(1)}$, $P^{(2)}$, and P^{r+1} yields:

$$P^{(1)} = (X + \rho I)^{-1}G^n - (X + \rho I)^{-1}(N_x - \rho I)P^{(r)}$$

$$\begin{aligned}
P^{(2)} &= (Y + \rho I)^{-1}G^n - (Y + \rho I)^{-1}(N_y - \rho I)P^{(1)} \\
&= (Y + \rho I)^{-1}G^n - (Y + \rho I)^{-1}(N_y - \rho I)(X + \rho I)^{-1}G^n \\
&\quad + (Y + \rho I)^{-1}(N_y - \rho I)(X + \rho I)^{-1}(N_x - \rho I)P^r
\end{aligned}$$

$$\begin{aligned}
P^{(r+1)} &= (Z + \rho I)^{-1}G^n - (Z + \rho I)^{-1}(N_z - \rho I)P^{(2)} \\
&= (Z + \rho I)^{-1}G^n - (Z + \rho I)^{-1}(N_z - \rho I)(Y + \rho I)^{-1}G^n \\
&\quad + (Z + \rho I)^{-1}(N_z - \rho I)(Y + \rho I)^{-1}(N_y - \rho I)(X + \rho I)^{-1}G^n \\
&\quad - (Z + \rho I)^{-1}(N_z - \rho I)(Y + \rho I)^{-1}(N_y - \rho I)(X + \rho I)^{-1}(N_x - \rho I)P^r
\end{aligned}$$

$$P^{(r+1)} = H(\rho)P^r + K(\rho)G^n, \tag{7.147}$$

with the definition of the iteration matrix:

$$H(\rho) = -(Z + \rho I)^{-1}(N_z - \rho I)(Y + \rho I)^{-1}(N_y - \rho I)(X + \rho I)^{-1}(N_x - \rho I), \tag{7.148a}$$

and of:

$$K(\rho) =(Z+\rho I)^{-1} - (Z+\rho I)^{-1}(N_z - \rho I)(Y+\rho I)^{-1} + (Z+\rho I)^{-1}(N_z - \rho I)(Y+\rho I)^{-1}(N_y - \rho I)(X+\rho I)^{-1}. \tag{7.148b}$$

Again, the definition of the matrix $K(\rho)$ is not essential for the stability analysis.

One may try to vary the parameter ρ in (7.148) and correspondingly to calculate the spectral radius $S[H(\rho)]$. If this is smaller than one, the algorithm is convergent, and the convergence rate (CR) is given by:

$$\text{CR} = -\log_n(S[H(\rho)]) = -\log_n(|\lambda \max|) \tag{7.149}$$

where $|\lambda\max|$ is the modulus of the maximum eigenvalue of the iteration matrix.

Alternatively, one can proceed as follows: Let λ_x, λ_y, and λ_z be eigenvalues of the matrix X, Y, and Z, respectively, and ν_x, ν_y, and ν_z be eigenvalues of the matrix N_x, N_y, and N_z, respectively. Corresponding to this set of eigenvalues, the iteration matrix $H(\rho)$ has the eigenvalue:

$$\lambda_H(\rho) = -\frac{(\nu_z - \rho)(\nu_y - \rho)(\nu_x - \rho)}{(\lambda_z + \rho)(\lambda_y + \rho)(\lambda_x + \rho)}. \tag{7.150}$$

Equating to zero the derivative of $\lambda_H(\rho)$ with respect to ρ, namely:

$$\frac{d}{dr}\lambda_H(\rho) = 0, \tag{7.151}$$

one derives a fourth-degree polynomial in ρ:

$$\rho^4(S_1+S_2) + 2\rho^3(-a_1+a_2) + \rho^2[3(P_1+P_2) - a_1S_2 - a_2S_1] + 2\rho(P_1S_2 - S_1P_2) + a_1P_2 + a_2P_1 = 0 \tag{7.152}$$

with:

$$S_1 = \nu_x + \nu_y + \nu_z, \tag{7.153a}$$

$$S_2 = \lambda_x + \lambda_y + \lambda_z, \tag{7.153b}$$

$$P_1 = \nu_x\nu_y\nu_z, \tag{7.153c}$$

$$P_2 = \lambda_x\lambda_y\lambda_z, \tag{7.153d}$$

$$a_1 = \nu_x\nu_y + \nu_x\nu_z + \nu_y\nu_z, \tag{7.153e}$$

$$a_2 = \lambda_x\lambda_y + \lambda_x\lambda_z + \lambda_y\lambda_z. \tag{7.153f}$$

Equation (7.152) can be solved numerically to find its four roots ρ_i $(i = 1, 2, 3, 4)$, and correspondingly, one finds out the root for which $\lambda_H(\rho)$ has a minimum.

7.5.2 Advanced Variants of the ADI Method

Recently, advanced variants of the Alternating Direction Implicit (ADI) scheme have been proposed, making use of convergence accelerating parameters and based on algorithms vectorizable to a high degree. We explain in this section the approach followed in these advanced variants.

(i) *The Messina-Londrillo (ML-ADI) Variant with Constant Optimization Parameter*

For the numerical solution of the Poisson equation, Messina and Londrillo (1988) proposed a new ADI variant summarized by the following equations:

$$(Z + rI)P^{(1)} = G - (Z + Y + X)P^r, \tag{7.154a}$$

$$(Y + rI)P^{(2)} = rP^{(1)}, \tag{7.154b}$$

$$(X + rI)P^{(3)} = rP^{(2)}, \tag{7.154c}$$

$$P^{r+1} = P^r + 2P^{(3)}. \tag{7.154d}$$

$P^{(1)}$, $P^{(2)}$, $P^{(3)}$ are auxiliary vectors which tend to the zero vector at convergence, as P^{r+1} and P^r both converge to the updated solution P^{n+1}. Here, r is the optimization parameter, while, as superscript, it is an iteration index. The right-hand side of Eq. (7.154a) gives the residuals which thus need not be computed separately. The main achievement with respect to previous schemes consists in minimizing matrix by vector multiplications. Only vector operations are considered at the right-hand side, in steps two and three of the above scheme.

The evaluation of an optimum parameter r is made as follows. Eliminating the auxiliary unknowns $P^{(1)}$ to $P^{(3)}$ in the above equations, one obtains the following relationship between two consecutive iterative values of the vector P:

$$P^{r+1} = H(r)P^r + K, \tag{7.155}$$

with:

$$H(r) = I - 2r^2(X + rI)^{-1}(Y + rI)^{-1}(Z + rI)^{-1}(X + Y + Z), \tag{7.156}$$

$$K = 2r^2(X + rI)^{-1}(Y + rI)^{-1}(Z + rI)^{-1}G. \tag{7.157}$$

The algorithm can be optimized by minimizing the spectral radius of the iteration matrix $H(r)$. The functional dependence of the eigenvalues of $H(r)$ from the corresponding eigenvalues $\lambda_x^j, \lambda_y^i, \lambda_z^k$ of matrices X, Y, Z is:

$$H_{\Lambda}\left(r,\lambda_x^i,\lambda_y^j,\lambda_z^k\right)=1-\frac{2r^2\left(\lambda_y^i+\lambda_x+\lambda_z^k\right)}{\left(\lambda_y^i+r\right)\left(\lambda_x^i+r\right)\left(\lambda_z^k+r\right)}\qquad(i,j,k=1,2,\ldots,N). \tag{7.158}$$

A minimum of H_{Λ} as function of r is obtained by computing the derivative:

$$\frac{dH_{\Lambda}(r)}{dr}=0, \tag{7.159}$$

which yields a fourth-degree polynomial:

$$r^4-r^2a-2mr=0. \tag{7.160}$$

Beyond the irrelevant solution $r_1=0$, there are the three roots of the algebraic equation:

$$r^3+3pr+2q=0, \tag{7.161}$$

with $p=-a/3$, $q=-m$. If $p<0$ and $d=q^2+p^3\leq 0$, the three roots are real and given by:

$$r_2=-2c\cos\left(\delta/3\right), \tag{7.162a}$$

$$r_3=-2c\cos\left(\frac{\pi}{3}-\frac{\delta}{3}\right), \tag{7.162b}$$

$$r_4=-2c\cos\left(\frac{\pi}{3}+\frac{\delta}{3}\right), \tag{7.162c}$$

with:

$$\delta=arc\cos\left(\frac{q}{c^3}\right) \tag{7.163}$$

$$c=\pm\sqrt{|p|}\quad\text{(with the sign equal to that of } q\text{)}. \tag{7.164}$$

These roots correspond to extrema of $H_{\Lambda}(r)$. Because $H_{\Lambda}(r=0)=1$ and $\lim_{r\to\infty}[H_{\Lambda}(r)]=1$ and because for r sufficiently large $|H_{\Lambda}(r)|<1$, there is a minimum of $H_{\Lambda}(r)$ for some $r_i(i=2,3,4)$ positive. Let $r^{\max}$ be the minimum corresponding to the maxima of the eigenvalues $\lambda_x^j,\lambda_y^i,\lambda_z^k$ and $r^{\min}$ be the minimum corresponding to the minima. Then a reasonable evaluation of the parameter r is given by:

$$\tilde{r} = \sqrt{r^{\max} r^{\min}}. \tag{7.165}$$

Notice that this formula gives only an order of magnitude and not an exact value. However, this ADI variant with constant r is much inferior in the rate of convergence to that obtained by letting r vary with the iterations. If the maximum and minimum eigenvalues of the matrix of the coefficients $A = X + Y + Z$ can be computed or approximated, the sequence explained in the following subsection can be used.

(ii) *Evaluation of an Optimum Sequence of the Optimization Parameter r_j*
It is customary to use an optimization parameter r_j which sweeps the range:

$$r_o = \frac{|\lambda_{\min}|}{\mu} \leq r \leq r_M = \frac{|\lambda_{\max}|}{\mu} \tag{7.166}$$

of eigenvalues of the iteration matrix $A = X + Y + Z$ with a geometric sequence. Here $\lambda_{\min}$, $\lambda_{\max}$ are the minimum and maximum eigenvalues of the matrix of the coefficients, and μ $(0 \leq \mu \leq 1)$ is a chosen number (usually $\mu = 0.1 - 0.3$). The geometric sequence of r_j is then built up as follows:

(i) Fix an integer M $(M \cong 30 - 80)$.
(ii) Compute a number v $(v > 1)$ from the formula:

$$\left(\frac{v}{\mu}\right)^{M-1} = \frac{\lambda_{\max}}{\lambda_{\min}}. \tag{7.167}$$

Verify that $v > 1$. If not, increase μ.

(iii) Compute successive values of the parameter r:

$$r_j = \left(\frac{v}{\mu}\right)^{j-1} r_0. \qquad (j = 1, 2, \ldots, M) \tag{7.168}$$

If M steps are not sufficient to reach convergence (which is usually the case), the sequence (7.168) is repeated cyclically.

7.5.3 Two-Dimensional Model Problem

Let us recall the model Poisson equation treated with several variants in Chap. 6:

$$\frac{\partial^2 u}{\partial x^2} + \frac{\partial^2 u}{\partial y^2} = 6xy(xy + x + y - 3)e^{(x+y)}(=f), \tag{7.169}$$

defined in the square domain ($0 \leq x, y \leq 1$) with the condition $u = 0$ on its boundary. Is solution is:

$$u = 3\left(x - x^2\right)\left(y - y^2\right)e^{(x+y)}. \tag{7.170}$$

We partition the definition domain with grids parallel to the coordinate axes and spaced $\Delta x = 1/\text{NI}$ and $\Delta y = 1/\text{NJ}$, respectively. The total number of cells in the domain is then NE = NI × NJ, and the total number of nodes (including the boundary nodes) is NP = (NI + 1) (NJ + 1). Since, in this case, we are interested in the performance of the ADI method, and not in the accuracy of the discretization technique, we discretize Eq. (7.169) with the simple 5-point difference operator, but retaining the option of different mesh lengths in the two coordinate directions. Thus, one obtains:

$$\frac{1}{\Delta x^2}\left(u_{i-1,j} - 2u_{ij} + u_{i+1,j}\right) + \frac{1}{\Delta y^2}\left(u_{i,j-1} - 2u_{ij} + u_{i,j+1}\right) = f_{ij}. \tag{7.171}$$

The system of Eq. (7.171) can be written in matrix form as:

$$\left(A_x + A_y\right)U = F. \tag{7.172}$$

F is the known vector; the tridiagonal matrices A_x and A_y are denoted symbolically by:

$$A_x = \left\{\frac{1}{\Delta x^2}, -\frac{2}{\Delta x^2}, \frac{1}{\Delta x^2}\right\}, \tag{7.173}$$

$$A_y = \left\{\frac{1}{\Delta y^2}, -\frac{2}{\Delta y^2}, \frac{1}{\Delta y^2}\right\}. \tag{7.174}$$

The ML-ADI technique is given, in the 2D case, by the following algorithm:

$$(A_x + rI)\Phi^{(1)} = F - \left(A_x + A_y\right)U^r \tag{7.175a}$$

$$\left(A_y + rI\right)\Phi^{(2)} = r\Phi^{(1)} \tag{7.175b}$$

$$U^{r+1} = U^r + 2\Phi^{(2)}. \tag{7.175c}$$

In Eqs. (7.175), I is the identity matrix; r is a convergence accelerating parameter and, as a superscript, is an iteration index. $\Phi^{(1)}$ and $\Phi^{(2)}$ are auxiliary vectors which at convergence tend to the zero vector, while the vector U^{r+1} tends to the solution

U^{n+1} at time level t^{n+1}. The advantage of this scheme with respect to the classical ones is its full vectorization capability.

The iteration matrix of the scheme (7.175) is:

$$H(r) = I - 2r\left(A_y + rI\right)^{-1}\left(A_x + rI\right)^{-1}\left(A_x + A_y\right). \tag{7.176}$$

The maximum convergence rate of the algorithm (7.175) is obtained by varying cyclically the parameter r in geometrical sequence between the minimum and the maximum of the eigenvalues of matrices A_x and A_y. The eigenvalues of $H(r)$ are:

$$\Lambda^{i,j}(r) = 1 - 2r\frac{\lambda_x^i + \lambda_y^i}{\left(\lambda_x^i + r\right)\left(\lambda_y^j + r\right)} = \frac{\left(\lambda_x^i - r\right)\left(\lambda_y^j - r\right)}{\left(\lambda_x^i + r\right)\left(\lambda_y^j + r\right)}. \tag{7.177}$$

The eigenvalues of A_x and A_y are:

$$\lambda_x^i = -\frac{4}{\Delta x^2}\sin^2\left[\frac{i\pi}{2(N+1)}\right], \quad [i = 1, 2, \ldots, N = (\mathrm{NI}+1)] \tag{7.178}$$

$$\lambda_y^j = -\frac{4}{\Delta y^2}\sin^2\left[\frac{j\pi}{2(M+1)}\right]. \quad [j = 1, 2, \ldots, M = (\mathrm{NJ}+1)] \tag{7.179}$$

For the applicability of formula (7.177), see the remark at the end of this section.

Because $|\Lambda^{i,j}(r) < 1|$, algorithm (7.175) is convergent. Using formulas (7.177) through (7.179), the range of variation of r which minimizes $\Lambda^{i,j}(r)$ can be found. For three-dimensional problems, this is however not the case, and the solution of an eigenvalue problem would be an arduous and time-consuming task. Thus, Bottoni and Dorr (1991) devised a heuristic approach to determine the optimum range of the parameter r. This approach is now explained for the 2D case of the model problem although the model problem could be treated analytically and would not require a heuristic approach. After analyzing the 2D case, its extrapolation to 3D problems can be better understood.

Heuristic approach: First we compute numerically the matrix $H(r)$, which implies two matrix inversions and three matrix multiplications. Then, we calculate the eigenvalues of $H(r)$, say Λ_k (k = 1, 2, ..., NP) (see, for instance, Press et al., 1986). The convergence behavior of the ADI algorithm depends not only on the spectral radius of the iteration matrix $H(r)$ but also on the clustering of the eigenvalues and on their mean value. Useful information can be derived by calculating their mean value and the mean value of their moduli and taking their difference:

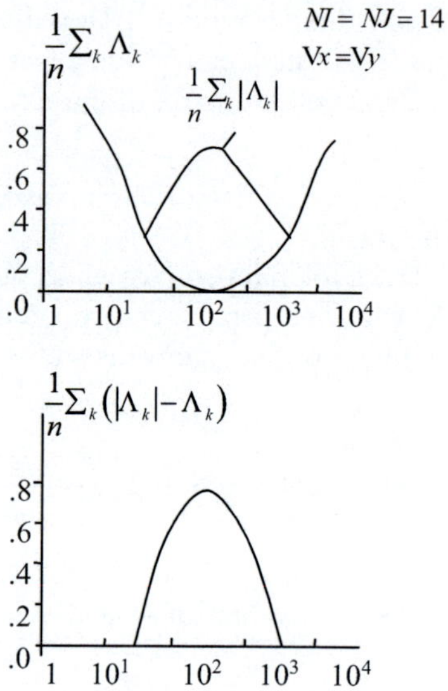

Fig. 7.3 Distribution of the eigenvalues of the iteration matrix $H(r)$, Eq. (7.176), for the 2D model problem: mean value, mean value of the moduli (upper plots), and their difference (lower plot). The lowermost plot gives the range of the convergence accelerating parameter r for optimum convergence

$$\overline{\Lambda}(r) = \frac{1}{n}\sum_{k=1}^{n}\Lambda_k(r), \tag{7.180}$$

$$\left|\overline{\Lambda}(r)\right| = \frac{1}{n}\sum_{k=1}^{n}|\Lambda_k(r)|, \tag{7.181}$$

$$\Delta\Lambda(r) = \left|\overline{\Lambda}(r)\right| - \overline{\Lambda}(r). \tag{7.182}$$

Repeating these calculations for several values of r, one derives the distributions plotted in Fig. 7.3 for the case of rough discretization of the definition domain with NI = NJ = 14 and $\Delta x = \Delta y$.

The mean value of the moduli $\left|\overline{\Lambda}(r)\right|$ indicates the clustering of the eigenvalues of the iteration matrix. The stronger the clustering, the better is the convergence behavior of the numerical scheme. The distribution of the difference $\Delta\Lambda(r)$ displays the range of r in which the clustering is stronger and therefore the optimum range of variation of the sequence of r-values. In case of the 2D model problem, the range of r, evaluated in this way, coincides exactly with the range (λ_{min}, λ_{max}) of the eigenvalues of the matrix of coefficients.

Table 7.1 shows the performance of the ML-ADI algorithm in comparison with a direct vectorized Gaussian elimination technique (labelled GAUSS) for a medium-size problem with NI = NJ = 100, hence with NP = (NI + 1)(NJ + 1) = 10,201 computational nodes. The accuracy of the computed results is given in terms of the Hamming distance σ_{H} between the analytical solution U, Eq. (7.170), and the computed solution U^{C}. By the same accuracy, the ML-ADI algorithm reduced the computational time by a factor of 2.

Table 7.1 Comparison between Gaussian elimination and ML-ADI algorithm for a 2D model problem

	GAUSS	ML-ADI
$\sigma_H = \sum_{k=1}^{NP} \lvert U_k - U_k^C \rvert$	3.9×10^{-5}	3.9×10^{-5}
ε (maximum of residuals)	3.2×10^{-10}	1.6×10^{-10}
ITER (number of iterations to obtain $\Delta u/u < 10^{-8}$)	–	40
CPU [s][a]	2.72	1.46

[a]A Siemens-Fujitsu VP 50 computer was used at Kernforschungszentrum Karlsruhe (KfK) in 1990

Remark on the applicability of formula (7.177): The matrix theory (Zurmühl & Falk, 1984) states that, under given hypotheses for the coefficients of matrices, the eigenvalues of the matrix:

$$P = A_1 A_2 \ldots A_t, \tag{7.183}$$

product of matrices A_j ($j = 1, 2, \ldots, t$) of order N, are given by the product of the eigenvalues of matrices A_j:

$$P_{\Lambda k} = \lambda_{1k}\lambda_{2k} \ldots \lambda_{tk} \quad (k = 1, 2, \ldots, N). \tag{7.184}$$

These products must be made observing strictly the sequence of the eigenvalues determined by the condition:

$$\left(A_j - \lambda_{jk} I\right) X_k = 0 \quad (j = 1, 2, \ldots, t), \quad (k = 1, 2, \ldots, N). \tag{7.185}$$

The set of eigenvalues X_k is the same for the t eigenvalue problems. If the hypothesis under which this theorem holds is satisfied, the eigenvalues of the iteration matrix $K(r)$ are given exactly by formula (7.177) (for $k = 1, 2, \ldots, N$), by observing strictly the sequence of eigenvalues. However, even if the hypotheses were satisfied, one could not afford in real cases to solve t eigenvalue problems to determine the right sequence of the eigenvalues. Therefore, in the treatment outlined above, the rigorous theoretical way was abandoned, and the heuristic way was followed.

7.5.4 Three-Dimensional Problem from Nuclear Reactor Safety Analysis

In this section, we recall the problem outlined in Sect. 7.3 for the numerical simulation of two-phase flow through the bundle shown in Fig. 7.2. For uniformity of notations with the preceding sections, we use in the following the same coordinate directions (x, y, z), where x corresponds to the radial direction of the bundle, y to the azimuthal, and z to the axial one.

In the tree-dimensional case, the ML-ADI algorithm is given in matrix form by the following equations:

$$(Z + rI)\Phi^{(1)} = F - (Z + Y + X)U^r, \tag{7.186a}$$

$$(Y + rI)\Phi^{(2)} = r\Phi^{(1)}, \tag{7.186b}$$

$$(X + rI)\Phi^{(3)} = r\Phi^{(2)}, \tag{7.186c}$$

$$U^{r+1} = U^r + 2\Phi^{(3)}. \tag{7.186d}$$

The right-hand side of Eq. (7.186a) gives the residuals of the Poisson equation to be solved. The main improvement of this algorithm with respect to the classical variant of Douglas (1962) consists in avoiding matrix-by-vector multiplications in the second and third steps above. Instead, only vector operations, which are much faster, are needed at the right-hand side of the respective equations. From the mathematical viewpoint, however, the two algorithms are equivalent, having the same iteration matrix:

$$H(r) = I - 2r^2(A_x + rI)^{-1}\left(A_y + rI\right)^{-1}(A_z + rI)^{-1}\left(A_x + A_y + A_z\right). \tag{7.187}$$

Let $\lambda^{\alpha}_{\min}$, $\lambda^{\alpha}_{\max}$, and $\overline{\lambda}^{\alpha}$ ($\alpha = x, y, z$) be the minimum, maximum, and mean value of the eigenvalues of matrices X, Y, and Z. They are approximated using the same formulas (7.178) and (7.179) and a third one formally similar for the z direction, using average values of mesh lengths. These formulas are correct for equal mesh lengths. Using them with unequal mesh lengths implies making recourse to the heuristic approach, instead of relying upon a strict, but unknown, theory. It has been ascertained with numerical tests that the minimum of $\lambda^{\alpha}_{\min}$ and the maximum of $\lambda^{\alpha}_{\max}$ ($\alpha = x, y, z$) approach closely the minimum and maximum, respectively, of the eigenvalues of the matrix of coefficients $A = X + Y + Z$. The nine numbers $\lambda^{\alpha}_{\min}$, $\lambda^{\alpha}_{\max}$, $\overline{\lambda}^{\alpha}$, when combined in triples, give 84 combinations, but only 27 of them are "legal," in the sense that every coordinate direction is represented in the triple only once. One can use these 27 legal combinations to compute, from Eq. (7.187), approximate values of the corresponding eigenvalues of the iteration matrix $H(r)$:

$$\Lambda_k(r) = 1 - 2r^2 \frac{\lambda^x_k + \lambda^y_k + \lambda^z_k}{\left(\lambda^x_k + r\right) + \left(\lambda^y_k + r\right) + \left(\lambda^z_k + r\right)} \qquad (k = 1, 2, \ldots, 27) \tag{7.188}$$

These 27 eigenvalues $\Lambda_k(r)$ can be considered as a representative selection of the total number of eigenvalues of the iteration matrix. Then, we use the same formulas (7.180)–(7.182), as in the 2D case, and repeat the calculations for several values of the parameter r, obtaining the distributions shown in Fig. 7.4. Similar to the 2D case, the lowermost plot displays the optimum range of r, where clustering of the

Fig. 7.4 Distribution of estimated eigenvalues of the iteration matrix $H(r)$, Eq. (7.187), in the three-dimensional case: mean value, mean value of moduli, and their difference (lowermost plot). The latter gives the optimum range of the iteration parameter r

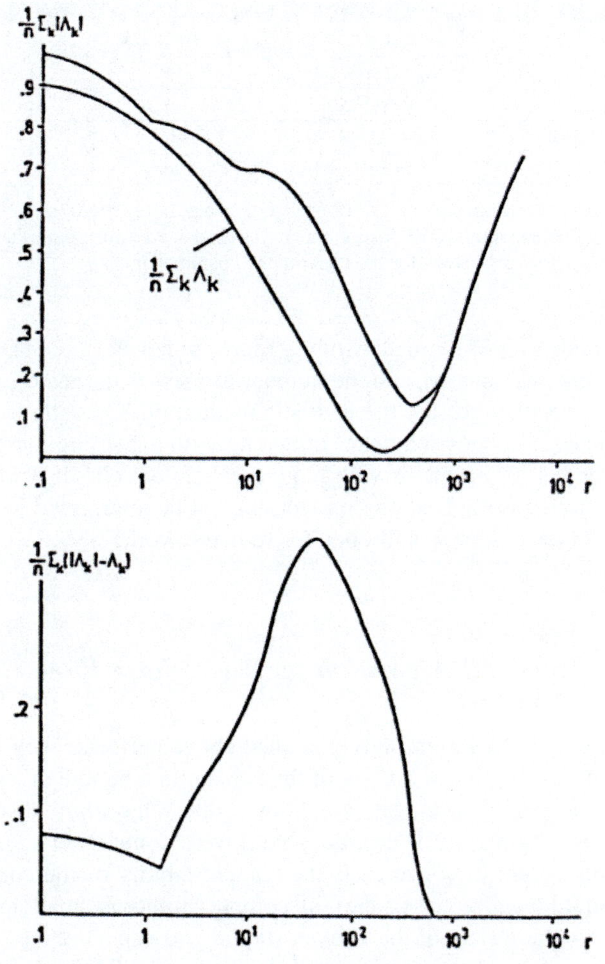

eigenvalues is stronger. After having estimated the optimum r-range, there remain in the computational praxis two free parameters: the number of cycles of r, within one outer iteration step, and the number of steps within the cycle.

Some computational results are given in Table 7.2 for two cases: a 127-pin bundle discretized in NP = (MC × NTH × NC) = 41 × 12 × 7 = 3444 control cells and a 169-pin bundle discretized in NP = 40 × 12 × 8 = 3840 cells. Results are given for the numerical solution of the Poisson equation for pressure, obtained by combining mass and momentum equation for the liquid coolant, as explained in Sect. 7.3.5. As shown in the table, the improvement of convergence when using the heuristic approach is remarkable in comparison with the classical GAUSS variant. In case of large bundles, or in case of finer discretization of small bundles, the difference in performance between the GAUSS algorithm and the heuristic method becomes stronger. The CPU time for the GAUSS method increases with the square of the

Table 7.2 Comparison between Gaussian elimination and ML-ADI algorithm for a three-dimensional problem

	127-pin bundle		169-pin bundle	
	GAUSS	ML-ADI	GAUSS	ML-ADI
Range of r	–	0.2–700	–	0.25–700
CPU [s][a]	513	309 (4 × 40)[b]	1728	1116 (4 × 25)
	346 (5 × 40)		1545 (6 × 30)	

[a]A Siemens-Fujitsu VP 50 computer was used at Kernforschungszentrum Karlsruhe (KfK) in 1990
[b]4 cycles of 40 steps for the convergence parameter r

bandwidth of the matrix of coefficients. For the ML-ADI algorithm, the CPU time increases linearly with the number of meshes in the coordinate directions.

Needless to say, the heuristic method outlined above is just one method among many other possible ways of coping with the strong increase of computational time needed for problems with large numbers of cells. It is not implied that the method outlined is the best one; in fact, it must be interpreted just as an attempt to attack a problem which is still open for research work.

7.5.5 ADI Algorithm for Two-Phase Flow Calculations

The ML-ADI algorithm explained above performs well for single-phase flow conditions. In case of boiling of the coolant in a bundle of a nuclear reactor, simulating hypothetical accident conditions, the algorithm discussed above, given by Eqs. (7.186a)–(7.186d), does not give a convergent solution. The main reason for this failure to converge is the compressibility of the coolant. When the coolant is liquid (single-phase state), its compressibility is small and can be neglected. In the two-phase state of the coolant (liquid and vapor), the compressibility is strong and cannot be disregarded any longer. A modification of the algorithm (7.186) is therefore necessary for simulating correctly the different and more complex physical conditions. It can be remarked that, as so often happens in mathematical modelling of physical problems, a more complex physical situation requires, to be described correctly, more sophisticated mathematical approach.

In two-phase flow calculations, the central coefficient α_{ijk} of the Poisson equation for the pressure distribution depends strongly on the time derivative of the coolant mixture density. This time derivative can be calculated by differentiating formally the equation of state $\rho_{\mathrm{m}} = \rho_{\mathrm{m}}(p, h_{\mathrm{m}})$ of the liquid-vapor mixture (index m), which relates the density to pressure and enthalpy:

$$\left(\frac{\partial \rho_{\mathrm{m}}}{\partial t}\right)^{n+1} = \left(\frac{\partial \rho_{\mathrm{m}}}{\partial p}\right)_{h_{\mathrm{m}}}^{n} \frac{p^{n+1} - p^{n}}{\Delta t} + \left(\frac{\partial \rho_{\mathrm{m}}}{\partial h_{\mathrm{m}}}\right)_{p}^{n} \frac{h_{\mathrm{m}}^{n+1} - h_{\mathrm{m}}^{n}}{\Delta t}. \tag{7.189}$$

The superscript n refers to time discretization. The computation of the derivatives in Eq. (7.189) has been explained in Sect. 7.3.5. The term $\left(\frac{\partial \rho_{\mathrm{m}}}{\partial p}\right)_{h_{\mathrm{m}}}^{n}$ at the right-hand side of Eq. (7.189), which multiplies the unknown pressure p^{n+1}, adds to the central coefficient of the Poisson equation for pressure, while all other terms contribute to the right-hand side. Let us call K the diagonal matrix formed with the above compressibility terms for all cells of the computational domain. According to Eq. (7.92), the coefficients of the matrix K are:

$$K_{ijk} = \frac{\varepsilon_{ijk}}{\Delta t}\left(\frac{\partial \rho_{\mathrm{m}}}{\partial p}\right) \tag{7.190}$$

The Poisson equation can therefore be written in matrix form as:

$$(Z + Y + X + K)P = G. \tag{7.191}$$

The contribution of the diagonal elements k_{ijk} must be taken into account by means of an additional contribution to the convergence optimization parameter r. Let k^M be the maximum of the compressibility terms k_{ijk} and r_ℓ be the optimization parameter for single-phase flow calculation. We then define:

$$r^m = r_\ell + \frac{nk^M}{3} \tag{7.192}$$

the optimization parameter for two-phase flow calculation. In Eq. (7.192), n is a multiplier which has been determined as follows. In single-phase flow calculations, one can set $n = 0$ because the compressibility term is negligibly small and all eigenvalues of the iteration matrix are in modulus smaller than 1. In two-phase flow, however, because of the compressibility terms, the eigenvalues of the iteration matrix may be larger than unity. Numerical investigation has shown that, in all test cases, the eigenvalues of the iteration matrix were kept below unity using $n > 2$, when two-phase flow was simulated. Optimization of this free parameter may be investigated as a research problem.

The ML-ADI algorithm can then be applied in the variant:

$$\begin{aligned}
&\left(Z + \frac{1}{3}K + r^m I\right)P^{(1)} = G - (AZ + Y + X + K)P^r, \\
&\left(Y + \frac{1}{3}K + r^m I\right)P^{(2)} = r^m P^{(1)}, \\
&\left(X + \frac{1}{3}K + r^m I\right)P^{(3)} = r^m P^{(2)}, \\
&P^{r+1} = P^r + 2P^{(3)}.
\end{aligned} \tag{7.193}$$

7.6 Historical Note About the Computer Codes BACCHUS-3D/TP and COMMIX-2

(i) *The BACCHUS Code*

In this chapter, after introducing the "pressure method" from the theoretical viewpoint (Sect. 7.1), we showed how it can be used in practice for deriving a Poisson-like equation (Eq. 7.21) from the continuity and momentum equations of a single-phase flow (Sect. 7.2).

In Sect. 7.3, we introduced the conservation equations for mass, momentum, and energy of a two-phase flow considered as a mixture of liquid and vapor phases. The approach followed is applicable not only when the two phases are of the same chemical component (for instance, water and water vapor, liquid sodium and its vapor, etc.) but also when they are different chemical components (for instance, water and air). We called this approach SLIP model, if different flow velocities are computed for the two phases of the fluid, or homogeneous equilibrium model (HEM), if it is assumed that the phases have the same velocity. In the frame of the slip model, we derived the Poisson-like equation for the pressure distribution in the fluid and the Poisson-like equation for the enthalpy distribution in the fluid. The numerical algorithms have been used in the "pseudo-cylindrical" geometrical configuration shown in Fig. 7.2, but they are obviously applicable in different geometries. In the particular case of cylindrical coordinates, the treatment of the geometry becomes simpler because of the simpler definitions of volumes and surfaces of the cells.

Thus, summarizing, we showed the numerical treatment to obtain the following equations:

(i) Poisson-like equation (Eq. 7.21) for the pressure distribution in a single-phase flow, derived from continuity and momentum equations
(ii) Poisson-like equation (Eq. 7.100) for the pressure distribution in the two-phase fluid, derived from continuity and momentum equations of a two-phase fluid mixture
(iii) Poisson-like equation (Eq. 7.112) for the enthalpy distribution in the two-phase fluid mixture

All these numerical techniques have been applied and coded in the computer program BACCHUS-3D/TP (three-dimensional, two phase). In this program, these techniques have also been applied to the numerical simulation of a two-phase flow based on the separated phases model (SPM). The SPM is based on five conservation equations for the flow: two continuity equations for the liquid and vapor (or gas) phases, two momentum equations, and one energy equation. The latter is the same as for the slip model (Eq. 7.36).

The computer program BACCHUS-3D/TP has been developed at the nuclear research center KfK (now KIT) of Karlsruhe, Germany, in the years 1980 through 1992 starting from a steady-state, two-dimensional version of the code kindly provided to KfK from colleagues of the CENG (Centre d'Etudes Nucleaires de Grenoble) in the frame of a cooperation for development and safety analysis of fast breeder reactors, with particular reference to the reactor designs of SNR-300 in Germany and of Superphenix in France. One of the very first publications on the BACCHUS code is by Basque et al. (1982). An intermediate state of development of the code was documented in (Bottoni et al., 1987). The final documentation of the code followed in the already quoted reference (Bottoni et al., 1992). This latter reference has a large part dedicated to the validation of the code against results obtained in in-pile and out-of-pile experiments.

A short synopsis of the code BACCHUS-3D/TP is given below.

The program is based on a three-dimensional representation of a fuel bundle of a fast nuclear reactor, with hexagonal geometrical configuration, by means of the "porous-body" model. The thermal hydraulic calculation for the reactor coolant, sodium in a fast breeder reactor, is coupled to a pin model describing the temperature distribution in fuel and cladding. An option is provided for the simulation of central or displaced blockages of arbitrary thickness. The program can be applied to the description of in-pile bundles of out-of-pile experiments. As mentioned above, the two-phase flow model is based on (i) a slip model or (ii) the model of separated phases. The program user can choose between these two options or a third option which, in the course of the development of a hypothetical accident, can simulate a transition between the slip model, used at boiling inception, and the model of separated phases applied when the volume fraction has exceeded a given threshold. Needless to say, this third option allows for the most realistic simulation.

The numerical solution of the Poisson-like equations discussed in this chapter is made with the direct Cholesky method when the number of unknowns does not exceed about 50,000; otherwise, the computational time would be too large. For larger problems, one can choose between Gaussian eliminations or the ADI technique, in its variant for two-phase flow calculations. The ADI method is for most problems faster by a factor of 1.5 with respect to the Gaussian method. The code is available in a sequential version and in a vectorized version. The latter allows for a speedup factor of about 5. For two-phase flow calculations, double precision is mandatory. The code has been written in FORTRAN IV and consists of about 30,000 statements, divided in 175 subroutines.

(ii) *The COMMIX-2 Code*

The development of the COMMIX-2 program has proceeded in parallel with that of the BACCHUS code versions. The first version of the code was developed during a 1-year delegation of the author from KfK to the Argonne National Laboratory (ANL) at Argonne, Illinois. The first publication about this program (Bottoni et al., 1985) documented the theoretical basis of the code, while code verification and improvement from the numerical viewpoint were made and documented in the following years at KfK.

The fast code development of only 1 year, starting from a single-phase code version available at ANL, was possible mainly for three reasons: (i) The experience already gained with the development of BACCHUS; (ii) the support given by a colleague Dr. Hank Domanus for the structure to be given to the code; and (iii) the theoretical support given by Dr. Robert Lyczkowski for all problems concerning two-phase flows and especially for the calculation of the physical properties of liquid sodium and sodium vapor. The fast development of the code would not have been possible without Dr. Domanus' and Dr. Lyczkowski's contributions. In this respect, the author wishes to acknowledge that Dr. Domanus and Dr. Lyczkowski were both mentors and friends. An account of the cooperation between the author and Dr. Lyczkowski for the development of the COMMIX-2 code is given in (Lyczkowski, 2018).

In subsequent years, a multiphase version of the code was tried to establish, named COMMIX-M, which had to be structured to simulate up to ten different phases. However, for failure of the ANL management, this code version, though applicable in some particular cases (see, for instance, Bottoni et al., 1994; Bottoni & Ninokata, 1998), never became fully operational.

A short synopsis of the COMMIX-2 program is given below.

The singe-phase flow version COMMIX-1A has provided the basis for the development of the COMMIX-2 version. The modelling of the geometrical configuration relies upon the concepts of volume porosity, surface porosity, distributed resistances, and heat sources. Two two-phase flow models are incorporated into COMMIX-2.

The first two-phase flow model is a three-equation slip model (SM) which provides for either a constant slip ratio or a relative slip velocity normalized by the mixture velocity in each of the coordinate directions. The limiting case of equal phase velocities, known as the homogeneous equilibrium model (HEM), is a sub-case of the slip model. In the SM version, the conservation equations for momentum and enthalpy of the two-phase mixture are formally identical with those for a single-phase flow, with the inclusion of additional terms which take into account the so-called momentum slip and enthalpy slip between the phases.

The second two-phase flow model is a five-equation unequal phase velocity, equal phase temperature (UVET) two-fluid continuum which is referred to as the separated phase model (SPM).

For both phase models, the same as in the BACCHUS code versions, the pressure solution method is applied by deriving, through combination of the continuity and momentum equations, a Poisson-like equation which describes the pressure distribution in the fluid. The numerical solution of the Poisson-like equation is obtained either with the iterative SOR method or, for sufficiently small problems, with the direct matrix inversion based on the Cholesky matrix decomposition.

In the frame of the SPM, the code describes a space and time variation of the thickness of a liquid film wetting structural surfaces. The liquid phase is thus split into two components, the liquid film and slugs or liquid droplets entrained by the flow.

The code, written in FORTRAN IV, consists of about 52,000 statements organized in 157 subroutines.

(iii) *The Fate of the Codes*

Both codes required for their developments about 20 man-years. In the case of the BACCHUS-3D/TP code, it was the work of a group of three persons for about 7 or 8 years, and the estimated time includes the development of intermediate code versions. In the case of the COMMIX-2 code, the estimated time includes the development and verification of the single-phase COMMIX-1C (Domanus et al., 1990) code that was the basis from which the two-phase flow version was started. In Germany, at KfK, both codes were applied for the safety analysis of the German SNR-300 prototype of fast breeder reactor. In the USA, COMMIX-2 had to be applied for the safety analysis of the Clinch River Fast Breeder Reactor (CRFBR). Both projects turned out to be ill-fated and were abandoned around 1990, not for technical but for political reasons. A detailed account of the history of the SNR-300 fast reactor development is given in (Marth, 1992). Most politicians in both countries either were unaware of the inherent safety characteristics of fast reactors or found it more convenient for their political carrier to ride the rising way of antinuclear attitudes. This kind of antinuclear consciousness, real or simulated, was likely to assure more votes by the next elections. The damage to the environment due to the non-use of the nuclear power started becoming visible in recent years. Future generations, all over the world, will have to pay a very high price both in terms of money and of human lives.

At KfK, the code COMMIX-2 was downgraded to a single-phase flow program by eliminating all parts simulating the two-phase flow. The new single-phase flow version was renamed FLUTAN, and subsequent development work, made after departure of the author from KfK, consisted in introducing in the code new turbulence models, for instance, the Anisotropic Turbulence Model (ATM) (Cfr. Sect. 8.7), in addition to the already available $k - \varepsilon$ model. The COMMIX-2 program was also delivered to the research center PNC (Power Reactor and Nuclear Fuel Company) at Mito, Ibaraki-ken, Japan, where it was renamed into AQUA. Renaming the codes, with little regard for their origin, was a blatant violation of professional ethics.

In the years 1990–1995 from COMMIX-2, the most ambitious COMMIX-M, designed to treat up to ten different phases or components, was developed, but, as mentioned above, it never became fully operational. It was however applied for safety analysis of the containment of the AP-600 thermal reactor (Bottoni et al., 1993).

References

Basque, B., Grand, D., Delapierre, L., Bottoni, M., Dorr, B., Homann, C., & Struwe, D. (1982). Thermohydraulic analysis of LMFBR subassemblies with the BACCHUS programme. In *Proceedings ANS/ENS International Topical Meeting on LMFBR Safety and Related Design and Operational Aspects*, Lyon, France, July 19–23, 1982.

Bird, R. B., Stewart, W. E., & Lightfoot, E. N. (1960). *Transport phenomena*. Wiley.

Bottoni, M., & Ninokata, H. (1998). An algorithm for attenuation of turbulence in particulate flow linked to the fluid-dynamic code COMMIX-M. *Journal of Nuclear Science and Technology, 35* (2), 101–112.

Bottoni, M., & Sengpiel, W. (1992). *Review of mathematical and physical basis of two-phase flow modeling* (Report No. KfK 4759).

Bottoni, M., et al. (1985, September). *COMMIX-2: A three-dimensional transient computer program for thermal-hydraulic analysis of two-phase flow* (Report NUREG/CR-4371, ANL-85-47).

Bottoni, M., Dorr, B., Homann, C., & Struwe, D. (1987). State of development of the computer programme BACCHUS-3D/TP for the description of transient two-phase flow conditions in LMFBR fuel pin bundles. *Nuclear Engineering and Design, 100*, 321–349.

Bottoni, M., Dorr, B., & Homann, C. (1992, April). *The three-dimensional transient two-phase flow computer programme BACCHUS-3D/TP* (Report No. KfK 4760).

Bottoni, M., Chang, F. C., Chien, T. H., Ding, J., & Sha, W. T. (1993). *Liquid film tracking model for the passive containment cooling system of the AP-600 reactor* (Report NUREG/CR-5985, ANL-92-22, R7).

Bottoni, M., Chang, F. C., & Ding, J. (1994, May). *First assessment of computation of turbulent bubbly flow and particulate flow with the COMMIX-M program* (Report No. ANL-95/6).

Bridgman, P. W. (1961). *The thermodynamics of electrical phenomena in metals and a condensed collection of thermodynamics formulas*. Dover Publications Inc.

Domanus, H. M., et al. (1990, November). *COMMIX-1C: A three-dimensional transient single-phase computer program for thermal-hydraulic analysis of single-components and multicomponent engineering systems. Volume I: Equations and numerics: Volume II: User's guide manual* (Report NUREG/CR-5649, Argonne National Laboratory Report ANL-90/33).

Douglas, J. (1962). Alternating direction methods for three space variables. *Numerische Mathematik, 4*, 41–63.

Fehlberg, E. (1968, October). *Classical fifth-, sixth-, seventh- and eigth-order Runge-Kutta formulas with stepwise control* (Report No. NASA-TR-R-287).

Lyczkowski, R. W. (2018). *The history of multiphase science and computational fluid dynamics – A personal memoir*. Springer International Publishing.

Marth, W. (1992). Der schnelle Bräter SNR-300 im auf und ab seiner Geschichte, KfK 4666, März.

Messina, A., & Londrillo, P. (1988). *A N-body code for elliptical galaxies, super-computing in astrophysics*. Astronet Special Publications.

Peaceman, D. W., & Rachford, H. H. (1955). The numerical solution of parabolic and elliptic differential equations. *Journal of the Society for Industrial and Applied Mathematics, 3*, 28–41.

Press, W. H., Flannery, B. P., Teukolsky, S. A., & Vetterling, W. T. (1986). *Numerical recipes*. Cambridge University Press.

Zurmühl, R., & Falk, S. (1984). *Matrizen und Ihre Anwendungen*. Springer Verlag.

Chapter 8
Numerical Treatment of the Transport Equations of Turbulence

This chapter explains the numerical treatment of transport equations used in turbulence models. The explanation is preceded by an introductory section in which the main concepts of turbulence theory are presented and an overview is given about the most frequently used turbulence models. More details about the theoretical foundations can be found in fundamental textbooks, like Lesieur (1990), McComb (1990), and Wilcox (1998).

8.1 Introduction to Turbulence

In this introductory section, we give a general description of turbulent flows by defining the main parameters, like length scales, which characterize them, and defining physical quantities, like turbulence intensity and dissipation. In the following sections, we present the transport equations of turbulence from the rigorous theoretical viewpoint, their physical modelling and numerical treatment.

8.1.1 Kolmogorov Scales of Turbulence

Let us consider a plate, as shown in Fig. 8.1, referring to a coordinate system with axis x along the plate and origin at the edge and axis z normal to the plate. The y-axis is along the depth of the plate as to form a right-handed system. We assume that a fluid flows past the plate forming a boundary layer of thickness growing from the edge of the plate to a maximum value δ. The velocity profile in the fluid grows from zero, at the surface of the plate, to a maximum value U at the top of the boundary layer, while the fluid velocity remains constant in the region outside of the boundary layer. As the boundary layer forms along the plate, eddies are generated, whirling

M. Bottoni, *Physical Modeling and Computational Techniques for Thermal and Fluid-dynamics*, Mechanical Engineering Series,
https://doi.org/10.1007/978-3-030-79717-1_8

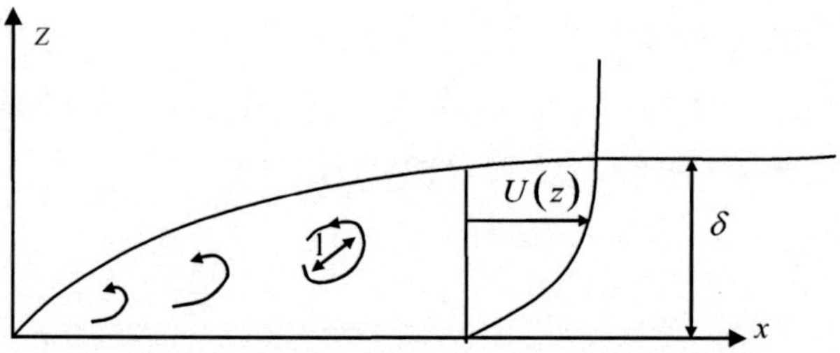

Fig. 8.1 Sketch of turbulent flow forming a boundary layer along a horizontal plate

with characteristic velocities u_e (the subscript "e" refers to eddies) which in general will be about 0.1–0.2 times the value of the velocity U of the main flow. Let us call ℓ the maximum size of these whirls, which in general will be of the order of magnitude of 0.8 δ. We refer to ℓ as the macroscopic length scale of turbulence.

As the flow goes along the plate, kinetic energy of turbulence associated with the largest eddies is gradually transferred to smaller eddies and from these again to smaller ones, until finally all kinetic energy of turbulence is dissipated through viscosity into inner energy of the fluid. This process of energy transfer from the largest to the smallest eddies is generally referred to as "cascade process". Eddies originated at the beginning of the plate, where the boundary layer forms, may preserve their identities for a length of about 20–30 times the thickness of the boundary layer, until they eventually decade and dissipate. The minimum length scale of the eddies which finally decay into viscosity is referred to as the microscopic length scale of turbulence or Kolmogorov's length scale η .

We define the turbulent Reynolds number by

$$\mathrm{Re}_t = \frac{u_e \ell}{\nu} = \frac{\nu_t}{\nu}, \tag{8.1}$$

where ν is the kinematic viscosity of the fluid and $\nu_t = u_e \ell$ is, by definition, the turbulent viscosity. This is a parameter which characterizes the turbulence intensity, but it is not a physical property of the fluid. Its definition is however consistent with the fact that the molecular kinematic viscosity is proportional to the average velocity of the molecules and to their mean free path λ_m. As representative of the average velocity of the molecules one can take the speed of sound $c = (\partial p/\partial \rho)^{1/2}$. Therefore, one has:

$$\nu = c\lambda_m. \tag{8.2}$$

Accordingly, the turbulent (or eddy) viscosity v_t is defined as the product of a velocity characterizing the fluctuating motion with a typical length, also referred to as "mixing length."

The turbulence Mach number and the Mach number are defined, with reference to the speed of sound, by:

$$Ma_t = \frac{u_e}{c}, \tag{8.3}$$

$$Ma = \frac{U}{c}. \tag{8.4}$$

An alternative definition of the Kolmogorov's length scale, which will be justified later, is the length which corresponds to a turbulent Reynolds number of one unity; hence:

$$\mathrm{Re}_t = \frac{u_e \eta}{\nu} = 1. \tag{8.5}$$

In numerical applications aiming at describing fluid flows, one defines an effective or equivalent viscosity as the sum of the molecular and of the turbulent viscosity, namely:

$$\tilde{\nu} = \nu + \nu_t. \tag{8.6}$$

The ratio of the microscopic Kolmogorov's length scale to the macroscopic length scale is a function of the turbulent Reynolds number. As it will be shown later [see Eq. (8.51)], this function is given by:

$$\frac{\eta}{\ell} = \frac{1}{\mathrm{Re}_t^{3/4}}. \tag{8.7}$$

With increasing turbulent Reynolds number, the ratio of the microscopic to the macroscopic length scale decreases. Because the macroscopic length scale is generally fixed by the geometric size of the problem, for high values of the turbulent Reynolds number, the Kolmogorov's length scale decreases and may reach values of the order of magnitude of 10^{-5} or 10^{-6}m. To get a feeling of the magnitude of the microscopic length scale of turbulence, we may compare it to other physical parameters characterizing molecular sizes and dynamics.

The diameter of an air molecule, taken as a weighted mean value of the diameters of the oxygen and nitrogen components, is $d = 4.15 \times 10^{-10}$m. Referring to an atmosphere at 1 bar pressure and 300 Kelvin temperature, the mean average spacing between molecules is $\delta_m = 3.3 \times 10^{-9}$m, and the molecular mean free path is $\lambda_m = 4.9 \times 10^{-8}$m . With a good approximation, we can therefore write that the ratios of these parameters are given by d: δ_m: $\lambda_m = 1$: 10: 100. The Kolmogorov's length scale is approximately given, as an order of magnitude, by $\eta \approx 100\,\lambda_m$. As an example, in case of air with kinematic viscosity $\nu = 10^{-5}\mathrm{m}^2/\mathrm{s}$, assuming a characteristic velocity $u_e = 10$ m/s, one obtains, from Eq. (8.5), $\eta = \nu/u_e = 10^{-6}$m.

The ratios of the molecular mean free path to the microscopic and macroscopic lengths are defined as the microstructure Knudsen number and the free molecular Knudsen numbers. Using Eq. (8.2), one obtains:

$$\mathrm{Kn}_{\mathrm{micr}} = \frac{\lambda_m}{\eta} = \frac{\nu}{c\eta} = \frac{\nu}{c}\frac{R_{et}^{3/4}}{\ell} = \frac{\nu}{u_e/M_{at}}\frac{R_{et}^{3/4}}{\ell} = Ma_t\frac{R_{et}^{3/4}}{R_{et}} = \frac{Ma_t}{\mathrm{Re}_t^{1/4}}, \tag{8.8}$$

$$\mathrm{Kn} = \frac{\lambda_m}{\ell} = \frac{\nu}{c\ell} = \frac{U}{U}\frac{\nu}{c\ell} = \frac{Ma}{\mathrm{Re}}, \tag{8.9}$$

with the Reynolds number defined by:

$$\mathrm{Re} = \frac{U\ell}{\nu}. \tag{8.10}$$

In rarefied gases, the mean free path of molecules λ_m increases, and the Knudsen number Kn approaches 1. At the limit of extreme rarefaction (free-molecule limit), one has a collision-less motion, and the Knudsen number goes to infinity.

Numerical applications of the Navier-Stokes equations to turbulent flows would require discretization of the flow domain in cells of size equal to the Kolmogorov's length scale in order to simulate numerically all energy carrying whirls in the flow. This would be the so-called direct numerical simulation (DNS) of turbulence, which is not feasible at present, for high Reynolds numbers, because of the long computational time which would be involved. However, the conceptual question arises of whether a three-dimensional cell of size η could be treated as a continuum, which is the fundamental prerequisite for the application of the Navier-Stokes equations. The question is equivalent to ask whether a cube of size η contains a sufficient number of molecules as to be considered as a continuum in which macroscopic physical quantities like density and viscosity can be defined. Assuming $\eta = 10^{-6}$m, the volume of a computational cell would be $\eta^3 = 10^{-18}\mathrm{m}^3$.

The Loschmidt number, ratio of the Avogadro number and of the volume of a mole of gas, is:

$$n_0 = \frac{A_v}{V_{\mathrm{mole}}} = \frac{6.0022137 \times 10^{26}}{22.414\ \mathrm{m}^3} = 2.68684 \times 10^{25}\ \left[\mathrm{molecules/m}^3\right]. \tag{8.11}$$

From these numbers, it results that in the volume η^3, there are 10^7 molecules, which is a sufficiently large number to define macroscopic quantities. Therefore, from a conceptual viewpoint, the Navier-Stokes equations could be applied to a flow domain partitioned in cells of linear size equal to the Kolmogorov's length scale, but in practice, this would lead to computational times unacceptable for today's computers. In fact, computational praxis shows that the number of operations N_c to be made for simulation of turbulent flows varies as:

$$N_c = \mathrm{Re}^{9/4}. \tag{8.12}$$

Assuming a Reynolds number $\mathrm{Re} = 5 \times 10^5$, one has $N_c = 6.65 \times 10^{12}$. Assuming that one second of problem time must be simulated by performing 10^5 time steps of $\Delta t = 10^{-5}$ s, it turns out that 6.65×10^{17} operations must be performed. At the rate of 10^9 operations per second, this would require 6.65×10^8 s or about 20 years of computation. For these reasons DNS is thus far limited to flows with small Reynolds numbers. In case of medium or large Reynolds numbers, it is therefore necessary to apply turbulence models which allow the use of larger grid sizes.

Beside the microscopic length scale, Kolmogorov introduced the microscopic scales of time and velocity, generally denoted with τ and ν_k, respectively (subscript K is used for the microscopic scale of velocity to avoid confusion with the symbol of kinematic viscosity). Analytical expressions of these scales have been obtained on the basis of the so-called Kolmogorov's first hypothesis, based on the "universal equilibrium theory" (1941), which can be formulated as follows:

> For large Reynolds numbers, the small scale components (small eddies) of turbulence are statistically isotropic and independent of the size and distribution of the large scale eddies. At the smallest scales of turbulence the rate of dissipation of the turbulent kinetic energy into heat, which depends on the molecular viscosity ν, can be assumed to be the same as the rate at which dissipation $\varepsilon = -dK/dt$ [m^2/s^3] of turbulent kinetic energy K is transferred through the cascade process from the largest to the smallest eddies. For this reason all microscopic scales of length, time and velocity are functions of the two parameters ε and ν only.

As for the microscopic length scale, letting $\eta = \varepsilon^a \nu^b$ and applying dimensional analysis, one obtains $a = -1/4$ and $b = 3/4$, and similarly for the other microscopic length scales, which are then given by

$$\eta = \left(\frac{\nu^3}{\varepsilon}\right)^{1/4}, \tag{8.13}$$

$$\tau = \left(\frac{\nu}{\varepsilon}\right)^{1/2}, \tag{8.14}$$

$$v_K = (\nu\varepsilon)^{1/4} = \eta/\tau. \tag{8.15}$$

Using Eq. (8.13) and Eq. (8.15) (v_K replacing u_e) in Eq. (8.5) yields $\text{Re}_t = 1$, which justifies the reason for defining the Kolmogorov's length scale as the length scale which makes the Reynolds number of turbulence equal to one. Consistently, $\eta = \tau v_K$.

8.1.2 Definitions of Main Turbulence Parameters

Let us refer to a parameter λ, which we call "wavelength," as representative of the linear dimensions of eddies, from the largest scale ℓ, at which they are produced, through the cascade process, till the smallest scales η at which they are dissipated. The wavelength λ is associated with a "wavenumber":

$$k = \frac{2\pi}{\lambda}. \tag{8.16}$$

During the cascade process referred to above, the wavenumber k increases from a value $k_i = 2\pi/\ell$, associated with the macroscopic length scale, to a value $k_d = 2\pi/\eta$, corresponding to the microscopic length scale. To every wavenumber k in the interval (k_i, k_d) corresponds to an eddy of size $\lambda = 2\pi/k$ to which is associated a kinetic energy of turbulence, due to its whirling in the flow field. We define $E(k)$ $[m^3/s^2]$ as the turbulent kinetic energy per unit wavenumber k. For the time being, $E(k)$ is still an unknown function. It represents however the distribution of turbulent kinetic energy over the wavenumbers k and is referred to as "energy spectrum." As we will see later, the function $E(k)$ is not limited to the above given interval (k_i, k_d) but extends from zero to (theoretically) infinity. In practice there will be a cutoff at a maximum possible wavenumber $k_{\mathbf{max}}$. Integration of the distribution $E(k)$ over the full range of k, from zero to infinity, yields the (total) turbulent kinetic energy K $[m^2/s^2]$, thus defined by:

$$K = \int_0^\infty E(k)dk. \tag{8.17}$$

Similarly, we can define a "dissipation spectrum" $D(k)$ $[m^3/s^2]$ as the dissipation of turbulent kinetic energy per unit wavenumber k. It can be proved that the function $D(k)$ is related to the energy spectrum $E(k)$ by the relation (Tennekes & Lumley, 1972):

$$D(k) = 2\nu k^2 E(k). \tag{8.18}$$

Integration of the dissipation spectrum over the full range of k, from zero to infinity, yields the (total) dissipation, to which we already referred above:

$$\varepsilon = \int_0^\infty D(k)dk = 2\nu \int_0^\infty k^2 E(k)dk, \tag{8.19}$$

having assumed a constant value of the kinematic viscosity.

It can be proved that the dissipation ε is also given by:

$$\varepsilon = 2\nu\langle s_{ij}s_{ij}\rangle, \tag{8.20}$$

where the symbol $\langle\rangle$ denotes mean value and:

$$s_{ij} = \frac{1}{2}\left(\frac{\partial u_i}{\partial x_j} + \frac{\partial u_j}{\partial x_i}\right) \tag{8.21}$$

is the strain rate tensor, defined with the velocity components of the fluid.

During the cascade process, for every wavenumber k, the turbulent kinetic energy becomes in part dissipated and in part transferred to higher wavenumbers (smaller eddies). We define $T(k)[m^2/s^3]$ as the "spectrum of energy flux," which represents the turbulent kinetic energy which, in the cascade process, flows through the wavenumber k. Because the decrement of energy flux is equal to the dissipation rate spectrum, we define $T(k)$ by: means of the differential equation

$$\frac{dT(k)}{dk} = -D(k) = -2\nu k^2 E(k). \tag{8.22}$$

8.1.3 Analytical Expressions of the Energy Spectrum

The analytical expressions of the turbulence characteristic quantities $D(k)$ and $T(k)$ depend on the analytical expression of the energy spectrum $E(k)$. The first determination of the function $E(k)$ in the interval (k_i, k_d) has been made on the basis of the so-called Kolmogorov's second hypothesis which can be formulated as follows:

> On the k axis there is a subrange, called inertial range, in which dissipation of turbulent kinetic energy is negligible and the transfer of energy by inertia forces is the dominant process. In this subrange the energy spectrum is uniquely determined, beside k, by ε only and is therefore independent of viscosity.

According to Kolmogorov's second hypothesis, the energy spectrum must be of the form $E(k) = \varepsilon^a k^b$, and dimensional analysis yields $a = 2/3$, $b = -5/3$. The analytical expression of the energy spectrum thus becomes the well-known "Kolmogorov's spectrum" given by:

$$E(k) = \xi \varepsilon^{2/3} k^{-5/3}, \tag{8.23}$$

where the constant $\xi \approx 1.5$, called Kolmogorov's constant, has been inserted upon experimental information.

The dissipation spectrum in the inertial subrange, according to Eq. (8.18), becomes:

$$D(k) = 2\nu k^2 E(k) = 2\xi \nu \varepsilon^{2/3} k^{1/3}. \tag{8.24}$$

From the historical viewpoint, the energy spectrum has been determined first in the inertial subrange. Subsequently, the energy spectrum has been evaluated also in

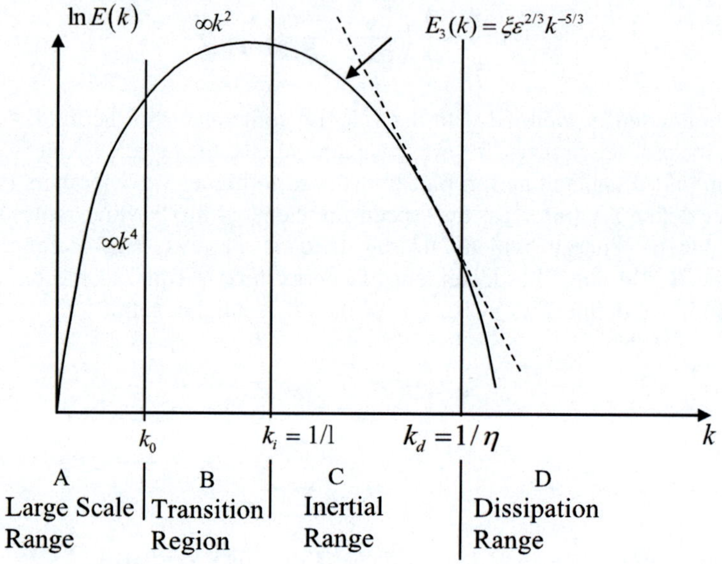

Fig. 8.2 Regions of the wavenumber $k = 2\pi/\lambda$ for the determination of the energy spectrum

the remaining part of the k axis. In the following, we summarize the results obtained, with reference to the four intervals of the k axis, sketched in Fig. 8.2:

(i) Large scale range, from $k = 0$ to an arbitrary limit generally set to $k_0 = 10$ (Region A);
(ii) transition region from $k_0 = 10$ to $k_i = 1/\ell$ (Region B);
(iii) inertial range, from $k_i = 1/\ell$ to $k_d = 1/\eta$ (Region C);
(iv) dissipation range (or equilibrium range) from $k_d = 1/\eta$ to infinity (in practice to a cutoff value) (Region D)

(i) **Region A, Large-Scale Range** ($0 \leq k \leq k_0 = 10$)

In this region the distributions of turbulent kinetic energy and of its dissipation per unit wavenumber are given by:

$$E_1(k) = \xi\varepsilon^{-3/4}\nu^{17/4}k^4, \tag{8.25}$$

$$D_1(k) = 2\nu k^2 E_1(k) = 2\xi\varepsilon^{-3/4}\nu^{21/4}k^6, \tag{8.26}$$

respectively.

(ii) **Region B, Transition Region** ($k_0 = 10 \leq k \leq k_i = 1/\ell$)

In this region is generally accepted a parabolic approximation:

$$E_2(k) = a_1 k^2 + b_1. \tag{8.27}$$

Constants a_1 and b_1 are determined by the conditions that for $k = k_0$ and $k = k_i = 1/\ell$, the distribution matches those of the regions A and C, respectively, the latter one being already known in the inertial subrange. Thus, one obtains:

$$a_1 = \frac{A_1 - B_1}{k_0^2 - k_i^2}, \tag{8.28}$$

$$b_1 = \frac{-A_1 k_i^2 + B_1 k_0^2}{k_0^2 - k_i^2}, \tag{8.29}$$

with:

$$A_1 = \xi \varepsilon^{-3/4} \nu^{17/4} k_0^4, \tag{8.30}$$

$$B_1 = \xi \varepsilon^{2/3} k_i^{-5/3}. \tag{8.31}$$

For the dissipation spectrum, we obtain:

$$D_2(k) = 2\nu k^2 E_2(k) = 2\nu\left(a_1 k^4 + b_1 k^2\right). \tag{8.32}$$

(iii) **Region C, Inertial Range,** $(k_i = 1/\ell \leq k \leq k_d = 1/\eta)$

The equations for the inertial range have been obtained above and are repeated here for completeness:

$$E_3(k) = \xi \varepsilon^{2/3} k^{-5/3}, \tag{8.33}$$

$$D_3(k) = 2\nu k^2 E_3(k) = 2\xi \nu \varepsilon^{2/3} k^{1/3}. \tag{8.34}$$

In the inertial range, for $k \leq 1/\eta$, the turbulent kinetic energy is transferred from the large eddies to smaller eddies without significant dissipation, and the energy flux is almost constant and can be approximated by (Tennekes & Lumley, 1972):

$$T_3(k) = \xi^{-3/2} k^{5/2} E_3(k)^{3/2}. \tag{8.35}$$

(iv) **Region D, Dissipation Range (or Equilibrium Range)** ($k_d = 1/\eta \leq k \leq \infty$)

This range extends to infinity. Its lower boundary is set by the Kolmogorov's wavenumber $k_d = 1/\eta$. It is assumed that in this range, Kolmogorov's first hypothesis holds. For large Reynolds numbers, the small-scale components (small eddies) of turbulence are statistically isotropic and independent of the structure of the large scale eddies. The motion of the flow associated with the small-scale components is determined, beside k, by the parameters ε and η (or ε and ν) only. The spectrum of turbulent kinetic energy is given in this range by:

$$E_4(k) = f(\varepsilon, \eta, k) = \xi\varepsilon^{2/3}k^{-5/3}\exp\left[-\frac{3}{2}\xi(k\eta)^{4/3}\right]. \tag{8.36}$$

The spectrum of energy dissipation is:

$$D_4(k) = 2\nu k^2 E_4(k) = 2\xi\nu\varepsilon^{2/3}k^{1/3}\exp\left[-\frac{3}{2}\xi(k\eta)^{4/3}\right]. \tag{8.37}$$

In the high-frequency range, for $k \geq 1/\eta$, the energy flux decreases with increasing wavenumber by an amount corresponding to the energy dissipation. It can be proved (see above reference) that:

$$T_4(k) = \frac{1}{\xi}\varepsilon^{1/3}k^{5/3}E_4(k). \tag{8.38}$$

8.1.4 Reynolds Stress Tensor and Anisotropy Tensor

The instantaneous value of the velocity of a particle of fluid, $\vec{V}(t)$, can be considered as the sum of a mean value $\overline{V} = (U, V, W)$ and of fluctuations (u', v', w') [or (u_1, u_2, u_3)] around the mean value. The mean values (U, V, W) can be considered as constant in time or varying slowly with respect to the time intervals of consideration for characterizing turbulent fluctuations. Conversely, the fluctuations of velocity (u', v', w') are, generally speaking, high-frequency fluctuations with mean values equal to zero, but with mean values of their cross products $u_i u_j$ $(i, j = 1, 2, 3)$ different from zero. (In the following, we will drop the apex denoting fluctuations, always using small letters to denote fluctuating velocity components and capital letters to denote mean values).

We define the Reynolds stress tensor by:

$$\tau_{ij} = -\rho\langle u_i u_j \rangle, \tag{8.39}$$

while the tensor $-\langle u_i u_j \rangle$ is referred to as the specific Reynolds stress tensor. Its diagonal components are used to define the turbulent kinetic energy:

$$K = \frac{1}{2}(\langle u_1 u_1 \rangle + \langle u_2 u_2 \rangle + \langle u_3 u_3 \rangle) = \frac{1}{2}\langle u_i u_i \rangle \quad (8.40)$$

where, according to Einstein's convention, summation is implied and mean values over time of the products $(u_i u_i)$ are assumed.

The definition (8.40) of the turbulent kinetic energy was given first by the theorists of turbulence. It however coincides with the definition given above in Eq. (8.17) which considers the problem of turbulence modelling from the viewpoint of its spectral distribution.

Knowledge of the turbulent kinetic energy given by Eq. (8.40) does not give information about the distribution of the turbulent intensities among the components of the velocity oscillations. In fact, the single-velocity components can be very different from each other. Let us refer again to the flow over the plate shown in Fig. 8.1. With reference to the bulk flow velocity U, we define the following normalized Reynolds stresses:

$$\widehat{u} = \frac{\sqrt{\langle u^2 \rangle}}{U}, \quad (8.41a)$$

$$\widehat{v} = \frac{\sqrt{\langle v^2 \rangle}}{U}, \quad (8.41b)$$

$$\widehat{w} = \frac{\sqrt{\langle w^2 \rangle}}{U}. \quad (8.41c)$$

Figure 8.3 shows the intensities of the normalized Reynolds stresses as function of the normalized distance z/δ from the plate. Close to the wall, the ratio of the normalized Reynolds stresses is roughly given by: $\widehat{u} : \widehat{v} : \widehat{w} = 4 : 3 : 2$. The plots display a maximum in proximity of the wall and then decay and become very close to each other at the abscissa $z/\delta \approx 1$. A similar plot holds also for a turbulent flow inside a pipe where the three coordinate directions, referring to a cylindrical coordinate system, are the axial, the azimuthal, and the radial one.

When the diagonal components of the Reynolds stress tensor are different from each other, which is in general the case, the turbulence is said to be anisotropic. Conversely, when the diagonal components of the Reynolds stress tensor are equal, the turbulence is isotropic. In this particular case, denoting with $u_\alpha u_\alpha$ the common value of the three products $u_i u_i$ one has (summation not implied):

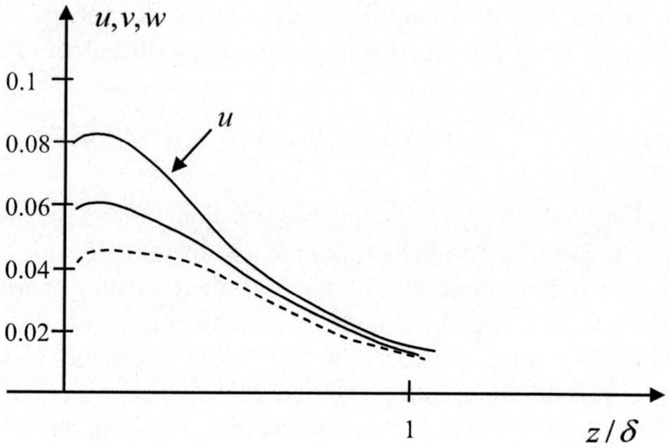

Fig. 8.3 Qualitative sketch of the normalized Reynolds stresses as function of the normalized distance z/δ from the plate of Fig. 8.1

$$K = \frac{3}{2}\langle u_\alpha u_\alpha \rangle, \tag{8.42}$$

hence

$$\langle u_\alpha u_\alpha \rangle = \frac{2}{3}K. \tag{8.43}$$

In isotropic turbulence, the value $2K/3$ is the common value of the diagonal components of the Reynolds stress tensor. Therefore, in anisotropic turbulence, it is straightforward to assume the difference $\langle u_i u_i \rangle - 2K/3$ as a measure of the anisotropy. Using the Kronecker delta δ_{ij}, one defines the Reynolds stress anisotropy tensor by:

$$a_{ij} = \frac{1}{K}\left(\langle u_i u_j \rangle - \frac{2}{3}K\delta_{ij}\right). \tag{8.44}$$

It is easy to verify that it always holds:

$$a_{ii} = a_{11} + a_{22} + a_{33} = 0, \tag{8.45}$$

while in isotropic turbulence, one has:

$$a_{11} = a_{22} = a_{33} = 0. \tag{8.46}$$

8.1.5 Taylor's Hypothesis and the Mixing Length Model

The ultimate goal in practical applications of turbulence theory is to determine the turbulent viscosity ν_t, so that Eq. (8.6) can be applied to calculate the effective viscosity as sum of the laminar (or molecular) and of the turbulent contributions. To this purpose in 1935, Taylor's assumption was made, which is still accepted in elementary models of turbulence. Because $\nu_t = u_e \ell$, by definition, and the characteristic velocity u_e can be identified with the root mean square of the velocity fluctuations, thus setting $u_e = K^{1/2}$, we can write:

$$\nu_t = u_e \ell = K^{1/2} \ell. \tag{8.47}$$

Taylor's hypothesis consists in assuming that during a turnover time ℓ/u_e the eddies loose a fraction u_e^2 of their kinetic energy per unit mass. Therefore, the dissipation of turbulent kinetic energy is proportional to the third power of the characteristic velocity u_e and inversely proportional to the macroscopic length of turbulence:

$$\varepsilon = \frac{u_e^2}{\ell/u_e} = \frac{u_e^3}{\ell} = \frac{K^{3/2}}{\ell}, \tag{8.48}$$

from which the macroscopic length scale or "mixing length" is derived:

$$\ell = \frac{K^{3/2}}{\varepsilon}. \tag{8.49}$$

Inserting Eq. (8.49) into Eq. (8.47), one obtains the most celebrated formula of turbulence modelling, namely:

$$\nu_t = \frac{K^2}{\varepsilon}. \tag{8.50}$$

As shown in the following [see Eq. (8.126)], numerical coefficients, suggested by experimental information, are introduced in this formula. According to the way in which the values of the turbulent kinetic energy K and of its dissipation ε are calculated in practical applications, one distinguishes several so-called "turbulence models," which will be summarized thereafter.

The ratios of the Kolmogorov's scales of turbulence (η, τ, v_K) to the macroscopic scales $(\ell, \ell/u_e, u_e)$ can be written as functions of the turbulent Reynolds number only. Using Eq. (8.13) and Eq. (8.48) for the macroscopic length (from which one derives $\varepsilon^{3/4} = u_e^{9/4}/\ell^{3/4}$), one obtains:

$$\frac{\eta}{\ell} = \frac{\nu^{3/4}}{\varepsilon^{1/4}} \frac{\varepsilon}{u_e^3} = \frac{\nu^{3/4}\varepsilon^{3/4}}{u_e^3} = \frac{\nu^{3/4}}{u_e^3} \frac{u_e^{9/4}}{\ell^{3/4}} = \left(\frac{\nu}{u_e \ell}\right)^{3/4} = \frac{1}{\mathrm{Re}_t^{3/4}}, \tag{8.51}$$

Similarly, from Eqs. (8.14) and (8.15), one obtains:

$$\frac{\tau}{\ell/u_e} = \left(\frac{\nu}{u_e \ell}\right)^{1/2} = \frac{1}{\mathrm{Re}_t^{1/2}}, \qquad (8.52)$$

$$\frac{\nu}{u_e} = \left(\frac{\nu}{u_e \ell}\right)^{1/4} = \frac{1}{\mathrm{Re}_t^{1/4}}. \qquad (8.53)$$

The simplest model of turbulence, and from a historical viewpoint the first one to have been proposed, consists in assuming that the length scale ℓ represents a mixing length over which fluid parcels of different turbulent intensities mix ($\ell = \ell_m$). This mixing length is considered to be proportional to the distance from the wall; hence:

$$\ell_m = \chi z, \qquad (8.54)$$

where $\chi = 0.435$ is the von Karman constant.

Thus, the turbulent viscosity becomes:

$$\nu_t = u_e \ell_m = u_e \chi z \qquad (8.55)$$

It is further assumed that:

$$u_e = \ell_m \left|\frac{\partial U}{\partial z}\right|, \qquad (8.56)$$

from which one obtains:

$$\nu_t = \ell_m^2 \left|\frac{\partial U}{\partial z}\right|. \qquad (8.57)$$

The turbulent shear stress becomes, in the two-dimensional case,

$$\tau_t = \mu_t \frac{\partial U}{\partial z} = \rho \nu_t \frac{\partial U}{\partial z} = \rho \ell_m^2 \left|\frac{\partial U}{\partial z}\right| \frac{\partial U}{\partial z}. \qquad (8.58)$$

In a tree-dimensional case, one has:

$$\nu_t = \ell_m^2 \frac{\partial U_i}{\partial x_j} \left(\frac{\partial U_i}{\partial x_j} + \frac{\partial U_j}{\partial x_i}\right). \qquad (8.59)$$

The above treatment has general validity. In the particular case of the Earth's atmosphere the mixing length model can be applied to the Prandtl sublayer, extending from the earth's surface (theoretically from 1 to 10 cm from the earth surface) to a height of about 100 m. In this layer, the shear stress:

$$\tau = \tau_{xz} = -\rho u'w' = \mathrm{c}. \tag{8.60}$$

can be considered as constant. In practice, one observes that the turbulent shear stress τ_{xz} decreases, proceeding form the Earth's surface upward. The ground value, denoted by a subscript 0 or w (wall), is:

$$\tau_0 = \tau_w = -\rho u_0' w_0'. \tag{8.61}$$

By convention, the shear stress $w'u'$ is considered negative ($w'u' < 0$) when momentum is transferred to the surface.

The shear stress at the surface allows to define the friction velocity

$$u^* = \sqrt{-w_0' u_0'} = \sqrt{\frac{\tau_w}{\rho}}. \tag{8.62}$$

We assume at first that there is not thermal exchange between Earth's surface and Prandtl sub-layer. Under this assumption, the square of the friction velocity can be written as:

$$u^{*2} = |w_0' u_0'| = K_m \frac{\partial u}{\partial z}, \tag{8.63}$$

which is a definition of the turbulent diffusion coefficient for momentum, K_m (in case of atmospheric turbulence it is customary to use the symbol K_m, rather than ν_t). This diffusion coefficient can be better specified in analogy with the molecular viscosity, which depends on the product of a characteristic velocity (the square mean root of the molecular velocity) with the free mean path of the molecules. Thus, using the Prandtl mixing length ℓ and a characteristic velocity:

$$u_c = \ell \left| \frac{\partial u}{\partial z} \right|, \tag{8.64}$$

the turbulent diffusion coefficient for momentum K_m can be written as:

$$K_m = \ell u_c = \ell^2 \left| \frac{\partial u}{\partial z} \right|. \tag{8.65}$$

Inserting Eq. (8.65) in Eq. (8.63) yields:

$$u^{*2} = \ell^2 \left(\frac{\partial u}{\partial z} \right)^2 \tag{8.66}$$

hence:

$$\frac{\partial u}{\partial z} = \frac{u^*}{\ell}. \tag{8.67}$$

Using Eq. (8.54), Eq. (8.67) becomes:

$$\frac{\partial u}{\partial z} = \frac{u^*}{\ell(z)} = \frac{u^*}{\chi z}. \tag{8.68}$$

Integrating Eq. (8.68) with respect to z, with the boundary condition $u = 0$ for $z = z_0$, yields:

$$u = \frac{u^*}{\chi} \ln\left(\frac{z}{z_0}\right). \tag{8.69}$$

Equation (8.69) is consistent with the experimental observation that in the Prandtl sublayer there is only one wind component, in the x direction, while it is well-known that in the Ekman layer, extending from the top of the Prandtl sublayer to a height of 1 to 2 Km, there are two main wind components. The length z_0 is called "rugosity length" and is the height above the earth's surface at which one can assume $u = 0$. It depends on the kind of surface and goes from a value of about 0.001 cm above smooth ice to about 10 cm above grass. In general a value of $z_0 = 1$ cm can be assumed.

From Eqs. (8.54), (8.65) and (8.68), finally the turbulent diffusion coefficient for momentum K_m can be written as:

$$K_m = \ell u_c = \ell^2 \left|\frac{\partial u}{\partial z}\right| = (\chi z)^2 \frac{u^*}{\chi z} = u^* \chi z = u^* \ell(z). \tag{8.70}$$

Equation (8.70) shows that in the Prandtl sublayer, the turbulent diffusion coefficient for momentum K_m increases linearly with height. Comparing with Eq. (8.65), one sees that the friction velocity u^* takes the role of the characteristic velocity u_c.

8.1.6 Dimensionless Quantities

Besides the friction velocity:

$$u^* = \sqrt{\frac{\tau_w}{\rho}}, \tag{8.71}$$

defined above in Eq. (8.62) and repeated here for clarity, one usually refers to characteristic length and time defined by:

$$\ell^* = \frac{\nu}{u^*}, \tag{8.72}$$

$$t^* = \frac{\ell^*}{u^*} = \frac{\nu}{u^{*2}}. \tag{8.73}$$

The quantities u^*, ℓ^* and, t^* are called "wall units," because they are used to render dimensionless physical quantities generally defined in boundary layers in proximity of surfaces. Velocities, coordinates, and time are usually made dimensionless dividing by the respective wall units:

$$u^+ = \frac{u}{u^*}, \tag{8.74}$$

$$x^+ = \frac{x}{\ell^*} = \frac{xu^*}{\nu}, \tag{8.75}$$

$$t^+ = \frac{t}{t^*} = \frac{x}{u}\frac{u^{*2}}{\nu} = \frac{xu^*}{\nu}\frac{u^*}{u} = \frac{x^+}{u^+}. \tag{8.76}$$

The kinetic energy of turbulence and its dissipation are made dimensionless by:

$$K^+ = \frac{K}{u^{*2}}, \tag{8.77}$$

$$\varepsilon^+ = \frac{\varepsilon\nu}{u^{*4}}. \tag{8.78}$$

In case of heat transfer problems, as, for instance, the problem of energy adsorbed or released from the Earth's surface, one defines a characteristic temperature T^* and a dimensionless temperature T^+ as follows:

$$T^* = \frac{\alpha^2 q_w}{\lambda u^*} = \frac{q_w}{\rho c_p u^*}, \tag{8.79}$$

$$T^+ = \frac{T_w - T_\infty}{T^*}, \tag{8.80}$$

where $\alpha^2 = \lambda/(\rho c_p)$ is a diffusion coefficient, λ is the thermal conductivity of the fluid at the surface, c_p is the specific heat at constant pressure, $q_w = -\lambda \nabla T$ is the heat flux at the surface, T_w is the surface temperature and T_∞ is the bulk temperature of the fluid, distant from the surface.

8.1.7 Overview of Turbulence Models

Complex turbulent flows occur in many engineering applications as well as in natural environments. For example, turbulent flows occur in the wake of ships, in the boundary layers around the wings of airplanes, in combustion processes, in the Earth boundary layer, and in marine environments. As long as (gas) flows are not rarefied, the Navier-Stokes equations are in principle suitable for properly describing turbulence, but their application would be too costly in most cases at high Reynolds numbers. Ensemble-averaged Navier-Stokes equations could be applied to describe turbulent phenomena, but in the averaging process, the details of a turbulent flow, which is unsteady, irregular, and rotational, are lost. To cope with this disadvantage, a turbulence model becomes necessary.

The simplest and historically speaking the first turbulence model, the **mixing length model**, was the only one used for practical purposes before the advent of the computer. In the past decades, as computing facilities became steadily stronger, more advanced turbulence models were adopted in the computational praxis. Among these, the **K-ε model** was the most successful one applied since the 1970s and became the standard of the so-called two-equation models although several other models have been tried. Among these, the most frequently used was the K-ω model, where ω (s^{-1}) is interpreted as the ratio of $\boldsymbol{\varepsilon}$ to K. The reciprocal of ω is the time scale at which dissipation of turbulent kinetic energy occurs. For details about the **K-ω model**, see Wilcox (1998).

These two-equation models assume isotropy of turbulence, assumption which, in most practical cases, and especially when the turbulent flow forms whirls, as behind obstacles, is not satisfied. In these cases it is necessary, for obtaining more precise predictions of the turbulent flow pattern and of the turbulent viscosity, to apply the complete Reynolds model of turbulence. In this model, the six components of the symmetric Reynolds stress tensor, $\tau_{ij} = -\rho\langle u_i u_j\rangle$, Eq. (8.39), are modelled by means of six transport equations to be solved simultaneously. Because the kinetic energy of turbulence is obtained by means of Eq. (8.40), in this case, the K- equation is not necessary, but still the **ε**-equation must be solved. The full model, called **Reynolds stress model (RSM)**, therefore consists, in case of unheated fluids, of seven transport equations.

Around the 1970s and 1980s it became evident that the K-**ε** model was not adequate to predict well turbulence in anisotropic flows, but the application of the full RSM was still beyond the computational capabilities and also outside of the experience of most research groups. In these times, alternatives were sought to the K-ε model, and the **algebraic stress model (ASM)** provided one of these alternatives. The key point of the problem can be stated as follows: in the production term of the K-ε model [see Eq. (8.83)] appear the Reynolds stresses $\langle u_i u_j\rangle$; these also appear in the production and diffusion terms of the transport equations for scalar fluxes [see Eqs. (8.121) and (8.122)]. These stresses should be computed rigorously, by solving the relative transport equations. If these solutions are however too demanding, one can still achieve an improvement with respect to the simple K-ε

model by introducing algebraic equations which give a reasonably good approximation of the stresses. One approximation of the Reynolds stresses, justified in Rodi (1980), is given in Sect. 8.1.7.2.

In case of heated fluids, one must also consider the transport equations for the turbulent heat fluxes $\langle \phi u_i \rangle$, where ϕ is the temperature fluctuation. To close the system, one more equation is necessary for the variance of the temperature fluctuations $\langle \phi^2 \rangle$. Thus, four additional equations are necessary for unheated fluids. When considering these four additional equations in combination with the K-ε , one has the so-called **anisotropic turbulence model (ATM)**. Details about the modelling and numerical treatment of the ATM are given in Chang et al. (1992). The use of the ATM became almost obsolete when the RSM was introduced in the computational praxis.

When the four equations (three equations for the turbulent heat fluxes $\langle \phi u_i \rangle$ and 1 equation for the variance of the temperature fluctuations $\langle \phi^2 \rangle$) are used in combination with the six equations for the Reynolds stresses and one equation for the dissipation ε of the turbulent kinetic energy, one obtains a self-closed system of 11 equations, which is referred to as **Reynolds stress model (RSM-HF)** for heated fluids.

Considerable success has obtained in the past decades the **large eddy Simulation (LES)** model. In this model, large whirls can be resolved by the numerical simulation of the turbulent flow. The turbulence intensity associated to whirls smaller than the grid used in the calculation is accounted for by a suitable definition of turbulent viscosity, which compensates for the unresolved small eddies.

A summary of this overview is given in Table 8.1.

A very comprehensive treatment of turbulent flows is given in the Lecture Notes by Chen (1990), issued form the Iowa University, USA. From the viewpoint of the statistical theory of turbulence, the lecture note by Orszag (1977) is recommended.

In the following, we give a brief review of the following turbulence models:

- The K-ε turbulence model
- The algebraic stress model (ASM)
- The large eddies simulation (LES) model

The complete Reynolds stress model (RSM) for heated fluids, consisting, as explained, of 11 transport equations, is treated in all analytical and numerical details in the following Sects. 8.2, 8.3, 8.4, and 8.5. All transport equations are finally transformed in algebraic equations of Poisson-like type, which can be solved numerically with the methods explained in Chapter 7 of this monograph.

8.1.7.1 The K-ε Turbulence Model

The transport equation describing the time and space distribution of the turbulent kinetic energy K is:

Table 8.1 Overview of the turbulence model introduced in Sect. 8.1.7

Name of model	Number of transport Equations	Characteristics of model
K-ε	2	Standard 2 Equations model used since the 1970's
K-ω	2	Good alternative to the standard K-ε model
ASM	2 + 6 algebraic Equations	Combines K and ε Equations with 6 algebraic equations for the Reynolds stresses
ATM	6	Combines K and ε Equations with 3 Equations $\langle \phi u_i \rangle$ and $\langle \phi^2 \rangle$ equation
RSM	7	6 Equations $\tau_{ij} = -\rho\langle u_i u_j \rangle$ for the components of the Reynolds stress tensor and ε equation
RSM-HF	11	6 Equations $\tau_{ij} = -\rho\langle u_i u_j \rangle$ for the components of the Reynolds stress tensor, ε equation, 3 Equations $\langle \phi u_i \rangle$ and $\langle \phi^2 \rangle$ equation
LES	(Navier-Stokes Equations)	Direct simulation of large eddies + turbulent viscosity to account for small unresolved eddies
DNS	(Navier-Stokes Equations)	Applicable for small Reynolds numbers

$$\rho \frac{\partial K}{\partial t} + \rho U_j \frac{\partial K}{\partial x_j} = \frac{\partial}{\partial x_j}\left[\left(\mu_\ell + \frac{\mu_t}{\sigma_k}\right)\frac{\partial K}{\partial x_j}\right] + \mu_t \frac{\partial U_i}{\partial x_j}\left(\frac{\partial U_i}{\partial x_j} + \frac{\partial U_j}{\partial x_i}\right) + \beta g_i \frac{\mu_t}{\mathrm{Pr}_t}\frac{\partial T}{\partial x_i} - \rho\varepsilon. \tag{8.81}$$

The first term at the right-hand side is the diffusive (laminar plus turbulent) transport of the turbulent kinetic energy. The second term is the production of K by shear. The third term is the production or suppression of turbulence due to buoyancy forces. Pr_t is the turbulent Prandtl number defined as the ratio of the diffusivity for momentum transfer and the diffusivity for heat transfer. β is the volumetric expansion coefficient of the fluid, given by:

$$\beta = -\frac{1}{\rho}\left(\frac{\partial \rho}{\partial T}\right)_p. \tag{8.82}$$

We define:

$$P_K = \mu_t \frac{\partial U_i}{\partial x_j}\left(\frac{\partial U_i}{\partial x_j} + \frac{\partial U_j}{\partial x_i}\right) = \rho\nu_t \frac{\partial U_i}{\partial x_j}\left(\frac{\partial U_i}{\partial x_j} + \frac{\partial U_j}{\partial x_i}\right) = -\rho \frac{\partial U_i}{\partial x_j}\langle u_i u_j \rangle, \tag{8.83}$$

$$G_K = \beta g_i \frac{\mu_t}{\mathrm{Pr}_t}\frac{\partial T}{\partial x_i}, \tag{8.84}$$

and compute the Reynolds stresses by:

$$-\langle u_i u_j \rangle = \nu_t \left(\frac{\partial U_i}{\partial x_j} + \frac{\partial U_j}{\partial x_i} \right). \tag{8.85}$$

With these definitions, Eq. (8.81) can be written in the compact form:

$$\rho \frac{\partial K}{\partial t} + \rho U_j \frac{\partial K}{\partial x_j} = \frac{\partial}{\partial x_j} \left[\left(\mu_\ell + \frac{\mu_t}{\sigma_k} \right) \frac{\partial K}{\partial x_j} \right] + P_K + G_K - \rho \varepsilon. \tag{8.86}$$

Equation (8.86) is treated numerically in the same way as each one of the six transport equations for the Reynolds stresses (see Sect. 8.4) to obtain a Poisson-like discretized equation, which can be solved with the methods outlined in Chap. 7. The ε-equation is given, modelled, and treated numerically in Sect. 8.5.

8.1.7.2 The Algebraic Stress Model (ASM)

The K- and ε-equations are the same as for the k-ε model. As explained above, instead of computing rigorously the components of the Reynolds stress tensor, by solving numerically the respective transport equations, an approximation is made using algebraic equations:

$$\langle u_i u_j \rangle = K \left[\frac{2}{3} \delta_{ij} + \frac{(1 - c_{2A})\left(P_{ij} - \frac{2}{3} P \delta_{ij}\right) + (1 - c_{3A})\left(G_{ij} - \frac{2}{3} G \delta_{ij}\right)}{c_{1A} + P + G - \varepsilon} \right], \tag{8.87}$$

where:

$$P_{ij} = -\langle u_i u_\ell \rangle \frac{\partial U_j}{\partial x_\ell} - \langle u_j u_\ell \rangle \frac{\partial U_i}{\partial x_\ell}, \tag{8.88}$$

$$G_{ij} = -\beta g_i \langle u_j \phi \rangle - \beta g_j \langle u_i \phi \rangle, \tag{8.89}$$

$$P = -\langle u_i u_j \rangle \frac{\partial U_i}{\partial x_j} = \frac{1}{2} P_{ii}, \tag{8.90}$$

$$G = -\beta g_i \langle u_i \phi \rangle = \frac{1}{2} G_{ii}. \tag{8.91}$$

The values of the constants in Eq. (8.87), suggested by experimental information, are: $c_{1A} = 1.5 - 2.2$, $c_{2A} = 0.55 - 0.6$, $c_{3A} = 0.5$. The last four equations are re-introduced in Sect. 8.4.

8.1.7.3 The Large Eddies Simulation (LES) Model

The ideal simulation of turbulent flows would require meshes so fine that every whirl could be simulated up to the smallest scale of turbulence, which is the Kolmogorov microscopic length scale η. In such an ideal case, the whole cascade range could be simulated numerically, from the large-scale range to the dissipation range, by solving the Navier-Stokes equations without any need for additional equations constituting what is called a turbulence model. This would be the goal of the direct numerical simulation (DNS) of turbulence. This refined simulation is impossible with today's computational power and may never be possible for large domains. This approach is thus far limited to flows with low Reynolds number. For this reason an alternative approach is necessary. It is provided by the **large eddy simulation (LES)**, based on the following ideas.

In view of a correct modelling it is important to realize that large eddies and small eddies have very different characteristics and behavior. Structure and time life of large eddies are strongly dependent on the geometrical configuration of the flow domain. They interact with the main flow, are strongly anisotropic, and are responsible for most of the transport of mass, momentum, and energy of the flow. Conversely, small eddies are nearly isotropic, have small interactions with the main flow, and tend to display universal characteristics independent of the main flow.

Recalling that the wavelength λ is representative of the dimensions of the eddies, the wave number $k = 2\pi/\lambda$ which corresponds to the inertial subrange (Range C, defined in Sect. 8.1.3) determines the separation between large and small eddies. In the inertial subrange, there is not a significant production of turbulent kinetic energy, nor dissipation of energy. Eddies larger than those in the inertial subrange are considered to be "large eddies" whilst eddies smaller than those in the inertial subrange are considered to be "small eddies". At small Reynolds numbers, the inertial subrange almost vanishes, and the distinction between large and small eddies becomes blurred.

Eddies larger than the meshes used for the discretization of the definition domain of the problem are captured and simulated by the numerical solution of the Navier-Stokes equations; eddies smaller than the meshes cannot be captured by the numerical simulation, and their presence is taken into account by means of a proper definition of the turbulent viscosity. The way in which this turbulent viscosity is defined makes the difference between several variants of the LES.

If the Navier-Stokes equations are solved by finite differences using grids of length Δx, the numerical scheme introduces a fictive "approximation filter" with a cut-off wave number $k_c = 1/\Delta x$. All wave numbers smaller than k_c, roughly covering the "large eddies" domain, namely, the regions A to C described in Sect. 8.1.3, are well simulated by the numerical scheme, while the wave numbers larger than k_c in the dissipation range cannot be simulated and must be accounted for by introducing an additional turbulent viscosity. This is accomplished by means of a **subgrid-scale**

(SGS) model. The purpose of a well-defined LES scheme is to simulate directly the eddies carrying at least 90% of the total turbulent energy $E(k)$.

The above qualitative considerations can be formulated with the following physical and mathematical approach.

Let us consider at first the Navier-Stokes equation for an incompressible flow:

$$\frac{\partial u_i}{\partial t} + \frac{\partial \left(u_i u_j\right)}{\partial x_j} = -\frac{1}{\rho}\frac{\partial p}{\partial x_i} + \nu \frac{\partial^2 u_i}{\partial x_j \partial x_j}, \tag{8.92}$$

with the continuity equation:

$$\frac{\partial u_i}{\partial x_i} = 0. \tag{8.93}$$

We average these equations over the discrete cells in which the definition domain has been partitioned. With the averaging process, the physical quantities become constant over one mesh and are represented by their instantaneous, but time-dependent, value. Averaging over the mesh volumes implies separating the eddies larger than the mesh size from eddies smaller than the mesh size. The first ones are resolved by the meshes, while the second ones are lost and shall be modelled. Thus, the process of averaging is the simplest case of a so-called filtering process. An overview of filtering techniques is given in Ferziger (1983). Upon averaging, the Navier-Stokes and the continuity equation assume the form:

$$\frac{\partial \langle u_i \rangle}{\partial t} + \frac{\partial \left\langle u_i u_j \right\rangle}{\partial x_j} = -\frac{1}{\rho}\frac{\partial \langle p \rangle}{\partial x_i} + \nu \frac{\partial^2 \langle u_i \rangle}{\partial x_j \partial x_j}, \tag{8.94}$$

$$\frac{\partial \langle u_i \rangle}{\partial x_i} = 0. \tag{8.95}$$

The instantaneous value of the velocity components is considered as the sum of its mean value and of a fluctuation, denoted by an apex as superscript:

$$u_i = \langle u_i \rangle + u_i'. \tag{8.96}$$

The averaging process, applied to the product of velocity components, yields

$$\langle u_i u_j \rangle = \langle \langle u_i \rangle \langle u_j \rangle \rangle + \left\langle \langle u_i \rangle u_j' \right\rangle + \langle u_i' \langle u_j \rangle \rangle + \left\langle u_i' u_j' \right\rangle, \tag{8.97}$$

which can be written as:

$$\langle u_i u_j \rangle = \langle \langle u_i \rangle \langle u_j \rangle \rangle + R_{ij}, \tag{8.98}$$

with the definition:

$$R_{ij} = \left\langle \langle u_i \rangle u'_j \right\rangle + \langle u'_i \langle u_j \rangle \rangle + \left\langle u'_i u'_j \right\rangle. \tag{8.99}$$

Equations (8.98) and (8.99) are simplified by the assumptions:

$$\langle \langle u_i \rangle \langle u_j \rangle \rangle = \langle u_i \rangle \langle u_j \rangle, \tag{8.100}$$

$$\langle u'_i \rangle = 0. \tag{8.101}$$

Equation (8.101) states that the mean value of the velocity fluctuations is zero and implies a simplification of Eq. (8.99) to:

$$R_{ij} = \left\langle u'_i u'_j \right\rangle. \tag{8.102}$$

Furthermore one can write:

$$R_{ij} = \left(R_{ij} - \frac{1}{3}\delta_{ij} R_{kk} \right) + \frac{1}{3} R_{kk}\delta_{ij} = \tau_{ij} + \frac{1}{3} R_{kk}\delta_{ij} = \tau_{ij} + \frac{2}{3} K\delta_{ij}, \tag{8.103}$$

having used:

$$K = R_{kk}/2, \tag{8.104}$$

and defined the SGS Reynolds stress tensor by:

$$\tau_{ij} = R_{ij} - \frac{1}{3} R_{kk}\delta_{ij}. \tag{8.105}$$

Using Eqs. (8.98), (8.100), and (8.103) one obtains:

$$\langle u_i u_j \rangle = \langle u_i \rangle \langle u_j \rangle + \tau_{ij} + \frac{1}{3} R_{kk}\delta_{ij}. \tag{8.106}$$

Inserting Eq. (8.106) in the second term of Eq. (8.94) yields:

$$\frac{\partial \langle u_i \rangle}{\partial t} + \frac{\partial}{\partial x_j}\left[\langle u_i \rangle \langle u_j \rangle + \tau_{ij} + \frac{1}{3} R_{kk}\delta_{ij} \right] = -\frac{1}{\rho}\frac{\partial \langle p \rangle}{\partial x_i} + \nu \frac{\partial^2 \langle u_i \rangle}{\partial x_j \partial x_j}. \tag{8.107}$$

The last term in the square brackets is the gradient of the scalar $R_{kk}\delta_{ij}/3$, which can be formally associated to the pressure term $\langle p\rangle/\rho$ by defining a generalized pressure:

$$P = \frac{\langle p\rangle}{\rho} + \frac{R_{kk}}{3}. \tag{8.108}$$

With this definition, Eq. (8.107) becomes:

$$\frac{\partial\langle u_i\rangle}{\partial t} + \frac{\partial}{\partial x_j}\left[\langle u_i\rangle\langle u_j\rangle\right] = -\frac{\partial P}{\partial x_i} + \nu\frac{\partial^2\langle u_i\rangle}{\partial x_j \partial x_j} - \frac{\partial \tau_{ij}}{\partial x_j}. \tag{8.109}$$

Equation (8.109), together with the averaged continuity Eq. (8.95), forms the basis of the LES method. Because of the "approximation filter" mentioned above, due to the discrete grids used, the higher wave numbers are not simulated and must be replaced by a proper definition of the turbulent viscosity to be inserted in the last term at the right-hand side of Eq. (8.109), which is approximated by letting:

$$\tau_{ij} = 2\nu_t\langle S_{ij}\rangle = \nu_t\left(\frac{\partial\langle u_i\rangle}{\partial x_j} + \frac{\partial\langle u_j\rangle}{\partial x_i}\right), \tag{8.110}$$

where S_{ij} is the strain rate tensor, equivalent to (8.21), but defined with reference to mean values of the velocity components.

According to the definition of the turbulent viscosity ν_t one can distinguish between several variants of the LES model. We recall some of the most frequently used:

(i) Eddy viscosity model (Smagorinsky model)

The turbulent viscosity is approximated by:

$$\nu_t = (C_S\Delta)^2|\langle S\rangle|, \tag{8.111}$$

where Δ is the grid size, C_S is the Smagorinsky coefficient, and:

$$|\langle S\rangle| = \left(\langle S_{ij}\rangle\langle S_{ij}\rangle\right)^{1/2}. \tag{8.112}$$

If the grid lengths Δ_x, Δ_y, and Δ_z are not uniform, one may take $\Delta = (\Delta_x\Delta_y\Delta_z)^{1/3}$. The Smagorinsky coefficient has been often adjusted from flow to flow in the range ($0.1 \leq C_S \leq 0.24$) to better match experimental data.

(ii) First model based on the macroscopic length scale L

The turbulent viscosity is approximated by:

$$\nu_t = C_D \Delta^{4/3} L^{2/3} |\langle S \rangle|, \tag{8.113}$$

where C_D is again a "fitting coefficient" to be evaluated by experimental information and the macroscopic length scale is:

$$L = c_D \frac{k^{3/2}}{\varepsilon}. \tag{8.114}$$

(iii) Second model based on the macroscopic length scale L

The turbulent viscosity is approximated by:

$$\nu_t = C_D \sqrt{K} L. \tag{8.115}$$

(iv) Model based on the definition of "cutoff velocity"

Let $k_c = 1/\Delta_c$ be the cutoff wavelength. The turbulent viscosity is approximated by:

$$\nu_t = u_c \Delta_c \tag{8.116}$$

and the cutoff velocity is defined with reference to the width of the energy spectrum in the domain of the small eddies which cannot be simulated:

$$u_c = \left[\int_{k_c}^{\infty} E(k) dk \right]^{1/2}. \tag{8.117}$$

The reader can find many more explanations and details in textbooks. For its clarity and exhaustiveness, the text of Wilcox (1998) is strongly recommended.

8.1.7.4 Historical Development of Turbulence Models at the Argonne National Laboratory, USA

Before proceeding with the analytical and numerical details of the following Sects. 8.2, 8.3, 8.4, and 8.5, we recall the historical development of the models used by the author and his co-operators at the Argonne National Laboratory (ANL) in the frame of the safety analysis of nuclear reactors.

At the Argonne Nation Laboratory (ANL) located at Argonne, 30 miles southwest from Chicago, Illinois, theoretical and numerical work was sponsored by the US Department of Energy (DOE) in frame of the safety analysis of nuclear reactors, both thermal and fast reactors. Inherent safe operations were one of the most important requirements in the frame of the development of the US Integral Fast Reactor (IFR) (Wade & Chang, 1988). The reactor design was characterized by a passive safety feature achieved by natural circulation as a result of thermal stratification in the reactor system. Even in the worst case of loss of power to the primary pumps, natural circulation of the liquid coolant would provide, through intermediate heat exchangers, a heat sink which could prevent coolant temperatures from reaching saturation and triggering local boiling with subsequent reduction of coolant capabilities and loss of core integrity. Therefore, accurate prediction of temperature distribution in a reactor system in a natural circulation environment was of paramount importance for the IFR.

Because natural circulation phenomena are not isotropic, there was the need to develop an anisotropic turbulence model (ATM) that could be used to predict accurately the effects of natural circulation on the reactor system. It was also necessary to model the impact of stratification of coolant temperature on turbulence, which can be enhanced or suppressed according to stable or unstable stratification patterns in a reactor pool. At ANL one of the most commonly used computer programs for reactor safety analysis was the code COMMIX-1C (Domanus et al., 1990). The ATM was implemented in a new version of this code, named COMMIX-1C/ATM (Chang et al., 1992). The development of the ATM helped refine the interaction between turbulence and buoyancy forces and accounted for the effect of temperature and velocity fluctuations on transport of turbulence. However, in the course of the verification of the ATM against experimental data, it became evident that a rigorous computation of Reynolds stresses was necessary. These stresses appear in the equations for the turbulent scalar fluxes and also replace the turbulent viscosity in the momentum equations of the fluid flow. Likewise, the turbulent scalar fluxes replace the turbulent conductivity in the enthalpy equation of the fluid.

In the frame of reactor safety analysis, in the simulation of a hypothetical loss of flow (LOF) accident, as a consequence of loss of power to the primary pumps, one critical event is the transition from forced to natural convection. The proper description of this transition phase requires a refined evaluation of the interaction between turbulence and buoyancy forces. These buoyancy forces strongly affect compressible fluids where density depends on temperature and pressure. The impact of buoyancy forces on production or dissipation of turbulence is represented by the term

$$G = -\rho\beta g_i \langle u_i\phi\rangle = \left(\frac{\partial\rho}{\partial T}\right)_p g_i \langle u_i\phi\rangle \tag{8.118}$$

which appears both in the K and ε equations.

Numerical calculations made with the computer program COMMIX-2, a two-phase flow version of the COMMIX series of codes (Bottoni & Willerding, 1987), proved that the performance of the K-ε model of turbulence is poor when the space distribution of fluid temperature is stratified. In such a case, it is necessary to compute rigorously the distribution of temperature fluxes $\langle u_i\phi\rangle$ $(i = 1, 2, 3)$ by solving numerically the three transport equations involved. These, in turn, have a term which depends on the temperature variance ϕ^2, which is computed by solving a fourth transport equation. This required the application of the ATM. With this model, the discrepancy between experimental data and computational results was reduced by approximately two thirds with respect to earlier calculations, but it also became evident that further improvements required the rigorous calculation of the Reynolds stresses which appear in the transport equations for the scalar fluxes. Thus, the full RSM with 11 transport equations was developed and applied for reactor safety analysis, using a new code variant called COMMIX-1C/RSM (Chang & Bottoni, 1995). The analytical and numerical treatments of the RSM, explained in the following Sects. 8.2, 8.3, 8.4, and 8.5 of this monograph, rely heavily on the above reference.

8.2 Transport Equations for Scalar Fluxes

In this section are presented the analytical and numerical treatments of the transport equations for the scalar fluxes.

I. **Modelling the Transport Equations for Scalar Fluxes**

We assume that the instantaneous fluid temperature can be written as the sum of a slowly varying term T and a fluctuation ϕ. Following a literature review, we adopted the following form of the transport equations for the scalar fluxes:

$$\begin{aligned}\frac{\partial}{\partial t}\langle u_i\phi\rangle + U_j\frac{\partial}{\partial x_j}\langle u_i\phi\rangle &= -\frac{\partial}{\partial x_j}\left(\langle u_i u_j\phi\rangle + \frac{1}{\rho}\langle p\phi\rangle\delta_{ij}\right)\\ &-\left(\langle u_i u_j\rangle\frac{\partial T}{\partial x_j} + \langle u_j\phi\rangle\frac{\partial U_i}{\partial x_j}\right) - \beta g_i\langle\phi^2\rangle + \frac{1}{\rho}\left\langle p\frac{\partial\phi}{\partial x_i}\right\rangle\end{aligned} \tag{8.119}$$

Equation (8.119) can be written in the more compact form:

$$\frac{\partial}{\partial t}\langle u_i\phi\rangle + U_j\frac{\partial}{\partial x_i}\langle u_i\phi\rangle = Diff\,\langle u_i\phi\rangle + P_{i\phi} + G_{i\phi} + \pi_{i\phi}, \tag{8.120}$$

with the definitions:

$$Diff\,\langle u_i\phi\rangle = -\frac{\partial}{\partial x_j}\left(\langle u_i u_j\phi\rangle + \frac{1}{\rho}\langle p\phi\rangle\delta_{ij}\right), \qquad \text{(diffusive transport)} \tag{8.121}$$

$$P_{i\phi} = -\left(\langle u_i u_j\rangle\frac{\partial T}{\partial x_j} + \langle u_j\phi\rangle\frac{\partial U_i}{\partial x_j}\right), \qquad \text{(mean-field production)} \tag{8.122}$$

$$G_{i\phi} = -\beta g_i \langle \phi^2 \rangle, \qquad \text{(buoyancy production)} \tag{8.123}$$

$$\pi_{i\phi} = \frac{1}{\rho}\left\langle p \frac{\partial \phi}{\partial x_i} \right\rangle. \qquad \text{(pressure-scalar gradient correlation)} \tag{8.124}$$

Diffusive transport is modelled as follows:

$$\begin{aligned} Diff \langle u_i \phi \rangle &= -\frac{\partial}{\partial x_j}\left(\langle u_i u_j \phi \rangle + \frac{1}{\rho} \langle p \phi \rangle \delta_{ij} \right) \\ &\simeq \frac{\partial}{\partial x_j}\left[\left(\nu_\ell + \frac{\nu_t}{\sigma_\phi} \right) \frac{\partial \langle u_i \phi \rangle}{\partial x_j} \right] \simeq \frac{\partial}{\partial x_j}\left[\left(\nu_\ell + \frac{c_\mu}{\sigma_\phi} \frac{K^2}{\varepsilon} \right) \frac{\partial \langle u_i \phi \rangle}{\partial x_j} \right] \\ &\simeq \frac{\partial}{\partial x_j}\left[\left(\nu_\ell + c_{s\phi} \frac{K^2}{\varepsilon} \right) \frac{\partial \langle u_i \phi \rangle}{\partial x_j} \right] \simeq \frac{\partial}{\partial x_j}\left[\Gamma_\phi^\nu \frac{\partial \langle u_i \phi \rangle}{\partial x_j} \right] \end{aligned} \tag{8.125}$$

having used:

$$\nu_t = c_\mu \frac{K^2}{\varepsilon}, \tag{8.126}$$

$$c_{s\phi} = \frac{c_\mu}{\sigma_\phi}, \tag{8.127}$$

$$\Gamma_\phi^\nu = \nu_\ell + c_{s\phi} \frac{K^2}{\varepsilon}. \tag{8.128}$$

The Reynolds stresses $\langle u_i u_j \rangle$ which appear in the production and diffusive terms of Eq. (8.120) are computed solving numerically the respective transport equations, as explained in Sect. 8.4. The pressure-scalar gradient correlation is modelled by:

$$\pi_{i\phi} = \frac{1}{\rho}\left\langle p \frac{\partial \phi}{\partial x_i} \right\rangle = \pi_{i\phi}^1 + \pi_{i\phi}^2 + \pi_{i\phi}^3 + \pi_{i\phi}^w, \tag{8.129}$$

with:

$$\pi_{i\phi}^1 = -c_{1\phi} \frac{\varepsilon}{K} \langle u_i \phi \rangle, \text{ (turbulence contribution)} \tag{8.130a}$$

$$\pi_{i\phi}^2 = c_{2\phi} \langle u_j \phi \rangle \frac{\partial U_i}{\partial x_j}, \qquad \text{(mean strain contribution)} \tag{8.130b}$$

$$\pi_{i\phi}^3 = c_{3\phi} \beta g_i \langle \phi^2 \rangle = -c_{3\phi} G_{i\phi}, \text{ (buoyancy contribution)} \tag{8.130c}$$

$$\pi_{i\phi}^{w} = -c_{1\phi}' \frac{\varepsilon}{K} \langle u_n \phi \rangle \delta_{in} f\left(\frac{L}{x_n}\right). \qquad \text{(wall effect)} \tag{8.130d}$$

Equation (8.130d) represents the effect of the wall: n is the direction normal to the wall, and x_n is the distance from the wall. L is a suitable length scale, whereas the function $f(L/x_n)$ needs further specification. Its calculation depends on the boundary conditions. The explanation of the choice of the boundary conditions goes beyond the purpose of this monograph, which is focused on the numerical treatment of the transport equations. Details about this issue can be found in (Chang & Bottoni, 1995).

The value of the constants used in the above equations is as follows: $\sigma_\phi = 1.3$; $c_\mu = 0.09$; $c_{s\phi} = c_\mu/\sigma_\phi = 0.09/1.3 = 0.07$; $c_{1\phi} = 3.1$; $c_{2\phi} = 0.4$, $c_{3\phi} = 0.5$; $c_{1\phi}' = 0.5$. Introducing Eqs. (8.121)–(8.130) into Eq. (8.119), one obtains:

$$\frac{\partial}{\partial t}\langle u_i \phi \rangle + U_j \frac{\partial}{\partial x_j}\langle u_i \phi \rangle = \frac{\partial}{\partial x_j}\left(\Gamma_\phi^\nu \frac{\partial \langle u_i \phi \rangle}{\partial x_j}\right) + P_{i\phi} + G_{i\phi}(1 - c_{3\phi})$$
$$- c_{1\phi}\frac{\varepsilon}{K}\langle u_i \phi \rangle + c_{2\phi}\langle u_j \phi \rangle \frac{\partial U_i}{\partial x_j} + \pi_{i\phi}^{w}. \tag{8.131}$$

Using the identity:

$$\nabla \cdot \left[\overline{U}\langle u_i \phi \rangle\right] = \langle u_i \phi \rangle\left(\nabla \cdot \overline{U}\right) + \overline{U} \cdot \nabla \langle u_i \phi \rangle \tag{8.132}$$

and assuming that the flow is divergence-free $\left(\nabla \cdot \overline{U} = 0\right)$, the second term at the left-hand side of Eq. (8.131) can be written as:

$$U_j \frac{\partial}{\partial x_i}\langle u_i \phi \rangle = \frac{\partial}{\partial x_i}\left(U_j \langle u_i \phi \rangle\right). \tag{8.133}$$

Introducing Eq. (8.133) into Eq. (8.131), one obtains:

$$\frac{\partial}{\partial t}\langle u_i \phi \rangle + \frac{\partial}{\partial x_j}\left[U_j \langle u_i \phi \rangle - \Gamma_\phi^\nu \frac{\partial \langle u_i \phi \rangle}{\partial x_j}\right]$$
$$= P_{i\phi} + G_{i\phi}(1 - c_{3\phi}) - c_{1\phi}\frac{\varepsilon}{K}\langle u_i \phi \rangle + c_{2\phi}\langle u_j \phi \rangle \frac{\partial U_i}{\partial x_j} + \pi_{i\phi}^{w} \tag{8.134}$$

Let us define convective, diffusive, and total (convective plus diffusive) fluxes by:

$$J_{\phi ji}^{c} = U_j \langle u_i \phi \rangle, \tag{8.135}$$

$$J^d_{\phi ji} = -\Gamma^{v}_{\phi} \frac{\partial \langle u_i \phi \rangle}{\partial x_j}, \tag{8.136}$$

$$\overline{J}_{\phi ji} = J^c_{\phi ji} + J^d_{\phi ji} = U_j \langle u_i \phi \rangle - \Gamma^{v}_{\phi} \frac{\partial \langle u_i \phi \rangle}{\partial x_j}. \tag{8.137}$$

Furthermore, we define the source term by:

$$S_{\phi i} = P_{i\phi} + G_{i\phi}(1 - c_{3\phi}) - c_{1\phi} \frac{\varepsilon}{K} \langle u_i \phi \rangle + c_{2\phi} \langle u_j \phi \rangle \frac{\partial U_i}{\partial x_j} + \pi^{w}_{i\phi}. \tag{8.138}$$

With these definitions, Eq. (8.134) can be written in compact form:

$$\frac{\partial}{\partial t} \langle u_i \phi \rangle + \frac{\partial}{\partial x_j} \left[\overline{J}_{\phi ji} \right] = S_{i\phi}. \tag{8.139}$$

Let us denote the scalar transported quantity as:

$$\Phi = \langle u_i \phi \rangle. \tag{8.140}$$

Notice that the capital Φ denotes anyone of the three scalar fluxes $\langle u_i \phi \rangle$ and therefore subscript i can be dropped. Equation (8.139) can be written in the more general form of a transport equation for the scalar as:

$$\frac{\partial \Phi}{\partial t} + \frac{\partial}{\partial x_j} \left[\overline{J}_{\phi j} \right] = S_{\Phi}. \tag{8.141}$$

Multiplying Eq. (8.141) by the fluid density ρ and recalling Eq. (8.137), one has:

$$\rho \frac{\partial \Phi}{\partial t} + \rho \frac{\partial (U_j \Phi)}{\partial x_j} - \rho \frac{\partial}{\partial x_j} \left(\Gamma^{v}_{\phi} \frac{\partial \Phi}{\partial x_j} \right) = \rho S_{\Phi}. \tag{8.142}$$

Using:

$$\frac{\partial (\rho \Phi)}{\partial t} = \rho \frac{\partial \Phi}{\partial t} + \Phi \frac{\partial \rho}{\partial t}, \tag{8.143}$$

$$\frac{\partial (\rho U_i \Phi)}{\partial x_j} = \rho \frac{\partial (U_i \Phi)}{\partial x_j} + U_i \Phi \frac{\partial \rho}{\partial x_j}, \tag{8.144}$$

Equation (8.142) becomes:

$$\frac{\partial(\rho\Phi)}{\partial t} - \Phi\frac{\partial\rho}{\partial t} + \frac{\partial(\rho U_j\Phi)}{\partial x_j} - U_j\Phi\frac{\partial\rho}{\partial x_j} - \frac{\partial}{\partial x_j}\left(\rho\Gamma_\phi^\nu\frac{\partial\Phi}{\partial x_j}\right) = \rho S_\Phi. \quad (8.145)$$

Using the continuity equation:

$$\frac{\partial\rho}{\partial t} + \frac{\partial(\rho U_j)}{\partial x_j} = 0, \quad (8.146)$$

the sum of the second and fourth terms at the left-hand side of Eq. (8.145) becomes:

$$\begin{aligned} -\Phi\frac{\partial\rho}{\partial t} - U_j\Phi\frac{\partial\rho}{\partial x_j} &= -\Phi\left(\frac{\partial\rho}{\partial t} + U_j\frac{\partial\rho}{\partial x_j}\right) \\ &= -\Phi\left(-\frac{\partial(\rho U_i)}{\partial x_j} + U_j\frac{\partial\rho}{\partial x_j}\right) = \rho\Phi\frac{\partial U_i}{\partial x_j}. \end{aligned} \quad (8.147)$$

Inserting Eq. (8.147) in Eq. (8.145) yields:

$$\frac{\partial(\rho\Phi)}{\partial t} + \frac{\partial(\rho U_j\Phi)}{\partial x_j} + \rho\Phi\frac{\partial U_j}{\partial x_j} - \frac{\partial}{\partial x_j}\left(\rho\Gamma_\phi^\nu\frac{\partial\Phi}{\partial x_j}\right) = \rho S_\Phi. \quad (8.148)$$

Because of the divergence-free assumption of the fluid, $\partial U_j/\partial x_j = 0$, and Eq. (8.148) simplifies to:

$$\frac{\partial(\rho\Phi)}{\partial t} + \frac{\partial}{\partial x_j}(\rho U_j\Phi) - \frac{\partial}{\partial x_j}\left(\rho\Gamma_\phi^\nu\frac{\partial\Phi}{\partial x_j}\right) = \rho S_\Phi, \quad (8.149)$$

or:

$$\frac{\partial(\rho\Phi)}{\partial t} + \frac{\partial J_{\phi j}}{\partial x_j} = \rho S_\Phi \quad (8.150)$$

with:

$$J_{\phi j} = \rho\bar{J}_{\phi j} = \rho U_j\Phi - \rho\Gamma_\phi^\nu\frac{\partial\Phi}{\partial x_j} = \rho U_j\Phi - \Gamma_\phi\frac{\partial\Phi}{\partial x_j}, \quad (8.151)$$

$$\Gamma_\phi = \rho\Gamma_\phi^\nu = \mu_\ell + c_{s\phi}\rho\frac{K^2}{\varepsilon}. \quad (8.152)$$

II. Numerical Treatment of the Transport Equations for Scalar Fluxes

We integrate Eq. (8.150) over the volume V_f of the fluid in a cell and obtain:

$$\int_{V_f} \frac{\partial(\rho\Phi)}{\partial t} dV + \int_{V_f} \frac{\partial J_{\phi j}}{\partial x_j} dV = \int_{V_f} \rho S_\phi dV. \tag{8.153}$$

Using the Gauss theorem, we replace the volume integral of the divergence of vector $J_{\phi j}$ with the fluxes through the boundary surfaces of the cell. Indexing the center node of a cell with (i,j,k), we have:

$$\begin{aligned} &\int_{V_f} \frac{\partial(\rho\Phi)}{\partial t} dV + \int_{A_{f,,i+1/2}} J_{\phi j} dA - \int_{A_{f,,i-1/2}} J_{\phi j} dA + \int_{A_{f,j+1/2}} J_{\phi j} dA \\ &- \int_{A_{f,j-1/2}} J_{\phi j} dA + \int_{A_{f,k+1/2}} J_{\phi j} dA - \int_{A_{f,,k-1/2}} J_{\phi j} dA \\ &= \int_{V_f} \rho S_\phi dV \end{aligned} \tag{8.154}$$

$A_{f,\, i\,+\,1/2}$, $A_{f,\, i\,-\,1/2}$, etc. are the areas of the cell surfaces normal to the coordinate directions. Denoting with the brackets $\langle\rangle$ mean values over volumes or over surfaces, as appropriate, and using superscript n for the time level, we discretize Eq. (8.154) as:

$$\begin{aligned} &\frac{V_f}{\Delta t}\left[(\rho\Phi)^{n+1} - (\rho\Phi)^n\right]_{ijk} + \left\langle J_\phi^c A_f \right\rangle_{i+1/2.j.k}^{n+1} - \left\langle J_\phi^c A_f \right\rangle_{i-1/2.j.k}^{n+1} + \left\langle J_\phi^c A_f \right\rangle_{i,j+1/2.k}^{n+1} \\ &- \left\langle J_\phi^c A_f \right\rangle_{i,j-1/2.k}^{n+1} + \left\langle J_\phi^c A_f \right\rangle_{i,j,k+1/2}^{n+1} - \left\langle J_\phi^c A_f \right\rangle_{i,j,k-1/2}^{n+1} + A_{f,i+1/2}\left\langle \Gamma_\phi \frac{\partial \Phi}{\partial x} \right\rangle_{i+1/2,j,k}^{n+1} \\ &- A_{f,i-1/2}\left\langle \Gamma_\phi \frac{\partial \Phi}{\partial x} \right\rangle_{i-1/2,j,k}^{n+1} + A_{f,j+1/2}\left\langle \Gamma_\phi \frac{\partial \Phi}{\partial x} \right\rangle_{i,j+1/2,k}^{n+1} - A_{f,j-1/2}\left\langle \Gamma_\phi \frac{\partial \Phi}{\partial x} \right\rangle_{j-1/2,j,k}^{n+1} \\ &+ A_{f,k+1/2}\left\langle \Gamma_\phi \frac{\partial \Phi}{\partial x} \right\rangle_{i,j,k+1/2}^{n+1} - A_{f,k-1/2}\left\langle \Gamma_\phi \frac{\partial \Phi}{\partial x} \right\rangle_{i,j,k-1/2}^{n+1} = V_f \langle \rho S_\Phi \rangle^{n+1}. \end{aligned} \tag{8.155}$$

In Eq. (8.155), we have separated the convective (superscript c) from the diffusive terms. The convective terms are computed as follows:

$$\left\langle J_\phi^c A_f \right\rangle_{\alpha+1/2.}^{n+1} = \left\langle \rho U \Phi A_f \right\rangle_{\alpha+1/2}^{n+1} = \langle F\Phi \rangle_{\alpha+1/2}^{n+1}$$
$$= \left[0, F_{\alpha+1/2}\right]\Phi_\alpha^{n+1} - \left[0, -F_{\alpha+1/2}\right]\Phi_{\alpha+1}^{n+1}, \tag{8.156a}$$

$$\left\langle J_\phi^c A_f \right\rangle_{\alpha-1/2.}^{n+1} = \left\langle \rho U \Phi A_f \right\rangle_{\alpha-1/2}^{n+1} = \langle F\Phi \rangle_{\alpha-1/2}^{n+1}$$
$$= \left[0, F_{\alpha-1/2}\right]\Phi_{\alpha-1}^{n+1} - \left[0, -F_{\alpha-1/2}\right]\Phi_{\alpha}^{n+1}, \tag{8.156b}$$

with:

$$F_{\alpha\pm1/2} = (\rho U A)_{\alpha\pm1/2}. \ (\alpha = i, j, k) \tag{8.157}$$

The diffusive terms are computed by:

$$A_{f,\alpha+1/2}\left\langle \Gamma_\Phi \frac{\partial \Phi}{\partial x_\alpha} \right\rangle_{\alpha+1/2}^{n+1} = \frac{A_{f,\alpha+1/2}(\Phi_{\alpha+1} - \Phi_\alpha)^{n+1}}{\frac{1}{2}\left(\frac{\Delta x_\alpha}{\Gamma_{\phi,\alpha}} + \frac{\Delta x_{\alpha+1}}{\Gamma_{\phi,\alpha+1}}\right)}$$
$$= D_{\alpha+1/2}^{\Phi,n+1}(\Phi_{\alpha+1} - \Phi_\alpha)^{n+1}, \tag{8.158a}$$

$$A_{f,\alpha-1/2}\left\langle \Gamma_\Phi \frac{\partial \Phi}{\partial x_\alpha} \right\rangle_{\alpha-1/2}^{n+1} = \frac{A_{f,\alpha-1/2}(\Phi_{\alpha} - \Phi_{\alpha-1})^{n+1}}{\frac{1}{2}\left(\frac{\Delta x_{\alpha-1}}{\Gamma_{\phi,\alpha-1}} + \frac{\Delta x_{\alpha}}{\Gamma_{\phi,\alpha}}\right)}$$
$$= D_{\alpha-1/2}^{\Phi,n+1}(\Phi_{\alpha} - \Phi_{\alpha-1})^{n+1}, \tag{8.158b}$$

with the definitions:

$$D_{\alpha+1/2}^{\Phi,} = \frac{A_{f,\alpha+1/2}}{\frac{1}{2}\left(\frac{\Delta x_\alpha}{\Gamma_{\phi,\alpha}} + \frac{\Delta x_{\alpha+1}}{\Gamma_{\phi,\alpha+1}}\right)}, \tag{8.159a}$$

$$D_{\alpha-1/2}^{\Phi} = \frac{A_{f,\alpha-1/2}}{\frac{1}{2}\left(\frac{\Delta x_{\alpha-1}}{\Gamma_{\phi,\alpha-1}} + \frac{\Delta x_{\alpha}}{\Gamma_{\phi,\alpha}}\right)}. \tag{8.159b}$$

Introducing Eqs. (8.156), (8.157), and (8.158) into Eq. (8.155) and rearranging we obtain:

$$
\begin{aligned}
&\Phi_{i,j,k}^{n+1}\left\{\frac{\left(V_f\rho^{n+1}\right)_{i,j,k}}{\Delta t}+\left[0,F_{i+1/2}\right]+\left[0,F_{j+1/2}\right]+\left[0,F_{k+1/2}\right]+\left[0,-F_{i-1/2}\right]\right.\\
&\left.+\left[0,-F_{j-1/2}\right]+\left[0,-F_{k-1/2}\right]+D_{i+1/2}^{\Phi}+D_{i-1/2}^{\Phi}+D_{j+1/2}^{\Phi}+D_{j-1/2}^{\Phi}+D_{k+1/2}^{\Phi}+D_{k-1/2}^{\Phi}\right\}\\
&\quad-\Phi_{i+1}^{n+1}\left\{\left[0,-F_{i+1/2}\right]+D_{i+1/2}^{\Phi}\right\}-\Phi_{i-1}^{n+1}\left\{\left[0,F_{i-1/2}\right]+D_{i-1/2}^{\Phi}\right\}\\
&\quad-\Phi_{j+1}^{n+1}\left\{\left[0,-F_{j+1/2}\right]+D_{j+1/2}^{\Phi}\right\}-\Phi_{j-1}^{n+1}\left\{\left[0,F_{j-1/2}\right]+D_{j-1/2}^{\Phi}\right\}\\
&\quad-\Phi_{k+1}^{n+1}\left\{\left[0,-F_{k+1/2}\right]+D_{k+1/2}^{\Phi}\right\}-\Phi_{k-1}^{n+1}\left\{\left[0,F_{k-1/2}\right]+D_{k-1/2}^{\Phi}\right\}\\
&\quad=V_{f,i,j,k}\frac{(\rho\Phi)_{i,j,k}^{n}}{\Delta t}+V_f\langle\rho S_{\Phi}\rangle^{n+1}.
\end{aligned}
\tag{8.160}
$$

With the convention of indexing with:

$$
\begin{array}{llllllll}
0 & \text{for} & i,j,k & 1 & \text{for} & i-1,j,k, & 2 & \text{for} & i+1,j,k,\\
 & & & 3 & \text{for} & i,j-1,k, & 4 & \text{for} & i,j+1,k,\\
 & & & 5 & \text{for} & i,j,k-1, & 6 & \text{for} & i,j,k+1,
\end{array}
$$

Equation (8.160) can be written in the compact form:

$$
a_0^{\Phi}\Phi_0^{n+1}-\sum_{\beta=1}^{6}a_{\beta}^{\Phi}\Phi_{\beta}^{n+1}=b_0^{\Phi},
\tag{8.161}
$$

with the definition of the coefficients:

$$
\begin{aligned}
a_0^{\Phi}=&\frac{\left(V_f\rho^{n+1}\right)_{i,j,k}}{\Delta t}+\left[0,F_{i+1/2}\right]+\left[0,F_{j+1/2}\right]+\left[0,F_{k+1/2}\right]\\
&+\left[0,-F_{i-1/2}\right]+\left[0,-F_{j-1/2}\right]+\left[0,-F_{k-1/2}\right]+D_{i+1/2}^{\Phi}+D_{i-1/2}^{\Phi}\\
&+D_{j+1/2}^{\Phi}+D_{j-1/2}^{\Phi}+D_{k+1/2}^{\Phi}+D_{k-1/2}^{\Phi},
\end{aligned}
\tag{8.162a}
$$

$$
a_1^{\Phi}=\left[0,F_{i-1/2}\right]+D_{i-1/2}^{\Phi},
\tag{8.162b}
$$

$$
a_2^{\Phi}=\left[0,-F_{i+1/2}\right]+D_{i+1/2}^{\Phi},
\tag{8.162c}
$$

$$a_3^{\Phi} = \left[0, F_{j-1/2}\right] + D^{\Phi}_{j-1/2}, \tag{8.162d}$$

$$a_4^{\Phi} = \left[0, -F_{j+1/2}\right] + D^{\Phi}_{j+1/2}, \tag{8.162e}$$

$$a_5^{\Phi} = \left[0, F_{k-1/2}\right] + D^{\Phi}_{k-1/2}, \tag{8.162f}$$

$$a_6^{\Phi} = \left[0, -F_{k+1/2}\right] + D^{\Phi}_{k+1/2}, \tag{8.162g}$$

$$b_0^{\Phi} = V_{f0}\frac{(\rho\Phi)_0^n}{\Delta t} + V_{f0}\rho_0^{n+1}\left\langle S_{\Phi}^{n+1}\right\rangle. \tag{8.162h}$$

Equation (8.160) is a Poisson-like equation which can be solved with the methods explained in Chap. 7.

8.3 Transport Equation for Variance of Temperature Fluctuations [g Equation]

In this section, we present the transport equation for the variance of temperature fluctuations $\langle\phi^2\rangle$ and its numerical treatment.

I. **Modelling the Transport Equation for Variance of Temperature Fluctuations**

The transport equation for the variance $\langle\phi^2\rangle$ of temperature fluctuations is:

$$\frac{\partial\langle\phi^2\rangle}{\partial t} + U_j\frac{\partial\langle\phi^2\rangle}{\partial x_j} = -\frac{\partial\langle u_j\phi^2\rangle}{\partial x_j} + \frac{\partial}{\partial x_j}\left(\frac{\lambda}{\rho c_p}\frac{\partial\langle\phi^2\rangle}{\partial x_j}\right) - 2\langle u_j\phi\rangle\frac{\partial T}{\partial x_j}$$
$$- 2\frac{\lambda}{\rho c_p}\left\langle\frac{\partial\phi}{\partial x_j}\frac{\partial\phi}{\partial x_j}\right\rangle. \tag{8.163}$$

Equation (8.163) can be written in the more compact form:

$$\frac{\partial\langle\phi^2\rangle}{\partial t} + U_j\frac{\partial\langle\phi^2\rangle}{\partial x_j} = D_t\langle\phi^2\rangle + D_\ell\langle\phi^2\rangle + P_\phi - \varepsilon_\phi, \tag{8.164}$$

where:

$$D_t\langle\phi^2\rangle = -\frac{\partial\langle u_j\phi^2\rangle}{\partial x_j} = \frac{\partial}{\partial x_j}\left(c'_\phi \frac{K^2}{\varepsilon}\frac{\partial\langle\phi^2\rangle}{\partial x_j}\right), \quad \text{(turbulent diffusion)} \tag{8.165}$$

$$D_\ell\langle\phi^2\rangle = \frac{\partial}{\partial x_j}\left(\frac{\lambda}{\rho c_p}\frac{\partial\langle\phi^2\rangle}{\partial x_j}\right), \quad \text{(molecular diffusion)} \tag{8.166}$$

$$P_\phi = -2\langle u_j\phi\rangle\frac{\partial T}{\partial x_j}, \quad \text{(mean-field production)} \tag{8.167}$$

$$\varepsilon_\phi = 2\frac{\lambda}{\rho c_p}\left\langle\frac{\partial\phi}{\partial x_j}\frac{\partial\phi}{\partial x_j}\right\rangle = \frac{\varepsilon}{K}\frac{\langle\phi^2\rangle}{R}. \quad \text{(dissipation of fluctuations)} \tag{8.168}$$

The constants used in the above equations have been determined experimentally (Rodi, 1980) as $c'_\phi = 0.13$ and R = 0.5.

By defining one-half of the variance of temperature fluctuations as:

$$g = \frac{1}{2}\langle\phi^2\rangle, \tag{8.169}$$

the transport equation for fluctuations of temperature, referred to as "g-equation", can be written, for a divergence-free flow ($\partial U_j/\partial x_j = 0$), as:

$$\frac{\partial g}{\partial t} + \frac{\partial(U_j g)}{\partial x_j} = \frac{\partial}{\partial x_j}\left(c'_\phi\frac{K^2}{\varepsilon}\frac{\partial g}{\partial x_j} + \frac{\lambda}{\rho c_p}\frac{\partial g}{\partial x_j}\right) - \langle u_j\phi\rangle\frac{\partial T}{\partial x_j} - \frac{\varepsilon}{K}\frac{g}{R}. \tag{8.170}$$

By defining:

$$\Gamma_g = \Gamma_{gt} + \Gamma_{g\ell} = c'_\phi\frac{K^2}{\varepsilon} + \frac{\lambda}{\rho c_p}, \tag{8.171}$$

$$P_g = \frac{1}{2}P_\phi = -\langle u_j\phi\rangle\frac{\partial T}{\partial x_j}, \tag{8.172}$$

$$\varepsilon_g = \frac{1}{2}\varepsilon_\phi = \frac{\varepsilon}{K}\frac{g}{R}, \tag{8.173}$$

Equation (8.170) can be shortened to:

$$\frac{\partial g}{\partial t} + \frac{\partial (U_j g)}{\partial x_j} = \frac{\partial}{\partial x_j}\left(\Gamma_g \frac{\partial g}{\partial x_j}\right) + P_g - \varepsilon_g. \tag{8.174}$$

We define the convective, diffusive, and total (convective plus diffusive) fluxes of temperature fluctuations and a generalized source term by:

$$J^c_{gj} = U_j g, \tag{8.175}$$

$$J^d_{gj} = -\Gamma_g \frac{\partial g}{\partial x_j}, \tag{8.176}$$

$$J_{gj} = J^c_{gj} + J^d_{gj} = U_j g - \Gamma_g \frac{\partial g}{\partial x_j}, \tag{8.177}$$

$$S_g = P_g - \varepsilon_g. \tag{8.178}$$

With these definitions, Eq. (8.174) can be written in the compact form:

$$\frac{\partial g}{\partial t} + \frac{\partial J_{gj}}{\partial x_j} = S_g. \tag{8.179}$$

II. **Numerical Treatment of the Transport Equations for Variance of Temperature Fluctuations**

Applying the same procedure used for the scalar fluxes, the differential Eq. (8.179) is transformed to the linear algebraic equation:

$$a^g_0 g^{n+1}_0 - \sum_{\beta=1}^{6} a^g_\beta g^{n+1}_\beta = b^g_0, \tag{8.180}$$

where the coefficients a^g_0 and a^g_β have the same form of the coefficients a^Φ_0 and a^Φ_β, given by Eqs. (8.162a–8.162g). The right-hand side of Eq. (8.180) is:

$$\begin{aligned} b^g_0 &= \frac{V_{f0}(\rho g)^n_0}{\Delta t} + V_{f0}\rho^{n+1}_0 \langle s_g \rangle^{n+1} \\ &= \frac{V_{f0}(\rho g)^{n+1}_0}{\Delta t} + V_{f0}\rho^{n+1}_0 \left[-\langle u_j \phi \rangle \frac{\partial T}{\partial x_j} - \left(\frac{\varepsilon}{K}\right)^{n+1} \frac{g^{n+1}}{R} \right]. \end{aligned} \tag{8.181}$$

8.4 Transport Equations for Reynolds Stress Model (RSM)

I. Modelling of Transport Equations for Reynolds Stresses

The transport equations for the Reynolds stresses have been derived by Chou (1954) in the form:

$$\frac{\partial\langle u_i u_j\rangle}{\partial t} + U_\ell \frac{\partial\langle u_i u_j\rangle}{\partial x_\ell} = -\frac{\partial\langle u_\ell u_i u_j\rangle}{\partial x_\ell} - \frac{1}{\rho}\left(\frac{\partial\langle u_j p\rangle}{\partial x_i} + \frac{\partial\langle u_i p\rangle}{\partial x_j}\right) + P_{ij} + G_{ij} + \pi_{ij} - \varepsilon_{ij}, \tag{8.182}$$

with:

$$P_{ij} = -\langle u_i u_\ell\rangle \frac{\partial U_j}{\partial x_\ell} - \langle u_j u_\ell\rangle \frac{\partial U_i}{\partial x_\ell}, \tag{8.183}$$

$$G_{ij} = -\beta g_i \langle u_j \phi\rangle - \beta g_j \langle u_i \phi\rangle, \tag{8.184}$$

$$\pi_{ij} = \left\langle \frac{p}{\rho}\left(\frac{\partial u_i}{\partial x_j} + \frac{\partial u_j}{\partial x_i}\right)\right\rangle, \tag{8.185}$$

$$\varepsilon_{ij} = 2\nu \left\langle \frac{\partial u_i}{\partial x_\ell}\frac{\partial u_j}{\partial x_\ell}\right\rangle. \tag{8.186}$$

The first two terms at the right-hand side of Eq. (8.182) are diffusion terms: the first depends only on velocity fluctuations, while the second depends on velocity-pressure correlations. P_{ij} is the stress production due to the mean flow; G_{ij} is the stress production due to buoyancy forces; π_{ij} is the pressure strain tensor; and ε_{ij} is the viscous dissipation tensor.

The diffusion terms are modelled by a gradient law:

$$\begin{aligned} Diff\langle u_i u_j\rangle &= -\frac{\partial\langle u_\ell u_i u_j\rangle}{\partial x_\ell} - \frac{1}{\rho}\left(\frac{\partial\langle u_j p\rangle}{\partial x_i} + \frac{\partial\langle u_i p\rangle}{\partial x_j}\right) \\ &= \frac{\partial}{\partial x_\ell}\left[\nu_\ell \frac{\langle u_i u_j\rangle}{\partial x_\ell} + c_s \frac{\partial}{\partial x_\ell}\left(\frac{K}{\varepsilon}\langle u_k u_\ell\rangle \frac{\partial\langle u_i u_j\rangle}{\partial x_k}\right)\right] \\ &= \frac{\partial}{\partial x_\ell}\left[\nu_\ell \frac{\langle u_i u_j\rangle}{\partial x_\ell} + c_s \frac{K^2}{\varepsilon}\frac{\partial\langle u_i u_j\rangle}{\partial x_l}\right]. \end{aligned} \tag{8.187}$$

The pressure strain tensor is modelled as the sum of three contributions:

$$\pi_{ij} = \left\langle \frac{p}{\rho}\left(\frac{\partial u_i}{\partial x_j} + \frac{\partial u_j}{\partial x_i}\right)\right\rangle = \pi^1_{ij} + \pi^2_{ij} + \pi^3_{ij}, \tag{8.188}$$

with:

$$\pi^1_{ij} = -c_1 \frac{\varepsilon}{K}\left(\langle u_i u_j\rangle - \frac{2}{3}K\delta_{ij}\right) = -c_1 \varepsilon a_{ij}, \tag{8.189}$$

$$\pi^2_{ij} = -c_2\left(P_{ij} - \frac{2}{3}P\delta_{ij}\right), \tag{8.190}$$

$$\pi^3_{ij} = -c_3\left(G_{ij} - \frac{2}{3}G\delta_{ij}\right). \tag{8.191}$$

In Eq. (8.189), a_{ij} is the Reynolds stress anisotropy tensor, which was introduced in Sect. 8.1.4 and is repeated here for the sake of clarity:

$$a_{ij} = \frac{1}{K}\left(\langle u_i u_j\rangle - \frac{2}{3}K\delta_{ij}\right). \tag{8.192}$$

Furthermore, we define:

$$P = -\langle u_i u_j\rangle \frac{\partial U_i}{\partial x_j} = \frac{1}{2}P_{ii}, \tag{8.193}$$

$$G = -\beta g_i \langle u_i \phi\rangle = \frac{1}{2}G_{ii}. \tag{8.194}$$

As remarked in Sect. 8.1.4, it holds always:

$$a_{ii} = a_{11} + a_{22} + a_{33} = 0, \tag{8.195}$$

but in case of isotropic turbulence, one has $u_\alpha u_\alpha = (2/3)K)$, $(\alpha = 1,2,3)$ (summation not implied) and therefore $a_{\alpha\alpha} = 0$, $(\alpha = 1,2,3)$. Thus also $\pi^1_{\alpha\alpha} = 0$ in isotropic turbulence. In the general case, the components of the tensor π^1_{ij} are a measure of the anisotropy of turbulence.

Similarly, the second contribution π^2_{ij} is a measure of the anisotropy of the production of the stress $\langle u_i u_j\rangle$ due to the mean flow. In the case of isotropic turbulence, $P_{\alpha\alpha} = (2/3)P$, and therefore $\pi^2_{\alpha\alpha} = 0$. The third contribution is a measure of the anisotropy of stress production due to buoyancy forces. In the case of isotropic turbulence, $G_{\alpha\alpha} = (2/3)G$; hence $\pi^3_{\alpha\alpha} = 0$.

Assuming that there is local isotropy at high wave numbers where dissipation occurs, the viscous dissipation tensor is modelled by:

$$\varepsilon_{ij} = 2\nu\left\langle \frac{\partial u_i}{\partial x_\ell}\frac{\partial u_j}{\partial x_\ell}\right\rangle = \frac{2}{3}\varepsilon\delta_{ij}, \tag{8.196}$$

with:

$$\varepsilon = \nu\left\langle \frac{\partial u_n}{\partial x_m}\frac{\partial u_n}{\partial x_m}\right\rangle. \tag{8.197}$$

In the computational praxis ε is obtained by solving the ε-equation, as explained in Sect. 8.5.

Inserting Eqs. (8.187), (8.189), (8.190), (8.191), and (8.196) into Eq. (8.182), one obtains the modelled transport equation for Reynolds stresses:

$$\begin{aligned}\frac{\partial\langle u_iu_j\rangle}{\partial t} + U_\ell\frac{\partial\langle u_iu_j\rangle}{\partial x_\ell} = {} & \frac{\partial}{\partial x_\ell}\left[\nu_\ell\frac{\langle u_iu_j\rangle}{\partial x_\ell} + c_s\frac{K^2}{\varepsilon}\frac{\partial\langle u_iu_j\rangle}{\partial x_l}\right] + P_{ij} + G_{ij}\\ & - c_1\frac{\varepsilon}{K}\left(\langle u_iu_j\rangle - \frac{2}{3}K\delta_{ij}\right) - c_2\frac{\varepsilon}{K}\left(P_{ij} - \frac{2}{3}P\delta_{ij}\right)\\ & - c_3\frac{\varepsilon}{K}\left(G_{ij} - \frac{2}{3}G\delta_{ij}\right) - \frac{2}{3}\varepsilon\delta_{ij}.\end{aligned} \tag{8.198}$$

Using Eqs. (8.183), 8.184), (8.193), and (8.194) and rearranging, one obtains:

$$\begin{aligned}\frac{\partial\langle u_iu_j\rangle}{\partial t} + U_\ell\frac{\partial\langle u_iu_j\rangle}{\partial x_\ell} = {} & \frac{\partial}{\partial x_\ell}\left[\nu_\ell\frac{\partial\langle u_iu_j\rangle}{\partial x_\ell} + c_s\frac{K^2}{\varepsilon}\frac{\partial\langle u_iu_j\rangle}{\partial x_l}\right]\\ & -(1-c_2)\left(\langle u_iu_\ell\rangle\frac{\partial U_j}{x_\ell} + \langle u_ju_\ell\rangle\frac{\partial U_i}{x_\ell}\right)\\ & -(1-c_3)\left(\beta g_i\langle u_j\phi\rangle + \beta g_j\langle u_i\phi\rangle\right)\\ & -c_1\frac{\varepsilon}{K}\left(\langle u_iu_j\rangle - \frac{2}{3}K\delta_{ij}\right) - c_2\frac{2}{3}\langle u_nu_m\rangle\frac{\partial U_n}{\partial x_m}\delta_{ij}\\ & -c_3\frac{2}{3}\beta g_\ell\langle u_\ell\phi\rangle\delta_{ij} - \frac{2}{3}\varepsilon\delta_{ij}.\end{aligned} \tag{8.199}$$

The numerical values of the constants are $c_s = 0.09$, $c_1 = 2.8$, $c_2 = 0.47$, and $c_3 = 0.47$.

Summing up the three Eq. (8.198) for $i = j$, one obtains the transport equation for $K = \langle u_iu_j\rangle/2$:

$$\frac{\partial K}{\partial t} + U_j\frac{\partial K}{\partial x_j} = \frac{\partial}{\partial x_j}\left[\nu_\ell\frac{\partial K}{\partial x_j} + \nu_t\frac{\partial K}{\partial x_j}\right] - \langle u_iu_j\rangle\frac{\partial U_i}{\partial x_j} - \beta g_i\langle u_j\phi\rangle - \varepsilon, \tag{8.200}$$

where [Cfr. Eq. (8.126)]:

$$\nu_t = c_s \frac{K^2}{\varepsilon} \tag{8.201}$$

is the turbulent diffusion coefficient.

II. **Numerical Treatment of Transport Equations for Reynolds Stresses**

Equation (8.199) can be written in the compact form:

$$\frac{\partial \Phi_{ij}}{\partial t} + \frac{\partial J_\ell(\Phi_{ij})}{\partial x_\ell} = S_{ij}^{\Phi}, \tag{8.202}$$

with the following definitions:

$$\Phi_{ij} = \langle u_i u_j \rangle, \tag{8.203}$$

$$J_\ell(\Phi_{ij}) = U_\ell \Phi_{ij} - \Gamma_{\Phi_{ij}} \frac{\partial \Phi_{ij}}{\partial x_\ell}, \tag{8.204}$$

$$\Gamma_{\Phi_{ij}} = \nu_\ell + c_s \frac{K^2}{\varepsilon} = \nu_\ell + \nu_t, \tag{8.205}$$

$$\begin{aligned} S_{ij}^{\Phi} = & -(1-c_2)\left(\langle u_i u_\ell \rangle \frac{\partial U_j}{x_\ell} + \langle u_j u_\ell \rangle \frac{\partial U_i}{x_\ell}\right) - (1-c_3) \\ & \times \left(\beta g_i \langle u_j \phi \rangle + \beta g_j \langle u_i \phi \rangle\right) - c_1 \frac{\varepsilon}{K}\left(\langle u_i u_j \rangle - \frac{2}{3} K \delta_{ij}\right) - c_2 \frac{2}{3} \langle u_n u_m \rangle \\ & \times \frac{\partial U_n}{\partial x_m} \delta_{ij} - c_3 \frac{2}{3} \beta g_\ell \langle u_\ell \phi \rangle \delta_{ij} - \frac{2}{3} \varepsilon \delta_{ij}, \end{aligned} \tag{8.206}$$

Equation (8.202) is formally identical with Eq. (8.150) and can be treated likewise to obtain a linear algebraic equation, again a Poisson-like equation, of the form:

$$a_0^{\Phi_{ij}} \left(\Phi_{ij}\right)_0^{n+1} - \sum_{\beta=1}^{6} a_\beta^{\Phi_{ij}} \left(\Phi_{ij}\right)_\beta^{n+1} = b_0^{\Phi_{ij}}, \tag{8.207}$$

with:

$$\begin{aligned} a_0^{\Phi_{ij}} = & \frac{\left(V_f \rho^{n+1}\right)_0}{\Delta t} + c_1 \frac{2}{3} \left(V_f \rho^{n+1}\right)_0 + \left[0, F_{i+1/2}\right] + \left[0, F_{j+1/2}\right] \\ & + \left[0, F_{k+1/2}\right] + \left[0, -F_{i-1/2}\right] + \left[0, -F_{j-1/2}\right] + \left[0, -F_{k-1/2}\right] \\ & + D_{i+1/2}^{\Phi_{ij}} + D_{i-1/2}^{\Phi_{ij}} + D_{j+1/2}^{\Phi_{ij}} + D_{j-1/2}^{\Phi_{ij}} + D_{k+1/2}^{\Phi_{ij}} + D_{k-1/2}^{\Phi_{ij}}, \end{aligned} \tag{8.208a}$$

$$a_1^{\Phi_{ij}} = [0, F_{i-1/2}] + D_{i-1/2}^{\Phi_{ij}}, \quad (8.208b)$$

$$a_2^{\Phi_{ij}} = [0, -F_{i+1/2}] + D_{i+1/2}^{\Phi_{ij}}, \quad (8.208c)$$

$$a_3^{\Phi_{ij}} = [0, F_{j-1/2}] + D_{j-1/2}^{\Phi_{ij}}, \quad (8.208d)$$

$$a_4^{\Phi_{ij}} = [0, -F_{j+1/2}] + D_{j+1/2}^{\Phi_{ij}}, \quad (8.208e)$$

$$a_5^{\Phi_{ij}} = [0, F_{k-1/2}] + D_{k-1/2}^{\Phi_{ij}}, \quad (8.208f)$$

$$a_6^{\Phi_{ij}} = [0, -F_{k+1/2}] + D_{k+1/2}^{\Phi_{ij}}, \quad (8.208g)$$

$$b_0^{\Phi_{ij}} = V_{f0} \frac{(\rho \Phi_{ij})_0^n}{\Delta t} + V_{f0} \rho_0^{n+1} \left\langle S_{ij}^R \right\rangle_0^{n+1}. \quad (8.208h)$$

The diffusion coefficients are given by equations formally identical to Eq. (8.159). The Poisson-like Eq. (8.207) can be solved numerically with the methods explained in Chap. 7.

8.5 Transport Equation for Dissipation of Turbulent Kinetic Energy [ε Equation]

I. Modelling the Transport Equation for Dissipation of Turbulent Kinetic Energy

The transport equation which describes the space and time distribution of the dissipation of the turbulent kinetic energy ε is:

$$\begin{aligned} \frac{\partial(\rho\varepsilon)}{\partial t} + \frac{\partial(\rho U_j \varepsilon)}{\partial x_j} &= c_{1\varepsilon} \frac{\varepsilon}{K} (P_k + G_k)(1 + c_{3\varepsilon} R_f) - c_{2\varepsilon} \rho \frac{\varepsilon^2}{K} \\ &+ \frac{\partial}{\partial x_j} \left[\mu_\ell \frac{\partial \varepsilon}{\partial x_j} + c_\varepsilon \rho \frac{K}{\varepsilon} \langle u_j u_m \rangle \frac{\partial \varepsilon}{\partial x_m} \right]. \end{aligned} \quad (8.209)$$

In Eq. (8.209):

$$R_f = -\frac{G_k}{P_k} \tag{8.210}$$

is the Richardson number. k as subscript refers to the production of turbulent kinetic energy. The numerical values of the constants are $c_{1\varepsilon} = 1.44$, $c_{2\varepsilon} = 1.92$, $c_{3\varepsilon} = 0.8$, and $c_\varepsilon = 0.15$.

Using the definition of the source term:

$$S_\varepsilon = c_{1\varepsilon}\frac{\varepsilon}{K}(P_k + G_k)\left(1 + c_{3\varepsilon}R_f\right) - c_{2\varepsilon}\rho\frac{\varepsilon^2}{K}, \tag{8.211}$$

Equation (8.209) can be re-written as:

$$\frac{\partial(\rho\varepsilon)}{\partial t} + \frac{\partial}{\partial x_j}\left[\rho U_j\varepsilon - \mu_\ell\frac{\partial\varepsilon}{\partial x_j} - c_\varepsilon\rho\frac{K}{\varepsilon}\langle u_j u_m\rangle\frac{\partial\varepsilon}{\partial x_m}\right] = S_\varepsilon. \tag{8.212}$$

We define the convective, diffusive laminar, diffusive turbulent, and total (convective plus diffusive) fluxes by:

$$J^c_{\varepsilon j} = \rho U_j\varepsilon, \tag{8.213}$$

$$J^{d\ell}_{\varepsilon j} = -\mu_\ell\frac{\partial\varepsilon}{\partial x_j}, \tag{8.214}$$

$$J^{dt}_{\varepsilon j} = c_\varepsilon\rho\frac{K}{\varepsilon}\langle u_j u_m\rangle\frac{\partial\varepsilon}{\partial x_m} = -\Gamma_{\varepsilon jm}\frac{\partial\varepsilon}{\partial x_m}, \tag{8.215}$$

$$J_{\varepsilon j} = J^c_{\varepsilon j} + J^{d\ell}_{\varepsilon j} + J^{dt}_{\varepsilon j} = \rho U_j\varepsilon - \mu_\ell\frac{\partial\varepsilon}{\partial x_j} - \Gamma_{\varepsilon jm}\frac{\partial\varepsilon}{\partial x_m}, \tag{8.216}$$

with:

$$\Gamma_{\varepsilon jm} = c_\varepsilon\rho\frac{K}{\varepsilon}\langle u_j u_m\rangle. \tag{8.217}$$

With these definitions, Eq. (8.212) can be written in the compact form:

$$\frac{\partial(\rho\varepsilon)}{\partial t} + \frac{\partial J_{\varepsilon j}}{\partial x_j} = S_\varepsilon. \tag{8.218}$$

Integration of Eq. (8.218) over the volume V_f of the fluid in a computational cell and application of the Gauss theorem yields:

$$\int_{V_f} \frac{\partial(\rho\varepsilon)}{\partial t} dV + \int_{A_{f,i+1/2}} J_{\varepsilon i} dA - \int_{A_{f,i-1/2}} J_{\varepsilon i} dA + \int_{A_{f,j+1/2}} J_{\varepsilon i} dA - \int_{A_{f,j-1/2}} J_{\varepsilon i} dA$$
$$+ \int_{A_{f,k+1/2}} J_{\varepsilon i} dA - \int_{A_{f,k-+1/2}} J_{\varepsilon i} dA = \int_{V_f} S_\varepsilon dV \tag{8.219}$$

II. **Numerical Treatment of the Transport Equation for Dissipation of Turbulent Kinetic Energy**

In this section, once more full details are given about the derivation of a Poisson-like equation from Eq. (8.219), which is discretized as follows:

$$\begin{aligned}&\frac{V_{f0}}{\Delta t}\left[(\rho\varepsilon)_0^{n+1} - (\rho\varepsilon)_0^n\right] + \left\langle J_\varepsilon^c A_f\right\rangle_{\alpha+1/2}^{n+1} - \left\langle J_\varepsilon^c A_f\right\rangle_{\alpha-1/2}^{n+1}\\ &+ \left\langle (J_\varepsilon^{dl} + J^{dl})A_f\right\rangle_{\alpha+1/2}^{n+1} - \left\langle (J_\varepsilon^{dt} + J_\varepsilon^{dt})A_f\right\rangle_{\alpha-1/2}^{n+1}\\ &= V_f\left\langle S_\varepsilon^{n+1}\right\rangle_0,\end{aligned} \tag{8.220}$$

where summation for $\alpha = i, j, k$ is implied.

The convective terms are computed by:

$$\begin{aligned}\left\langle J_\varepsilon^c A_f\right\rangle_{\alpha+1/2}^{n+1} &= \left\langle \rho U \varepsilon A_f\right\rangle_{\alpha+1/2}^{n+1} = \left\langle F\varepsilon\right\rangle_{\alpha+1/2}^{n+1}\\ &= \left[0, F_{\alpha+1/2}\right]\varepsilon_\alpha^{n+1} - \left[0, -F_{\alpha+1/2}\right]\varepsilon_{\alpha+1}^{n+1},\end{aligned} \tag{8.221a}$$

$$\left\langle J_\varepsilon^c A_f\right\rangle_{\alpha-1/2}^{n+1} = \left[0, F_{\alpha-1/2}\right]\varepsilon_{\alpha-1}^{n+1} - \left[0, -F_{\alpha-1/2}\right]\varepsilon_\alpha^{n+1}. \tag{8.221b}$$

Letting $Z = c_\varepsilon \rho K/\varepsilon$, diffusive terms are computed by:

$$\begin{aligned}\left\langle (J_\varepsilon^{d\ell} + J_\varepsilon^{dt})A_f\right\rangle_{i+1/2} &= \left\langle \left(-\mu_\ell \frac{\partial\varepsilon}{\partial x_j} - \Gamma_{\varepsilon jm}\frac{\partial\varepsilon}{\partial x_m}\right)A_f\right\rangle_{i+1/2}\\ &= -A_{f,i+1/2}\left[(Z\langle u_1u_1\rangle + \mu_\ell)\frac{\partial\varepsilon}{\partial x_1} + (Z\langle u_1u_2\rangle + \mu_\ell)\frac{\partial\varepsilon}{\partial x_2} + (Z\langle u_1u_3\rangle + \mu_\ell)\frac{\partial\varepsilon}{\partial x_3}\right]_{i+1/2},\end{aligned} \tag{8.222a}$$

$$\begin{aligned}&\left\langle\left(J_\varepsilon^{d\ell}+J_\varepsilon^{dt}\right)A_f\right\rangle_{j+1/2}\\&=-A_{f,j+1/2}\left[\left(Z\langle u_2u_1\rangle+\mu_\ell\right)\frac{\partial\varepsilon}{\partial x_1}+\left(Z\langle u_2u_2\rangle+\mu_\ell\right)\frac{\partial\varepsilon}{\partial x_2}+\left(Z\langle u_2u_3\rangle+\mu_\ell\right)\frac{\partial\varepsilon}{\partial x_3}\right]_{j+1/2},\end{aligned}\tag{8.222b}$$

$$\begin{aligned}&\left\langle\left(J_\varepsilon^{d\ell}+J_\varepsilon^{dt}\right)A_f\right\rangle_{k+1/2}\\&=-A_{f,k+1/2}\left[\left(Z\langle u_3u_1\rangle+\mu_\ell\right)\frac{\partial\varepsilon}{\partial x_1}+\left(Z\langle u_3u_2\rangle+\mu_\ell\right)\frac{\partial\varepsilon}{\partial x_2}+\left(Z\langle u_3u_3\rangle+\mu_\ell\right)\frac{\partial\varepsilon}{\partial x_3}\right]_{k+1/2}\end{aligned}\tag{8.222c}$$

and similarly for $i-1/2$, $j-1/2$, and $k-1/2$.

Inserting Eqs. (8.221) and (8.222) into Eq. (8.220) and replacing x, y, and z for x_i $(i=1,2,3)$, one obtains:

$$\begin{aligned}&\frac{V_{f0}}{\Delta t}\left[(\rho\varepsilon)_0^{n+1}-(\rho\varepsilon)_0^n\right]+\left[0,F_{i+1/2}\right]\varepsilon_{ijk}^{n+1}-\left[0,-F_{i+1/2}\right]\varepsilon_{i+1}^{n+1}-\left[0,F_{i-1/2}\right]\varepsilon_{i-1}^{n+1}+\left[0,-F_{i-1/2}\right]\varepsilon_{ijk}^{n+1}\\
&+\left[0,F_{j+1/2}\right]\varepsilon_{ijk}^{n+1}-\left[0,-F_{j+1/2}\right]\varepsilon_{j+1}^{n+1}-\left[0,F_{j-1/2}\right]\varepsilon_{j-1}^{n+1}+\left[0,-F_{j-1/2}\right]\varepsilon_{ijk}^{n+1}\\
&+\left[0,F_{k+1/2}\right]\varepsilon_{ijk}^{n+1}-\left[0,-F_{k+1/2}\right]\varepsilon_{k+1}^{n+1}-\left[0,F_{k-1/2}\right]\varepsilon_{k-1}^{n+1}+\left[0,-F_{k-1/2}\right]\varepsilon_{ijk}^{n+1}=V_{f0}\langle S_\varepsilon\rangle_{ijk}^{n+1}\\
&+\frac{A_{f,i+1/2}}{\Delta x_{i+1/2}}\left(Z\langle u_1u_1\rangle+\mu_\ell\right)_{i+1/2}\left(\varepsilon_{i+1}^{n+1}-\varepsilon_i^{n+1}\right)+A_{f,i+1/2}\left(Z\langle u_1u_2\rangle\frac{\partial\varepsilon}{\partial y}+Z\langle u_1u_3\rangle\frac{\partial\varepsilon}{\partial z}\right)_{i+1/2}\\
&-\frac{A_{f,i-1/2}}{\Delta x_{i-1/2}}\left(Z\langle u_1u_1\rangle+\mu_\ell\right)_{i-1/2}\left(\varepsilon_i^{n+1}-\varepsilon_{i-1}^{n+1}\right)+A_{f,i-1/2}\left(Z\langle u_1u_2\rangle\frac{\partial\varepsilon}{\partial y}+Z\langle u_1u_3\rangle\frac{\partial\varepsilon}{\partial z}\right)_{i-1/2}\\
&+\frac{A_{f,j+1/2}}{\Delta y_{j+1/2}}\left(Z\langle u_2u_2\rangle+\mu_\ell\right)_{j+1/2}\left(\varepsilon_{j+1}^{n+1}-\varepsilon_j^{n+1}\right)+A_{f,j+1/2}\left(Z\langle u_2u_1\rangle\frac{\partial\varepsilon}{\partial x}+Z\langle u_2u_3\rangle\frac{\partial\varepsilon}{\partial z}\right)_{j+1/2}\\
&-\frac{A_{f,j-1/2}}{\Delta y_{j-1/2}}\left(Z\langle u_2u_2\rangle+\mu_\ell\right)_{j-1/2}\left(\varepsilon_j^{n+1}-\varepsilon_{j-1}^{n+1}\right)+A_{f,j-1/2}\left(Z\langle u_2u_1\rangle\frac{\partial\varepsilon}{\partial x}+Z\langle u_2u_3\rangle\frac{\partial\varepsilon}{\partial z}\right)_{j-1/2}\\
&+\frac{A_{f,k+1/2}}{\Delta z_{k+1/2}}\left(Z\langle u_3u_3\rangle+\mu_\ell\right)_{k+1/2}\left(\varepsilon_{k+1}^{n+1}-\varepsilon_k^{n+1}\right)+A_{f,k+1/2}\left(Z\langle u_3u_1\rangle\frac{\partial\varepsilon}{\partial x}+Z\langle u_3u_2\rangle\frac{\partial\varepsilon}{\partial y}\right)_{k+1/2}\\
&-\frac{A_{f,k-1/2}}{\Delta z_{k-1/2}}\left(Z\langle u_3u_3\rangle+\mu_\ell\right)_{k-1/2}\left(\varepsilon_k^{n+1}-\varepsilon_{k-1}^{n+1}\right)+A_{f,k-1/2}\left(Z\langle u_3u_1\rangle\frac{\partial\varepsilon}{\partial x}+Z\langle u_3u_2\rangle\frac{\partial\varepsilon}{\partial y}\right)_{k-1/2}.\end{aligned}\tag{8.223}$$

Using the definitions:

$$D^{\varepsilon}_{i+1/2} = \frac{A_{f,i+1/2}}{\Delta x_{i+1/2}} \left(Z\langle u_1 u_1 \rangle + \mu_\ell\right)_{i+1/2}, \tag{8.224a}$$

$$D^{\varepsilon}_{\ j+1/2} = \frac{A_{f,j+1/2}}{\Delta y_{\ j+1/2}} \left(Z\langle u_2 u_2 \rangle + \mu_\ell\right)_{\ j+1/2}, \tag{8.224b}$$

$$D^{\varepsilon}_{k+1/2} = \frac{A_{f,k+1/2}}{\Delta z_{k+1/2}} \left(Z\langle u_3 u_3 \rangle + \mu_\ell\right)_{k+1/2}, \tag{8.224c}$$

and similar ones for $i - 1/2$, $j - 1/2$, and $k - 1/2$, and rearranging, one obtains:

$$\begin{aligned}
&\varepsilon_0^{n+1}\left\{\frac{V_{f}\rho_0}{\Delta t}+\left[0,F_{i+1/2}\right]+\left[0,F_{\ j+1/2}\right]+\left[0,F_{k+1/2}\right]+\left[0,-F_{i-1/2}\right]+\left[0,-F_{\ j-1/2}\right]+\left[0,-F_{k-1/2}\right]\right.\\
&\left.+D^{\varepsilon}_{i+1/2}+D^{\varepsilon}_{\ j+1/2}+D^{\varepsilon}_{k+1/2}+D^{\varepsilon}_{i-1/2}+D^{\varepsilon}_{\ j-1/2}+D^{\varepsilon}_{k-1/2}\right\}-\varepsilon_1^{n+1}\left\{\left[0,F_{i-1/2}\right]+D^{\varepsilon}_{i-1/2}\right\}\\
&-\varepsilon_2^{n+1}\left\{\left[0,-F_{i+1/2}\right]+D^{\varepsilon}_{i+1/2}\right\}-\varepsilon_3^{n+1}\left\{\left[0,F_{\ j-1/2}\right]+D^{\varepsilon}_{\ j-1/2}\right\}-\varepsilon_4^{n+1}\left\{\left[0,-F_{\ j+1/2}\right]+D^{\varepsilon}_{\ j+1/2}\right\}\\
&-\varepsilon_5^{n+1}\left\{\left[0,F_{k-1/2}\right]+D^{\varepsilon}_{k-1/2}\right\}-\varepsilon_6^{n+1}\left\{\left[0,-F_{k+1/2}\right]+D^{\varepsilon}_{k+1/2}\right\}=\frac{V_{f0}}{\Delta t}(\rho\varepsilon)_0^n+V_{f0}\langle S_\varepsilon\rangle_0^{n+1}\\
&+A_{f,i+1/2}\left(Z\langle u_1u_1\rangle\frac{\partial\varepsilon}{\partial y}+Z\langle u_1u_3\rangle\frac{\partial\varepsilon}{\partial z}\right)_{i+1/2}-A_{f,i-1/2}\left(Z\langle u_1u_2\rangle\frac{\partial\varepsilon}{\partial y}+Z\langle u_1u_3\rangle\frac{\partial\varepsilon}{\partial z}\right)_{i-1/2}\\
&+A_{f,j+1/2}\left(Z\langle u_2u_1\rangle\frac{\partial\varepsilon}{\partial x}+Z\langle u_2u_3\rangle\frac{\partial\varepsilon}{\partial z}\right)_{\ j+1/2}-A_{f,j-1/2}\left(Z\langle u_2u_1\rangle\frac{\partial\varepsilon}{\partial x}+Z\langle u_2u_3\rangle\frac{\partial\varepsilon}{\partial z}\right)_{\ j-1/2}\\
&+A_{f,k+1/2}\left(Z\langle u_3u_1\rangle\frac{\partial\varepsilon}{\partial x}+Z\langle u_3u_2\rangle\frac{\partial\varepsilon}{\partial y}\right)_{k+1/2}-A_{f,k-1/2}\left(Z\langle u_3u_1\rangle\frac{\partial\varepsilon}{\partial x}+Z\langle u_3u_2\rangle\frac{\partial\varepsilon}{\partial y}\right)_{k-1/2}.
\end{aligned} \tag{8.225}$$

Equation (8.225) can be written in the compact form of a Poisson-like equation:

$$a_0^{\varepsilon}\varepsilon_0^{n+1} - \sum_{\beta=1}^{6} a_\beta^{\varepsilon}\varepsilon_\beta^{n+1} = b_0^{\varepsilon}, \tag{8.226}$$

with the definition of the coefficients:

$$a_0^\varepsilon = \frac{V_f \rho_0^{n+1}}{\Delta t} + [0, F_{i+1/2}] + [0, F_{j+1/2}] + [0, F_{k+1/2}] + [0, -F_{i-1/2}]$$
$$+ [0, -F_{j-1/2}] + [0, -F_{k-1/2}] + D^\varepsilon_{i+1/2} + D^\varepsilon_{i-1/2} + D^\varepsilon_{j+1/2}$$
$$+ D^\varepsilon_{j-1/2} + D^\varepsilon_{k+1/2} + D^\varepsilon_{k-1/2}, \tag{8.227a}$$

$$a_1^\varepsilon = [0, F_{i-1/2}] + D^\varepsilon_{i-1/2}, \tag{8.227b}$$

$$a_2^\varepsilon = [0, -F_{i+1/2}] + D^\varepsilon_{i+1/2}, \tag{8.227c}$$

$$a_3^\varepsilon = [0, F_{j-1/2}] + D^\varepsilon_{j-1/2}, \tag{8.227d}$$

$$a_4^\varepsilon = [0, -F_{j+1/2}] + D^\varepsilon_{j+1/2}, \tag{8.227e}$$

$$a_5^\varepsilon = [0, F_{k-1/2}] + D^\varepsilon_{k-1/2}, \tag{8.227f}$$

$$a_6^\varepsilon = [0, -F_{k+1/2}] + D^\varepsilon_{k+1/2}, \tag{8.227g}$$

$$b_0^\varepsilon = \frac{V_{f0}}{\Delta t}(\rho\varepsilon)_0^n + V_{f0}\langle S_\varepsilon\rangle_0^{n+1} + A_{f,i+1/2}\left(Z\langle u_1u_1\rangle\frac{\partial\varepsilon}{\partial y} + Z\langle u_1u_3\rangle\frac{\partial\varepsilon}{\partial z}\right)_{i+1/2}$$
$$-A_{f,i-1/2}\left(Z\langle u_1u_2\rangle\frac{\partial\varepsilon}{\partial y} + Z\langle u_1u_3\rangle\frac{\partial\varepsilon}{\partial z}\right)_{i-1/2} + A_{f,j+1/2}\left(Z\langle u_2u_1\rangle\frac{\partial\varepsilon}{\partial x} + Z\langle u_2u_3\rangle\frac{\partial\varepsilon}{\partial z}\right)_{j+1/2}$$
$$-A_{f,j-1/2}\left(Z\langle u_2u_1\rangle\frac{\partial\varepsilon}{\partial x} + Z\langle u_2u_3\rangle\frac{\partial\varepsilon}{\partial z}\right)_{j-1/2} + A_{f,k+1/2}\left(Z\langle u_3u_1\rangle\frac{\partial\varepsilon}{\partial x} + Z\langle u_3u_2\rangle\frac{\partial\varepsilon}{\partial y}\right)_{k+1/2}$$
$$-A_{f,k-1/2}\left(Z\langle u_3u_1\rangle\frac{\partial\varepsilon}{\partial x} + Z\langle u_3u_2\rangle\frac{\partial\varepsilon}{\partial y}\right)_{k-1/2}. \tag{8.227h}$$

8.6 Synopsis of Governing Equations, Constants, Transport Quantities and Source Terms

The 11 transport equations used for the RSM are summarized in Table 8.2.

The constants used in the RSM are summarized in Table 8.3:

All transport equations have been transformed to the generalized form:

$$\frac{\partial \Phi}{\partial t}+\frac{\partial J_{\Phi i}}{\partial x_i}=S_{\Phi}, \tag{8.228}$$

with proper definitions of the transported quantity Φ, of the flux J_{Φ}, and of the source term S_{Φ}. These definitions are summarized in Table 8.4:

8.7 Reynolds Equations for Homogeneous Turbulence

I. General Reynolds Equations of Homogeneous Turbulence

Although ideal gases do not exist in nature, the theory of ideal gases is of paramount importance for understanding the behavior of real gases and their deviations from the ideal case. Similarly, the theory of homogeneous turbulence, which does not occur in nature, is important for understanding real turbulence.

The conditions for the realization of homogeneity in turbulent flows would require an infinite spatial domain which cannot be realized, nor can be simulated numerically by a finite domain with slip-free boundary conditions. Early experiments on nearly homogeneous turbulence were performed in wind tunnels with an almost unilateral flow with a constant velocity gradient. Turbulence generated by grids at the tunnel entrance becomes almost homogeneous in the center of the tunnel at some distance from the inlet. Among the most accurate and best documented experiments on homogeneous turbulence are those by Champagne et al. (1970).

The conditions which must be realized in a homogeneous turbulent shear flow are as follows. With reference to the sketch of Fig. 8.4, let x_1 be a coordinate along the horizontal axis of the wind tunnel, x_2 the vertical, and x_3 the transverse coordinate. In a homogeneous turbulence, the mean velocity components of the flow satisfy the conditions:

$$U_1=U_1(x_2), \tag{8.229a}$$

$$U_2=U_3=0, \tag{8.229b}$$

$$\partial U_1/\partial x_2=\text{const.}, \tag{8.229c}$$

$$\partial U_i/\partial x_j=0. \qquad (i\neq 1, j\neq 2) \tag{8.229d}$$

Let H be the height of the wind tunnel and $U_c=U_1(H/2)$ be the center velocity.

Table 8.2 Summary of transport equations used for the Reynolds stress model (RSM) of turbulence

1. Scalar fluxes (3 equations)

Rigorous

$$\frac{\partial}{\partial t}\langle u_i\phi\rangle + U_j\frac{\partial}{\partial x_i}\langle u_i\phi\rangle = -\frac{\partial}{\partial x_j}\left(\langle u_iu_j\phi\rangle + \frac{1}{\rho}\langle p\phi\rangle\delta_{ij}\right) - \left(\langle u_iu_j\rangle\frac{\partial T}{\partial x_i} + \langle u_j\phi\rangle\frac{\partial U_i}{\partial x_j}\right) - \beta g_i\langle\phi^2\rangle + \frac{1}{\rho}\left\langle p\frac{\partial\phi}{\partial x_i}\right\rangle \quad (8.119)$$

Modelled

$$\frac{\partial}{\partial t}\langle u_i\phi\rangle + \frac{\partial}{\partial x_j}\left[U_j\langle u_i\phi\rangle - \Gamma_\phi^v\frac{\partial\langle u_i\phi\rangle}{\partial x_j}\right] = P_{i\phi} + G_{i\phi}(1-c_{3\phi}) - c_{1\phi}\frac{\varepsilon}{K}\langle u_i\phi\rangle + c_{2\phi}\langle u_j\phi\rangle\frac{\partial U_i}{\partial x_j} + \pi_{i\phi}^w \quad (8.134)$$

2. Variance of temperature fluctuations [g equation] (1 equation)

Rigorous

$$\frac{\partial\langle\phi^2\rangle}{\partial t} + U_j\frac{\partial\langle\phi^2\rangle}{\partial x_j} = -\frac{\partial\langle u_j\phi^2\rangle}{\partial x_j} + \frac{\partial}{\partial x_j}\left(\frac{\lambda}{\rho c_p}\frac{\partial\langle\phi^2\rangle}{\partial x_j}\right) - 2\langle u_j\phi\rangle\frac{\partial T}{\partial x_j} - 2\frac{\lambda}{\rho c_p}\left\langle\frac{\partial\phi}{\partial x_j}\frac{\partial\phi}{\partial x_j}\right\rangle. \quad (8.163)$$

Modelled

$$\frac{\partial g}{\partial t} + \frac{\partial(U_jg)}{\partial x_j} = \frac{\partial}{\partial x_j}\left(c'_\phi\frac{K^2}{\varepsilon}\frac{\partial g}{\partial x_j} + \frac{\lambda}{\rho c_p}\frac{\partial g}{\partial x_j}\right) - \langle u_j\phi\rangle\frac{\partial T}{\partial x_j} - \frac{\varepsilon}{K}\frac{g}{R}. \quad (8.170)$$

3. Reynolds stresses (6 equations)

Rigorous

$$\frac{\partial\langle u_iu_j\rangle}{\partial t} + U_\ell\frac{\partial\langle u_iu_j\rangle}{\partial x_\ell} = -\frac{\partial\langle u_\ell u_iu_j\rangle}{\partial x_\ell} - \frac{1}{\rho}\left(\frac{\partial\langle u_jp\rangle}{\partial x_i} + \frac{\partial\langle u_ip\rangle}{\partial x_j}\right) + P_{ij} + G_{ij} + \pi_{ij} - \varepsilon_{ij}, \quad (8.182)$$

Modelled

$$\begin{aligned}\frac{\partial\langle u_iu_j\rangle}{\partial t} + U_\ell\frac{\partial\langle u_iu_j\rangle}{\partial x_\ell} &= \frac{\partial}{\partial x_\ell}\left[\nu_\ell\frac{\partial\langle u_iu_j\rangle}{\partial x_\ell} + c_s\frac{K^2}{\varepsilon}\frac{\partial\langle u_iu_j\rangle}{\partial x_l}\right]\\ &-(1-c_2)\left(\langle u_iu_\ell\rangle\frac{\partial U_j}{x_\ell} + \langle u_ju_\ell\rangle\frac{\partial U_i}{x_\ell}\right) - (1-c_3)\left(\beta g_i\langle u_j\phi\rangle + \beta g_j\langle u_i\phi\rangle\right) \quad (8.199)\\ &-c_1\frac{\varepsilon}{K}\left(\langle u_iu_j\rangle - \frac{2}{3}K\delta_{ij}\right) - c_2\frac{2}{3}\langle u_nu_m\rangle\frac{\partial U_n}{\partial x_m}\delta_{ij} - c_3\frac{2}{3}\beta g_\ell\langle u_\ell\phi\rangle\delta_{ij} - \frac{2}{3}\varepsilon\delta_{ij}.\end{aligned}$$

4. Dissipation of turbulent kinetic energy [ε equation] (1 equation)

Rigorous

$$\frac{\partial(\rho\varepsilon)}{\partial t} + \frac{\partial(\rho U_j\varepsilon)}{\partial x_j} = c_{1\varepsilon}\frac{\varepsilon}{K}(P_k + G_k)(1 + c_{3\varepsilon}R_f) - c_{2\varepsilon}\rho\frac{\varepsilon^2}{K} + \frac{\partial}{\partial x_j}\left[\mu_\ell\frac{\partial\varepsilon}{\partial x_j} + c_\varepsilon\rho\frac{K}{\varepsilon}\langle u_ju_m\rangle\frac{\partial\varepsilon}{\partial x_m}\right] \quad (8.209)$$

Modelled

$$\frac{\partial(\rho\varepsilon)}{\partial t} + \frac{\partial}{\partial x_j}\left[\rho U_j\varepsilon - \mu_\ell\frac{\partial\varepsilon}{\partial x_j} - c_\varepsilon\rho\frac{K}{\varepsilon}\langle u_ju_m\rangle\frac{\partial\varepsilon}{\partial x_m}\right] = S_\varepsilon \quad (8.212)$$

Table 8.3 Summary of constants used in the 11 equations of the Reynolds stress model for heated fluids (RSM-HF)

Transport equation	Transported quantity	Constant
Scalar fluxes	$\langle u_i\phi\rangle$	$\sigma_\phi = 1.3$
		$c_\mu = 0.09$
		$c_{s\phi} = c_\mu/\sigma_\phi = 0.07$
		$c_{1\phi} = 3.1$
		$c_{2\phi} = 0.4$
		$c_{3\phi} = 0.5$
		$c'_{1\phi} = 0.5$
Variance of temperature fluctuations	$g = \langle\phi^2\rangle/2$	$c'_\phi = 0.13$
		$R = 0.5$
Reynolds stresses	$\langle u_i u_j\rangle$	$c_s = 0.09$
		$c_1 = 2.8$
		$c_2 = 0.47$
		$c_3 = 0.47$
Dissipation of turbulent kinetic energy	ε	$c_{1\varepsilon} = 1.44$
		$c_{2\varepsilon} = 1.92$
		$c_{3\varepsilon} = 0.8$
		$c_\varepsilon = 0.15$

The kinetic energy of the flow per unit of mass is:

$$E = \frac{1}{2} U_j U_j. \tag{8.230}$$

Under these conditions, the components of the stress production tensor, given by Eq. (8.183) $P_{ij} = -\langle u_i u_\ell\rangle \frac{\partial U_j}{\partial x_\ell} - \langle u_j u_\ell\rangle \frac{\partial U_i}{\partial x_\ell}$, become:

$$P_{11} = -2\langle u_1 u_2\rangle \frac{\partial U_1}{\partial x_2}, \tag{8.231a}$$

$$P_{22} = 0, \tag{8.231b}$$

$$P_{33} = 0, \tag{8.231c}$$

$$P_{12} = -\langle u_2 u_2\rangle \frac{\partial U_1}{\partial x_2}, \tag{8.231d}$$

Table 8.4 Definitions of the transported quantity $\boldsymbol{\Phi}$, flux $J_{\boldsymbol{\Phi}}$, and source term $S_{\boldsymbol{\Phi}}$ in the generalized transport equation $\frac{\partial \Phi}{\partial t} + \frac{\partial J_{\Phi i}}{\partial x_i} = S_{\Phi}$ of the governing equations of the Reynolds stress model for heated fluids (RSM-HF)

Scalar fluxes (3 equations)
Transported quantity $\boldsymbol{\Phi}$: $\boldsymbol{\Phi}_i = \rho\langle u_i\phi\rangle$ (8.140)
Flux $J_{\boldsymbol{\Phi}}$: $\overline{J}_{\phi ji} = J^c_{\phi ji} + J^d_{\phi ji} = U_j\langle u_i\phi\rangle - \Gamma^{\nu}_{\phi}\frac{\partial\langle u_i\phi\rangle}{\partial x_j}$ (8.137)
Source term $S_{\boldsymbol{\Phi}}$: $S_{\phi i} = P_{i\phi} + G_{i\phi}(1 - c_{3\phi}) - c_{1\phi}\frac{\varepsilon}{K}\langle u_i\phi\rangle + c_{2\phi}\langle u_j\phi\rangle\frac{\partial U_i}{\partial x_j} + \pi^w_{i\phi}$ (8.138)
2. Variance of temperature fluctuations (1 equation)
Transported quantity $\boldsymbol{\Phi}$: $\boldsymbol{\Phi}_g = g = \langle\phi^2\rangle/2$ (8.169)
Flux $J_{\boldsymbol{\Phi}}$: $J_{gj} = J^c_{gj} + J^d_{gj} = U_j g - \Gamma_g\frac{\partial g}{\partial x_j} = U_j g - \left(\frac{\lambda}{\rho c_p} + c'_{\phi}\frac{K^2}{\varepsilon}\right)\frac{\partial g}{\partial x_j}$ (8.171), (8.177)
Source term $S_{\boldsymbol{\Phi}}$: $S_g = P_g - \varepsilon_g$ (8.178)
3. Reynolds stresses (6 equations)
Transported quantity $\boldsymbol{\Phi}$: $\boldsymbol{\Phi}_{ij} = \langle u_i u_j\rangle$ (8.203)
Flux $J_{\boldsymbol{\Phi}}$: $J_\ell(\boldsymbol{\Phi}_{ij}) = U_\ell\langle u_i u_j\rangle - \Gamma_{\Phi_{ij}}\frac{\partial\Phi_{ij}}{\partial x_\ell} = U_\ell\langle u_i u_j\rangle - \left(\nu_\ell + c_s\frac{K^2}{\varepsilon}\right)\frac{\partial\langle u_i u_j\rangle}{\partial x_\ell}$ (8.204), (8.205)
Source term $S^{\boldsymbol{\Phi}}_{ij} = -(1 - c_2)\left(\langle u_i u_\ell\rangle\frac{\partial U_j}{x_\ell} + \langle u_j u_\ell\rangle\frac{\partial U_i}{x_\ell}\right) - (1 - c_3)\left(\beta g_i\langle u_j\phi\rangle + \beta g_j\langle u_i\phi\rangle\right) - c_1\frac{\varepsilon}{K}\left(\langle u_i u_j\rangle - \frac{2}{3}K\delta_{ij}\right) - c_2\frac{2}{3}\langle u_n u_m\rangle\frac{\partial U_n}{\partial x_m}\delta_{ij} - c_3\frac{2}{3}\beta g_\ell\langle u_\ell\phi\rangle\delta_{ij} - \frac{2}{3}\varepsilon\delta_{ij}$ (8.206)
4. Dissipation of turbulent kinetic energy (1 equation)
Transported quantity $\boldsymbol{\Phi}$: $\rho\varepsilon$ (8.218)
Flux $J_{\boldsymbol{\Phi}}$: $J_{\varepsilon j} = J^c_{\varepsilon j} + J^{d\ell}_{\varepsilon j} + J^{dt}_{\varepsilon j} = \rho U_j\varepsilon - \mu_\ell\frac{\partial\varepsilon}{\partial x_j} - \Gamma_{\varepsilon jm}\frac{\partial\varepsilon}{\partial x_m} = \rho U_j\varepsilon - \mu_\ell\frac{\partial\varepsilon}{\partial x_j} - c_\varepsilon\rho\frac{K}{\varepsilon}\langle u_j u_m\rangle\frac{\partial\varepsilon}{\partial x_m}$ (8.215)–(8.217)
Source term $S_{\boldsymbol{\Phi}}$: $S_\varepsilon = c_{1\varepsilon}\frac{\varepsilon}{K}(P_k + G_k)(1 + c_{3\varepsilon}R_f) - c_{2\varepsilon}\rho\frac{\varepsilon^2}{K}$ (8.211)

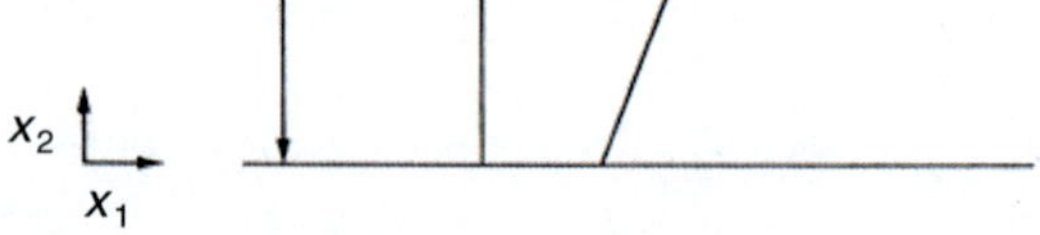

Fig. 8.4 Sketch of homogeneous shear flow

$$P_{13} = -\langle u_2 u_3 \rangle \frac{\partial U_1}{\partial x_2}, \tag{8.231e}$$

$$P_{23} = 0. \tag{8.231f}$$

The turbulent kinetic energy of the flow is:

$$K = \langle u_k u_k \rangle /2 = \langle u_1 u_1 \rangle /2 + \langle u_2 u_2 \rangle /2 + \langle u_3 u_3 \rangle /2 = K_1 + K_2 + K_3, \tag{8.232}$$

with the definition:

$$K_j = \langle u_j u_j \rangle /2, \ (\mathrm{j} = 1, 2, 3). \tag{8.233}$$

The Reynolds stress equations, given in the original form by Eq. (8.182), simplify in homogeneous turbulence. With reference to the above figure, all derivatives with respect to the coordinates x_1 and x_3 vanish because of the uniformity of the flow in these directions. Moreover, the averages of the dyadic products of the velocity fluctuations, and of their interactions with pressure, are not spatial dependent so that their spatially derivatives also vanish. Therefore, taking also in account conditions (8.229), Eq. (8.182) simplify to the following system of equations:

$$\begin{aligned} \frac{\partial \langle u_1 u_1 \rangle}{\partial t} &= P_{11} + \pi_{11} - \varepsilon_{11} \\ &= -2\langle u_1 u_2 \rangle \frac{\partial U_1}{\partial x_2} + 2\left\langle \frac{p}{\rho} \frac{\partial u_1}{\partial x_1} \right\rangle - 2\nu \left\langle \frac{\partial u_1}{\partial x_\ell} \frac{\partial u_1}{\partial x_\ell} \right\rangle, \end{aligned} \tag{8.234a}$$

$$\frac{\partial \langle u_2 u_2 \rangle}{\partial t} = \pi_{22} - \varepsilon_{22} = 2\left\langle \frac{p}{\rho} \frac{\partial u_2}{\partial x_2} \right\rangle - 2\nu \left\langle \frac{\partial u_2}{\partial x_\ell} \frac{\partial u_2}{\partial x_\ell} \right\rangle, \tag{8.234b}$$

$$\frac{\partial \langle u_3 u_3 \rangle}{\partial t} = \pi_{33} - \varepsilon_{33} = 2\left\langle \frac{p}{\rho} \frac{\partial u_3}{\partial x_3} \right\rangle - 2\nu \left\langle \frac{\partial u_3}{\partial x_\ell} \frac{\partial u_3}{\partial x_\ell} \right\rangle, \tag{8.234c}$$

$$\begin{aligned} \frac{\partial \langle u_1 u_2 \rangle}{\partial t} &= P_{12} + \pi_{12} - \varepsilon_{12} \\ &= -\langle u_2 u_2 \rangle \frac{\partial U_1}{\partial x_2} + \left\langle \frac{p}{\rho} \left(\frac{\partial u_1}{\partial x_2} + \frac{\partial u_2}{\partial x_1} \right) \right\rangle - 2\nu \left\langle \frac{\partial u_1}{\partial x_\ell} \frac{\partial u_2}{\partial x_\ell} \right\rangle, \end{aligned} \tag{8.234d}$$

$$\frac{\partial\langle u_1u_3\rangle}{\partial t}=P_{13}+\pi_{13}-\varepsilon_{13}$$

$$=-\langle u_2u_3\rangle\frac{\partial U_1}{\partial x_2}+\left\langle\frac{p}{\rho}\left(\frac{\partial u_1}{\partial x_3}+\frac{\partial u_3}{\partial x_1}\right)\right\rangle-2\nu\left\langle\frac{\partial u_1}{\partial x_\ell}\frac{\partial u_3}{\partial x_\ell}\right\rangle, \quad (8.234e)$$

$$\frac{\partial\langle u_2u_3\rangle}{\partial t}=\pi_{23}-\varepsilon_{23}=\left\langle\frac{p}{\rho}\left(\frac{\partial u_2}{\partial x_3}+\frac{\partial u_3}{\partial x_2}\right)\right\rangle-2\nu\left\langle\frac{\partial u_2}{\partial x_\ell}\frac{\partial u_3}{\partial x_\ell}\right\rangle. \quad (8.234f)$$

Summing the first three of the above equations, Eqs. (8.234a), (8.234b) and (8.234c), using $\pi_{ii}=0$ (summation implied) and:

$$P=\frac{P_{ii}}{2}=\frac{P_{11}}{2}=-\langle u_1u_2\rangle\frac{\partial U_1}{\partial x_2}, \quad (8.235)$$

one obtains the simplified K-equation for the case of homogeneous turbulence:

$$\frac{\partial K}{\partial t}=P-\varepsilon=-\langle u_1u_2\rangle\frac{\partial U_1}{\partial x_2}-\nu\left\langle\frac{\partial u_i}{\partial x_\ell}\frac{\partial u_i}{\partial x_\ell}\right\rangle. \quad (8.236)$$

II. **Modelled Reynolds Equations of Homogeneous Turbulence**

The pressure-strain production term π_{ij} is modelled for isothermal flow, without buoyancy effects, according to Eqs. (8.188) through (8.190), by:

$$\pi_{ij}=\left\langle\frac{p}{\rho}\left(\frac{\partial u_i}{\partial x_j}+\frac{\partial u_j}{\partial x_i}\right)\right\rangle=\pi_{ij}^1+\pi_{ij}^2$$

$$=-c_1\frac{\varepsilon}{K}\left(\langle u_iu_j\rangle-\frac{2}{3}K\delta_{ij}\right)-c_2\left(P_{ij}-\frac{2}{3}P\delta_{ij}\right). \quad (8.237)$$

Using Eqs. (8.231) and (8.235) and the definition (8.192) of the anisotropy tensor a_{ij}, one obtains the following components of the tensor π_{ij}:

$$\pi_{11}=-c_1\frac{\varepsilon}{K}\left(\langle u_1u_1\rangle-\frac{2}{3}K\right)+\frac{4}{3}c_2\langle u_1u_2\rangle\frac{\partial U_1}{\partial x_2}=-c_1\varepsilon a_{11}-c_2\frac{2}{3}P_{11}, \quad (8.238a)$$

$$\pi_{22}=-c_1\frac{\varepsilon}{K}\left(\langle u_2u_2\rangle-\frac{2}{3}K\right)-\frac{2}{3}c_2\langle u_1u_2\rangle\frac{\partial U_1}{\partial x_2}=-c_1\varepsilon a_{22}+c_2\frac{P_{11}}{3}, \quad (8.238b)$$

$$\pi_{33} = -c_1 \frac{\varepsilon}{K}\left(\langle u_3 u_3\rangle - \frac{2}{3}K\right) - \frac{2}{3}c_2\langle u_1 u_2\rangle\frac{\partial U_1}{\partial x_2} = -c_1\varepsilon a_{33} + c_2\frac{P_{11}}{3}, \tag{8.238c}$$

$$\pi_{12} = -c_1\frac{\varepsilon}{K}\langle u_1 u_2\rangle + c_2\langle u_2 u_2\rangle\frac{\partial U_1}{\partial x_2} = -c_1\varepsilon a_{12} - c_2 P_{12}, \tag{8.238d}$$

$$\pi_{13} = -c_1\frac{\varepsilon}{K}\langle u_1 u_3\rangle + c_2\langle u_2 u_3\rangle\frac{\partial U_1}{\partial x_2} = -c_1\varepsilon a_{13} - c_2 P_{13}, \tag{8.238e}$$

$$\pi_{23} = -c_1\frac{\varepsilon}{K}\langle u_2 u_3\rangle = -c_1\varepsilon a_{23}. \tag{8.238f}$$

Introducing Eq. (8.238) into Eq. (8.234), one obtains the modelled Reynolds equations for homogeneous turbulence:

$$\begin{aligned}\frac{\partial\langle u_1 u_1\rangle}{\partial t} &= -2\langle u_1 u_2\rangle\frac{\partial U_1}{\partial x_2} - c_1\varepsilon a_{11} + \frac{4}{3}c_2\langle u_1 u_2\rangle\frac{\partial U_1}{\partial x_2} \\ &\quad - 2\nu\left\langle\frac{\partial u_1}{\partial x_\ell}\frac{\partial u_1}{\partial x_\ell}\right\rangle,\end{aligned} \tag{8.239a}$$

$$\frac{\partial\langle u_2 u_2\rangle}{\partial t} = -c_1\varepsilon a_{22} - \frac{2}{3}c_2\langle u_1 u_2\rangle\frac{\partial U_1}{\partial x_2} - 2\nu\left\langle\frac{\partial u_2}{\partial x_\ell}\frac{\partial u_2}{\partial x_\ell}\right\rangle, \tag{8.239b}$$

$$\frac{\partial\langle u_3 u_3\rangle}{\partial t} = -c_1\varepsilon a_{33} - \frac{2}{3}c_2\langle u_1 u_2\rangle\frac{\partial U_1}{\partial x_2} - 2\nu\left\langle\frac{\partial u_3}{\partial x_\ell}\frac{\partial u_3}{\partial x_\ell}\right\rangle, \tag{8.239c}$$

$$\begin{aligned}\frac{\partial\langle u_1 u_2\rangle}{\partial t} &= -\langle u_2 u_2\rangle\frac{\partial U_1}{\partial x_2} - c_1\frac{\varepsilon}{K}\langle u_1 u_2\rangle + c_2\langle u_2 u_2\rangle\frac{\partial U_1}{\partial x_2} \\ &\quad - 2\nu\left\langle\frac{\partial u_1}{\partial x_\ell}\frac{\partial u_2}{\partial x_\ell}\right\rangle,\end{aligned} \tag{8.239d}$$

$$\frac{\partial\langle u_1u_3\rangle}{\partial t} = -\langle u_2u_3\rangle\frac{\partial U_1}{\partial x_2} - c_1\frac{\varepsilon}{K}\langle u_1u_3\rangle + c_2\langle u_2u_3\rangle\frac{\partial U_1}{\partial x_2} - 2\nu\left\langle\frac{\partial u_1}{\partial x_\ell}\frac{\partial u_3}{\partial x_\ell}\right\rangle, \tag{8.239e}$$

$$\frac{\partial\langle u_2u_3\rangle}{\partial t} = -c_1\frac{\varepsilon}{K}\langle u_2u_3\rangle - 2\nu\left\langle\frac{\partial u_2}{\partial x_\ell}\frac{\partial u_3}{\partial x_\ell}\right\rangle. \tag{8.239f}$$

III. Reynolds Equations of Homogeneous Isotropic Turbulence

If we assume that turbulence is not only homogeneous but is also isotropic in the dissipation spectral range, the above equations can be further simplified. The assumption of isotropic turbulence implies that Eq. (8.196) hold. Using Eq. (8.196) the above system of Eq. (8.239) becomes:

$$\frac{\partial\langle u_1u_1\rangle}{\partial t} = -2\langle u_1u_2\rangle\frac{\partial U_1}{\partial x_2} - c_1\varepsilon a_{11} + \frac{4}{3}c_2\langle u_1u_2\rangle\frac{\partial U_1}{\partial x_2} - \frac{2}{3}\varepsilon, \tag{8.240a}$$

$$\frac{\partial\langle u_2u_2\rangle}{\partial t} = -c_1\varepsilon a_{22} - \frac{2}{3}c_2\langle u_1u_2\rangle\frac{\partial U_1}{\partial x_2} - \frac{2}{3}\varepsilon, \tag{8.240b}$$

$$\frac{\partial\langle u_3u_3\rangle}{\partial t} = -c_1\varepsilon a_{33} - \frac{2}{3}c_2\langle u_1u_2\rangle\frac{\partial U_1}{\partial x_2} - \frac{2}{3}\varepsilon, \tag{8.240c}$$

$$\frac{\partial\langle u_1u_2\rangle}{\partial t} = -c_1\frac{\varepsilon}{K}\langle u_1u_2\rangle - (1-c_2)\langle u_2u_2\rangle\frac{\partial U_1}{\partial x_2}, \tag{8.240d}$$

$$\frac{\partial\langle u_1u_3\rangle}{\partial t} = -c_1\frac{\varepsilon}{K}\langle u_1u_3\rangle - (1-c_2)\langle u_2u_3\rangle\frac{\partial U_1}{\partial x_2}, \tag{8.240e}$$

$$\frac{\partial\langle u_2u_3\rangle}{\partial t} = -c_1\frac{\varepsilon}{K}\langle u_2u_3\rangle. \tag{8.240f}$$

IV. Integration of Modelled Reynolds Equations with Runge-Kutta Algorithm

In a homogeneous isotropic turbulence, the averages are spatially independent and depend only on time. Therefore, the partial derivatives of the above equations can be replaced by ordinary derivatives, and the system of Eq. (8.240) becomes a system of ordinary differential equations. These equations can be written in the compact form:

$$\frac{d\langle u_1u_1\rangle}{dt} = \left(1 - \frac{2}{3}c_2\right)P_{11} - c_1\varepsilon a_{11} - \frac{2}{3}\varepsilon, \tag{8.241a}$$

$$\frac{d\langle u_2u_2\rangle}{dt} = \frac{c_2}{3}P_{11} - c_1\varepsilon a_{22} - \frac{2}{3}\varepsilon, \tag{8.241b}$$

$$\frac{d\langle u_3u_3\rangle}{dt} = \frac{c_2}{3}P_{11} - c_1\varepsilon a_{33} - \frac{2}{3}\varepsilon \tag{8.241c}$$

$$\frac{d\langle u_1u_2\rangle}{dt} = (1 - c_2)P_{12} - c_1\varepsilon a_{12}, \tag{8.241d}$$

$$\frac{d\langle u_1u_3\rangle}{dt} = (1 - c_2)P_{13} - c_1\varepsilon a_{13}, \tag{8.241e}$$

$$\frac{d\langle u_2u_3\rangle}{dt} = -c_1\varepsilon a_{23}. \tag{8.241f}$$

Under the assumptions of homogeneous isotropic turbulence, the ε-equation [Eq. (8.209)] simplifies to:

$$\frac{\partial\varepsilon}{\partial t} = c_{1\varepsilon}\frac{\varepsilon}{K}\frac{P_{11}}{2} - c_{2\varepsilon}\frac{\varepsilon^2}{K}. \tag{8.242}$$

Equations (8.241) are not linear because the components of the anisotropic tensor a_{ij} [see. Eq. (8.192)] are functions of the dependent variables. These equations have been integrated numerically with a Runge-Kutta algorithm of the order four with an adaptive time step (Press et al., 1986). The experiments reported by Champagne et al. (1970) were performed with a center velocity of the flow in the wind tunnel of $U_c = 12.4$ m/s and a velocity gradient $\partial U_1/\partial x_2 = 13s^{-1}$. Measurements were taken in the region defined by $8.5H < x_1 < 10.5H$. The value of H is irrelevant as long as the velocity gradient is realized. In Table 8.5, experimental results are compared with the results of two calculations: the first calculation made with the above mentioned Runge-Kutta algorithm, using the program **PROG_VIII.1 [HOMRSM]**, given in **Appendix A.VIII**, and the second calculation made with the computer program COMMIX-1/RSM (Chang & Bottoni, 1995). It is important to notice that in the Runge-Kutta calculation, the system of ordinary differential equations, i.e., the system of simplified equations (Eq. (8.241)), was used, while in the calculations made with the computer program COMMIX-1/RSM, the original set of Eq. (8.199)

Table 8.5 Experimental and computed turbulence values in homogeneous isotropic shear flow

		Calculations	
Turbulence value	Experiment	Runge-Kutta	COMMIX-1/RSM
K $[m^2/s^2]$	0.053	0.030	0.033
$\sqrt{K}/U_c$	0.0185	0.0139	0.0147
$\sqrt{\langle u_1 u_1 \rangle}/U_c$	0.0178	0.0140	0.0161
$\sqrt{\langle u_2 u_2 \rangle}/U_c$	0.0132	0.0097	0.0093
$\sqrt{\langle u_3 u_3 \rangle}/U_c$	0.0141	0.0097	0.0093
$\sqrt{\langle u_1 u_2 \rangle}/U_c$	0.0104	0.0101	0.0100
$P = -u_1 u_2 \frac{\partial U_1}{\partial x_2}$ $[m^2/s^3]$	0.218	0.202	0.201

was used. In other words, the computer program COMMIX-1/RSM was not given instructions about the simplifications implied by homogeneous isotropic turbulence.

In the given experimental conditions, the quantities $\langle u_1 u_3 \rangle$ and $\langle u_2 u_3 \rangle$ are negligibly small and have not been given in the above table. The table shows that the results computed with the two different codes are closer to each other than to the experimental values. This is a consequence of the difficulty of realizing in practice the theoretical conditions of homogeneous turbulence. Theoretically, it must be $\langle u_2 u_2 \rangle = \langle u_3 u_3 \rangle$, a condition satisfied by the results of the calculations but not strictly verified in the experiment.

References

Bottoni, M., & Willerding, G. (1987). Advanced solution algorithms for transient multidimensional thermohydraulics flow problems in complex geometries with the programme COMMIX-2/KfK. *Nuclear Engineering and Design, 100*, 351.

Champagne, F. H., Harris, V. G., & Corrsin, S. (1970). Experiments on nearly homogeneous turbulent shear flow. *Journal of Fluid Mechanics, 41*(Part 1), 81.

Chang, F. C., Bottoni, M., & Sha W. T. (1992, Jan). Development and validation of the k-epsilon two-equation turbulence model and the six-equation anisotropic turbulence model in the COMMIX code (COMMIX-1C/ATM). ANL/ATHRP-44.

Chang, F. C., & Bottoni, M. (1995, Aug). Implementation and validation of a Reynolds stress model in the COMMIX-1C/RSM and CAPS-3D/RSM codes. ANL-96/8.

Chen Ching Jen. (1990). *Prediction of turbulent flows*, Lecture Notes of the University of Iowa, Iowa City, Iowa.

Chou, P. Y. (1954). On the velocity correlations and the solution of the equations of turbulent fluctuations. *Quarterly of Applied Mathematics, 3*, 38.

Domanus, H. M. et al. (1990, Nov). *COMMIX-1C: A three-dimensional transient single-phase computer program for thermal-hydraulic analysis of single-components and multicomponent engineering systems. Volume I: Equations and numerics: Volume II: User's guide manual, NUREG/CR-5649*, Argonne National Laboratory Report ANL-90/33.

Ferziger, J. H. (1983). Higher-level Simulations of Turbulent Flows, In Essers J. A., Computational Methods for Turbulent, Transonic, and Viscous Flows, Hemisphere Publishing CO.

Lesieur, M. (1990). *Turbulence in fluids*. Kluwer Academic Publishers.

McComb, W. B. (1990). *The physics of fluid turbulence*. Clarendon Press.

Orszag, S. A. (1977). In R. Balian & J. L. Peube (Eds.), *Lectures on the statistical theory of turbulence, fluid dynamics*. Gordon and Breach Science Publishers.

Press, W. H., Flannery, B. P., Teukolsky, S. A., & Vetterling, W. T. (1986). *Numerical recipes*. Cambridge University Press.

Rodi, W. (1980). *Turbulence models and their application in hydraulics*. IAHR.

Tennekes, H., & Lumley, J. L. (1972). *A first course in turbulence*. The MIT Press.

Wade, D. C., & Chang, Y. I. (1988). The integral fast reactor concept: Physics of operation and safety. *Nuclear Science and Engineering, 100*.

Wilcox, D. C. (1998). *Turbulence modeling for CFD*. DCW Industries.

Appendix A: List of Computer Programs

The computer programs given in these lists, in FORTRAN V, were written only for didactic purposes. It has not been tried to optimize the coding style, but to preserve the clarity of the logic. Each program is self-sufficient and does not need any link. Computations have been made by the author with a personal computer with peak performance of 13 Gcycles.

All examples have been re-run at the MEEO (Meteorological and Environmental Earth Observation) with the following platform:

CPU: Intel(R) Xeon(R) CPU E5-2640 v3 @ 2.60GHz

Compiler: GNU Fortran 9.3.0.

All results matched well with previous calculations.

The following computer programs are listed in italics, while general information and results of the calculations are given in cursive characters.

Appendix A.I

PROG_I_1 [PEANO-PICARD] Numerical solution of the ordinary differential equation of first order, Eq. (1.214), with initial condition $x_0 = 0$, $y(x_0) = 0$.

```
c   APPENDIX A, Program: PROG_I_1 [PEANO-PICARD]

c   Last Change: 18.03.2020
c   ---------------------------------------------------
c   This program codes the numerical integration with Gaussian
    quadratures of ordinary differential equations of first order of the
    type
c   Y' = f(X,Y), based on the existence theorem in the form of Peano-Picard.
```


M. Bottoni, *Physical Modeling and Computational Techniques for Thermal and Fluid-dynamics*, Mechanical Engineering Series,
https://doi.org/10.1007/978-3-030-79717-1

```
c   Reference: Cugiani M., Metodi dell' Analisi Numerica, UTET, 1967.
c   Cfr.: pp. 314-316.

c   Example: Y' = 1 + X Y with Y(0) = 0.

c   OUTPUT:
c
c   1) SOLNUM = Numerical solution obtained integrating numerically the
c              above differential equation.
c   2) SOLSER = Approximate solution obtaines as sum of the series:

c   Y = X + X**3/3 + X**5/(3 x 5) + X**7/(3 x 5 x 7) + ...
c                                                X**(2N + 1)/[(2N + 1)!!]

c  3) SOLEXA = Exact solution:  Y = exp(x**2/2) x Int(0,x) [exp(-x**2/2)]
c   --------------------------------------------
c   LIST of Subroutines and Functions:
c  Subroutine FACODD  - Computes the double factorial: n!! = Product of all
c                       odd integers smaller than or equal to n.
c   Subroutine GAUSS40 - Performs numerical integration with Gaussian
c                         quadratures of order 40.
c   Function ANTEG1 - Codes function to be integrated.
c   Function ERFF   - Computes Error function.

   implicit real*8 (a-h, o-z)

   common/beta/ir

   open(60, file='OUTPUT' , status='unknown')
   open(10, file='OUTPL0' , status='unknown')  ! Data for plotting

   pi=4.d0*datan(1.d0)
c  -----------------------------------------------------
c   Solution 1) Get SOLNUM
   write(60,*) ' '
   write( *,*) ' '

   x0=0.d0              ! Initial conditions
   y0=0.d0
   solnum=0.d0
   solser=0.d0
   solexa=0.d0
   write(10,1010) istep,x0,solnum,solser,solexa

   ngauss=40
   istepmax=20 ! Number of intervals to be integrated
   irmax = 20 ! Number of terms of the integrand to be integrated
   irp1max= irmax + 1

   do 500 istep=1,istepmax
     xmax=float(istep) * 0.1d0
```

```
      do 100 ir=1,irmax
        call gauss40(x0,xmax,ngauss,aint1)
        solnum= y0 + aint1
100     continue

c   --------------------------------------------------
c   Get solution 2) SOLSER as sum of the series.

      do 200 irp1=1,irp1max
        ir=irp1-1   ! Ir is the number of terms considered
c                        in the series expansion

      if(ir.eq.0) then
           solser=xmax
           go to 195
      endif

      if(ir.eq.1) then
           solser=xmax + xmax**3/3.d0
           go to 195
      endif

      if(ir.eq.2) then
           solser=xmax + xmax**3/3.d0 + xmax**5/15.d0
           go to 195
      endif

      if(ir.ge.3) then ! See general term of Eq. I.1.3.7 of Text
           if(ir.eq.3) solser=xmax + xmax**3/3.d0 + xmax**5/15.d0
           call facodd(2*ir+1, faco)
           solser=solser + xmax**(2*ir+1)/faco
      endif

195     continue
200   continue
c   --------------------------------------------------

c   Solution 3) Exact solution SOLEXA
      solexa= dexp((xmax**2)/2.d0) * erff(xmax/dsqrt(2.d0))
     1                                 * dsqrt(pi/2.d0)
c   --------------------------------------------------
      write(60,*) ' '
      write( *,*) ' '
      error_solnum = DABS(solnum - solexa)/solexa
      error_solser = DABS(solser - solexa)/solexa
      write(60,*) 'istep,solnum/ser/exa', istep,solnum,solser,solexa
      write(60,*) '    error_solnum, error_solser',
     1                 error_solnum, error_solser
      write( *,*) 'istep,solnum/ser/exa', istep,solnum,solser,solexa
      write( *,*) '    error_solnum, error_solser',
     1                 error_solnum, error_solser
```

```
   write(10,1010) istep,xmax,solnum,solser,solexa
1010 FORMAT(1H ,i5,4e12.5)

500  continue

   close(60)

   stop
   end

   subroutine facodd(n,faco)
c   Computes the product FACO = 1 x 3 x 5 x (n-3)(n-1)  if n is even,
c                             = 1 x 3 x 5 x (n-2) x n  if n is odd.
   implicit real*8 (a-h,o-z)

   if(n.le.2) then
      faco=1.d0
      return
   endif

   if(n.eq.3 .or.n.eq.4) then
      faco=3.d0
      return
   endif

   if(n.ge.5) then
      faco=3.d0
      do 10 i=5,1000,2
         if(i.gt.n) go to 20
         faco=faco * float(i)
10    continue
20    continue
   endif

   return
   end

   subroutine gauss40(a,b,n,sum)
c   Computes a definite integral with Gaussian quadrature
c   input= a,b limits of integration (finite)
c            n= order of gaussian quadrature
c   output= sum= computed integral
c
   implicit real*8 (a-h, o-z)
   dimension r(40),w(40)

400  continue
c
     r(40)=0.998237709710559200350d0
     r(39)=0.990726238699457006450d0
     r(38)=0.977259949983774262660d0
```

```
      r(37)=0.957916819213791655 80d0
      r(36)=0.932812808278676533 36d0
      r(35)=0.902098806968874296 73d0
      r(34)=0.865959503212259503 82d0
      r(33)=0.824612230833311663 20d0
      r(32)=0.778305651426519387 69d0
      r(31)=0.727318255189927103 28d0
      r(30)=0.671956684614179548 38d0
      r(29)=0.612553889667980237 95d0
      r(28)=0.549467125095128202 08d0
      r(27)=0.483075801686178712 91d0
      r(26)=0.413779204371605001 52d0
      r(25)=0.341994090825758473 01d0
      r(24)=0.268152185007253681 14d0
      r(23)=0.192697580701371099 72d0
      r(22)=0.116084070675255208 48d0
      r(21)=0.038772417506050821 93d0
c
      w(40)=0.004521277098533191 26d0
      w(39)=0.010498284531152813 62d0
      w(38)=0.016421058381907888 71d0
      w(37)=0.022245849194166957 26d0
      w(36)=0.027937006980023401 10d0
      w(35)=0.033460195282547847 39d0
      w(34)=0.038782167974472017 64d0
      w(33)=0.043870908185673271 99d0
      w(32)=0.048695807635072232 06d0
      w(31)=0.053227846983936824 36d0
      w(30)=0.057439769099391551 37d0
      w(29)=0.061306242492928939 17d0
      w(28)=0.064804013456601038 07d0
      w(27)=0.067912045815233903 83d0
      w(26)=0.070611647391286779 70d0
      w(25)=0.072886582395804059 06d0
      w(24)=0.074723169057968264 20d0
      w(23)=0.076110361900626242 37d0
      w(22)=0.077039818164247965 59d0
      w(21)=0.077505947978424811 26d0
      do 405 i=1,20
         k=41-i
         r(i)=-r(k)
405   w(i)=w(k)
500   continue
c
c   Change the variable so that the integration limits range from (-1,+1).
c   this is equivalent to shift the zeros
c
    do 510 i=1,n
510   r(i)=(r(i)*(b-a)+a+b)/2.d0
c
c   Compute the function to be integrated in the shifted zeros and
    multiply by
c   respective weigths
```

```
c
   sum=0.d0
c
   do 520 i=1,n
     bet=anteg1(r(i))
     sum=sum+bet*w(i)
520  continue

   sum=sum*(b-a)/2.d0
c
   return
   end

   function anteg1(r)  ! Integrand in Eq. I.1.3.6d and following terms
   implicit real*8 (a-h,o-z)
   common/beta/ir
   if(ir.eq.1) anteg1=1.d0
   if(ir.eq.2) anteg1=1.d0 + r**2

   if(ir.ge.3) then
        anteg1=1.d0 + r**2
        do 10 i=3,ir
           call facodd(2*i-2,faco)
         anteg1=anteg1 + r**(2*i -2)/faco
10      continue
   endif

   return
   end

   function erff(x)         ! Reference: Press et al. 1986
   implicit real*8 (a-h, o-z)

   z=dabs(x)
   t=1./(1.+0.5*z)
   erfcc=t*dexp(-z*z-1.26551223+t*(1.00002368+t*(.37409196+
  *  t*(.09678418+t*(-.18628806+t*(.27886807+t*(-1.13520398+
  *  t*(1.48851587+t*(-.82215223+t*.17087277)))))))))
   if (x.lt.0.) erfcc=2.-erfcc
   erff=1.d0 -erfcc
   return
   end
```

OUTPUT From Program PROG_I.1 [PEANO-PICARD] (Only every fifth result was printed)

istep, solnum/ser/exa 1 0.10033400095344011 0.10033400095344011 0.10033403900242073
error_solnum, error_solser 3.79223053285731412E-007 3.79223053285731412E-007

```
istep,solnum/ser/exa      5 0.54382651945215066
0.54382651945215055    0.54382655214077436
   error_solnum, error_solser 6.01085466829175675E-008
6.01085468870677808E-008

istep,solnum/ser/exa     10 1.4106861346424477    1.4106861346424480
1.4106860830099526
   error_solnum, error_solser 3.66009814586787322E-008
3.66009816160805884E-008

istep,solnum/ser/exa     15 3.3446636845284083    3.3446636845284075
3.3446636644429240
   error_solnum, error_solser 6.00523293278782611E-009
6.00523266723693392E-009

istep,solnum/ser/exa     20 8.8394392409188676    8.8394392409190292
8.8394392681512315
   error_solnum, error_solser 3.08077957134196502E-009
3.08076128415569908E-009
```

PROG_I_2 [TEST_RK4] Test of Runge-Kutta method of order four.

```
c   APPENDIX A.I, Program: PROG_I_2 [TEST_RK4]

c   Last Change: 18.03.2020
c   --------------------------------------------
c   Test of Runge-Kutta integration
c   --------------------------------------------
c
c   The differential equation y'=cos(x), with analytical solu-
c   tion y(x)=sin(x), is considered in the interval A=0 to B=2 x pi
c   This subroutine performs :
c   i)  A forward integration in space from A to B with boun-
c         dary condition y(0)=0;
c   ii)  A backward integration in space from B to A with boun-
c         dary condition y(2 x pi)=0;
c   iii) Shows that the numerical results are identical in both
c          cases with the analytical solution.
c   Subroutine RK4 can be used for a large system of differential
c   equations, but it is used here for only one equation (neq = 1 )
c
c   YAN  = Analytical solution
c   YF   = Result of forward integration
c   YB   = Result of backward integration

c   LIST of Subroutines and Functions:
c   Subroutine RK4
c   Subroutine DER

      implicit real*8 (a-h,o-z)
      dimension y(1), yback(1), yf(100), yb(100), yan(100)
      open(60,file='OUTPUT',status='unknown')
      open(10,file='OUTPL0',status='unknown') ! Save data for plotting
```

```
c
   pi=4.d0*datan(1.d0)
   a=0.d0
   b=pi * 2.d0
   nmesh=100
   neq=1
   h=(b-a)/dfloat(nmesh)
c
c  Forward integration
c
   y(1)=0.d0     ! Initial conditions
   yan(1)=0.d0
c
   do 100 i=1,nmesh
      xi=dfloat(i)*h
      yan(i)=dsin(xi)
      call rk4(a,h,y,neq)
      yf(i)=y(1)
100  continue
c
c  Backward integration

   yback(1)=0.d0  ! Initial conditions
   yb(nmesh)=0.d0
   hback=-h
   do 200 i=1,nmesh
      ii=nmesh - i
      call rk4(b,hback,yback,neq)
      if(ii.ge.1) yb(ii)=yback(1)
200  continue
c
   write(60,1300)   ! Write on OUTPUT file
   write( *,1300)   ! Write on computer screen
   do 300 i=1,nmesh
      xi=float(i)*h
      write(60,1301) i,xi,yan(i),yf(i),yb(i)
      write( *,1301) i,xi,yan(i),yf(i),yb(i)
      write(10,1010) i,xi,yan(i),yf(i),yb(i)  ! Write data for plotting
300  continue

1010 FORMAT(1H ,i7,4e14.5)
1300 format(' i',9x,'xi',9x,'yan',9x,'yf ',9x,'yb')
1301 format(i5,4e12.5)

   close(60)
   close(10)
   stop
   end

   subroutine rk4(t,h,y,n)
   implicit real*8 (a-h,o-z)
   dimension y(100),yp1(100),yp2(100),yp3(100),yp4(100),aus(100)
```

```
c
c     t   independent variable
c     h   integration step (dt)
c     n   Number of equations in the diff. system  (here n=1)
c     y(n) dependent variables (initial values are here introduced
c     yp(n) auxiliary arrays
c     aus(n) auxiliary array
c
      t0 = t
      do 5 i=1,n
         aus(i) = y(i)
 5    continue
c
      call der(t,y,yp1)
      t     = t0+h*0.5
      do 10 i=1,n
         y(i) = aus(i)+h*yp1(i)*0.5
 10   continue
      call der(t,y,yp2)
      do 11 i=1,n
        y(i) = aus(i)+h*yp2(i)*0.5
 11   continue
      call der(t,y,yp3)
      t    = t0+h
      do 12 i=1,n
         y(i) = aus(i)+h*yp3(i)
 12   continue
      call der(t,y,yp4)
c
      do 20 i=1,n
         y(i) = aus(i)+h*(yp1(i)+2.0*yp2(i)+2.0*yp3(i)+yp4(i))/6.0
 20   continue
c
      return
      end

      subroutine der(x,y,yp)       ! Code derivative
      implicit real*8(a-h, o-z)
      dimension y(1),yp(1)
      yp(1)=dcos(x)
      return
      end
```

OUTPUT From Program PROG_I.2 [TEST_RK4]

i	*xi*	*yan*	*yf*	*yb*
1	*0.62832E-01*	*0.62791E-01*	*0.62791E-01*	*0.62791E-01*
2	*0.12566E+00*	*0.12533E+00*	*0.12533E+00*	*0.12533E+00*
3	*0.18850E+00*	*0.18738E+00*	*0.18738E+00*	*0.18738E+00*
4	*0.25133E+00*	*0.24869E+00*	*0.24869E+00*	*0.24869E+00*
5	*0.31416E+00*	*0.30902E+00*	*0.30902E+00*	*0.30902E+00*
6	*0.37699E+00*	*0.36812E+00*	*0.36812E+00*	*0.36812E+00*
7	*0.43982E+00*	*0.42578E+00*	*0.42578E+00*	*0.42578E+00*

```
 8 0.50265E+00 0.48175E+00 0.48175E+00 0.48175E+00
 9 0.56549E+00 0.53583E+00 0.53583E+00 0.53583E+00
10 0.62832E+00 0.58779E+00 0.58779E+00 0.58779E+00
11 0.69115E+00 0.63742E+00 0.63742E+00 0.63742E+00
12 0.75398E+00 0.68455E+00 0.68455E+00 0.68455E+00
13 0.81681E+00 0.72897E+00 0.72897E+00 0.72897E+00
14 0.87965E+00 0.77051E+00 0.77051E+00 0.77051E+00
15 0.94248E+00 0.80902E+00 0.80902E+00 0.80902E+00
16 0.10053E+01 0.84433E+00 0.84433E+00 0.84433E+00
17 0.10681E+01 0.87631E+00 0.87631E+00 0.87631E+00
18 0.11310E+01 0.90483E+00 0.90483E+00 0.90483E+00
19 0.11938E+01 0.92978E+00 0.92978E+00 0.92978E+00
20 0.12566E+01 0.95106E+00 0.95106E+00 0.95106E+00
21 0.13195E+01 0.96858E+00 0.96858E+00 0.96858E+00
22 0.13823E+01 0.98229E+00 0.98229E+00 0.98229E+00
23 0.14451E+01 0.99211E+00 0.99211E+00 0.99211E+00
24 0.15080E+01 0.99803E+00 0.99803E+00 0.99803E+00
25 0.15708E+01 0.10000E+01 0.10000E+01 0.10000E+01
26 0.16336E+01 0.99803E+00 0.99803E+00 0.99803E+00
27 0.16965E+01 0.99211E+00 0.99211E+00 0.99211E+00
28 0.17593E+01 0.98229E+00 0.98229E+00 0.98229E+00
29 0.18221E+01 0.96858E+00 0.96858E+00 0.96858E+00
30 0.18850E+01 0.95106E+00 0.95106E+00 0.95106E+00
31 0.19478E+01 0.92978E+00 0.92978E+00 0.92978E+00
32 0.20106E+01 0.90483E+00 0.90483E+00 0.90483E+00
33 0.20735E+01 0.87631E+00 0.87631E+00 0.87631E+00
34 0.21363E+01 0.84433E+00 0.84433E+00 0.84433E+00
35 0.21991E+01 0.80902E+00 0.80902E+00 0.80902E+00
36 0.22619E+01 0.77051E+00 0.77051E+00 0.77051E+00
37 0.23248E+01 0.72897E+00 0.72897E+00 0.72897E+00
38 0.23876E+01 0.68455E+00 0.68455E+00 0.68455E+00
39 0.24504E+01 0.63742E+00 0.63742E+00 0.63742E+00
40 0.25133E+01 0.58779E+00 0.58779E+00 0.58779E+00
41 0.25761E+01 0.53583E+00 0.53583E+00 0.53583E+00
42 0.26389E+01 0.48175E+00 0.48175E+00 0.48175E+00
43 0.27018E+01 0.42578E+00 0.42578E+00 0.42578E+00
44 0.27646E+01 0.36812E+00 0.36812E+00 0.36812E+00
45 0.28274E+01 0.30902E+00 0.30902E+00 0.30902E+00
46 0.28903E+01 0.24869E+00 0.24869E+00 0.24869E+00
47 0.29531E+01 0.18738E+00 0.18738E+00 0.18738E+00
48 0.30159E+01 0.12533E+00 0.12533E+00 0.12533E+00
49 0.30788E+01 0.62791E-01 0.62791E-01 0.62791E-01
50 0.31416E+01-0.32163E-15-0.15106E-14-0.17029E-13
51 0.32044E+01-0.62791E-01-0.62791E-01-0.62791E-01
52 0.32673E+01-0.12533E+00-0.12533E+00-0.12533E+00
53 0.33301E+01-0.18738E+00-0.18738E+00-0.18738E+00
54 0.33929E+01-0.24869E+00-0.24869E+00-0.24869E+00
55 0.34558E+01-0.30902E+00-0.30902E+00-0.30902E+00
56 0.35186E+01-0.36812E+00-0.36812E+00-0.36812E+00
57 0.35814E+01-0.42578E+00-0.42578E+00-0.42578E+00
58 0.36442E+01-0.48175E+00-0.48175E+00-0.48175E+00
59 0.37071E+01-0.53583E+00-0.53583E+00-0.53583E+00
60 0.37699E+01-0.58779E+00-0.58779E+00-0.58779E+00
```

```
 61 0.38327E+01-0.63742E+00-0.63742E+00-0.63742E+00
 62 0.38956E+01-0.68455E+00-0.68455E+00-0.68455E+00
 63 0.39584E+01-0.72897E+00-0.72897E+00-0.72897E+00
 64 0.40212E+01-0.77051E+00-0.77051E+00-0.77051E+00
 65 0.40841E+01-0.80902E+00-0.80902E+00-0.80902E+00
 66 0.41469E+01-0.84433E+00-0.84433E+00-0.84433E+00
 67 0.42097E+01-0.87631E+00-0.87631E+00-0.87631E+00
 68 0.42726E+01-0.90483E+00-0.90483E+00-0.90483E+00
 69 0.43354E+01-0.92978E+00-0.92978E+00-0.92978E+00
 70 0.43982E+01-0.95106E+00-0.95106E+00-0.95106E+00
 71 0.44611E+01-0.96858E+00-0.96858E+00-0.96858E+00
 72 0.45239E+01-0.98229E+00-0.98229E+00-0.98229E+00
 73 0.45867E+01-0.99211E+00-0.99211E+00-0.99211E+00
 74 0.46496E+01-0.99803E+00-0.99803E+00-0.99803E+00
 75 0.47124E+01-0.10000E+01-0.10000E+01-0.10000E+01
 76 0.47752E+01-0.99803E+00-0.99803E+00-0.99803E+00
 77 0.48381E+01-0.99211E+00-0.99211E+00-0.99211E+00
 78 0.49009E+01-0.98229E+00-0.98229E+00-0.98229E+00
 79 0.49637E+01-0.96858E+00-0.96858E+00-0.96858E+00
 80 0.50265E+01-0.95106E+00-0.95106E+00-0.95106E+00
 81 0.50894E+01-0.92978E+00-0.92978E+00-0.92978E+00
 82 0.51522E+01-0.90483E+00-0.90483E+00-0.90483E+00
 83 0.52150E+01-0.87631E+00-0.87631E+00-0.87631E+00
 84 0.52779E+01-0.84433E+00-0.84433E+00-0.84433E+00
 85 0.53407E+01-0.80902E+00-0.80902E+00-0.80902E+00
 86 0.54035E+01-0.77051E+00-0.77051E+00-0.77051E+00
 87 0.54664E+01-0.72897E+00-0.72897E+00-0.72897E+00
 88 0.55292E+01-0.68455E+00-0.68455E+00-0.68455E+00
 89 0.55920E+01-0.63742E+00-0.63742E+00-0.63742E+00
 90 0.56549E+01-0.58779E+00-0.58779E+00-0.58779E+00
 91 0.57177E+01-0.53583E+00-0.53583E+00-0.53583E+00
 92 0.57805E+01-0.48175E+00-0.48175E+00-0.48175E+00
 93 0.58434E+01-0.42578E+00-0.42578E+00-0.42578E+00
 94 0.59062E+01-0.36812E+00-0.36812E+00-0.36812E+00
 95 0.59690E+01-0.30902E+00-0.30902E+00-0.30902E+00
 96 0.60319E+01-0.24869E+00-0.24869E+00-0.24869E+00
 97 0.60947E+01-0.18738E+00-0.18738E+00-0.18738E+00
 98 0.61575E+01-0.12533E+00-0.12533E+00-0.12533E+00
 99 0.62204E+01-0.62791E-01-0.62791E-01-0.62791E-01
100 0.62832E+01 0.64326E-15-0.31039E-14 0.00000E+00]
```

Appendix A.II

Numerical solution of Laplace equations, Eq. (2.97) with boundary conditions (2.98).

(i) Program **PROG_II_1_LAPLMC** with Subroutine LAPLMC, based on the Monte Carlo technique.

```
c   Program: PROG_II_1_LAPLMC Last change: 22.03.2020  (14.35)

C   SOLUTIONS OF MODEL LAPLACE EQUATION WITH SUBROUTINE LAPLMC (Based on
Monte Carlo technique)

c   Test case  : xmax=ymax=4.
c                  u=0 for (x=0, y=0-ymax), and (x=0-xmax, y=0)
c                  u=x**3                  for y=ymax
c                  u=(y/ymax)*ymax**3 for x=xmax.

c   LIST of Subroutines and Functions:
c   Subroutine LAPLMC
c   Subroutine CPU_TIME
c   Function RF

    implicit real*8 (a-h,o-z)
    Parameter (idima=400)
    DIMENSION RESID(idima)

    open (60,file='OUTPUT', status='unknown')
    open (10,file='OUTPL0', status='unknown') ! Computed solution
c                                              and residuals for plotting
    open (12,file='OUTPL2', status='unknown') ! Computed solution for 2D
plot

    call CPU_TIME (cput0)

    write(60,*) 'LAPLMC - Solution of LAPLACE eq. with Monte Carlo '
    write( *,*) 'LAPLMC - Solution of LAPLACE eq. with Monte Carlo '

    call LAPLMC (RESID)

    resmin= 1.e30      ! Find minimun and maximum of residuals
    resmax=-1.e30
    do 10 i=1,idima
      if(dabs(resid(i)).le.resmin) resmin=dabs(resid(i))
      if(dabs(resid(i)).ge.resmax) resmax=dabs(resid(i))
10  continue

    call CPU_TIME(cpuend)
    cputot=cpuend - cput0
```

```
      write(60,*) ' CPUTOT, resmin, resmax ', cputot, resmin, resmax
      write( *,*) ' CPUTOT, resmin, resmax ', cputot, resmin, resmax

      close(60)
      close(10)
      close(12)

      STOP
      END

      subroutine LAPLMC (RESID)
      implicit real*8 (a-h,o-z)
c     Solution of Laplace equation with Monte Carlo method.
c     Example taken from G. D. Smith, "Numerische Loesung von
c     partiellen Differentialgleichungen", Vieweg, pag. 214 ff.
c
c     INPUT: ni = number of interval in x direction
c            nj = number of interval in y direction
c        xmin, xmax, ymin, ymax = coordinates delimiting the
c                                    integration domain.
c        nnodes = (ni+1)*(nj+1)
c
c     Method of solution:
c     Consider one starting point P0=(i=i0, j=j0) in which we want to find
c     the solution. Generate a random number 'r' = 1,2,3,4.
c
c     If r=1 set i = i-1
c     If r=2 set j = j+1
c     If r=3 set i = i+1
c     If r=4 set j = j-1

c     When the point P(i,j) reaches the boundary, take the boundary value
c     and assign to the point P0(i0,j0). Sum up the contributions from all
c     walks. In the end take mean value of the contributions summed in all
c     points P0.

c     N.B. Some "write" statements have been commented. They can be turned
c        on to get information on the computer screen about the advancement
c        of the calculation


c
      Parameter (idima=400)
      dimension a(idima,idima),u(idima),resid(idima)
      data xmin,xmax,ymin,ymax /0.d0, 4.d0, 0.d0, 4.d0/

      nip1=20
      njp1=20
      ni=nip1-1
      nj=njp1-1
      nnodes=nip1*njp1
```

```
c
   do 10 m=1,nnodes
      u(m)     =0.d0
      resid(m)=0.d0
10  continue
c
   dx=(xmax-xmin)/float(ni)
   dy=(ymax-ymin)/float(nj)
   write(60,*) ' ni,nj,nnodes ', ni,nj,nnodes
   write( *,*) ' ni,nj,nnodes ', ni,nj,nnodes
c
c   ---------------- Initialization ----------------------
c   Set initial and boundary values
   do 95 i=2,ni
   do 95 j=2,nj
      a(i,j)= 0.d0
95  continue
c
   do 100 i=1,nip1
c   On x-axis, lower boundary
     x=float(i-1)*dx
     m=(i-1)*nip1 + 1
     a(i,1) = 0.d0
     u(m)=a(i,1)
c
c   Upper boundary
     m=(i-1)*nip1 + njp1
     a(i,njp1)=x**3
     u(m)=a(i,njp1)
100 continue
c
   do 105 j=1,njp1
c   On y-axis, left boundary
     y=float(j-1)*dy
     a(1,j)=0.d0
     u(j)=0.d0
c
c   Right boundary
     m=(nip1-1)*nip1 + j
     a(nip1,j)=(y/ymax)*ymax**3
     u(m)=a(nip1,j)
105 continue
c   ---------------------- End initialization ----------------
c
c   ============= Main loop for Monte Carlo walk =============
c
   r=rf(-1)        ! Initialization of Function RF
   iterx = 1 000 000 ! Number of sweeps for every node
c
   do 200 i0=2,ni
   do 200 j0=2,nj
```

```
c    write(60,*) 'iterx,i0,j0 in progress', iterx,i0,j0
c    write( *,*) 'iterx,i0,j0 in progress', iterx,i0,j0

      do 150 iter=1,iterx
        i=i0
        j=j0
c
       do 130 iwalk=1,10000
         ir=1 + rf(0)*4.d0
c
         if(ir.eq.1) i = i-1
         if(ir.eq.2) j = j+1
         if(ir.eq.3) i = i+1
         if(ir.eq.4) j = j-1

    if(i.eq.1 .or. i.eq.nip1 .or. j.eq.1 .or. j.eq.njp1) go to 140
130      continue
    write(60,*) ' LAPLMC - Warning: Walk not completed, STOP'
    write( *,*) ' LAPLMC - Warning: Walk not completed, STOP'
    STOP
c
140      continue
c    if(mod(iter,10 000) .eq. 0) then
c      write(60,*) 'i0,j0 iwalk',i0,j0,iwalk
c      write( *,*) 'i0,j0 iwalk',i0,j0,iwalk
c    endif

      a(i0,j0)=a(i0,j0) + a(i,j)
150     continue
c
     a(i0,j0) = a(i0,j0)/float(iterx)
200  continue
c    ========================== End Monte Carlo Walks ================
c
c    Check maximum residual
    resmax=0.d0
    do 210 i=2,ni
    do 210 j=2,nj
       m= (i-1)*nip1 + j
       u(m)=a(i,j)
    resid(m)=a(i-1,j) + a(i+1,j) + a(i,j-1) + a(i,j+1) -4.d0*a(i,j)
    if(dabs(resid(m)).gt.resmax) resmax=dabs(resid(m))
210  continue

    do 300 m=1,nnodes  ! Fill OUTPL0 (solution and residuals)
      absres=dabs(resid(m))
      write(10,3020) m, absres,u(m)
3020    format(i6,2e14.5)
300  continue
c
c
    do 400 i=1,nip1    ! Fill OUTPL2 for plotting in 2D
    do 400 j=1,njp1
```

```
      x=float(i-1)*dx
      y=float(j-1)*dy
      m=(i-1)*nip1 + j
      write(12,1012) i,j,x,y,u(m)
1012 format(1H ,2i5,3e14.5)
400  continue
c
     return
     end

     function rf(IDUM)
*--generates a uniformly distributed random fraction between 0 and 1
*----IDUM will generally be 0, but negative values may be used to
*------re-initialize the seed
c    Cfr. function RUN3 of Numerical Recipes, pag. 199.
     implicit real*8 (a-h,o-z)
     SAVE MA,INEXT,INEXTP
     PARAMETER (MBIG=1000000000,MSEED=161803398,MZ=0,FAC=1.E-9)
     DIMENSION MA(55)
     DATA IFF/0/
     IF (IDUM.LT.0.OR.IFF.EQ.0) THEN
      IFF=1
      MJ=MSEED-IABS(IDUM)
      MJ=MOD(MJ,MBIG)
      MA(55)=MJ
      MK=1
      DO 50 I=1,54
       II=MOD(21*I,55)
       MA(II)=MK
       MK=MJ-MK
       IF (MK.LT.MZ) MK=MK+MBIG
       MJ=MA(II)
50     CONTINUE
      DO 100 K=1,4
       DO 60 I=1,55
        MA(I)=MA(I)-MA(1+MOD(I+30,55))
        IF (MA(I).LT.MZ) MA(I)=MA(I)+MBIG
60     CONTINUE
100    CONTINUE
      INEXT=0
      INEXTP=31
     END IF
200  INEXT=INEXT+1
     IF (INEXT.EQ.56) INEXT=1
     INEXTP=INEXTP+1
     IF (INEXTP.EQ.56) INEXTP=1
     MJ=MA(INEXT)-MA(INEXTP)
     IF (MJ.LT.MZ) MJ=MJ+MBIG
     MA(INEXT)=MJ
     RF=MJ*FAC
     IF (RF.GT.1.E-8.AND.RF.LT.0.99999999) RETURN
     GO TO 200
     END
```

```
OUTPUT from PROG_II_1_LAPLMC (when using 100,000 sweeps for every node)

LAPLMC - Solution of LAPLACE eq. with Monte Carlo
 ni,nj,nnodes       19      19      400
 CPUTOT, resmin, resmax  44.631886000000002     0.0000000000000000
0.65570365359326033

OUTPUT from PROG_II_1_LAPLMC (when using 1000,000 sweeps for every
node)

LAPLMC - Solution of LAPLACE eq. with Monte Carlo
ni,nj,nnodes       19      19      400
CPUTOT, resmin, resmax  446.89606400000002     0.0000000000000000
0.19412083627742582

OUTPUT from PROG_II_1_LAPLMC (when using 10,000,000 sweeps for every
node)

LAPLMC - Solution of LAPLACE eq. with Monte Carlo
 ni,nj,nnodes       19      19      400
 CPUTOT, resmin, resmax  4467.6814380000005     0.0000000000000000
7.3760358329398201 2E-002
```

(ii) Program **PROG_II_2_LAPLFI** with Subroutine LAPLFI, based on the 5-Point formula.

```
c   Program: PROG_II_2_LAPLFI    Last change: 20.03.2020 (17.54)

c   SOLUTIONS OF MODEL LAPLACE EQUATION WITH SUBROUTINE LAPLFI (Five
c   Point Formula)

c   Test case  : xmax=ymax=4.
c                u=0 for (x=0, y=0-ymax), and (x=0-xmax, y=0)
c                u=x**3               for y=ymax
c                u=(y/ymax)*ymax**3 for x=xmax.

c   LIST of Subroutines and Functions:
c   Subroutine LAPLFI
c   Subroutine MINV
c   Subroutine CPU_TIME

    implicit real*8 (a-h,o-z)
    Parameter (idima=400)
    DIMENSION RESID(idima)

    open (60,file='OUTPUT', status='unknown')
    open (10,file='OUTPL0', status='unknown') ! Computed solution and
residuals
    open (12,file='OUTPL2', status='unknown') ! Computed solution for 2D
plot
```

```
      call CPU_TIME (cput0)

      write(60,*) ('LAPLFI - Solves LAPLACE Eq. with 5-point formula')
      write( *,*) ('LAPLFI - Solves LAPLACE Eq. with 5-point formula')

      call LAPLFI (RESID)

      resmin= 1.e30       ! Find minimun and maximum of residuals
      resmax=-1.e30
      do 10 i=1,idima
      if(dabs(resid(i)).le.resmin) resmin=dabs(resid(i))
      if(dabs(resid(i)).ge.resmax) resmax=dabs(resid(i))
10     continue

      call CPU_TIME(cpuend)
      cputot=cpuend - cput0

      write(60,*) ' CPUTOT, resmin, resmax ', cputot, resmin, resmax
      write( *,*) ' CPUTOT, resmin, resmax ', cputot, resmin, resmax
      close(60)
      close(10)
      close(12)

      STOP
      END

      Subroutine LAPLFI (RESID)
      implicit real*8 (a-h,o-z)

c     Example taken from G. D. Smith, "Numerische Loesung von
c     partiellen Differentialgleichungen", Vieweg, pag. 214 ff.
c
c     INPUT: ni = number of interval in x direction
c            nj = number of interval in y direction
c            xmin, xmax, ymin, ymax = coordinates delimiting the
c                                       integration domain.
c            nnodes = (ni+1)*(nj+1)
c
      Parameter (idima=400)
      dimension a(idima,idima),am1(idima,idima),gamma(idima,idima),
     1      b(idima),z1(idima),z2(idima),u(idima),resid(idima)
      data xmin,xmax,ymin,ymax /0.d0, 4.d0, 0.d0, 4.d0/

      nip1=20
      njp1=20
      ni=nip1-1
      nj=njp1-1
      nnodes=nip1*njp1
c
      do 5 i=1,idima
      do 5 j=1,idima
       a(i,j)=0.d0
```

```
5   continue
c
    do 10 m=1,nnodes
      b(m)   =0.d0
      u(m)   =0.d0
      resid(m)=0.d0
10  continue
c
    dx=(xmax-xmin)/float(ni)
    dy=(ymax-ymin)/float(nj)
c
    do 200 m=1,nnodes          ! Set initial and boundary values
      m0=mod(m,nip1)

c   Subroutine is coded as if dx, dy were different
c   Boundary nodes are skipped and treated later
      if(m.ge.1 .and. m.le.njp1) go to 200
      if(m.ge.(nnodes-ni) .and. m.le.nnodes) go to 200
      if(m0.eq.0 .or. m0.eq.1) go to 200

      a(m,m-1)   = 1.d0
      a(m,m+1)   = 1.d0
      a(m,m-nip1) = 1.d0
      a(m,m+nip1) = 1.d0
      a(m,m)     = -4.d0
      b(m)=0.d0
200 continue
c
c   Boundary nodes
    do 100 m=1,njp1     ! For x = 0, left boundary
      a(m,m) =1.d0
      b(m)   =0.d0
100 continue
c
    do 102 i=1,nip1     ! For y = ymax, upper boundary
      m=(i-1)*nip1 + njp1
      x=float(i-1)*dx
      b(m)=x**3
      a(m,m)=1.d0
102 continue
c
    do 105 j=1,njp1     ! For y= 0, lower boundary
      m=(j-1)*njp1 + 1
      a(m,m)=1.d0
      b(m)  =0.d0
105 continue
c
    do 107 j=1,njp1     ! For x = xmax, rigth boundary
      m=(nip1-1)*nip1 + j
      y=float(j-1)*dy
      b(m)=(y/ymax)*ymax**3
      a(m,m)=1.d0
107 continue
```

```
   do 234 m=1,nnodes
     if(a(m,m) .eq. 0.d0) then
       write(60,*) ' STOP a(m,m) equal to zero for m ', m
       write( *,*) ' STOP a(m,m) equal to zero for m ', m
       stop
       endif
234  continue
c    ==================  Solution ================================
   ires=1
   call MINV (a,am1,u,B,RESID,z1,z2,nnodes,IRES,KOD,IDIMA,IDIMA)
c    ==============================================================
c
c    Check maximum residual. Residuals at the boundaries are zero.
   resmax=0.d0
   do 210 m=1,nnodes
     if(dabs(resid(m)).gt.resmax) resmax=dabs(resid(m))
210  continue

c    Fill OUTPL0 (computed solution and residuals)
   do 300 m=1,nnodes
     absres=dabs(resid(m))
     write(10,3020) m, absres,u(m)
3020   format(i6,2e14.5)
300  continue
c
   do 400 i=1,nip1    ! Fill OUTPL2 for plotting
   do 400 j=1,njp1
      x=float(i-1)*dx
      y=float(j-1)*dy
      m=(i-1)*nip1 + j
      write(12,1012) i,j,x,y,u(m)
1012   format(1H ,2i5,5e14.5)
400  continue
c
   return
   end

   SUBROUTINE MINV (ZA,ZC,Y,B,RESID,ZB,ZM,N,IRES,KOD,IDIMZA,IDIMZC)
   implicit real*8 (a-h,o-z)
C       MINV COMPUTES THE INVERSE OF A MATRIX
C       N IS THE DIMENSION OF THE INPUT/OUTPUT MATRICES
C       ZA(N,N) INPUT MATRIX NOT MODIFIED BY THE COMPUTATION.
C       ZC(N,N) OUTPUT MATRIX ZC=ZA**-1
C    MATRIX ZC DOES NOT NEED TO BE ZEROED
C       ZB(N)   AND  ZM(N) ARE  VECTORS FOR INTERMEDIATE CALCULUS
C     Y = SOLUTION VECTOR OF THE MATRICIAL EQUATION  A*Y=B (Y=ZC*B)
C    RESID= RESIDUALS VECTOR = A*Y-B
C    IRES= -1 ONLY INVERSE (NOT Y) IS CALCULATED
C    IRES=0 SOLUTION NOT REFINED
C    IRES=1 SOLUTION REFINED WITH THE RESIDUALS
C    KOD=0 NORMAL RESULTS;  KOD.NE.0 SINGULAR CASE, ZC IS NOT COMPUTED
C    IDIMZA= DIMENSION OF MATRIX ZA IN CALLING PROGRAMME
C    IDIMZC= DIMENSION OF MATRIX ZC IN CALLING PROGRAMME
```

```
c     DIMENSION ZA(IDIMZA,1),ZC(IDIMZC,1)
      DIMENSION ZA(IDIMZA,idimza),ZC(IDIMZC,idimzc)
      DIMENSION Y(1),B(1),RESID(1),ZB(1),ZM(1)
      KOD=0
      EPS=1.D-12
      DO 80 I=2,N
      DO 80 J=1,N
      ZC(I,J)=0.0D0
   80 CONTINUE
      DO 2 I=1,N
    2 ZC(I,I)=1./ZA(I,I)
      ITER=2
      ZM(1)=ZC(1,1)
      ZB(1)=0.D0
      DO 50 L=2,N
      ZB(L)=ZA(1,L)*ZC(L,L)
   50 ZC(1,L)=-ZB(L)*ZM(1)
   30 ITERP=ITER+1
      ITERM=ITER-1
      DO 33 I=1,N
   33 ZM(I)=ZC(I,ITER)
      IF(ITER.EQ.N) GOTO 6
      DO 36 L=1,N
      ZB(L)=0.D0
      DO 3 K=1,ITERM
    3 ZB(L)=ZB(L)+ZA(ITER,K)*ZC(K,L)
      DO 4 K=ITERP,N
    4 ZB(L)=ZB(L)+ZA(ITER,K)*ZC(K,L)
   36 CONTINUE
      GOTO 5
    6 CONTINUE
      DO 37 L=1,N
      ZB(L)=0.D0
      DO 37 K=1,ITERM
      ZB(L)=ZB(L)+ ZA(ITER,K)*ZC(K,L)
   37 CONTINUE
    5 ZIJ=1.0+ZB(ITER)
      IF(ABS(ZIJ).GT.EPS) GOTO 39
        KOD=1
        WRITE(60,8040) KOD
8040    FORMAT(' KOD',I10)
        RETURN
39     ZIJ=1.0/ZIJ
       DO 40 I=1,N
40     ZB(I)=ZB(I)*ZIJ
       DO 42 L=1,N
       ZIJ=ZB(L)
       DO 41 I=1,N
41     ZC(I,L)=ZC(I,L)-ZIJ*ZM(I)
42     CONTINUE
       ITER=ITER+1
       IF(ITER.LE.N) GOTO 30
43     CONTINUE
```

```
    IF(IRES.EQ.-1) RETURN
    DO 100 I=1,N
    Y(I)=0.0D0
    DO 95 J=1,N
    Y(I)=Y(I)+ZC(I,J)*B(J)
95  CONTINUE
100 CONTINUE
    IREP=0
120 CONTINUE
    DO 150 I=1,N
    RESID(I)=0.0D0
    DO 145 J=1,N
145 RESID(I)=RESID(I)+ ZA(I,J)*Y(J)
    RESID(I)=RESID(I)-B(I)
150 CONTINUE
    IF(IRES.EQ.0.OR.IREP.EQ.1) GO TO 250
    DO 200 I=1,N
    ZB(I)=0.0D0
    DO 195 J=1,N
    ZB(I)=ZB(I)+ZC(I,J)*RESID(J)
195 CONTINUE
    Y(I)=Y(I)-ZB(I)
200 CONTINUE
    IREP=1
    GO TO 120
250 CONTINUE
    RETURN
    END
```

OUTPUT From PROG_II_2_LAPLFI

```
LAPLFI - Solves LAPLACE Eq. with 5-point formula
CPUTOT, resmin, resmax 0.18720099999999998     0.0000000000000000
1.77635683940025046E-014
```

(iii) Program **PROG_II_3_LAPLUN** with Subroutine LAPLUN, based on under-relaxation.

```
c   Program: <PROG_II_3_LAPLUN.f>    Last change: 21.03.2020

C   SOLUTIONS OF MODEL LAPLACE EQUATION WITH SUBROUTINE LAPLUN

c   Test case  : xmax=ymax=4.
c                u=0 for (x=0, y=0-ymax), and (x=0-xmax, y=0)
c                u=x**3        for y=ymax
c                u=(y/ymax)*ymax**3 for x=xmax.

c   LIST of Subroutines and Functions:
c   Subroutine LAPLUN
c   Subroutine CPU_TIME
```

```
      implicit real*8 (a-h,o-z)
      Parameter (idima=400)
      DIMENSION RESID(idima)

      open (60,file='OUTPUT', status='unknown')
      open (10,file='OUTPL0', status='unknown') ! File for plotting
c                          omputed solution and residuals
      open (12,file='OUTPL2', status='unknown') ! For 2D plot of solution

      call CPU_TIME (cput0)

      write(60,*) 'Solution of LAPLACE equation with under-relaxation'
      write( *,*) 'Solution of LAPLACE equation with under-relaxation'

      call LAPLUN (RESID)

      call CPU_TIME(cpuend)
      cputot=cpuend - cput0
      write(60,*) ' cputot = ', cputot
      write( *,*) ' cputot = ', cputot

      close(60)
      close(10)
      close(12)

      STOP
      END

      Subroutine LAPLUN (RESID)
      implicit real*8 (a-h,o-z)

c     Solution of Laplace equation with underelaxation method.
c     Example taken from G. D. Smith, "Numerische Loesung von
c     partiellen Differentialgleichungen", Vieweg, pag. 214 ff.
c
c     INPUT: ni = number of interval in x direction
c            nj = number of interval in y direction
c            xmin, xmax, ymin, ymax = coordinates delimiting the
c                                    integration domain.
c            nnodes = (ni+1)*(nj+1)
c
      Parameter (idima=400)
      dimension a(idima,idima),am1(idima,idima),u(idima),resid(idima)
      data xmin,xmax,ymin,ymax /0.d0, 4.d0, 0.d0, 4.d0/

      nip1=20
      njp1=20
      ni=nip1-1
      nj=njp1-1
      nnodes=nip1*njp1
c
```

```
      do 10 m=1,nnodes
      u(m)   =0.d0
      resid(m)=0.d0
10    continue
c
      dx=(xmax-xmin)/float(ni)
      dy=(ymax-ymin)/float(nj)
c
c     ---------------- Initialization ----------------------
      do 95 i=2,ni   Set initial and boundary values
      do 95 j=2,nj
       a(i,j)= 0.d0
95    continue

      do 100 i=1,nip1
       x=float(i-1)*dx
       m=(i-1)*nip1 + 1
       a(i,1)  =0.d0
       u(m)=a(i,1)
c
       m=(i-1)*nip1 + njp1
       a(i,njp1)=x**3
       u(m)=a(i,njp1)
100   continue
c
      do 105 j=1,njp1
       y=float(j-1)*dy
       a(1,j)=0.d0
       u(j)=0.d0
c
       m=(nip1-1)*nip1 + j
       a(nip1,j)=(y/ymax)*ymax**3
       u(m)=a(nip1,j)
105   continue
c
c     Initial residuals
      do 110 i=2,ni
      do 110 j=2,nj
       am1(i,j)=a(i-1,j) +a(i+1,j) +a(i,j-1) + a(i,j+1) -4.d0*a(i,j)
110   continue
c
c     ---------------------- End initialization ----------------
c
c     Main loop to reduce residuals and to update values of residuals
c     at neighbouring points.
      omega=0.7d0
      iterx=10000
      do 250 iter=1,iterx
c
      do 200 i=2,ni
      do 200 j=2,nj
       acorr =omega * am1(i,j)*0.25d0
       a(i,j)=a(i,j) + acorr
```

```
      am1(i,j)=am1(i,j) -4.d0*acorr
      am1(i+1,j)=am1(i+1,j) + acorr
      am1(i-1,j)=am1(i-1,j) + acorr
      am1(i,j+1)=am1(i,j+1) + acorr
      am1(i,j-1)=am1(i,j-1) + acorr
200   continue                 ! End Under-relaxation
c

c
    resmax=0.d0
    do 210 i=2,ni              ! Check maximum residual
    do 210 j=2,nj
      m=(i-1)*nip1 + j
      u(m)=a(i,j)
      resid(m)=a(i-1,j) +a(i+1,j) +a(i,j-1) + a(i,j+1) -4.d0*a(i,j)
      if(dabs(resid(m)).gt.resmax) resmax=dabs(resid(m))
210   continue

    if(resmax.lt.1.d-12) go to 280

250   continue

280   continue
    write(60,*) ' LAPLUN - iter, resmax ', iter, resmax
    write( *,*) ' LAPLUN - iter, resmax ', iter, resmax

    do 300 m=1,nnodes  ! Fill OUTPL0 (computed solution and residuals)
      absres=abs(resid(m))
      write(10,3020) m, absres,u(m)
3020    format(i6,2e14.5)

300   continue

    do 400 i=1,nip1  ! Fill OUTPL2 with i,j,x,y,u(m) for 2D plot
    do 400 j=1,njp1
      x=float(i-1)*dx
      y=float(j-1)*dy
      m=(i-1)*nip1 + j
      write(12,1012) i,j,x,y,u(m)
1012    format(1H ,2i5,3e14.5)
400   continue
c
    return
    end
```

OUTPUT from PROG_II_3 LAPLUN

```
Solution of LAPLACE equation with under-relaxation
LAPLUN - iter, resmax      1993 9.98312543742940761E-013
cputot =  4.67999999999999944E-002
```

Appendix A.III

PROG_III_1 [FRONT1] Numerical solution of model Eq. (3.97), with a choice of initial and boundary conditions, with orthogonal collocation methods and with finite differences.

```
c
c   APPENDIX A.III, Program: PROG_III_1_FRONT1 - Last change 24.02.2020
-

c   -----------------------------------------------------------------
C   NUMERICAL SIMULATION OF WAWE-FRONT PROPAGATION
c   ---------------------------------------------  ----------------
c   MAIN
c   LIST OF SUBROUTINES:
c   Subroutine FRONT1 - Driving subroutine. A main loop runs over
C              METH=0     SUBDOMAIN COLLOCATIONS
C              METH=1     POINTS COLLOCATIONS
C              METH=2     GALERKIN
C              METH=3     FINITE DIFFERENCES:
C                     isolfd=1 Explicit Point Jacobi
c                            =2 Explicit Point Gauss
c                            =3 Implicit
c   Subroutine MINVIS - Solves for Y a system of algebraic equations
c              written in matrix form [A] Y = B
c   Subroutine TRIDI -  Solves a system of algebraic equations with
c              tri-diagonal matrix of coefficients.
c   Subroutine CERROR - Computes an error based on the discrepancy
c              between numerical and theoretical results.

   implicit real*8 (a-h,o-z)
   open (60,file='OUTPUT', status='unknown')
   open (10,file='OUTPL0', status='unknown') ! Subdomain Collocations
   open (11,file='OUTPL1', status='unknown') ! Point Collocations
   open (12,file='OUTPL2', status='unknown') ! Galerking method
   open (13,file='OUTPL3', status='unknown') ! Point Jacobi
   open (14,file='OUTPL4', status='unknown') ! Point Gauss
   open (15,file='OUTPL5', status='unknown') ! Implicit scheme

c   itest=1: Front wave equal to 1 for z < Z0, equal to 0 for z > Z0.
C        *******************************
C        TEST CASE DU/DT + D(U**2/2)/DZ=0
C             U(Z,0)=1       (Z.LT.Z0)
C             U(Z,0)=0       (Z.GE.Z0)
C        *******************************
c   itest=2: u(0,z)=sin(pai*z) for z < Z0,
c        u(0,z)=0        for z > Z0.
c        The wave breakes at t=1/pai = 0.318 sec
c
   itest=1 ! Default. User may change it to itest =2.
c
```

```
      write(60,*) ' MAIN PROG_III_1_FRONT1>' ! Write on OUTPUT
      write( *,*) ' MAIN PROG_III_1_FRONT1>' ! Write on computer screen

      CALL FRONT1(itest)
c
      close(60)
      close(10)
      close(11)
      close(12)
      close(13)
      close(14)
      close(15)
      stop
      end

      SUBROUTINE FRONT1(itest)
      implicit real*8 (a-h,o-z)

c     List of main FORTRAN variables

c     AM   = Matrix used for solution with subroutine MINVIS
c     AMI  = Auxiliary matrix used for solution with subroutine MINVIS
c     ATRI = Tri-dimensional matrix
c     B    = Known vector in the matrix equation AM * Y = B
c     B1   = Auxiiary vector used in subroutine TRIDI
c     UO   = Initial value of wave
c     UR   = Wave amplitude at iteration step r
c     URM1 = Wave amplitude at iteration step r-1
c     U1   = Wave amplitude at time level n

       parameter (IDIMZ=162) ! Number of nodes (must be divisible by 3)
c
       DIMENSION AM(IDIMZ,IDIMZ),AMI(IDIMZ,IDIMZ),ATRI(IDIMZ,3)
       DIMENSION UO(IDIMZ),U1(IDIMZ),UR(IDIMZ),URM1(IDIMZ),B(IDIMZ),
      1     BI(IDIMZ)
c
       DATA ZERO,ONE,TWO,TRI,FOUR,FIVE,SIX/0.d0, 1.d0, 2.d0, 3.d0,
      1                 4.d0, 5.d0, 6.d0/
       PAI=4.d0*datan(1.d0)
       DT=1.D-03
       if(itest.eq.1) ITMAX=1001
       if(itest.eq.2) ITMAX= 320

       if(itest.eq.1) write(60,9001), DT
       if(itest.eq.1) write( *,9001), DT
       if(itest.eq.2) write(60,9002), DT
       if(itest.eq.2) write( *,9002), DT
9001  format(/,' FRONT1 - Test case 1: Front wave, DT =',e12.5,/)
9002  format(/,' FRONT1 - Test case 2: u(0,z)=sin(pai*z),DT=',e12.5,/)

       MZ=(IDIMZ-1)/2    ! Number of meshes
       MZM1=MZ-1
```

```
      MZD2=MZ/2
      Z0=1.D0
      DZ=2.D0*Z0/float(MZ)
      IAMMAX = 2 * MZ + 1

      OMEGA = 0.5d0         ! Under-relaxation parameter
      THETA = 0.5D0         ! Time discretization parameter
      TEDT=THETA*DT
      THHH=THETA*DT - (ONE-THETA)*DT
      TEDTDS=THETA*DT/SIX

      DO 510 METH = 0,3       ! Loop over all solution methods
      T=0.D0
      IF(METH.EQ.3) GO TO 600  ! Go to separate branch

c     ------------------------- Initialisation --------------------
      DO 2 I=1,IAMMAX        ! Blank matrix AM and vector B
         B(I)=0.
      DO 2 J=1,IAMMAX
         AM(I,J)=0.D0
2     continue

c
      DO 5 I=1,IAMMAX
         MZM=IAMMAX-I
         if(itest.eq.1) U1(I)=1.D0
c
         if(itest.eq.2) then
         z=0.5d0*dz*float(i)
         u1(i)=dsin(pai*z)
         endif

        IF(I.GT.(IAMMAX/2)) U1(I)=0.d0
          UO(I)=U1(I)
          if(METH .eq. 0) write(60,888) i, uo(i)  ! Write initial values
          if(METH .eq. 0) write( *,888) i, uo(i)
888        format(' FRONT1 - Initial values i, uo(i)' i5,e12.5)
5      CONTINUE
c     ----------------------------------------------------------------
      IF(METH.EQ.0) WRITE(60,1000)
      IF(METH.EQ.0) WRITE( *,1000)
      IF(METH.EQ.1) WRITE(60,1001)
      IF(METH.EQ.1) WRITE( *,1001)
      IF(METH.EQ.2) WRITE(60,1002)
      IF(METH.EQ.2) WRITE( *,1002)

1000 FORMAT(/,' METH = 0 *** METHOD OF SUBDOMAIN COLLOCATIONS',/)
1001 FORMAT(/,' METH = 1 *** METHOD OF POINTS COLLOCATIONS',/)
1002 FORMAT(/,' METH = 2 *** METHOD OF GALERKIN',/)
c
      DO 500 IT=1,ITMAX
```

```
      do 380 i=1,IAMMAX  ! Blank Matrix AM
      do 380 j=1,IAMMAX
        AM(i,j)=0.d0
380   continue
c
     T=T+DT
     IMUM=0
c
     DO 400 M=1,MZ
        IMUM=IMUM+2
        IMUI=IMUM-1
        IMUO=IMUM+1
        DUDZN0=-3.D0*U1(IMUI)+4.D0*U1(IMUM)-U1(IMUO)
        DUDZ12=U1(IMUO)-U1(IMUI)
        DUDZN1=U1(IMUI)-4.D0*U1(IMUM)+3.D0*U1(IMUO)
        IF(METH.EQ.1) GO TO 100
        IF(METH.EQ.2) GO TO 200

C       METH = 0, SUBDOMAIN COLLOCATIONS - Use Eq.s (2.22) of text

       AM(IMUM,IMUI)=DZ/SIX - TEDT*U1(IMUI)
       AM(IMUM,IMUM)=2.D0*DZ/3.D0
       AM(IMUM,IMUO)=DZ/SIX + TEDT*U1(IMUO)
       B(IMUM)=DZ*(U1(IMUI)+4.D0*U1(IMUM)+U1(IMUO))/SIX
   1   +(U1(IMUI)*DUDZN0+4.D0*U1(IMUM)*DUDZ12+U1(IMUO)*DUDZN1)*
   2   THHH/SIX

     GO TO 230
c    ============================================================
100  continue
C     METH = 1, POINTS COLLOCATIONS - Use Eq.s (3.27) of text

     AM(IMUM,IMUI)= DZ +TEDT*(-6.d0*U1(IMUI) + 3.d0*U1(IMUM) )
     AM(IMUM,IMUM)= DZ +TEDT*(3.d0*(U1(IMUI) - U1(IMUO))   )
     AM(IMUM,IMUO)= DZ +TEDT*(-3.d0*U1(IMUM) + 6.d0*U1(IMUO) )
     B(IMUM)=DZ*(U1(IMUI) + U1(IMUM) + U1(IMUO)) +
   1  (U1(IMUI)*DUDZN0 + U1(IMUM)*DUDZ12 + U1(IMUO)*DUDZN1)*THHH

     GO TO 230
c    =============================================================
200  CONTINUE
C     METH = 2, GALERKIN METHOD - Use Eq.s (4.29) of text

     AM(IMUM,IMUI)= DZ/SIX - TEDT*U1(IMUI)
     AM(IMUM,IMUM)= FOUR*DZ/SIX
     AM(IMUM,IMUO)= DZ/SIX + TEDT*U1(IMUO)
     B(IMUM)= DZ*(U1(IMUI)+ FOUR*U1(IMUM)+ U1(IMUO))/SIX
   1  +THHH*(U1(IMUO)**2-U1(IMUI)**2)/TWO

230   CONTINUE
C    ADD DISSIPATIVE TERM: See: Beam R. M., Warming R. F.,
c    Journal of Computational Physics, 22, pp. 87-110 (1976).
```

```
      IF(M.GE.3.AND.M.LE.(MZ-2)) B(IMUM)=B(IMUM)
     1  - OMEGA*(U1(IMUM-2)-4.*U1(IMUM-1)+6.*U1(IMUM)-4.*U1(IMUM+1)
     2  +U1(IMUM+2))/8.D0 *DZ
C
C    For METH=0/2: Impose continuity of derivative at mesh boundaries
      IF(M.EQ.1) GO TO 250
        AM(IMUI,IMUI )=SIX
        AM(IMUI,IMUI -2)=1.D0
        AM(IMUI,IMUI -1)=-4.D0
        AM(IMUI,IMUI +1)=-4.D0
        AM(IMUI,IMUI +2)=1.D0
250     CONTINUE

C    SET BOUNDARY CONDITIONS
c
     if(M.eq.1) THEN
       DO 270 IAM=1,IAMMAX
         AM(1,IAM)=ZERO
270     CONTINUE
       AM(1,1)=ONE
       if(itest.eq.1) B(1)=1.D0
       if(itest.eq.2) B(1)=0.D0
     ENDIF
c
     if(M.eq.MZ) THEN
       DO 290 IAM=1,IAMMAX
        AM(IMUO,IAM)=ZERO
290    CONTINUE
        AM(IMUO,IMUO)=ONE
        B(IMUO)=0.D0
     ENDIF
c
400    CONTINUE        ! End loop over meshes
c
c    ------------------- SOLUTION ----------------------------------
     IRESMV=0
     call MINVIS(AM,AMI,UR,B,IAMMAX,IDIMZ,IRESMV)
c
-----------------------------------------------------------------

     if(IT.eq.ITMAX) THEN   ! Print Results  for METH = 0/1/2
       IMUM=0
       DO 475 M=1,MZ
         IMUM=IMUM + 2
         Z=(M-1)*DZ  + DZ/2.d0
         if(METH.eq.0) WRITE(10,8010) Z ,UR(IMUM)  ! Store results
         if(METH.eq.1) WRITE(11,8010) Z ,UR(IMUM)
         if(METH.eq.2) WRITE(12,8010) Z ,UR(IMUM)
8010        format(2f11.3)
475     CONTINUE

       IF(ITEST.eq.1) CALL CERROR(METH,MZ,Z0,T,DZ,UR)
     ENDIF
```

```
c   ------------------------------------------------
   DO 480 IAM=1,IAMMAX
     UO(IAM)=U1(IAM)
     U1(IAM)=UR(IAM)
480  CONTINUE

500  CONTINUE   ! End Loop IT = 1, ITMAX
510  CONTINUE   ! End Loop METH = 0,3
   Go TO 900
c  ========= End solution with Orthogonal CCollocations ==========

600  CONTINUE   ! FINITE DIFFERENCES (METH = 3)

   DO 800 ISOLFD = 1,3
   T=0.d0
c
------------------------------------------------------------
   if(ISOLFD.eq.1) THEN
     write(60,*) ' '
     write( *,*) ' '
     write(60,*) ' Expl. Point Jacobi, isolfd=', isolfd
     write( *,*) ' Expl. Point Jacobi, isolfd=', isolfd
    ENDIF
   if(ISOLFD.eq.2) THEN
      write(60,*) '/'
     write( *,*) '/'
     write(60,*) ' Expl. Point Gauss,  isolfd=', isolfd
     write( *,*) ' Expl. Point Gauss,  isolfd=', isolfd
    ENDIF
   if(ISOLFD.eq.3) THEN
      write(60,*) '/'
     write( *,*) '/'
     write(60,*) ' Implicit, isolfd=      ', isolfd
     write( *,*) ' Implicit, isolfd=      ', isolfd
   ENDIF

c  ------------------------------------------------------------

   DO 705 M=1,MZ
     if(ITEST.eq.1) U1(M)=1.d0
     if(ITEST.eq.2) THEN
       Z=(M-1)*DZ + 0.5d0*DZ
       U1(M)=dsin(PAI*Z)
     ENDIF
    IF(M.GT.MZD2) U1(M)=1.D-20
705  CONTINUE

   IF(ISOLFD.eq.1 .or. ISOLFD.eq.2) THEN
c  Explicit ( POINT JACOBI or POINT GAUSS)
   EPSI= 1.d-08
   DTD4DZ=DT*0.25d0/DZ
```

```
    DO 741 M=1,MZ
      URM1(M)=U1(M)
741  CONTINUE

    DO 745 IT=1,ITMAX
    T = T + DT
c   -------------------------- Start iterations ------------------
    DO 728 ITREXP = 1,100
    UR(1)  = U1(1)
    UR(mz) = U1(mz)
      DO 720 M=2,MZM1
        IF(ISOLFD . eq. 1)
   1     ur(m)=u1(m)-dtd4dz*(urm1(m+1)**2-urm1(m-1)**2)

         if(ISOLFD . eq. 2)
   1     ur(m)=u1(m)-dtd4dz*(urm1(m+1)**2-ur(m-1)**2)

c        Add dissipative term
        IF(m.ge.3 .and. m.le. (mz-2)) ur(m) = ur(m)
   1     - OMEGA*(u1(m-2)-4.d0*u1(m-1)+6.d0*u1(m)
   2     -4.d0*u1(m+1)+ u1(m+2))/8.d0
720     continue

      diffmx=0.d0
      do 725 m=1,mz
        diff=dabs(ur(m)-urm1(m))/dabs(ur(m))
        if(diff .gt. diffmx) diffmx= diff
725     continue
      if(diffmx .lt. epsi) go to 738

      do 726 m=1,mz
726     urm1(m) = ur(m)
728  continue
c   -------------------------- End iterations -------------------
738  continue
c    write(*,*) ' ITREXP ', ITREXP
       do 743 m=1,mz
743     u1(m) = ur(m)
745     continue

c------------------- Print results for METH = 3 ---------------------
c
    if(IT.ge.ITMAX) THEN
      DO 752 M=1,MZ
        Z=(M-1)*DZ+ DZ/2.d0
        IF(ISOLFD.eq.1) WRITE(13,8010) Z,UR(M)
        if(ISOLFD.eq.2) WRITE(14,8010) Z,UR(M)
752     CONTINUE
      IF(ITEST.eq.1) CALL CERROR(METH,MZ,Z0,T,DZ,UR)
    ENDIF
```

```
   ENDIF
c  ------------------------------------------------------------
   if(ISOLFD.eq.3) then  ! IMPLICIT (BEAM,WARMING)
c
   do 760 it=1,itmax
   t = t + dt
   ATRI(1,1)  = 0.d0
   ATRI(1,3)  = 0.d0
   ATRI(mz,1) = 0.d0
   ATRI(mz,3) = 0.d0
   ATRI(1,2)  = 1.d0
   ATRI(mz,2) = 1.d0
   if(itst .eq.1) B(1)=1.d0
   if(itst .eq.2) B(1)=0.d0
   B(mz) = 0.d0

   do 753 m=2,mzm1
    ATRI(m,1) = -u1(m-1)*dt/(4.d0*dz)
    ATRI(m,2) = 1.d0
    ATRI(m,3) = u1(m+1)*dt/(4.d0*dz)
    B(m) = u1(m)
c    Add dissipative term.
    if(m.ge.3 .and. m. le. mz-2) B(m) = B(m)
  1    - OMEGA*(u1(m-2)-4.d0*u1(m-1)+6.d0*u1(m)
  2    -4.d0*u1(m+1)+ u1(m+2))/8.d0
753 continue
c
--------------------------------------------------------------
   call TRIDI(ATRI,UR,B,BI,IDIMZ,MZ,KOD)
c
--------------------------------------------------------------
   if(it.ge.itmax) then
    DO 755 M=1,MZ
     Z=(M-1)*DZ+ DZ/2.
     if(ISOLFD.eq.3) WRITE(15,8010) Z,UR(M)
755   CONTINUE
    IF(ITEST.eq.1) CALL CERROR(METH,MZ,Z0,T,DZ,UR)
   endif

c-------------------------------------------
--------------------
   do 758 M=1,MZ
    u1(M) = ur(M)
758 CONTINUE

760 CONTINUE
c-----------------------------------------------
----------------
   endif
800 continue        ! End Loop ISOLFD= 1/2/3

900 continue
```

```
      close(10)
      close(11)
      close(12)
      close(13)
      close(14)
      close(15)
c
      open (10,file='OUTPL0', status='old') ! Subdomain Collocations
      open (11,file='OUTPL1', status='old') ! Point Collocations
      open (12,file='OUTPL2', status='old') ! Galerking method
      open (13,file='OUTPL3', status='old') ! Point Jacobi
      open (14,file='OUTPL4', status='old') ! Point Gauss
      open (15,file='OUTPL5', status='old') ! Implicit scheme

      write(60,*) (' ')
      write( *,*) (' ')
      write(60,*) 'Table: z, UR 6 solutions'
      write( *,*) 'Table: z, UR 6 solutions'

      DO 910 m=1,mz
           read(10,8010) Z ,  a1
           read(11,8010) dummy, a2
           read(12,8010) dummy, a3
           read(13,8010) dummy, a4
           read(14,8010) dummy, a5
           read(15,8010) dummy, a6
        write(60,8015) m,z,a1,a2,a3,a4,a5,a6
        write( *,8015) m,z,a1,a2,a3,a4,a5,a6
8015  format(i5, 7e11.4)
910   CONTINUE
c    Information about results. If dt=0.001 with MZ=80:
c
c    Error (computed if itest=1):
c        error = 0.0532 with METH=0 (Subdomain collocations)
c        error = 0.0468 with METH=1 (Points collocations)
c        error = 0.0532 with METH=2 (Galerkin method)
c        error = 0.0655 with METH=3 (FD, explicit, Jacobi scheme)
c         error = 0.0655 with METH=3 (FD, explicit, Gauss scheme)
c        error = 0.0654 with METH=3 (FD, implicit)

      RETURN
      END

      subroutine cerror(meth,mz,z0,t,dz,ur)
c    Compute LER = Length of error vector. This subroutine is
c    used only for itest=1, because only in this case the reference
c    values are known.
      implicit real*8 (a-h, o-z)
      dimension ur(1)
c
      error=0.d0
      zend= z0 + t*0.5d0
c
```

```
      if(meth.lt.3) then
      imum=0
      do 90 m=1,mz
        imum=imum+2
        z=(m-1)*dz + dz/2.d0
        if(z.lt.zend) uref=1.d0
        if(z.ge.zend) uref=0.d0
        error=error + (ur(imum)-uref)**2
90    continue
        error=error/float(imum/2)
      else
      do 95 m=1,mz
        z=(m-1)*dz + dz/2.d0
        if(z.lt.zend) uref=1.d0
        if(z.ge.zend) uref=0.d0
        error=error+ (ur(m)-uref)**2
95    continue
      error=error/float(mz)
      endif
c
      error = dsqrt(error)
      write(60,1000) t, error
      write( *,1000) t, error
1000 format(' Sub. CERROR - t, error ',2e12.5)
c
      return
      end

      SUBROUTINE MINVIS(B,A,Y,BM,N,IDIMA,IRES)
C     USED FOR MATRIX INVERSION
C     MATRIX B IS STORED IN A
C     MATRIX B IS REPLACED BY THE INVERSE
C     IRES=-1 ONLY ONVERSE (NOT Y) IS CALCULATED
C     IRES= 0 Y IS CALCULATED
      implicit real*8 (a-h,o-z)
      DIMENSION B(IDIMA,1),A(IDIMA,1)
      DIMENSION Y(1),BM(1)
c
      DO 30 I=1,N
      IF(B(I,1)) 33,32,33
   32 WRITE(6,1001) I
      STOP
   33 RESU=1./B(I,1)
      DO 34 J=2,N
   34 B(I,J-1)=B(I,J)*RESU
      B(I,N)=RESU
      DO 30 I1=1,N
      IF(I1-I) 35,30,35
   35 RESU=B(I1,1)
      DO 36 K=2,N
   36 B(I1,K-1)=B(I1,K)-RESU*B(I,K-1)
      B(I1,N)=-RESU*B(I,N)
   30 CONTINUE
```

```
c
   IF(IRES.EQ.-1) RETURN
   DO 120 I=1,N
   Y(I)=0.
   DO 120 J=1,N
   Y(I)=Y(I)+ B(I,J)*BM(J)
120  CONTINUE

125  CONTINUE

500  continue
c
 1001 FORMAT('0     NULL PIVOT I=',I5)
   RETURN
   END

   SUBROUTINE TRIDI(AM,Y,B,B1,IA,N,KOD)
   implicit real*8 (a-h,o-z)
C   SOLUTION OF A SYSTEM OF N ALGEBRAIC LINEAR EQUATIONS WITH TRIDIAGO
C   NAL MATRIX OF COEFFICIENTS.
C   REFERENCE: G.E.FORSYTHE, W.R.WASOW-FINITE DIFFERENCE METHODS FOR
C   PARTIAL DIFFERENCE EQUATIONS (PAGE 103-105).
C   PIVOTING NOT PERFORMED
C   N= NUMBER OF EQUATIONS TO BE SOLVED
C   AM(N,3) CONTAINS ONLY THE TRIDIAGONAL MATRIX ELEMENS (MATRIX COEFF
C        ICIENTS ARE PRESERVED)
C   MATRIX AM(N,3) MUST BE FILLED AS FOLLOWS
C          A*X(I-1) + B*X(I) + C*X(I+1)
C          0       X      X
C          X       X      X
C         ----------
C          X       X      X
C          X       X      0
C   Y(N) IS THE SOLUTION VECTOR
C   B(N) CONTAINS THE KNOWN TERMS OF THE MATRIX EQUATION A.Y= B
C   B1=  WORKING SPACE
C   IA= NUMBER OF ROWS OF MATRIX AM IN CALLING PROGRAMME  (MAX. DIMENS
C   IF KOD=0 NORMAL END OF THE CALCULATION
C   IF KOD=1 WARNING
C   IF KOD=2 ABNORMAL END OF THE CALCULATION
   DIMENSION AM(IA,3),B(1),Y(1),B1(1)
   KOD=0
   ZERO=0.D0
   Y(1) =AM(1,3)/AM(1,2)
   B1(1)=B(1)/AM(1,2)
   DO 10 I=2,N
   DEN= AM(I,2) - AM(I,1)*Y(I-1)
   AEN= ABS(DEN)
   IF(AEN.EQ.ZERO) GO TO 102
   IF(AEN.LT.1.D-06) KOD=1
   IF(I.EQ.N) GO TO 5
   Y(I)= AM(I,3)/DEN
  5 B1(I)= (B(I)- AM(I,1)*B1(I-1))/DEN
```

```
  10 CONTINUE
   Y(N)=B1(N)
   DO 20 I=2,N
   II=N-I+1
   Y(II)=B1(II)-Y(II)*Y(II+1)
  20 CONTINUE
   IF(KOD.NE.0) WRITE(6,1000) KOD
1000 FORMAT('  ******** KOD FROM TRIDI ',I5)
   RETURN
 102 KOD=2
   WRITE(6,1002) I,KOD
1002 FORMAT(' SOLUTION OF ALGEBRAIC SYSTEM HAS TO BE CHANGED-KOD=',2I5)
   RETURN
   END
```

```
OUPTPUT from FRONT1 for itest=1:

MAIN PROG_III_1_FRONT1>
[Nodes 1-70, where all wave values are u=1, and nodes 85- 161, where all
wave values are u=0, were omitted]

 FRONT1 - Initial values i, uo(i)  71 0.10000E+01
 FRONT1 - Initial values i, uo(i)  72 0.10000E+01
 FRONT1 - Initial values i, uo(i)  73 0.10000E+01
 FRONT1 - Initial values i, uo(i)  74 0.10000E+01
 FRONT1 - Initial values i, uo(i)  75 0.10000E+01
 FRONT1 - Initial values i, uo(i)  76 0.10000E+01
 FRONT1 - Initial values i, uo(i)  77 0.10000E+01
 FRONT1 - Initial values i, uo(i)  78 0.10000E+01
 FRONT1 - Initial values i, uo(i)  79 0.10000E+01
 FRONT1 - Initial values i, uo(i)  80 0.10000E+01
 FRONT1 - Initial values i, uo(i)  81 0.00000E+00
 FRONT1 - Initial values i, uo(i)  82 0.00000E+00
 FRONT1 - Initial values i, uo(i)  83 0.00000E+00
 FRONT1 - Initial values i, uo(i)  84 0.00000E+00

METH = 0 *** METHOD OF SUBDOMAIN COLLOCATIONS

 Sub. CERROR - t, error 0.10010E+01 0.53222E-01

 METH = 1 *** METHOD OF POINTS COLLOCATIONS

 Sub. CERROR - t, error 0.10010E+01 0.46829E-01

 METH = 2 *** METHOD OF GALERKIN

 Sub. CERROR - t, error 0.10010E+01 0.53222E-01

  Expl. Point Jacobi, isolfd=       1
 Sub. CERROR - t, error 0.10010E+01 0.65504E-01
```

```
 Expl. Point Gauss, isolfd=      2
Sub. CERROR - t, error 0.10010E+01 0.65504E-01

 Implicit, isolfd=           3
Sub. CERROR - t, error 0.10010E+01 0.65458E-01

Table: z, UR 6 solutions
[Meshes 1-70, where all wave values are u=1, and meshes 85- 161, where all
wave values are u=0, were omitted]

 51 0.1263E+01 0.1000E+01 0.1000E+01 0.1000E+01 0.9780E+00 0.9780E+00
0.9780E+00
 52 0.1288E+01 0.1002E+01 0.1000E+01 0.1002E+01 0.9800E+00 0.9800E+00
0.9790E+00
 53 0.1313E+01 0.1003E+01 0.1000E+01 0.1003E+01 0.9950E+00 0.9950E+00
0.9940E+00
 54 0.1338E+01 0.9990E+00 0.1000E+01 0.9990E+00 0.1027E+01 0.1027E+01
0.1026E+01
 55 0.1363E+01 0.9850E+00 0.1001E+01 0.9850E+00 0.1071E+01 0.1071E+01
0.1071E+01
 56 0.1388E+01 0.9830E+00 0.9980E+00 0.9830E+00 0.1112E+01 0.1112E+01
0.1113E+01
 57 0.1413E+01 0.1032E+01 0.9910E+00 0.1032E+01 0.1117E+01 0.1117E+01
0.1120E+01
 58 0.1438E+01 0.1111E+01 0.1024E+01 0.1111E+01 0.1050E+01 0.1050E+01
0.1053E+01
 59 0.1463E+01 0.1019E+01 0.1058E+01 0.1019E+01 0.8840E+00 0.8840E+00
0.8860E+00
 60 0.1488E+01 0.5590E+00 0.5940E+00 0.5590E+00 0.6320E+00 0.6320E+00
0.6320E+00
 61 0.1513E+01 0.4700E-01-0.2200E-01 0.4700E-01 0.3490E+00 0.3490E+00
0.3480E+00
 62 0.1538E+01-0.1140E+00-0.7400E-01-0.1140E+00 0.1090E+00 0.1090E
+00 0.1070E+00
 63 0.1563E+01-0.5000E-01 0.1800E-01-0.5000E-01-0.4200E-01-0.4200E-
01-0.4400E-01
 64 0.1588E+01 0.1100E-01 0.1200E-01 0.1100E-01-0.1000E+00-0.1000E
+00-0.1010E+00
 65 0.1613E+01 0.1700E-01-0.5000E-02 0.1700E-01-0.9300E-01-0.9300E-
01-0.9300E-01
 66 0.1638E+01 0.4000E-02-0.2000E-02 0.4000E-02-0.5800E-01-0.5800E-
01-0.5800E-01
 67 0.1663E+01-0.3000E-02 0.1000E-02-0.3000E-02-0.2100E-01-0.2100E-
01-0.2100E-01
 68 0.1688E+01-0.2000E-02 0.0000E+00-0.2000E-02 0.4000E-02 0.4000E-
02 0.4000E-02
 69 0.1713E+01-0.0000E+00-0.0000E+00-0.0000E+00 0.1500E-01 0.1500E-
01 0.1500E-01
 70 0.1738E+01 0.1000E-02-0.0000E+00 0.1000E-02 0.1500E-01 0.1500E-01
0.1500E-01
```

Appendix A.IV

PROG_IV_1 [TEST_RNG] Test for real randomness of subroutines used as random number generators.

```
c      APPENDIX A.IV, Program; PROG_PREF_IV_1 [TEST_RNG] - Stand:
       24.02.2020

c==================================================================
C      Test of random number generators RANDOM and RF
c =================================================================

c      LIST OF SUBROUTINES and FUNTIONS
c      Subroutine CHECK_RANDOM: Checks random number generator "RANDOM",
       based on
c              modulus function, by computing number PI (Greek pai)
c      Subroutine CHECK_RF: Checks random number generator "RF" from Bird
       library.
c       Subroutine RW  : Simulation of random walk by simple sampling.
c       Subroutine CPU_TIME
c       Function RANDOM : Random number generator based on modulus
        function.
c       Function RF    : Random number generators taken from Bird library.
c              Reference: G. A. Bird, Molecular Gas Dynamics and
c              the Simulation of Gas Flows, Clarandon Press,
c              Oxford 1994.
c       MAIN
c       Simulation of random walk by simple sampling.
c       Reference: K. Binder, D. W. Heermann, "Monte Carlo simulation
c       in statistical physics", Springer Verlag, 1988. See section 3.2,
        p. 73 ff.

      call CPU_TIME (cput0)

      open(60,file='OUTPUT',     status='unknown')
      open(10,file='OUTPL0_RANDOM',status='unknown') ! Write NSTEPX, EDZ2
      open(11,file='OUTPL0_RF'    ,status='unknown') ! Write NSTEPX, EDZ2
      write (60,1000)          ! Write on OUTPUT file
      write (* ,1000)          ! Write on Computer screen
1000  format(' Program PROG_PREF_IV_1 [TEST_RNG] ')
c
      call CHECK_RANDOM        ! Compute Greek Pai with function RANDOM
      call CHECK_RF          ! Compute Greek Pai with function RF

      call CPU_TIME(T1)
      TIME1= T1 - CPUT0
      write(60,*) ' Time for calc. with CHECH_RANDOM + _RF)', TIME1
      write( *,*) ' Time for calc. with CHECH_RANDOM + _RF)', TIME1
c     ---------------------------------------------------------------
c     iran=1: Use Function "RANDOM" for random number generation.
c     iran=2: Use Function "RF"    for random number generation.
```

```
c   The code is organized to treat both cases in a single run.
c   --------------------------------------------------------------

    call RW
c
    close(60)
    close(10)
    close(11)

    call CPU_TIME (cpuend)
    cpu=cpuend - cput0
    write(60,*) ' END - Total CPU time =     ', cpu
    write( *,*) ' END - Total CPU time =     ', cpu

    close(60)
    close(10)
    close(11)

    stop
    end
c
c
    subroutine RW
    implicit real*8 (a-h, o-z)
c
c    ======================================================
c    Definition of main variables
c    edx2= mean value of the square of final distance in
c       x-direction
c    edy2= mean value of the square of final distance in
c       y-direction
c    edz = mean value of the final distance in z-direction
c    edz2= mean value of the square of final distance.
c
c    iseed = seed used for the random function (must be odd)
c
c    nsampx = number of samples taken before averaging
c    nstepx = number of steps for every random walk
c    ====================================================
c
    do 900 IRAN= 1,2
    write(60,*) ' IRAN = ', IRAN
    write( *,*) ' IRAN = ', IRAN

    if(iran.eq.1) THEN
      iseed=1
      call CPU_TIME(T2)
    ENDIF

    if(iran.eq.2) THEN
      arf=rf(-1)
      call CPU_TIME(T3)
       TIME_RANDOM = T3 - T2
     ENDIF
```

```
      nsampx = 100

      do 800 nstepx=1,10 000   ! Number of steps for every random walk
      edx2=0.d0
      edy2=0.d0
      edz2=0.d0
      edz =0.d0
c
      do 500 nsamp=1,nsampx ! Number of samples taken before averaging
c
      x=0.d0
      y=0.d0
c
      if(iran.eq.1) then
c
      do 100 nstep=1,nstepx
      ir=random(iseed)*4
      if(ir.eq.0) x= x + 1.d0
      if(ir.eq.1) y= y + 1.d0
      if(ir.eq.2) x= x - 1.d0
      if(ir.eq.3) y= y - 1.d0
100   continue
c
       else
c
       do 200 nstep=1,nstepx
      ir=rf(0)*4
      if(ir.eq.0) x= x + 1.d0
      if(ir.eq.1) y= y + 1.d0
      if(ir.eq.2) x= x - 1.d0
      if(ir.eq.3) y= y - 1.d0
200   continue
c
      endif
c
      x2=x**2
      y2=y**2
      z2= x2 +y2
      z = dsqrt(z2)
      rdist=dsqrt(x2 + y2)
c
      edx2=edx2 + x2
      edy2=edy2 + y2
      edz2=edz2 + z2
      edz =edz  + z


500   continue
       edx2 = edx2/float(nsampx)
       edy2 = edy2/float(nsampx)
       edz2 = edz2/float(nsampx)
       edz  = edz /float(nsampx)
       if(mod(nstepx,10).eq.0.and.IRAN.eq.1) write(10,1020) nstepx,edz2
       if(mod(nstepx,10).eq.0.and.IRAN.eq.2) write(11,1020) nstepx,edz2
1020   format(i12, e12.5)
```

```
800  continue

900  CONTINUE

      call CPU_TIME(T4)
      TIME_RF = T4 - T3
      write(60,*) ' Time for calc. with Function RANDOM)', TIME_RANDOM
      write( *,*) ' Time for calc. with Function RANDOM)', TIME_RANDOM

      write(60,*) ' Time for calc. with Function RF)    ', TIME_RF
      write( *,*) ' Time for calc. with Function RF)    ', TIME_RF

      return
      end

      subroutine CHECK_RANDOM
C    Checks the random number generator random.
c    Couples of random numbers are considered as points (xi,yi) in
c    the square (0-1, 0-1).
C    Let ri=sqrt(xi**2 + yi**2). The sum of the number of the
C     points (xi,yi) falling in the circle of radius=1 is pi/4.
C
      implicit real *8 (a-h, o-z)

      iseed=1
      isum=0
      imax= 2 000 000 000
      ifreq=  20 000 000
      do 10 i=2,imax,2
        xi=random(iseed)
        yi=random(iseed)
        ri=dsqrt(xi**2 + yi**2)
        if(ri.le.1.d0) isum = isum + 1

        if(mod(i,ifreq) .eq. 0) then
           sumi =4.d0 *float(isum)/float(i/2)
           id2= i/2
        endif
10    continue
C
      sum =4.d0 *float(isum)/float(imax/2)
      pi=4.d0*atan(1.d0)
      perc_error = dabs(pi - sum)/pi
      write(60, 1000) pi, sum, perc_error
      write(* , 1000) pi, sum, perc_error
1000  format(' Sub. CHECK_RANDOM - pi, sum, perc_error', 3e17.9)
      return
      end

      subroutine CHECK_RF
C        Checks the random number generator RF from bird library.
C        are considered as points (xi,yi)in the square (0-1, 0-1).
```

```
C     Let ri=sqrt(xi**2 + yi**2). The sum of the number of the
C     points (xi,yi) falling in the circle of radius=1 is pi/4.
C
      implicit real *8 (a-h, o-z)
C
c     Initialize RF
      arf=rf(-1)
      iseed=1
      isum=0
      imax=2000 000 000
       do 10 i=1,imax,2
         xi=rf(0)
         yi=rf(0)
         ri=dsqrt(xi**2 + yi**2)
         if(ri.le.1.d0) isum = isum + 1
10     continue
C
      sum =4.d0 *float(isum)/float(imax/2)
      pi=4.d0*atan(1.d0)
      perc_error = dabs(pi - sum)/pi
      write(60, 1000) pi, sum, perc_error
      write(* , 1000) pi, sum, perc_error
1000 format(' Sub. CHECK_RF - pi, sum, perc_error', 3e17.9)

      return
      end

      double precision function RANDOM(ibm)
      implicit real*8 (a-h, o-z)
      mult=1277
      modo=2**17
      rmodo=modo
      ibm=ibm*mult
      ibm=mod(ibm,modo)
      random=ibm/rmodo
      return
      end

      function RF(IDUM)
*--generates a uniformly distributed random fraction between 0 and 1
*----IDUM will generally be 0, but negative values may be used to
*------re-initialize the seed
c   Cfr. function RUN3 of Numerical Recipes, p. 199.
      implicit real*8 (a-h,o-z)
      SAVE MA, INEXT, INEXTP
      PARAMETER (MBIG=1000000000,MSEED=161803398,MZ=0,FAC=1.E-9)
      DIMENSION MA(55)
      DATA IFF/0/
      IF (IDUM.LT.0.OR.IFF.EQ.0) THEN
        IFF=1
        MJ=MSEED-IABS(IDUM)
        MJ=MOD(MJ,MBIG)
        MA(55)=MJ
```

```
      MK=1
      DO 50 I=1,54
       II=MOD(21*I,55)
       MA(II)=MK
       MK=MJ-MK
       IF (MK.LT.MZ) MK=MK+MBIG
       MJ=MA(II)
50    CONTINUE
      DO 100 K=1,4
       DO 60 I=1,55
         MA(I)=MA(I)-MA(1+MOD(I+30,55))
         IF (MA(I).LT.MZ) MA(I)=MA(I)+MBIG
60     CONTINUE
100   CONTINUE
      INEXT=0
      INEXTP=31
     END IF
200  INEXT=INEXT+1
     IF (INEXT.EQ.56) INEXT=1
     INEXTP=INEXTP+1
     IF (INEXTP.EQ.56) INEXTP=1
     MJ=MA(INEXT)-MA(INEXTP)
     IF (MJ.LT.MZ) MJ=MJ+MBIG
     MA(INEXT)=MJ
     RF=MJ*FAC
     IF (RF.GT.1.E-8.AND.RF.LT.0.99999999) RETURN
     GO TO 200
     END
```

OUTPUT from PROG_IV_1 [TEST_RNG]

Sub. CHECK_RANDOM - pi, sum, perc_error 0.314159265E+01 0.314208983E +01 0.158255530E-03

Sub. CHECK_RF - pi, sum, perc_error 0.314159265E+01 0.314161063E+01 0.572143247E-05

Time for calc. with CHECH_RANDOM + _RF) 29.905392
IRAN = 1
IRAN = 2

Time for calc. with Function RANDOM) 95.363412000000011
Time for calc. with Function RF) 116.92274900000000

END - Total CPU time = 242.19156

Besides, the codes writes the files OUTPL0_RANDOM and OUTPL0_RF. Data of these files can be used with a plotting software to reproduce Fig.s 4.2 and 4.3 of the text.

Appendix A.V

PROG_V_1 [TEST_PROA] Test of a matrix for "Property A".

```
c   APPENDIX A.V, Program: [TEST_PROA] - Stand: 28.02.2020

C   ==========================================================
C   Checks a given matrix for PROPERTY A and CONSISTENT ORDER
C   ==========================================================
c   MAIN

c   LIST OF SUBROUTINES and FUNTIONS
c   Subroutine TSPROA
c   Subroutine PROPA

      open (60, file='OUTPUT', status='unknown')
      call TSPROA
      close(60)
      stop
      end

      SUBROUTINE TSPROA      ! DRIVING PROGRAMME
c    For the purpose of this program only the position (not the value)
c    of the non-zero elements of a matrix A is important. We set a value 4
c    on the main diagonal and set A(i,j) = -1 for every off-diagonal
c    non-zero element.
      PARAMETER(NN=14)
      DIMENSION AM(NN,NN),MATR(NN,NN),IYO(NN),IYSO(NN),IAA(NN),IDD(NN)

      DO 100 ICASE=1,7
         WRITE(60,1100) ICASE
         WRITE( *,1100) ICASE
         IF(ICASE.eq.1) N=6
         IF(ICASE.eq.2) N=6
         IF(ICASE.eq.3) N=4
         IF(ICASE.eq.4) N=9
         IF(ICASE.eq.5) N=3
         IF(ICASE.EQ.6) N=3
         IF(ICASE.EQ.7) N=14
         NM1=N-1

       DO 10 I=1,N
       DO 10 J=1,N
          AM(I,J)=0.0d0
10    CONTINUE

       DO 12 I=1,N
          AM(I,I)=4.0d0
12    CONTINUE
```

```
      GO TO (1,2,3,4,5,6,7), ICASE

1   CONTINUE  ! TEST CASE 1, EXAMPLE 1.A of text
c          with sequential numbering of nodes

c ****  MATRIX R  (We set R(i,j) = 1 if A(i,j) or A(j,i) are non-zero
c                           and set a star * on the main diagonal)
c  1  * 1 0 0 0 1
c  2  1 * 1 0 1 0
c  3  0 1 * 1 0 0
c  4  0 0 1 * 0 0
c  5  0 1 0 0 * 1
c  6  1 0 0 0 1 *

      AM(1,2)=-1.
      AM(1,6)=-1.
      AM(2,1)=-1.
      AM(2,3)=-1.
      AM(2,5)=-1.
      AM(3,2)=-1.
      AM(3,4)=-1.
      AM(4,3)=-1.
      AM(5,2)=-1.
      AM(5,6)=-1.
      AM(6,5)=-1.

      GO TO 90

2   CONTINUE ! TEST CASE 2, EXAMPLE 1.A of text
c             with diagonal numbering of nodes
c****  MATRIX R *******
c  1  * 1 1 0 0 0
c  2  1 * 0 1 1 0
c  3  1 0 * 0 1 0
c  4  0 1 0 * 0 1
c  5  0 1 1 0 * 0
c  6  0 0 0 1 0 *

      AM(1,2)=-1.
      AM(1,3)=-1.
      AM(2,1)=-1.
      AM(2,4)=-1.
      AM(2,5)=-1.
      AM(3,1)=-1.
      AM(3,5)=-1.
      AM(4,2)=-1.
      AM(4,6)=-1.
      AM(5,2)=-1.
      AM(5,3)=-1.
      AM(6,4)=-1.
      GO TO 90
```

```
C
3    CONTINUE     ! TEST CASE 3, EXAMPLE 1.B of Text

c ****  MATRIX R *******
c  1  * 1 1 0
c  2  1 * 0 1
c  3  1 0 * 1
c  4  0 1 1 *

      AM(1,2)=-1.
      AM(1,3)=-1.
      AM(2,1)=-1.
      AM(2,4)=-1.
      AM(3,1)=-1.
      AM(3,4)=-1.
      AM(4,2)=-1.
      AM(4,3)=-1.

      GO TO 90

4    CONTINUE   ! TEST CASE 4, Exampke 1.C of Text

c****  MATRIX R *******
c  1  * 0 0 0 0 0 0 0 0
c  2  0 * 1 0 0 0 0 0 0
c  3  0 1 * 0 0 0 0 0 0
c  4  0 0 0 * 0 0 1 0 0
c  5  0 0 0 0 * 1 0 1 0
c  6  0 0 0 0 1 * 0 0 1
c  7  0 0 0 1 0 0 * 0 0
c  8  0 0 0 0 1 0 0 * 1
c  9  0 0 0 0 0 1 0 1 *

      AM(2,3)=-1.0
      AM(3,2)=-1.0
      AM(4,7)=-1.0
      AM(5,6)=-1.0
      AM(5,8)=-1.0
      AM(6,5)=-1.0
      AM(6,9)=-1.0
      AM(7,4)=-1.0
      AM(8,5)=-1.0
      AM(8,9)=-1.0
      AM(9,6)=-1.0
      AM(9,8)=-1.0

      GO TO 90

5    CONTINUE     ! TEST CASE 5, Exmple 2.2.B of text

c ****  MATRIX R *******
c  1  * 1 1
```

```
c  2  1 * 0
c  3  1 0 *

      AM(1,2)=-1.0
      AM(1,3)=-1.0
      AM(2,1)=-1.0
      AM(3,1)=-1.0

      GO TO 90

6   CONTINUE    ! TEST CASE 6, Example 2.2.C of text
c****  MATRIX R *******
c  1  * 1 1
c  2  1 * 1
c  3  1 1 *
C
      AM(1,2)=-1.0
      AM(1,3)=-1.0
      AM(2,1)=-1.0
      AM(2,3)=-1.0
      AM(3,1)=-1.0
      AM(3,2)=-1.0

      GO TO 90

7   CONTINUE       !    TEST CASE 7
c ****  MATRIX R *******
c  1   * 1 0 0 0 0 0 0 0 0 0 0 0 0
c  2   1 * 1 1 0 0 0 0 0 0 0 0 0 0
c  3   0 1 * 0 0 1 0 0 0 0 0 0 0 0
c  4   0 1 0 * 1 0 0 0 0 0 0 0 0 0
c  5   0 0 0 1 * 0 0 1 1 0 0 0 0 0
c  6   0 0 1 0 0 * 1 1 0 0 0 0 0 0
c  7   0 0 0 0 0 1 * 0 0 0 0 0 0 1
c  8   0 0 0 0 1 1 0 * 0 0 0 1 0 0
c  9   0 0 0 0 1 0 0 0 * 1 0 0 0 0
c 10   0 0 0 0 0 0 0 0 1 * 1 0 0 0
c 11   0 0 0 0 0 0 0 0 0 1 * 1 0 0
c 12   0 0 0 0 0 0 0 1 0 0 1 * 1 0
c 13   0 0 0 0 0 0 0 0 0 0 0 1 * 1
c 14   0 0 0 0 0 0 1 0 0 0 0 0 1 *

      DO 301 I=2,NM1
        AM(I,I-1)=-1.
        AM(I,I+1)=-1.
301   CONTINUE
      AM(3,4)=0.
      AM(4,3)=0.
      AM(5,6)=0.
      AM(6,5)=0.
      AM(7,8)=0.
      AM(8,7)=0.
```

```
      AM(8,9)=0.
      AM(9,8)=0.
      AM(2,4)=-1.
      AM(4,2)=-1.

      AM(3,6)=-1.
      AM(6,3)=-1.
      AM(5,8)=-1.
      AM(8,5)=-1.
      AM(5,9)=-1.
      AM(9,5)=-1.
      AM(6,8)=-1.
      AM(8,6)=-1.
      AM(7,14)=-1.
      AM(14,7)=-1.
      AM(8,12)=-1.
      AM(12,8)=-1.
C
  90 CONTINUE
c
-----------------------------------------------------------------
      CALL PROPA(AM,NN,N,MATR,IYO,IYSO,IAA,IDD)
c
-----------------------------------------------------------------
100  CONTINUE

 1100 FORMAT(' *****  TEST CASE',I5)
 2233 FORMAT('      PERMUTATION P APPLIED TO GET CONSISTENT ORDER')
      RETURN
      END
C
      SUBROUTINE PROPA(AM,IDIM,N,MATR,IYO,IYSO,IAA,IDD)
C     CHECKS WHETHER THE MATRIX AM HAS PROPERTY A AND IS
C     CONSISTENTLY ORDERED
C
C     REFERENCE: D.M. YOUNG -ITERATIVE SOLUTION OF LARGE LINEAR
C     SYSTEMS, ACADEMIC PRESS 1971. (PP 159-161)

C     VECTOR IYO (ORDERING) : TESTS FOR MATRIX HAVING PROPERTY A
C     VECTOR IYSO (SUITABLE ORDERING) : TESTS FOR MATRIX BEING
c            CONSISTENTLY ORDERED

C     INDEX=0  MATRIX A HAS    PROPERTY A
C     INDEX=1  MATRIX A HAS NOT PROPERTY A - NO FURTHER INVESTIGATION

C     IALPHA=1 MATRIX A IS CONSISTENTLY ORDERED, HAS PROPERTY A
C     IALPHA=2 MATRIX A IS NOT CONSISTENTLY ORDERED, HAS PROPERTY A

C     KODO=0 ORDERING VECTOR IYO CHECKED - NO ERROR
C     KOD==1 ORDERING VECTOR IYO CHECKED - ERROR
C
C     KODSO=0 SUITABLE ORDERING VECTOR IYSO CHECKED - NO ERROR
```

```
C     KODSO=1 SUITABLE ORDERING VECTOR IYSO CHECKED - ERROR
C
      DIMENSION AM(IDIM,IDIM),MATR(N,N),IYO(N),IYSO(N),IAA(N),IDD(N)

       IPRINT=0  ! Turno to 1 to get more prnting
       INDEX=0
       KODSO=0
       KODO=0
       IALPHA=1

C     FIRST STEP - INITIALIZE MATRIX R (MATR)
      DO 10 I=1,N
      DO 10 J=1,N
         MATR(I,J)=0
 10 CONTINUE
C
      DO 20 J=1,N
      DO 20 I=1,N
         IF(I.EQ.J) GO TO 20
         IF(AM(I,J).NE.0.0.OR.AM(J,I).NE.0.0) MATR(I,J)=1
 20 CONTINUE
C
      IYO(1) =1
      IYSO(1)=1
      IAA(1)=1
      IDD(1)=1

      DO 30 J=2,N
        IYO (J)=0
        IYSO (J)=0
        IAA(J)=0
        IDD(J)=0
 30 CONTINUE

c-------------------------------------------------------
      IF(IPRINT.NE.0) THEN
         WRITE(60,1002)
         WRITE( *,1002)
        DO 220 I=1,N
           WRITE(60,1003) I,(AM(I,J),J=1,N)
           WRITE( *,1003) I,(AM(I,J),J=1,N)
 220   CONTINUE
C
        WRITE(60,1004)
        WRITE( *,1004)
        DO 230 I=1,N
          WRITE(60,1005) I,(MATR(I,J),J=1,N)
          WRITE( *,1005) I,(MATR(I,J),J=1,N)
 230   CONTINUE
      ENDIF
```

```
1002 FORMAT(//'  **** MATRIX A ****')
1003 FORMAT(I5,2X,1P,10E10.1/99(7X,10E10.1/))
1004 FORMAT(//' ****  MATRIX R *******')
1005 FORMAT(I5,2X,50I2/99(7X,50I2/))

      WRITE(60,'(//)')
      WRITE( *,'(//)')
c------------------------------------------------------------
      I=1
 100  CONTINUE
      J=1
 105  CONTINUE
      IF(MATR(I,J).EQ.0) GO TO 150
      IGAMBS=1- IYO(I)
      IF(IAA(J).EQ.0) GOTO 70
      IF(IGAMBS .EQ. IYO(J)) GO TO 200
      INDEX=1
      GO TO 600
  70  CONTINUE
      IYO(J)=IGAMBS
      IDD(J)=1
C
 200  CONTINUE
C     AS LONG AS IALPHA=1 TRY TO CONSTRUCT VECTOR IYSO
C     WHEN IALPHA=2 STOP TRYING TO CONSTRUCT VECTOR IYSO
      IF(IALPHA.EQ.2) GO TO 300
      IF(J.GT.I) IGAMS=IYSO(I) +1
      IF(J.LT.I) IGAMS=IYSO(I) -1
      IF(IAA(J).EQ.0) GO TO 80
      IF(IGAMS.EQ.IYSO(J)) GO TO 300
      IALPHA=2
      I1=I
      J1=J
      GO TO 300
C
  80 CONTINUE
      IYSO(J)=IGAMS
C
 300 CONTINUE
      IAA(J)=1
 150 CONTINUE
      IF(J.GE.N) GO TO 160
      J=J+1
      GO TO 105
 160 CONTINUE
      IDD(I)=0
C
      DO 165 K=1,N
      IF(IDD(K).NE.0) GO TO 170
 165 CONTINUE
      GO TO 190
C
 170 CONTINUE
```

```
      I=K
      GO TO 100
C
 190 CONTINUE
      DO 195 K=1,N
      IF(IAA(K).EQ.0) GO TO 210
 195 CONTINUE
      GO TO 600
C
C
 210 CONTINUE
      I=K
      IAA(I)=1
      IDD(I)=1
      IYO(I)=1
      IYSO (I)=1
      GO TO 100
 600 CONTINUE
C
C   WRITE RESULTS
      IF(INDEX.EQ.1)  THEN
        WRITE(60,1008)
        WRITE( *,1008)
1008    FORMAT(' MATRIX A HAS NOT PROPERTY A ')
        WRITE(60,'(//)')
        WRITE( *,'(//)')
        GO TO 900
      ENDIF

      IF(IALPHA.EQ.1) WRITE(60,1006)
      IF(IALPHA.EQ.1) WRITE( *,1006)
1006  FORMAT(' MATRIX A HAS PROPERTY A AND IS CONSISTENTLY ORDERED')

      IF(IALPHA.EQ.2) THEN
        WRITE(60,1007)
        WRITE( *,1007)
1007  FORMAT('MATRIX A HAS PROPERTY A BUT IS NOT CONSISTENTLY ORDERED')
      KODSO=1
        GO TO 650
      ENDIF

      DO 610 J=1,N   ! CHECK SUITABLE ORDERING VECTOY IYSO
        DO 606 I=1,N
          IF(I.EQ.J) GO TO 605
          IF(MATR(I,J).EQ.0.AND.MATR(J,I).EQ.0) GO TO 605
          IF(I.GT.J.AND.(IYSO(I)-IYSO(J)).EQ.1) GO TO 605
          IF(I.LT.J .AND.(IYSO(I)-IYSO(J)).EQ.-1) GO TO 605
          KODSO = 1
          GO TO 630
 605      CONTINUE
 606    CONTINUE
 610 CONTINUE
```

```
 630 CONTINUE

 650 CONTINUE
     WRITE(60,1009) KODSO
     WRITE( *,1009) KODSO
1009  FORMAT(' SUITABLE ORDERING VECTOR IYSO CHECKED, KODSO=',I5)

     DO 660 I=1,N   ! CHECK ORDERING VECTOR IYO
       DO 655 J=1,N
         IF(I.EQ.J) GO TO 654
         IF(MATR(I,J).EQ.0.AND.MATR(J,I).EQ.0) GO TO 654
         IF(IABS(IYO(I)- IYO(J)).EQ.1) GO TO 654
         KODO = 1
         GO TO 680
 654     CONTINUE
 655   CONTINUE
 660 CONTINUE

 680 CONTINUE
     WRITE(60,1010) KODO
     WRITE( *,1010) KODO
1010  FORMAT(' ORDERING VECTOR IYO CHECKED, KODO =',I5)

C
     IF(KODO .eq. 0 .and. KODSO .eq. 0) THEN ! WRITE ITO, IYSO
       WRITE(60,'(//)')
       WRITE( *,'(//)')
       DO 820 J=1,N
         WRITE(60,1020) J,IYO(J),IYSO(J)
         WRITE( *,1020) J,IYO(J),IYSO(J)
1020     FORMAT(' J,IYO,IYSO  ',3I5)
820    CONTINUE
       WRITE(60,'(//)')
       WRITE( *,'(//)')
     ENDIF

     IF(KODO .eq. 0 .and. KODSO .ne. 0) THEN ! WRITE IYO only
       WRITE(60,'(//)')
       WRITE( *,'(//)')
       DO 830 J=1,N
         WRITE(60,1030) J,IYO(J)
         WRITE( *,1030) J,IYO(J)
1030     FORMAT(' J,IYO  ',2I5)
830    CONTINUE

       WRITE(60,'(//)')
       WRITE( *,'(//)')
     ENDIF

900  CONTINUE
     RETURN
     END
```

```
OUTPUT from PROGRAM A.V

*****  TEST CASE   1

MATRIX A HAS PROPERTY A BUT IS NOT CONSISTENTLY ORDERED
 SUITABLE ORDERING VECTOR IYSO CHECKED, KODSO=   1
 ORDERING VECTOR IYO CHECKED, KODO =   0

 J,IYO     1  1
 J,IYO     2  0
 J,IYO     3  1
 J,IYO     4  0
 J,IYO     5  1
 J,IYO     6  0

 *****  TEST CASE   2

 MATRIX A HAS PROPERTY A AND IS CONSISTENTLY ORDERED
 SUITABLE ORDERING VECTOR IYSO CHECKED, KODSO=   0
 ORDERING VECTOR IYO CHECKED, KODO =   0

 J,IYO,IYSO     1  1  1
 J,IYO,IYSO     2  0  2
 J,IYO,IYSO     3  0  2
 J,IYO,IYSO     4  1  3
 J,IYO,IYSO     5  1  3
 J,IYO,IYSO     6  0  4

 *****  TEST CASE   3

 MATRIX A HAS PROPERTY A AND IS CONSISTENTLY ORDERED
 SUITABLE ORDERING VECTOR IYSO CHECKED, KODSO=   0
 ORDERING VECTOR IYO CHECKED, KODO =   0

 J,IYO,IYSO     1  1  1
 J,IYO,IYSO     2  0  2
 J,IYO,IYSO     3  0  2
 J,IYO,IYSO     4  1  3

 *****  TEST CASE   4

 MATRIX A HAS PROPERTY A AND IS CONSISTENTLY ORDERED
 SUITABLE ORDERING VECTOR IYSO CHECKED, KODSO=   0
 ORDERING VECTOR IYO CHECKED, KODO =   0

 J,IYO,IYSO     1  1  1
 J,IYO,IYSO     2  1  1
 J,IYO,IYSO     3  0  2
 J,IYO,IYSO     4  1  1
 J,IYO,IYSO     5  1  1
 J,IYO,IYSO     6  0  2
```

```
J,IYO,IYSO    7  0  2
J,IYO,IYSO    8  0  2
J,IYO,IYSO    9  1  3

 *****  TEST CASE  5

MATRIX A HAS PROPERTY A AND IS CONSISTENTLY ORDERED
SUITABLE ORDERING VECTOR IYSO CHECKED, KODSO=  0
ORDERING VECTOR IYO CHECKED, KODO =  0

J,IYO,IYSO    1  1  1
J,IYO,IYSO    2  0  2
J,IYO,IYSO    3  0  2

 *****  TEST CASE  6

MATRIX A HAS NOT PROPERTY A

 *****  TEST CASE  7

MATRIX A HAS PROPERTY A BUT IS NOT CONSISTENTLY ORDERED
SUITABLE ORDERING VECTOR IYSO CHECKED, KODSO=  1
ORDERING VECTOR IYO CHECKED, KODO =  0

J,IYO    1  1
J,IYO    2  0
J,IYO    3  1
J,IYO    4  1
J,IYO    5  0
J,IYO    6  0
J,IYO    7  1
J,IYO    8  1
J,IYO    9  1
J,IYO   10  0
J,IYO   11  1
J,IYO   12  0
J,IYO   13  1
J,IYO   14  0
```

Appendix A.VI

PROG_VI_1 [POISSON] Numerical solutions of the model Poisson equation, Eqs. (6.1) and (6.2).

```
c   APPENDIX A.VI, Program: PROG_VI_1 [POISSON]
c
c   Last change: 11.03.2020

c   ------------------------------------------------------------
C   TEST FIVE SOLUTIONS OF MODEL POISSON EQUATION (See Chapter VI)
c   ------------------------------------------------------------
c   List of Subroutines and Functions

c   SUB. TESTPS - Driving subroutine for the program
c   SUB. FIVEPF - Solution of Poisson equation with the
c            five-point formula
c   SUB. NINEPF - Solution of Poisson equation with the
c            nine-point formula
c   SUB. CLRITZ - Solution of Poisson equation with the
c            classical Ritz method
c   SUB. GALERK - Solution of Poisson equation with the
c            Galerkin method
c   SUB. PORTCO - Solution of Poisson equation with the
c            point orthogonal collocations megthod
c   SUB. GAUSS2 - Gaussian umerical integration with a Legendre
c            polynomial of degree 20
c   SUB. MINV  - Solution of an algebraic system of equations
c            by means of matrix inversion
c   SUB. MINMAX - Finds minimum and maximum of the values in an array.

c   FUNCTION AN1N6 - Computes shape functions to be integrated
c   FUNCTION ANTEG2 - Computes function to be integrated
c   FUNCTION D2FXYZ - Computes a Laplacian used in Sub. NINEPF
c   FUNCTION FBOUND - Sets boundary conditions
c   FUNCTION FIJM  - Auxiliary function used in SUB. CLRITZ
c
C
      open (60,file='OUTPUT', status='unknown')
      open (11,file='OUTPL0_1',status='unknown') ! For data from FIVEPF
      open (12,file='OUTPL0_2',status='unknown') ! For data from NINEPF
      open (13,file='OUTPL0_3',status='unknown') ! For data from CLRITZ
c   Plotting data have not been provived for tests 4 and 5.
c
      call CPU_TIME (cput0)
c
-------------------------------------------------------------------
      CALL TESTPS
c
-------------------------------------------------------------------
c
```

```
      call CPU_TIME(cpuend)
      cputot=cpuend - cput0
      WRITE(60,1001)
      WRITE( *,1001)
1001  FORMAT(//)
      write(60,1000) cputot
      write( *,1000) cputot
1000  format(' TESTPS - cputot', e12.5)

      close(60)
      close(11)
      close(12)
      close(13)

      STOP
      END

      SUBROUTINE TESTPS ! Test numerical solutions of Poisson equation
      implicit real*8 (a-h,o-z)
      EXTERNAL ANTEG2,D2FXYZ,FBOUND
      Parameter (idima=400, idimi=400)
      DIMENSION A(idima,idima),AMI(idima,idima),U(idima),B(idima),
     1 Z1(idima),Z2(idima),RESID(idima),UEXACT(idima)

c   ---------------- Definition of main FORTRAN symbols ----------

c   A(idima,idima)  - Matrix of coefficients of the discretized
c                       model equation
c  AMI(idima,idima) - Inverse of matrix A
c  B(idima)      - Known right hand side of matrix equation A*Y = B
c  RESID(idima)    - Residuals R = A*Y - B after solution with Sub. MINVV
c  U(idima)      - Solution vector of the equation A*Y = B
c  UEXACT(idima)   - Exact solution of the model equation
c  Z1(idima)      _Auxiliary array used in Sub. MINV
c  Z2(idima)      _Auxiliary array used in Sub. MINV
c  -----------------------------------------------------
--------------

c   Indices and locations of nodes defining the 2D domain
c                          | Y
c                          |
c                        4 |--------------- 3
c                          |                |
c                          |                |
c                          |                |
c                          |                |
c                          |                |
c                          |                | 2
c                        1 |---------------- x
      X1=0.D0
      Y1=X1
      X2=1.D0
```

```
      Y2=0.D0
      X3=1.D0
      Y3=X3
      X4=0.D0
      Y4=1.D0

c   For solution of POISSON equation:
c
c   itest= 1 Use subroutine FIVEPF  - Five point formula
c   itest= 2 Use subroutine NINEPF  - Nine point formula
c   itest= 3 Use subroutine CLRITZ  - Classical Ritz method
c   itest= 4 Use subroutine GALERK  - Galerkin method
c   itest= 5 Use subroutine PORTCO  - Orthogonal (Domain) collocations

      do 500 itest=1,5

         WRITE(60,'(//)')
        WRITE( *,'(//)')
        write(60,1001) ITEST
        write( *,1001) ITEST
1001    FORMAT(' ITEST = ', i5)

        NI = 19
        if(itest.eq.4) ni= 9
        if(itest.eq.5) ni=11
        NJ=NI

      if(itest.eq.1) then
        WRITE(60,*) 'SOLUTION OF POISSON EQ. WITH FIVE POINT FORMULA'
        write( *,*) 'SOLUTION OF POISSON EQ. WITH FIVE POINT FORMULA'
c
-----------------------------------------------------------------
        CALL FIVEPF(NI,NJ,RESID,U)
        call minmax(idima,resid,resmin,resmax)
        call minmax(idima,u,umin,umax)
c
-----------------------------------------------------------------
        call CPU_TIME (cput1)
        time1 = cput1 - cput0
        write(60,*) 'Time1 elapsed,resmax,umax = ', time1,resmax,umax
        write( *,*) 'Time1 elapsed,resmax,umax = ', time1,resmax,umax
      endif

      if(itest.eq.2) then
        WRITE(60,*)'SOLUTION OF POISSON EQ. WITH NINE POINT FORMULA'
        write( *,*)'SOLUTION OF POISSON EQ. WITH NINE POINT FORMULA'
c
-----------------------------------------------------------------
        CALL NINEPF(NI,NJ,X1,Y1,X3,Y3,A,AMI,B,U,
   1        Z1,Z2,RESID,UEXACT,ANTEG2,D2FXYZ,FBOUND,IDIMA,IDIMI)
        call minmax(idima,resid,resmin,resmax)
        call minmax(idima,u,umin,umax)
```

```
c
-----------------------------------------------------------------
      call CPU_TIME (cput2)
      time2 = cput2 - cput1
      write(60,*) 'Time2 elapsed,resmax,umax=', time2, resmax,umax
      write( *,*) 'Time2 elapsed,resmax,umax=', time2, resmax,umax
     endif

      if(itest.eq.3) then
      WRITE(60,*)'SOLUTION OF POISSON EQ. WITH CLASS. RITZ METHOD'
      write( *,*)'SOLUTION OF POISSON EQ. WITH CLASS. RITZ METHOD'
c
-----------------------------------------------------------------
      CALL CLRITZ(NI,NJ,X1,Y1,X3,Y3,A,AMI,B,U,
   1                 Z1,Z2,RESID,IDIMA,IDIMI)
      call minmax(idima,resid,resmin,resmax)
      call minmax(idima,u,umin,umax)
c
-----------------------------------------------------------------
      call CPU_TIME (cput3)
      time3 = cput3 - cput2
      write(60,*)'Time3 elapsed,resmax,umax=', time3, resmax,umax
      write( *,*)'Time3 elapsed,resmax,umax=', time3, resmax,umax
      ENDIF

      if (itest.eq.4) then
       WRITE(60,*) ' GALERKIN METHOD-QUADRATIC SHAPE FUNC. N1-N6'
       write( *,*) ' GALERKIN METHOD-QUADRATIC SHAPE FUNC. N1-N6'
c
-----------------------------------------------------------------
       CALL GALERK(NI,NJ,X1,Y1,X3,Y3,A,AMI,B,U,
   1                Z1,Z2,RESID,IDIMA,IDIMI)
       call minmax(idima,resid,resmin,resmax)
       call minmax(idima,u,umin,umax)
c
-----------------------------------------------------------------
       call CPU_TIME (cput4)
       time4 = cput4 - cput3
       write(60,*)'Time4 elapsed,resmax,umax=',time4,resmax,umax
       write( *,*)'Time4 elapsed,resmax,umax=',time4,resmax,umax
      endif

      if (itest.eq.5) then
      WRITE(60,*) ' DOMAIN ORTHOGONAL COLLOCATIONS'
      write( *,*) ' DOMAIN ORTHOGONAL COLLOCATIONS'
c
-----------------------------------------------------------------
      CALL PORTCO(NI,NJ,X1,Y1,X3,Y3,A,AMI,B,U,
   1                Z1,Z2,RESID,IDIMA,IDIMI)
      call minmax(idima,resid,resmin,resmax)
      call minmax(idima,u,umin,umax)
c
-----------------------------------------------------------------
```

```
        call CPU_TIME (cput5)
        time5 = cput5 - cput4
       write(60,*)'Time5 elapsed,resmax,umax=',time5,resmax,umax
       write( *,*)'Time5 elapsed,resmax,umax=',time5,resmax,umax
   endif

 500 CONTINUE
C ----------------------------------------------------------------
      RETURN
      END

      SUBROUTINE FIVEPF(NI,NJ,RESID,U)
      implicit real*8 (a-h,o-z)

C     SOLUTION OF THE POISSON EQUATION BY MEANS OF THE FIVE POINT FORMULA
C     INPUT= NI=NUMBER OF INTERVALS IN X DIRECTION
C         = NJ=NUMBER OF INTERVALS IN Y DIRECTION
C         = XMIN,XMAX,YMIN,YMAX =COORDINATES DELIMITING THE INTEGRATION
C                            DOMAIN
C         =NNODES=(NI+1)*(NJ+1)
C    A= MATRIX OF COEFFICIENTS(TO BE INVERTED)
C    B= VECTOR OF KNOWN RIGHT-HAND TERMS
C    OUTPUT= U= SOLUTION VECTOR Y=A**(-1)*B
C    BOUNDARY FUNCTION MUST BE PROGRAMMED IN FBOUND
C
      PARAMETER (IDIMA=400)

      dimension A(IDIMA,IDIMA),AM1(IDIMA,IDIMA),
     1          B(IDIMA),U(IDIMA),Z1(IDIMA),Z2(IDIMA),RESID(IDIMA),
     2        UEXACT(IDIMA),WR(IDIMA),WI(IDIMA),VR(IDIMA),VI(IDIMA),
     3          AM3(IDIMA,3),B3(IDIMA)

      nip1=ni+1
      njp1=nj+1
c
      if(nip1**2 .gt. idima .or. njp1**2 .gt. idima**2) then
        write(60,*) ' FIVEPF - Increase IDIMA - stop '
        stop
      endif

      RDIAX = -2.0
      IERROR = 0
      XMIN  = 0.0
      YMIN  = 0.0
      XMAX  = 1.0
      YMAX  = 1.0
      PI = 4.0*ATAN(1.0)
      NNODES=(NIP1)*(NJP1)
C
      DO 10 I=1,NNODES
        B(I) = 0.0
        U(I) = 0.0
```

```
      DO 10 J=1,NNODES
        A(I,J) = 0.0
 10   CONTINUE

      NI = NIP1-1
      NJ = NJP1-1
      ANI=NI
      DX=(XMAX-XMIN)/ANI
      ANJ=NJ
      DY=(YMAX-YMIN)/ANJ
      DXY = DX*DY
      DXQ = 1.0/(DX*DX)
      DYQ = 1.0/(DY*DY)
C
      DO 20 M=1,NNODES
C     BOUNDARY NODES ARE SKIPPED AND TREATED LATER
        M0=MOD(M,NIP1)
        MM = M0
        IF(MM.EQ.0) MM=NIP1
        X=(MM-1)*DX
        MY=(M-1)/NIP1
        Y= MY*DY
        UEXACT(M)= 3.D0* EXP(X+Y)*(X-X**2)*(Y-Y**2)
        B(M)=     6.D0*X*Y* EXP(X+Y)*(X*Y+X+Y-3.D0)
        B3(M)= B(M)

C     BOUNDARY NODES ARE SKIPPED AND TREATED LATER
        IF(M.GE.1.AND.M.LE.NIP1) GO TO 20
        IF(M.GE.(NNODES-NI).AND.M.LE.NNODES) GO TO 20
        IF(M0.EQ.0.OR.M0.EQ.1) GO TO 20
        ID=M-1
        IF=M+1
        IB=M-NIP1
        IH=M+NIP1
        A(M,IB)=1.D0/(DY*DY)
        A(M,ID)=1.D0/(DX*DX)
        A(M,IF)=1.D0/(DX*DX)
        A(M,IH)=1.D0/(DY*DY)
        A(M,M) =-(A(M,IB)+A(M,ID)+A(M,IF)+A(M,IH))
 20   CONTINUE
C
C   BOUNDARY POINTS
      DO 25 M=1,NNODES
        WI(M) = 0.0
        WR(M) = 0.0
        VI(M) = 0.0
        VR(M) = 0.0
        IF(M.GE.1.AND.M.LE.NIP1)GO TO 22
        MM=MOD(M,NIP1)
        IF(M.GE.(NNODES-NI).AND.M.LE.NNODES)GO TO 22
        IF(MM.EQ.0.OR.MM.EQ.1)GO TO 22
        GO TO 25
 22   CONTINUE
```

```
      WR(M) =-UEXACT(M)
      B(M)  = 0.0
      B3(M) = UEXACT(M)
      A(M,M)=1.D0
 25  CONTINUE

c  ------------------------------------------
     ires=1
     CALL MINV(A,AM1,U,B3,RESID,Z1,Z2,NNODES,IRES,KOD,IDIMA,IDIMA)
c  ------------------------------------------

c  Fill OUTPL0_1 with information on i,j,x,y,absres,uexact,u for
c  plotting with MATLAB
     do 400 i=1,nip1
     do 400 j=1,njp1
      x=(i-1)*dx
      y=(j-1)*dy
      m=(i-1)*nip1 + j
      absres=dabs(resid(m))
      write(11,1012) i,j,m,x,y,absres,uexact(m),u(m)
1012  format(1H ,3i5,5e13.5)
400  continue

     RETURN
     END

     SUBROUTINE NINEPF(NI,NJ,XMIN,YMIN,XMAX,YMAX,A,AM1,B,U,
   1 Z1,Z2,RESID,UEXACT,FUNC,D2FUNC,FBOUND,IDIMA,IDIMI)
      implicit real*8 (a-h,o-z)

C   SOLUTION OF THE POISSON EQUATION BY MEANS OF THE NINE POINT FORMUL
C   KNOWN FUNCTION F MUST BE SPECIFIED IN FUNCTION ANTEG2
C   LAPLACIAN OF F (D2F) MUST BE GIVEN IN FUNCTION D2FXYZ
C   INPUT= NI=NUMBER OF INTERVALS IN X DIRECTION
C       = NJ=NUMBER OF INTERVALS IN Y DIRECTION
C       = XMIN,XMAX,YMIN,YMAX =COORDINATES DELIMITING THE INTEGRATION
C                    DOMAIN
C       =NNODES=(NI+1)*(NJ+1)
C   A= MATRIX OF COEFFICIENTS(TO BE INVERTED)
C   B= VECTOR OF KNOWN RIGHT-HAND TERMS
C   OUTPUT= U= SOLUTION VECTOR Y=A**(-1)*B
C   BOUNDARY FUNCTION MUST BE PROGRAMMED IN FBOUND

     DIMENSION A(IDIMA,IDIMA),AM1(IDIMA,IDIMA),B(IDIMA),U(IDIMA),
   1 Z1(IDIMA),Z2(IDIMA),RESID(IDIMA),UEXACT(IDIMA)
C
     nip1=ni+1
     njp1=nj+1
     NNODES=(NI+1)*(NJ+1)

     DO 10 I=1,NNODES
      B(I)=0.D0
```

```
        U(I)=0.D0
      DO 10 J=1,NNODES
        A(I,J)=0.D0
10    CONTINUE
      DX=(XMAX-XMIN)/float(ni)
      DY=(YMAX-YMIN)/float(nj)
      WRITE(60,*) 'NI,NJ,Dx,DY', ni, nj, dx, dy

      DO 200 M=1,NNODES
C     BOUNDARY NODES ARE SKIPPED AND TREATED LATER
        IF(M.GE.1.AND.M.LE.NIP1) GO TO 200
        IF(M.GE.(NNODES-NI).AND.M.LE.NNODES) GO TO 200
        MM=MOD(M,NIP1)
        IF(MM.EQ.0.OR.MM.EQ.1) GO TO 200
        IA=M-NIP1-1
        IB=IA+1
        IC=IA+2
        ID=M-1
        IE=M
        IFF=M+1
        IG=M+NIP1-1
        IH=IG+1
        II=IG+2
        A(M,IA)=1.D0
        A(M,IB)=4.D0
        A(M,IC)=1.D0
        A(M,ID)=4.D0
        A(M,M) =-20.D0
        A(M,IFf)=4.D0
        A(M,IG)=1.D0
        A(M,IH)=4.D0
        A(M,II)=1.D0

        X=(MM-1)*DX
        MY=M/NIP1
        Y= MY*DY
        UEXACT(M)= 3.D0* DEXP(X)* DEXP(Y)*(X-X**2)*(Y-Y**2)
        F=FUNC(X,Y)
        D2F=D2FUNC(X,Y)
        B(M)= 6.D0*DX**2*F +0.5D0*DX**4*D2F
 200  CONTINUE

      DO 250 M=1,NNODES     ! TREAT BOUNDARY POINTS
        IF(M.GE.1.AND.M.LE.NIP1)GO TO 220
        IF(M.GE.(NNODES-NI).AND.M.LE.NNODES)GO TO 220
        MM=MOD(M,NIP1)
        IF(MM.EQ.0.OR.MM.EQ.1)GO TO 220
        GO TO 250
220   CONTINUE
        X=(MM-1)*DX
        IF(MM.EQ.0) X=XMAX
        MY=M/NIP1
        IF(MM.EQ.0) MY=MY-1
```

```
      Y=MY*DY
      B(M)=FBOUND(X,Y)
      A(M,M)=1.D0
 250 CONTINUE

c   ---------------- Solution ---------------
      IRES=1
      CALL MINV (A,AM1,U,B,RESID,Z1,Z2,NNODES,IRES,KOD,IDIMA,IDIMI)
c   --------------------------------------------

c   Fill OUTPL0_2 with information on i,j,x,y,absres,uexact,u for
c   plotting with MATLAB
      do 400 i=1,nip1
      do 400 j=1,njp1
        x=(i-1)*dx
        y=(j-1)*dy
        m=(i-1)*nip1 + j
        absres=dabs(resid(m))
        write(12,1012) i,j,m,x,y,absres,uexact(m),u(m)
1012    format(1H ,3i5,5e13.5)
400   continue

      RETURN
      END

      SUBROUTINE CLRITZ(NI,NJ,XMIN,YMIN,XMAX,YMAX,AK,AKM1,B,U,
     1 Z1,Z2,RESID,IDIMA,IDIMI)
       implicit real*8 (a-h,o-z)

C   SOLUTION OF POISSON EQUATION BY CLASSICAL RITZ METHOD
      EXTERNAL FIJM
      COMMON/ALPHA/A1,A2,A3
      DIMENSION AK(IDIMA,1),AKM1(IDIMI,1),B(1),U(1),Z1(1),Z2(1),
     1 RESID(1)
C   NUMBER OF TRIANGLES =NI*NJ*2
      NNODES=(NI+1)*(NJ+1)          ! NUMBER OF NODES

      DO 10 I=1,NNODES
      B(I)=0.D0
      DO 10 J=1,NNODES
  10 AK(I,J)=0.D0

      NIP1=NI+1
      NJP1=NJ+1
      ANI=FLOAT(NI)
      DX=(XMAX-XMIN)/ANI
      DY=DX
      DELTA= DX*DY/2.D0

      DO 200 I=1,NI
      DO 200 J=1,NJ
        IA= (J-1)*(NI+1)+I
```

```
      IB= IA+1
      IC= IB + (NI+1)
      ID= IA + (NI+1)
      XA= (I-1) *DX
      YA= (J-1) *DY
      XB=XA+DX
      YB=YA
      XC=XB
      YC=YB+DY
      XD=XA
      YD=YC
     DO 200 MT=1,2
      XI=XA
      YI=YA
      IF(MT.EQ.2) GO TO 30
      XJ=XC
      YJ=YC
      XM=XD
      YM=YD
      II=IA
      JJ=IC
      MM=ID
      GO TO 40
30   CONTINUE
      XJ=XB
      YJ=YB
      XM=XC
      YM=YC
      II=IA
      JJ=IB
      MM=IC
40   CONTINUE
      AI= XJ*YM - XM*YJ
      AJ= XM*YI - XI*YM
      AM= XI*YJ - XJ*YI
      BI= YJ - YM
      BJ= YM - YI
      BM= YI - YJ
      CI= XM - XJ
      CJ= XI - XM
      CM= XJ - XI
C    FILL MATRIX AK
      AK(II,II) = AK(II,II) + (BI**2 +CI**2)
      AK(II,JJ) = AK(II,JJ) + (BI*BJ +CI*CJ)
      AK(II,MM) = AK(II,MM) + (BI*BM +CI*CM)
      AK(JJ,II) =AK(JJ,II) + BI*BJ+CI*CJ
      AK(JJ,JJ) =AK(JJ,JJ) + BJ**2+CJ**2
      AK(JJ,MM) =AK(JJ,MM) + BJ*BM+ CJ*CM
      AK(MM,II) =AK(MM,II) + BI*BM+ CI*CM
      AK(MM,JJ) =AK(MM,JJ) + BJ*BM+ CJ*CM
      AK(MM,MM) =AK(MM,MM) + BM**2+CM**2
      DO 50 IJM=1,3
        IF(IJM.EQ.2) GO TO 42
```

```
      IF(IJM.EQ.3) GO TO 43
      A1=AI
      A2=BI
      A3=CI
      GO TO 45
42    A1=AJ
      A2=BJ
      A3=CJ
      GO TO 45
43    A1=AM
      A2=BM
      A3=CM
45    CONTINUE
c    ----------------------------------------
      N=20
      CALL GAUSS2(XA,YA,XB,YB,XC,YC,XD,YD,N,MT,SUMXY,FIJM)
c    ----------------------------------------
      IF(IJM.EQ.1) B(II)=B(II)+SUMXY*2.D0
      IF(IJM.EQ.2) B(JJ)=B(JJ)+SUMXY*2.D0
      IF(IJM.EQ.3) B(MM)=B(MM)+SUMXY*2.D0
50    CONTINUE
 200 CONTINUE
C
C   SET VALUES CORRESPONDING TO BOUNDARY NODES
     DO 250 M=1,NNODES
     IF(M.GE.1.AND.M.LE.(NI+1)) GO TO 220
     IF(M.GE.(NNODES-NI).AND.M.LE.NNODES) GOTO 220
     MM=MOD(M,NIP1)
     IF(MM.EQ.0.OR.MM.EQ.1) GO TO 220
     GO TO 250
 220 CONTINUE
     DO 222 J=1,NNODES
222  AK(M,J)=0.D0
     IF(MM.EQ.0) X=XMAX
     IF(MM.EQ.1) X=XMIN
     IF(M.LE.NIP1) Y=YMIN
     IF(M.GE.(NNODES-NI))Y=YMAX
C
     AK(M,M)=1.D0
     B(M)=FBOUND(X,Y)
 250 CONTINUE

c   ------------------------------------
     IRES=1
    CALL MINV (AK,AKM1,U,B,RESID,Z1,Z2,NNODES,IRES,KOD,IDIMA,IDIMI)
c   ------------------------------------

c   Fill OUTPL0_3 with information on i,j,x,y,absres,u for
c   plotting with MATLAB
     dummy = 1.d0
     do 400 i=1,nip1
     do 400 j=1,njp1
       x=(i-1)*dx
```

```
       y=(j-1)*dy
       m=(i-1)*nip1 + j
       absres=dabs(resid(m))
       write(13,1012) i,j,m,x,y,absres,dummy,U(m)
1012   format(1H ,3i5,5e13.5)
400  continue

      RETURN
      END

      SUBROUTINE GALERK(NI,NJ,XMIN,YMIN,XMAX,YMAX,A,AM1,B,U,
    1 Z1,Z2,RESID,IDIMA,IDIMI)
       implicit real*8 (a-h,o-z)
C    SOLUTION OF POISSON EQUATION BY MEANS OF GALERKIN RESIDUAL METHOD
C    PARABOLIC INTERPOLATION WITH SHAPE FUNCTIONS N1 TO N6
      EXTERNAL AN1N6
      COMMON/N16/A1,B1,C1,A2,B2,C2,A3,B3,C3,DELTA
      COMMON/NN16/INN1N6
      DIMENSION COE(6),G(6,6),IND(6),CNODES(IDIMA,2)
      DIMENSION B(1),U(1),Z1(1),Z2(1),RESID(1),
    1      A(IDIMA,1),AM1(IDIMI,1)
C    NUMBER OF TRIANGLES = NI*NJ*2
C    NUMBER OF NODES     = (2NJ+1)*(2NI+1)  (Including mid-point of edges)
      NNODES=(2*NI+1)*(2*NJ+1)
      WRITE(60,1003) NI,NJ,NNODES
      WRITE( *,1003) NI,NJ,NNODES
1003 FORMAT(' NI,NJ,NNODES ',3I5)

c    Indices and locations of nodes

c              |y
c              |   G
c            D |--------- C
c              |         /|
c              |        / |
c              |       /  |
c              |      /   |
c            H |    I/    | F
c              |    /     |
c              |   /      |
c              |  /       |
c              | /        !
c              |--------------------- x
c              A    E    B

      DO 5 I=1,IDIMA
        RESID(i) = 0.d0
5    CONTINUE

      DO 10 I=1,NNODES
        B(I)=0.D0
      DO 10 J=1,NNODES
        A (I,J)=0.D0
10   CONTINUE
```

```
      SDELTA=0.D0
      NP=2*NI+1
      ANI=NI
      DX=(XMAX-XMIN)/ANI
      ANJ=NJ
      DY=(YMAX-YMIN)/ANJ
      DO 200 I=1,NI
      DO 200 J=1,NJ
      IA=2*(J-1)*(2*NI+1) +2*I-1
      IB=IA+2
      IC=IB+2*(2*NI+1)
      ID=IC-2
      IE=IA+1
      IF=IB+2*NI+1
      IG=ID+1
      IH=IA+2*NI+1
      II=IH+1
      XA=XMIN+(I-1)*DX
      YA=YMIN+(J-1)*DY
      XB=XA+DX
      YB=YA
      XC=XB
      YC=YB+DY
      XD=XA
      YD=YA+DY
      XE=(XA+XB)/2.D0
      YE=(YA+YB)/2.D0
      XF=(XB+XC)/2.D0
      YF=(YB+YC)/2.D0
      XG=(XC+XD)/2.D0
      YG=(YC+YD)/2.D0
      XH=(XA+XD)/2.D0
      YH=(YA+YD)/2.D0
      XI=(XA+XC)/2.D0
      YI=(YA+YC)/2.D0
      DO 200 MT=1,2
      X1=XA
      Y1=YA
      IF(MT.EQ.2) GO TO 30
      X2=XC
      Y2=YC
      X3=XD
      Y3=YD
      IND(1)=IA
      IND(2)=IC
      IND(3)=ID
      IND(4)=II
      IND(5)=IG
      IND(6)=IH
      GO TO 40
   30 CONTINUE
      X2=XB
      Y2=YB
```

```
      X3=XC
      Y3=YC
      IND(1)=IA
      IND(2)=IB
      IND(3)=IC
      IND(4)=IE
      IND(5)=IF
      IND(6)=II
   40 CONTINUE
      DELTA=0.5D0*(X2*Y3-X3*Y2+X1*(Y2-Y3)+Y1*(X3-X2))
      SDELTA=SDELTA+DELTA
      DELTA2=DELTA**2
      CNODES(IA,1)=XA
      CNODES(IA,2)=YA
      CNODES(IB,1)=XB
      CNODES(IB,2)=YB
      CNODES(IC,1)=XC
      CNODES(IC,2)=YC
      CNODES(ID,1)=XD
      CNODES(ID,2)=YD
      CNODES(IE,1)=XE
      CNODES(IE,2)=YE
      CNODES(IF,1)=XF
      CNODES(IF,2)=YF
      CNODES(IG,1)=XG
      CNODES(IG,2)=YG
      CNODES(IH,1)=XH
      CNODES(IH,2)=YH
      CNODES(II,1)=XI
      CNODES(II,2)=YI
      A1=X2*Y3-X3*Y2
      A2=X3*Y1-X1*Y3
      A3=X1*Y2-X2*Y1
      B1=Y2-Y3
      B2=Y3-Y1
      B3=Y1-Y2
      C1=X3-X2
      C2=X1-X3
      C3=X2-X1
C
      G(1,1)=(B1**2+C1**2)/(4.D0*DELTA)
      G(1,2)= -(B1*B2 + C1*C2)/(12.D0*DELTA)
      G(1,3)= -(B1*B3 + C1*C3)/(12.D0*DELTA)
      G(1,4)= (B1*B2 +C1*C2)/(3.D0*DELTA)
      G(1,5)= 0.D0
      G(1,6)= (B1*B3 +C1*C3)/(3.D0*DELTA)
      G(2,2)=(B2**2+C2**2)/(4.D0*DELTA)
      G(2,3)= -(B2*B3 +C2*C3)/(12.D0* DELTA)
      G(2,4)= (B1*B2+C1*C2)/(3.D0*DELTA)
      G(2,5)= (B2*B3 + C2*C3)/(3.D0*DELTA)
      G(2,6)= 0.D0
      G(3,3)=(B3**2+C3**2)/(4.D0*DELTA)
      G(3,4)= 0.D0
```

```
      G(3,5)= (B2*B3 +C2*C3)/(3.D0*DELTA)
      G(3,6)= (B1*B3 +C1*C3)/(3.D0*DELTA)
      G(4,4)= 2.D0*(B1**2 +B2**2 +B1*B2 +C1**2+C2**2+C1*C2)/(3.
D0*DELTA)
      G(4,5)= (B2*B1+B2**2 +B2*B3 +2.D0*B1*B3 +
   1      C2*C1+C2**2 +C2*C3 +2.D0*C1*C3)/(3.D0*DELTA)
      G(4,6)= (B1*B2+B1*B3+B1**2 +2.D0*B2*B3+
   1      C1*C2+C1*C3+C1**2 +2.D0*C2*C3)/(3.D0*DELTA)
     G(5,5)=2.D0*(B2**2+B3**2+B2*B3+C2**2+C3**2+C2*C3)/(3.D0*DELTA)
      G(5,6)= (B3*B1 +B3*B2+ B3**2+ 2.D0*B1*B2 +
   1      C3*C1 +C3*C2+ C3**2+ 2.D0*C1*C2)/(3.D0*DELTA)
     G(6,6)=2.D0*(B1**2+B3**2+B1*B3+C1**2+C3**2+C1*C3)/(3.D0*DELTA)

      DO 800 IG=1,6
      DO 800 JG=1,6
        IF(IG.LE.JG) GO TO 800
        G(IG,JG)= G(JG,IG)
 800 CONTINUE

      NG=20
      DO 100 INN1N6=1,6
C
------------------------------------------------------------------
        CALL GAUSS2(XA,YA,XB,YB,XC,YC,XD,YD,NG,MT,ANTNKF,AN1N6)
C
------------------------------------------------------------------
79     CONTINUE
        MM=IND(INN1N6)
        DO 80 M=1,6
          JJ=IND(M)
          A(MM,JJ)=A(MM,JJ)- G(INN1N6,M)
 80    CONTINUE
        B(MM)=B(MM) + ANTNKF
100  CONTINUE

 200 CONTINUE

C    SET VALUES CORRESPONDING TO BOUNDARY NODES
      DO 250 M=1,NNODES
      IF(M.GE.1.AND.M.LE.(2*NI+1))GOTO 220
      IF(M.GE.(NNODES-(2*NI+1)).AND.M.LE.NNODES) GO TO 220
      MM=MOD(M,NP)
      IF(MM.EQ.0.OR.MM.EQ.1) GO TO 220
      GO TO 250
 220 CONTINUE
      DO 222 J=1,NNODES
 222 A(M,J)=0.D0
      A(M,M)=1.D0
      IF(MM.EQ.0) X=XMAX
      IF(MM.EQ.1) X=XMIN
      IF(M.LE.NP) Y=YMIN
      IF(M.GE.(NNODES-NP+1)) Y=YMAX
      B(M)= FBOUND(X,Y)
```

```
 250 CONTINUE
C
c
-----------------------------------------------------------------------
      IRES=1
      CALL MINV (A,AM1,U,B,RESID,Z1,Z2,NNODES,IRES,KOD,IDIMA,IDIMI)
c
-----------------------------------------------------------------------
      RETURN
      END

      SUBROUTINE PORTCO(NI,NJ,XMIN,YMIN,XMAX,YMAX,A,AMI,B,U,
    1 Z1,Z2,RESID,IDIMA,IDIMI)
       implicit real*8 (a-h,o-z)

      EXTERNAL ANTEG2
      DIMENSION Z1(1),Z2(1),U(1),B(1),RESID(1)
      DIMENSION A(IDIMA,IDIMI),AMI(IDIMA,IDIMI)

      IAMMAX=3*NI*NJ + NI+NJ      ! TOTAL NUMBER OF EQATIONS
      WRITE(60,8563) NI,NJ, IAMMAX
      WRITE( *,8563) NI,NJ, IAMMAX
8563  FORMAT(' NI,NJ, IAMMMAX ',3I8)
      LBAND=1+2*(3*NI+1)

      DO 5 I=1,IDIMA
        RESID(i) = 0.d0
5    CONTINUE

      DO 10 I=1,IAMMAX
        B(I )=0.D0
      DO 10 J=1,IAMMAX
        A(I,J)=0.D0
10    CONTINUE

      ANI=NI
      DX=1.D0/ANI
      DY=DX
      DO 200 J=1,NJ
      DO 200 I=1,NI
        NP=NI+(J-1)*(3*NI+1)+2+2*(I-1)
        IPA=NP-1
        IPB=NP+1
        IPC=NP+2
        IPD=NP+3
        IPE=NP-(NI+I)
        IPF=IPE+3*NI+1
        IPG=NP+3*NI+1
        IPH=IPF+3*NI+1
        A(NP,NP)=-4.D0
        A(NP,IPA)=1.D0
        A(NP,IPB)=1.D0
```

```
      A(NP,IPE)=1.D0
      A(NP,IPF)=1.D0
      IF(I.EQ.1) A(IPA,IPA)=1.D0
      IF(I.EQ.NI) A(IPB,IPB)=1.D0
      IF(I.EQ.NI) GO TO 30
      A(IPB,IPB)=6.D0
      A(IPB,NP)=-4.D0
      A(IPB,IPA)=1.D0
      A(IPB,IPC)=-4.D0
      A(IPB,IPD)=1.D0
30   CONTINUE
      IF(J.EQ.1) A(IPE,IPE)=1.D0
      IF(J.EQ.NJ) A(IPF,IPF)=1.D0
      IF(J.EQ.NJ) GO TO 40
      A(IPF,IPF)=6.D0
      A(IPF,NP)=-4.D0
      A(IPF,IPE)=1.D0
      A(IPF,IPG)=-4.D0
      A(IPF,IPH)=1.D0
40   CONTINUE
150  CONTINUE
      XI=(I-1)*DX
      YJ=(J-1)*DY
      X1=XI
      Y1=YJ
      X2=X1+DX
      Y2=Y1
      X3=X2
      Y3=Y2+DY
      X4=X1
      Y4=Y3
      INDEX=0
      NG=20
c
---------------------------------------------------------------
      CALL GAUSS2(X1,Y1,X2,Y2,X3,Y3,X4,Y4,NG,INDEX,SUM,ANTEG2)
c
---------------------------------------------------------------
      B(NP)=SUM/4.D0
200  CONTINUE

c
---------------------------------------------------------------
     ires=1
     CALL MINV(A,AMI,U,B,RESID,Z1,Z2,IAMMAX,IRES,KOD,IDIMA,IDIMI)
c
---------------------------------------------------------------
     RETURN
     END

     SUBROUTINE GAUSS2(X1,Y1,X2,Y2,X3,Y3,X4,Y4,N,INDEX,SUMXY,ANTEG)
   implicit real*8 (a-h,o-z)
C   COMPUTES A DEFINITE INTEGRAL WITH GAUSSIAN QUADRATURE
```

```
C   INPUT= X1,Y1,X2,Y2,X3,Y3,X4,Y4 COORDINATES OF POINTS FORMING A
C       A RECTANGLE WITH LATES PARALLEL TO THE AXIS. POINTS 1-2-3-4
C       POINTS MUST BE GIVEN IN COUNTERCLOCKWISE SEQUENCE
C   INDEX=0  INTEGRATION OVER THE RECTANGLE HAVING THE POINTS (X1,Y1),
C        AS OPPOSITE VERTICES
C   INDEX=1  INTEGRATION OVER THE TRIANGLE WITH VERTICES 1-3-4
C   INDEX=11 INTEGRATION OVER THE TRIANGLE WITH VERTICES 4-2-3
C   INDEX=2  INTEGRATION OVER THE TRIANGLE 1-2-3
C   INDEX=22 INTEGRATION OVER THE TRIANGLE 1-2-4
C   INDEX=1/11 DOMAIN NORMAL TO THE Y AXIS
C   INDEX=2/22 DOMAIN NORMAL TO THE X AXIS
C       N= ORDER OF GAUSSIAN QUADRATURE
C       R,W= ARRAYS CONTAINING ZEROS AND WEIGTHS OF LEGENDRE POL.
C       ANTEG (EXTERNAL) FUNCTION TO BE INTEGRATED
C   OUTPUT= SUMXY= COMPUTED INTEGRAL
C   Y,YY AUXILIARY ARRAYS

c   If INDEX= 0 Integration over domain

c             4___3
c             |   |
c             |   |
c             1---2

c   If INDEX= 1 Integration over domain
c             _____
c             |   /
c             |  /
c             | /

c   If INDEX= 11 Integration over domain

c             _____
c             \   |
c              \  |
c               \ !

c   If INDEX=2 Integration over domain

c               /|
c              / |
c             /  |
c             ----

c   If INDEX=22 Integration over domain

c
c             |\
c             | \
c             |  \
c             ----\
```

```
      DIMENSION X(20),Y(20),YY(20),R(20),W(20)

      A=X1
      B=X2
      C=Y1
      D=Y4

      X(20)=0.993128599185094924790D0 ! Zeroes of Legendre Polynomial
      X(19)=0.963971927277913791270D0
      X(18)=0.912234428251325905870D0
      X(17)=0.839116971822218823390D0
      X(16)=0.746331906460150792610D0
      X(15)=0.636053680726515025450D0
      X(14)=0.510867001950827098000D0
      X(13)=0.373706088715419560670D0
      X(12)=0.227785851141645078080D0
      X(11)=0.076526521133497333750D0
C
      W(20)=0.017614007139152118310D0 ! Associated weights
      W(19)=0.040601429800386941330D0
      W(18)=0.062672048334109063570D0
      W(17)=0.083276741576704748730D0
      W(16)=0.101930119817240435040D0
      W(15)=0.118194531961518417310D0
      W(14)=0.131688638449176626900D0
      W(13)=0.142096109318382051330D0
      W(12)=0.149172986472603746790D0
      W(11)=0.152753387130725850700D0
C
      DO 205 I=1,10
       K=21-I
       X(I)=-X(K)
       W(I)=W(K)
205   CONTINUE
C
500   CONTINUE

      IF(INDEX.EQ.1.OR.INDEX.EQ.11) GO TO 601
      DO 510 I=1,N
      YY(I)=X(I)
510   R(I)=(X(I)*(B-A)+A+B)/2.D0
C     COMPUTE THE FUNCTION TO BE INTEGRATED IN THE SHIFTED ZEROS
C     AND MULTIPLY BY THE RESPECTIVE WEIGTHS
      SUM=0.D0
      DO 535 I=1,N
      IF(INDEX.EQ.0) GO TO 30
      IF(INDEX.EQ.22) GO TO 29
      C=Y1
      D=Y1+(Y3-Y1)*(R(I)-X1)/(X3-X1)
      GO TO 30
29    C=Y1
      D=Y1+ (R(I)-X2)*(Y4-Y1)/(X1-X2)
30    CONTINUE
```

```
      DO 520 J=1,N
520   Y(J)=(YY(J)* (D-C)+D+C)/2.D0
      SUMY=0.D0
      DO 525 J=1,N
      BETA=ANTEG(R(I),Y(J))
525   SUMY=SUMY+ BETA*W(J)
      SUMY=SUMY*(D-C)/2.D0
535   SUM=SUM+SUMY*W(I)
C
C   MULTIPLY BY (B-A)/2. (BECAUSE OF THE CHANGE OF DX)
      SUM=SUM*(B-A)/2.D0
600   CONTINUE
      SUMXY=SUM
      RETURN
C --------------------------------------------------
601   CONTINUE
      C=Y1
      D=Y4
      DO 610 J=1,N
      YY(J)=X(J)
610   Y(J)=(X(J)*(D-C)+D+C)/2.D0
      SUMXY=0.D0
      DO 635 J=1,N
      IF(INDEX.EQ.11) GO TO 629
      A=X1
      B=X1+(X3-X1)*(Y(J)-Y1)/(Y3-Y1)
      GO TO 630
629   A=X1+(Y(J)-Y4)*(X2-X1)/(Y1-Y4)
      B=X2
630   CONTINUE
      DO 620 I=1,N
620   R(I)=(YY(I)*(B-A)+A+B)/2.D0
      SUMX=0.D0
      DO 625 I=1,N
      BETA=ANTEG(R(I),Y(J))
625   SUMX=SUMX+BETA*W(I)
      SUMX=SUMX*(B-A)/2.D0
635   SUMXY=SUMXY+SUMX*W(J)
      SUMXY=SUMXY*(D-C)/2.D0
      RETURN
      END

     SUBROUTINE MINV (ZA,ZC,Y,B,RESID,ZB,ZM,N,IRES,KOD,IDIMZA,IDIMZC)
      implicit real*8 (a-h,o-z)
C       MINV COMPUTES THE INVERSE OF A MATRIX
C       N IS THE DIMENSION OF THE INPUT/OUTPUT MATRICES
C       ZA(N,N) INPUT MATRIX NOT MODIFIED BY THE COMPUTATION.
C       ZC(N,N) OUTPUT MATRIX ZC=ZA**-1
C    MATRIX ZC DOES NOT NEED TO BE ZEROED
C       ZB(N)   AND  ZM(N) ARE  VECTORS FOR INTERMEDIATE CALCULUS
C     Y = SOLUTION VECTOR OF THE MATRICIAL EQUATION  A*Y=B (Y=ZC*B)
C    RESID= RESIDUALS VECTOR = A*Y-B
C    IRES= -1 ONLY INVERSE (NOT Y) IS CALCULATED
```

```
C    IRES=0 SOLUTION NOT REFINED
C    IRES=1 SOLUTION REFINED WITH THE RESIDUALS
C    KOD=0 NORMAL RESULTS;   KOD.NE.0 SINGULAR CASE, ZC IS NOT COMPUTED
C    IDIMZA= DIMENSION OF MATRIX ZA IN CALLING PROGRAMME
C    IDIMZC= DIMENSION OF MATRIX ZC IN CALLING PROGRAMME
c     DIMENSION ZA(IDIMZA,1),ZC(IDIMZC,1)
      DIMENSION ZA(IDIMZA,idimza),ZC(IDIMZC,idimzc)
      DIMENSION Y(1),B(1),RESID(1),ZB(1),ZM(1)
      KOD=0
      EPS=1.D-12
      DO 80 I=2,N
      DO 80 J=1,N
      ZC(I,J)=0.0D0
   80 CONTINUE
      DO 2 I=1,N
    2 ZC(I,I)=1./ZA(I,I)
      ITER=2
      ZM(1)=ZC(1,1)
      ZB(1)=0.D0
      DO 50 L=2,N
      ZB(L)=ZA(1,L)*ZC(L,L)
   50 ZC(1,L)=-ZB(L)*ZM(1)
   30 ITERP=ITER+1
      ITERM=ITER-1
      DO 33 I=1,N
   33 ZM(I)=ZC(I,ITER)
      IF(ITER.EQ.N) GOTO 6
      DO 36 L=1,N
      ZB(L)=0.D0
      DO 3 K=1,ITERM
    3 ZB(L)=ZB(L)+ZA(ITER,K)*ZC(K,L)
      DO 4 K=ITERP,N
    4 ZB(L)=ZB(L)+ZA(ITER,K)*ZC(K,L)
   36 CONTINUE
      GOTO 5
    6 CONTINUE
      DO 37 L=1,N
      ZB(L)=0.D0
      DO 37 K=1,ITERM
      ZB(L)=ZB(L)+ ZA(ITER,K)*ZC(K,L)
   37 CONTINUE
    5 ZIJ=1.0+ZB(ITER)
      IF(ABS(ZIJ).GT.EPS) GOTO 39
        KOD=1
        WRITE(60,8040) KOD
8040    FORMAT(' KOD',I10)
        RETURN
39    ZIJ=1.0/ZIJ
      DO 40 I=1,N
40    ZB(I)=ZB(I)*ZIJ
      DO 42 L=1,N
      ZIJ=ZB(L)
      DO 41 I=1,N
```

```
41   ZC(I,L)=ZC(I,L)-ZIJ*ZM(I)
42   CONTINUE
      ITER=ITER+1
      IF(ITER.LE.N) GOTO 30
43   CONTINUE
      IF(IRES.EQ.-1) RETURN
      DO 100 I=1,N
      Y(I)=0.0D0
      DO 95 J=1,N
      Y(I)=Y(I)+ZC(I,J)*B(J)
95   CONTINUE
100  CONTINUE
      IREP=0
120  CONTINUE
      DO 150 I=1,N
      RESID(I)=0.0D0
      DO 145 J=1,N
145  RESID(I)=RESID(I)+ ZA(I,J)*Y(J)
      RESID(I)=RESID(I)-B(I)
150  CONTINUE
      IF(IRES.EQ.0.OR.IREP.EQ.1) GO TO 250
      DO 200 I=1,N
      ZB(I)=0.0D0
      DO 195 J=1,N
      ZB(I)=ZB(I)+ZC(I,J)*RESID(J)
195  CONTINUE
      Y(I)=Y(I)-ZB(I)
200  CONTINUE
      IREP=1
      GO TO 120
250  CONTINUE
      RETURN
      END

      subroutine minmax(len, field, amin,amax)
C     Given a 'field' of lenght 'len', this subroutine returns
C     the minimum (amin) and maximum (amax) values in the field.
      implicit real*8 (a-h, o-z)
      dimension field(*)
C
      amin= 1.e30
      amax=-1.e30
      do 10 i=1,len
      if(field(i).le.amin) amin=field(i)
      if(field(i).ge.amax) amax=field(i)
10   continue
      return
      end

      FUNCTION AN1N6(X,Y)
   implicit real*8 (a-h,o-z)
      COMMON/N16/A1,B1,C1,A2,B2,C2,A3,B3,C3,DELTA
      COMMON/NN16/INN1N6
```

```
      IERROR=1
      IERROR=0
      D2=2.D0*DELTA
      AL1= (A1+B1*X+C1*Y)/D2
      AL2= (A2+B2*X+C2*Y)/D2
      AL3= (A3+B3*X+C3*Y)/D2
      GO TO (1,2,3,4,5,6),INN1N6
    1 AN1N6=(2.D0*AL1-1.D0)*AL1
      GO TO 10
    2 AN1N6=(2.D0*AL2-1.D0)*AL2
      GO TO 10
    3 AN1N6=(2.D0*AL3-1.D0)*AL3
      GO TO 10
    4 AN1N6= 4.D0*AL1*AL2
      GO TO 10
    5 AN1N6=4.D0*AL2*AL3
      GO TO 10
    6 AN1N6=4.D0*AL1*AL3
C
   10 CONTINUE
      F=6.D0* DEXP(X)* DEXP(Y)*X*Y*(X*Y+ X+Y-3.D0)
      AN1N6=F*AN1N6
      RETURN
      END

      FUNCTION FIJM(X,Y)
      implicit real*8 (a-h,o-z)
      COMMON/ALPHA/A1,A2,A3
      F=- 6.D0* EXP(X)* EXP(Y)*X*Y*(X*Y+ X+Y -3.D0)
      FIJM=F*(A1+A2*X+A3*Y)
      RETURN
      END

      FUNCTION FBOUND(X,Y)
c   Sets boundry conditions. Entering the arguments X, Y allows for
c   setting boundary conditions different from those here coded.
      implicit real*8 (a-h,o-z)
      FBOUND= 0.D0
      RETURN
      END

      FUNCTION D2FXYZ(X,Y)
      implicit real*8 (a-h,o-z)
      F= 6.D0*X*Y* EXP(X)* EXP(Y)*(X*Y+X+Y -3.D0)
      D2FXYZ= F*(2.D0/X + 2.D0/Y +2.D0) +
     1 12.D0* EXP(X)* EXP(Y)*(X+1.D0)*(Y+1.D0)*(X+Y)
      RETURN
      END

      FUNCTION ANTEG2(X,Y)
      implicit real*8 (a-h,o-z)
      ANTEG2=6.D0* EXP(X)* EXP(Y)*X*Y*(X*Y+X+Y-3.D0)
      RETURN
      END
```

OUTPUT from Program PROG_VI_1 [POISSON]

```
 ITEST =    1
 SOLUTION OF POISSON EQ. WITH FIVE POINT FORMULA
 Time1 elapsed,resmax,umax =  0.20280099999999998
7.0110584005078 6355E-014  0.0000000000000000

 ITEST =    2
 SOLUTION OF POISSON EQ. WITH NINE POINT FORMULA
 NI,NJ,Dx,DY        19        19 5.2631578947368418 1E-002
5.2631578947368418 1E-002
 Sub. NINEPF - KOD   0
 Time2 elapsed,resmax,umax= 0.14040100000000003
1.0824674490095 2763E-015 0.57444791494296099

 ITEST =    3
 SOLUTION OF POISSON EQ. WITH CLASS. RITZ METHOD
 Time3 elapsed,resmax,umax= 0.26520100000000002
1.6985128174816 8575E-018 0.57299046250206420

 ITEST =    4
  GALERKIN METHOD-QUADRATIC SHAPE FUNC. N1-N6
  NI,NJ,NNODES     9   9 361
 Time4 elapsed,resmax,umax= 0.15600099999999995
3.7919970982680 5176E-016 0.57515066136819692

 ITEST =    5
  DOMAIN ORTHOGONAL COLLOCATIONS
  NI,NJ, IAMMMAX      11     11    385
 Time5 elapsed,resmax,umax= 0.12480099999999994
2.7755575615628 9135E-016 0.57274054937808871

 TESTPS - cputot 0.87360E+00
```

Appendix A.VIII

PROG_VIII.1 [HOMRSM]

```
c    Last Change: 12.03.2010

c   PROG_VIII.1 [HOMRSM - Computer program to simulate numerically the
c   simplified RSM (Reynolds Stress Model) for homogeneous, isotropic
c   turbulence. Integration of non linear ordinary differential
c   equations of first order is made with Runge-Kutta algorithm.

c   Subroutines used

c    subroutine HOMRSM - Driving Program
c    subroutine RK4MB  - Integration with Runge Kutta method of order 4
c    subroutine RKQC   - Runge Kutta of order 5 with variable time step,
c              adapted from Numerical Recipes.
c    subroutine RK4    - Called from rkqc
c
c    MAIN
      index=1           ! User must set index=1 or index=2
      call HOMRSM(index)
      stop
      end

      subroutine HOMRSM(index)
c    if index=1: Driving subroutine for Runge-Kutta of order 4 rk4mb.
c    if index=2: Driving subroutine for rkqc, Runge Kutta of order five
c          with adaptable time step.
c
c    LIST of main FORTRAN SYMBOLS:
c    a11,a22,a33,a12,a13,a23: components of anisotropic tensor
c    ak:  Turbulent kinetic energy [m**2 / s**2 ]
c    deru: Velocity gradient in Champagne experiments [1/s]
c    h:   Time step [s]
c    neq: Number of differential equayions to be integrated
c    nsteps: Number of time steps used for integration
c    t:   Time [s]
c    uc:  Central velocity in Champagne experiments [m/s]
c    y(1)-y(6): Components of Reynolds stress tensor [m**2/s**2]
c   y(7): Dissipation (epsilon) of turbulent kinetic energy [m**2/s**3]
c    yp(1)-yp(6): Time derivative of components of Reynolds stress
c       tensor [m**2/s**3]
c    y(7): Time derivative of dissipation (epsilon) of turbulent
c       kinetic energy [m**2/s**4]

      implicit real*8 (a-h,o-z)
      common/alpha/a11,a22,a33,a12,a13,a23,ak
      common/beta/c1,c2,c1e,c2e,deru
      parameter(n=7)
      dimension y(n),dydx(n),yscal(n),b(n),resid(n),yold(n)
      open(60,file='OUTPUT',status='unknown')
```

```
      open(10, file='OUT_TRAJ' ,  status='unknown')
c
c   Constants for calcultion of dissipation of turbulent energy
      c1e=1.44d0
      c2e=1.92d0
      c1=2.8d0
      c2=0.47d0

      deru=13.d0
      uc=12.4d0
c
c   Set initial values
      t=0.d0
      epsi=1.e-02
      eps3=1.e-10
      y(1)=epsi
      y(2)=epsi
      y(3)=epsi
      y(4)= - epsi
      y(5)=1.e-04
      y(6)=1.e-04
      y(7)=epsi
      neq=7
      nsteps=100 000 ! Problem: with increasing number of time steps the
c                          resuls do not approach a steady state.
c
      go to (1,2) index
c   ---------------------------------------------------
1     continue
      h=1.e-6
      write(60,999) h
      write( *,999) h
999   format(/,' RK4MB Integration step h = ',e12.5)
c
      do 100 i=1,nsteps
      y1o=y(1)
      y2o=y(2)
      y3o=y(3)
      y4o=y(4)
      y5o=y(5)
      y6o=y(6)
      y7o=y(7)
      call rk4mb(t,h,y,neq)
c
      if( mod(i,1000).eq.0) then ! Intermediat printing of data
        yp1=(-y1o+y(1))/h
        yp2=(-y2o+y(2))/h
        yp3=(-y3o+y(3))/h
        yp4=(-y4o+y(4))/h
        yp5=(-y5o+y(5))/h
        yp6=(-y6o+y(6))/h
        yp7=(-y7o+y(7))/h
        write(60,1000) i, y(1),y(2),y(3),y(4),y(5)
```

```
      write(60,1003)  y(6),y(7)
      write(60,1001) a11,a22,a33,ak
      write(60,1002) yp1,yp2,yp3,yp4,yp5
      write(60,1004) yp6,yp7
1000   format(/,' i,y1-5', i8,' ', 5e10.3)
1003   format('   y6-7',     2e10.3)
1001   format(' a11,a22,a33,k ', 4e12.4)
1002   format(' yp1-5',      5e10.3)
1004   format(' yp6-7',      2e10.3)
c
      y1norm= dsqrt(y(1))/uc   ! Compute normalized values
      y2norm= dsqrt(y(2))/uc
      y3norm= dsqrt(y(3))/uc
      y4norm= dsqrt(dabs(y(4)))/uc

      write(10,1050) i,y1norm,y2norm,y3norm,y4norm,y(7)
1050   format(1H ,i7,';',5(e16.8,';'))

      endif
c
      dymax=0.d0
      if(abs(y1o-y(1)).gt.dymax) dymax=abs(y1o-y(1))
      if(abs(y2o-y(2)).gt.dymax) dymax=abs(y2o-y(2))
      if(abs(y3o-y(3)).gt.dymax) dymax=abs(y3o-y(3))
      if(abs(y4o-y(4)).gt.dymax) dymax=abs(y4o-y(4))
      if(abs(y5o-y(5)).gt.dymax) dymax=abs(y5o-y(5))
      if(abs(y6o-y(6)).gt.dymax) dymax=abs(y6o-y(6))
      if(abs(y7o-y(7)).gt.dymax) dymax=abs(y7o-y(7))
      if(dymax.lt.eps3) go to 105
100   continue
105   continue
c
      go to 500
c   ---------------------------------------------
2   continue
       write(60,2009)
       write( *,2009)
2009   format(/, ' ***** TEST WITH SUBROUTINE RKQC ***',/)
      htry=1.d-06
      eps=1.e-08
      call der (y,dydx)
c
      do 11 i=1,n
      yscal(i)=1.0
11   continue
c
      do 20 i=1,nsteps
      call rkqc(y,dydx,n,t,htry,eps,yscal,hdid,hnext)
c
      if(mod(i,10000).eq.0) then
      write(60,1999) i,t,hdid
      write( *,1999) i,t,hdid
1999 format(' i,t,hdid ',i8,2e12.5)
```

```
      write(60,1000) i, y(1),y(2),y(3),y(4),y(5)
      write(60,1003)  y(6),y(7)
      write(60,1001) a11,a22,a33,ak
      write(60,1002) dydx(1),dydx(2),dydx(3),dydx(4)
      write(60,1004) dydx(5),dydx(6)
      endif
20   continue

c    -------------------------------------
500  continue                                  ! Final printing
      write(60,1000) i, y(1),y(2),y(3),y(4),y(5)
      write(60,1003)  y(6),y(7)
      write( *,1000) i, y(1),y(2),y(3),y(4),y(5)
      write( *,1003)  y(6),y(7)
      write(60,1001) a11,a22,a33,ak
      if(index.eq.1) then
        write(60,1002) yp1,yp2,yp3,yp4,yp5
        write(60,1004) yp6,yp7
      endif
      if(index.eq.2) then
        write(60,1002) dydx(1),dydx(2),dydx(3),dydx(4),dydx(5)
        write(60,1004) dydx(6),dydx(7)
      endif
c
      return
      end

      subroutine RK4MB(t,h,y,n)
      implicit real*8 (a-h,o-z)
      dimension y(7),yp1(7),yp2(7),yp3(7),yp4(7),aus(7)
c
c     t   independent variable
c     h   integration step (dt)
c     n   order of the diff. system
c     y(n) dependent variables (initial values are introduced here)
c     yp.(n) auxiliary arrays
c     aus(n) auxiliary array
c
      t0 = t
      do 5 i=1,n
      aus(i) = y(i)
 5   continue
c
      call der(y,yp1)
      t   = t0+h*0.5
      do 10 i=1,n
      y(i) = aus(i)+h*yp1(i)*0.5
 10  continue
      call der(y,yp2)
      do 11 i=1,n
      y(i) = aus(i)+h*yp2(i)*0.5
 11  continue
      call der(y,yp3)
```

```
      t   = t0+h
      do 12 i=1,n
      y(i) = aus(i)+h*yp3(i)
 12   continue
      call der(y,yp4)
c
      do 20 i=1,n
      y(i) = aus(i)+h*(yp1(i)+2.0*yp2(i)+2.0*yp3(i)+yp4(i))/6.0
 20   continue
c
      return
      end

      Subroutine DER(y,yp)
c
c   Derivatives coding the homogeneous Reynolds Stress Equations:
c   y(1) = u1u1; y(2)=u2u2; y(3)=u3u3; y(4)=u1u2; y(5)=u1u3,
c   y(6) = u2u3; y(7)=epsilon.
c
      implicit real*8 (a-h,o-z)
      common/alpha/a11,a22,a33,a12,a13,a23,ak
      common/beta/c1,c2,c1e,c2e,deru
      dimension y(7),yp(7)
      two3=2.d0/3.d0
c
      ak=0.5d0*(y(1) + y(2) + y(3))
      a11=y(1)/ak -two3
      a22=y(2)/ak -two3
      a33=y(3)/ak -two3
      a12=y(4)/ak
      a13=y(5)/ak
      a23=y(6)/ak
c
      p11=-2.d0*y(4)*deru
      p12=- y(2)*deru
      p13=- y(6)*deru
c
      yp(1)= p11*(1.d0 - two3*c2) - c1*y(7)*a11 - two3*y(7)
      yp(2)= p11*c2/3.d0      - c1*y(7)*a22 - two3*y(7)
      yp(3)= p11*c2/3.d0      - c1*y(7)*a33 - two3*y(7)
c
      yp(4)= p12*(1.d0 - c2) - c1*y(7)*a12
      yp(5)= p13*(1.d0 - c2) - c1*y(7)*a13
      yp(6)=           - c1*y(7)*a23
c
      yp(7)= 0.5d0*p11*c1e*y(7)/ak - c2e*y(7)**2/ak
c
      return
      end

c

      subroutine RKQC(y,dydx,n,x,htry,eps,yscal,hdid,hnext)
      implicit real*8(a-h,o-z)
```

```
c   Definition of variables:
c   Y    = Dependent variable (resut of integration)
c   DXDY = Derivative of Y with respect to X
c   N    = Number of equations to be integrated
c   X    = Independent variable
c   HTRY = Initial trial step size
c   EPX  = Required accuracy of the solution
c   YSCAL= Array for scalimg the error
c   HDID = Step size actually used
c   HNEXT= Estimated next step size
c
      parameter (nmax=10,fcor=.0666666667,
     *  one=1.,safety=0.9,errcon=6.e-4)
      dimension y(n),dydx(n),yscal(n),ytemp(nmax),ysav(nmax),dysav
      (nmax)
      pgrow=-0.20
      pshrnk=-0.25
      xsav=x
      do 11 i=1,n
       ysav(i)=y(i)
       dysav(i)=dydx(i)
11    continue
      h=htry
1     hh=0.5*h
      call rk4(ysav,dysav,n,xsav,hh,ytemp)
      x=xsav+hh
      call der(ytemp,dydx)
      call rk4(ytemp,dydx,n,x,hh,y)
      x=xsav+h
c     if(x.eq.xsav)pause 'stepsize not significant in rkqc.'
      if(x.eq.xsav) write(60,1010)
      if(x.eq.xsav) write( *,1010)
      if(x.eq.xsav) stop
1010  format(' stepsize not significant in rkqc - STOP')
      call rk4(ysav,dysav,n,xsav,h,ytemp)
      errmax=0.
      do 12 i=1,n
       ytemp(i)=y(i)-ytemp(i)
       errmax=max(errmax,abs(ytemp(i)/yscal(i)))
12    continue
      errmax=errmax/eps
      if(errmax.gt.one) then
       h=safety*h*(errmax**pshrnk)
       goto 1
      else
       hdid=h
       if(errmax.gt.errcon)then
        hnext=safety*h*(errmax**pgrow)
       else
        hnext=4.*h
       endif
      endif
      do 13 i=1,n
```

```
      y(i)=y(i)+ytemp(i)*fcor
13   continue
     return
     end

     subroutine RK4(y,dydx,n,x,h,yout)
     implicit real*8(a-h,o-z)
     parameter (nmax=10)
     dimension y(n),dydx(n),yout(n),yt(nmax),dyt(nmax),dym(nmax)
     hh=h*0.5
     h6=h/6.
     xh=x+hh
     do 11 i=1,n
      yt(i)=y(i)+hh*dydx(i)
11   continue
     call der(yt,dyt)
     do 12 i=1,n
      yt(i)=y(i)+hh*dyt(i)
12   continue
     call der(yt,dym)
     do 13 i=1,n
      yt(i)=y(i)+h*dym(i)
      dym(i)=dyt(i)+dym(i)
13   continue
     call der(yt,dyt)
     do 14 i=1,n
      yout(i)=y(i)+h6*(dydx(i)+dyt(i)+2.*dym(i))
14   continue
     return
     end
```

Formula Synopsis: Concepts and Definitions, Matrix Algebra, Determinants, and Vector and Tensor Operations

Concepts and Definitions

Hilbert space The Hilbert space is a set of functions, called the "elements" *of the space, defined in a given* domain *D*, dependent on the space coordinates and possibly on time, which possess some properties:

(i) A *sum* operation is defined, denoted with the usual symbol (+).With this operation, the functions form a linear set, so that if u and v are elements of the set, also $u + v$ and cu (where c is a constant) are elements of the set.
(ii) In the set of the functions, an operation *scalar product*, or *inner product*, is defined by:

$$(u, v) = \int_{\Omega} uv^* d\Omega, \quad \text{if } u, v \text{ are complex } (v^* \text{complex conjugate of } v),$$

$$(u, v) = \int_{\Omega} uvd\Omega, \quad \text{if } u, v \text{ are real,}$$

$$(\overline{u}, \overline{v}) = \int_{\Omega} (\overline{u} \cdot \overline{v}) d\Omega, \quad \text{if } \overline{u}, \overline{v} \text{ are vectors (the dot denotes vector scalar product).}$$

(iii) The scalar product has the following properties:

M. Bottoni, *Physical Modeling and Computational Techniques for Thermal and Fluid-dynamics*, Mechanical Engineering Series,
https://doi.org/10.1007/978-3-030-79717-1

$(u, v) = (v, u)^*$, if u, v are complex,
$(u, v) = (v, u)$, if u, v are real,
$(\overline{u}, \overline{v}) = (\overline{v}, \overline{u})$, if $\overline{u}, \overline{v}$ are vectors,
$(c_1 u_1 + c_2 u_2, v) = c_1(u_1, v) + c_2(u_2, v)$, (distributive property, also for vectors),
$(u, u) \geq 0$, $[(u, u) = 0$, if and only if $u = 0]$,
$(\overline{u}, \overline{u}) \geq 0$, $[(\overline{u}, \overline{u}) = 0$, if and only if $\overline{u} = 0]$.

Self-Adjoint Operator
An operator A is self-adjoint if, for any u and v functions, it satisfies:

$$(Au, v) = (u, Av).$$

If u and v are real functions, the self-adjoint operator is symmetric.

Matrix Algebra

A matrix is denoted in square brackets $[A] = [a_{ij}]$, which can be simplified to A, when there is no ambiguity with other symbols. Numbers of rows and columns of the matrix are denoted symbolically with $A(m, n)$.

Let $A(n, n) = [a_{ij}]$ be a square matrix with complex or real elements a_{ij}.

(i) A^*, the *conjugate* of A, is obtained replacing every element of A with its complex conjugate a_{ij}^*.
(ii) A^T, the *transpose* of A, is obtained exchanging in A rows and columns.
(iii) A^H, the *conjugate transpose* of A, is obtained from A taking the transpose and replacing every element with its complex conjugate. If A is real, $A^H = A^T$

It holds:

$$A^H = (A^T)^* = (A^*)^T,$$
$$(A^H)^H = (A^T)^T = (A^*)^* = A.$$

(iv) A^{-1}, the *inverse* of A, has elements $a_{rs} = (-1)^{r+s} \frac{A_{sr}}{\det(A)}$, where A_{sr} is the cofactor of a_{rs}, i.e., the determinant of the matrix obtained from the matrix A striking the s-th row and the r-th column.
(v) *Eigenvalues* of a square matrix A of order N are the solutions ($\lambda_i, i = 1, 2, \ldots N$) of the algebraic equation $\det(A - \lambda I) = 0$.

Spectral radius S of a matrix A is the maximum of the moduli of the eigenvalues:

$$S = \max{}_i |\lambda_i|.$$

The eigenvalues of a tri-diagonal matrix of rank N:

$$\begin{vmatrix} a & b & & & & \\ c & a & b & & & \\ & c & a & b & & \\ & & & \cdots\cdots & & \\ & & & c & a & b \\ & & & & c & a \end{vmatrix}$$

are $\lambda_i = a + 2b\sqrt{\frac{c}{b}}\cos\left(\frac{i\pi}{N+1}\right)$. $\quad (i = 1, 2 \ldots\ldots N)$

(vi) It holds: $\quad \det(A) = \prod_{i=1}^{N} \lambda_i,$

$$\mathrm{Trace}(A) = \sum_{i=1}^{N} a_{ii} = \sum_{i=1}^{N} \lambda_i.$$

(vii) *Eigenvectors* of A are the solutions of the equation $A\overline{x} = \lambda\overline{x}$, for every λ_i.
(viii) A matrix $[A]$ is:

Singular if	$\det(A) = 0,$	
Unitary if	$A = A^H = I,$	$(A^H = A^{-1}$ is the inverse$)$
Orthogonal if	$A = A^T = I,$	$(A^T = A^{-1}$ is the inverse$)$
Normal if	$AA^H = A^HA,$	
Hermitian if	$A = A^H$	(if A is real $A^H = A^T = A$, hence A is *symmetric*).

A unitary or Hermitian matrix is also normal.
Theorem 1: *All Eigenvalues of a Hermitian (hence of a real symmetric) matrix are real.*
Theorem 2: *The inverse and the transpose of an orthogonal matrix are orthogonal.*
Theorem 3: *The product of orthogonal matrices is orthogonal.*
Theorem 4: *The determinant of an orthogonal matrix is* ± 1.

(ix) *The rotation matrix* in two-dimensional space.

Let $P(x,y)$ be a point with coordinates in a reference frame (X,Y) with origin O. The coordinates of P in a frame (X',Y') with the same origin, rotated by an angle ϑ in a counterclockwise direction, are:

$$\begin{vmatrix} x' \\ y' \end{vmatrix} = \begin{vmatrix} \cos\vartheta & \sin\vartheta \\ -\sin\vartheta & cos\vartheta \end{vmatrix} \begin{vmatrix} x \\ y \end{vmatrix} = [R] \begin{vmatrix} x \\ y \end{vmatrix}$$

where $[R]$ is the rotation matrix and $\det(R) = 1$. It holds $[R][R^T] = [R]^T[R] = [I]$; hence $[R]^T = [R]^{-1}$ and the *t*matrix $[R]$ is orthogonal.

(x) It holds:

$$\begin{aligned} (AB)^H &= B^H A^H, \\ (AB)^{-1} &= B^{-1} A^{-1}, \\ (AB)^T &= B^T A^T, \\ (AB)^* &= A^* B^*. \end{aligned}$$

(xi) *Multiplication of matrices.* Multiplication (rows by columns) of matrices, say A and B, requires that the matrices be *comformable*: The number of columns of A (say r) must be equal to the number of rows of B. Symbolically, if m is the number of rows of A and n is the number of columns of B, one denotes: $(m, r)(r, n) = (m, n)$.

If $C = AB$, the elements of C are $c_{ij} = \sum_r a_{ir} b_{rj}$.

$$\begin{aligned} &AB \neq BA, \quad \text{(In general, the product of matrices is not commutative)} \\ &ABC = A(BC) = (AB)C, \quad \text{(Associative law)} \\ &A(B + C + D) = AB + AC + AD. \quad \text{(Distributive law)} \end{aligned}$$

(xii) A matrix A is *positive definite* if it is Hermitian ($A = A^H$) and $(v, Av) > 0$ for all $v \neq 0$. If $(v, Av) \geq 0$ for all v, A is nonnegative definite. Similar definitions apply for negative definite and non-positive definite.

(xiii) Theorem 5: *A matrix A is positive definite if and only if it is Hermitian and all its eigenvalues are positive.*

(xiv) Theorem 6: *A Hermitian matrix* ($A = A^H$) *of order N is positive definite if and only if every sub-determinant obtaining eliminating m* ($0 \leq m < N$) *rows and the corresponding columns is positive.*

In particular: $\det(A) > 0$.
All diagonal elements are positive ($a_{ii} > 0$).

(xv) Theorem 7: *If the matrix A is positive definite, there exists a unique positive definite matrix B such that* $B^2 = A$. One defines $B = A^{1/2.}$. If A is real, B is real.

(xvi) *Norms of a matrix.* For a matrix A of order N, the following norms can be defined:

$$\|A\| =N \max \sum_{j=1}^{N} |a_{ij}|, \quad (i = 1, 2, 3, \ldots N)$$

$$\|A\| =N \max \sum_{i=1}^{N} |a_{ij}|, \quad (j = 1, 2, 3, \ldots N)$$

$$\|A\| = \sqrt{S(A^H A)}, \quad \text{(Spectral norm)}$$

$$\|A\| =N \max |a_{ij}|,$$

$$\|A\| = \sqrt{\sum_{i,j=1}^{N} |a_{ij}|^2}.$$

Properties of Determinants

$$\det(AB) = \det(A)\det(B) = \det(BA),$$
$$\det\left(B^{-1}AB =\right) = \det\left(B^{-1}\right)\det(A)\det(B) = \det(A).$$

Vector and Tensor Operations

(i) *Definition and conventions.* In the following, we use:

$c, r, s\ldots$ as scalars (zero-order tensors),

$\overline{a}, \overline{b}, \ldots \overline{u}, \overline{v}, \overline{w}, \ldots \overline{i}, \overline{j}, \overline{k}, \overline{\delta}_i, \ldots (i = 1, 2, 3$ in 3D space) as vectors (first-order tensors) $\overline{\overline{\tau}}, \overline{\overline{\delta}}\ldots$ as second-order tensors.

We define the unit tensor $\overline{\overline{\delta}} = \begin{vmatrix} 1 & 0 & 0 \\ 0 & 1 & 0 \\ 0 & 0 & 1 \end{vmatrix}.$

In the following, we use the conventions of Bird et al. (1960) to denote scalars, vectors, or tensors:

Parentheses	( ) denote scalars.
Brackets	[ ] denote vectors.
Braces	{ } denote tensors.

The order of the product of multiplications on scalars, vectors, or tensors is given by the following rule ($\sum$ denotes the sum of the order of the factors):

Multiplication sign :	Order of result :	Example :
None	$\sum$	$(s)[\overline{v}] = [s\overline{v}]$
$\times$	$\sum -1$	$[\overline{a}] \times [\overline{b}] = [\overline{a} \times \overline{b}]$.
$\cdot$	$\sum -2$	$[\overline{a}] \cdot [\overline{b}] = (s)$.
$:$	$\sum -4$	$\{\overline{\overline{\sigma}}\} : \{\overline{\overline{\tau}}\} = (\overline{\overline{\sigma}} : \overline{\overline{\tau}})$.

Scalar product of two vectors : $\left(\overline{a} \cdot \overline{b}\right) = \sum_i a_i b_i$.

Cross product of two vectors : $\left[\overline{a} \times \overline{b}\right] = \begin{vmatrix} \overline{i} & \overline{j} & \overline{k} \\ a_1 & a_2 & a_3 \\ b_1 & b_2 & b_3 \end{vmatrix}$.

Triple product of vectors : $\left(\overline{u} \cdot \left[\overline{a} \times \overline{b}\right]\right) = \begin{vmatrix} u_1 & u_2 & u_3 \\ a_1 & a_2 & a_3 \\ b_1 & b_2 & b_3 \end{vmatrix} = uab \cos \phi \sin \vartheta$

where u, a, and b are the moduli of the vectors, ϑ is the angle between the vectors $\overline{b}$ and $\overline{c}$, and ϕ is the angle between $\overline{a}$ and the normal to the plane defined by $\overline{b}$ and $\overline{c}$.

$$[\nabla] = \overline{i}\frac{\partial}{\partial x} + \overline{j}\frac{\partial}{\partial y} + \overline{k}\frac{\partial}{\partial z},$$ (*Nabla operator* or "del" in rectangular coordinates)

$$\nabla^2 = \frac{\partial^2}{\partial x^2} + \frac{\partial^2}{\partial y^2} + \frac{\partial^2}{\partial z^2},$$ (*Laplacian operator* or "del square" in rectangular coordinates)

$$[\nabla s] = \overline{i}\frac{\partial s}{\partial x} + \overline{j}\frac{\partial s}{\partial y} + \overline{k}\frac{\partial s}{\partial z},$$ (Gradient of a scalar s, grad s)

$$[\nabla \times \overline{a}] = \text{curl}\overline{a} = \text{rot}\overline{a} = \begin{vmatrix} \overline{i} & \overline{j} & \overline{k} \\ \frac{\partial}{\partial x} & \frac{\partial}{\partial y} & \frac{\partial}{\partial z} \\ a_x & a_y & a_z \end{vmatrix}.$$ ("curl" or "rotational" of a vector)

(ii) *Summary of vector operations*

$$(\nabla \cdot \overline{v}) = \sum_i \frac{\partial v_i}{\partial x_i}, \quad \text{(Divergence of a vector field)}$$

$$(\nabla \cdot [\nabla \times \overline{a}]) = \operatorname{div}(\operatorname{rot}\overline{a}) = 0,$$

$$[\nabla \times \nabla s] = \operatorname{rot}(\operatorname{grad} s) = 0,$$

$$[\nabla \times s\overline{a}] = \nabla s \times \overline{a} + s[\nabla \times \overline{a}],$$

$$\begin{aligned}[\nabla(\overline{a} \cdot \overline{b}] &= \overline{b} \cdot \nabla\overline{a} + \overline{a} \cdot \nabla\overline{b} + \overline{a} \times [\nabla \times \overline{b}] + \overline{b} \times [\nabla \times \overline{a}] \\ &= (\overline{b} \cdot \nabla)\overline{a} + (\overline{a} \cdot \nabla)\overline{b} + \overline{a} \times (\nabla \times \overline{b}) + \overline{b} \times (\nabla \times \overline{a}),\end{aligned}$$

$$(\overline{a} \cdot (\overline{b} \times \overline{c})) = [\overline{a} \times \overline{b}] \cdot \overline{c} = [\overline{c} \times \overline{a}] \cdot \overline{b},$$

$$[\overline{a} \times (\overline{b} \times \overline{c})] = \overline{b}(\overline{a} \cdot \overline{c}) - \overline{c}(\overline{a} \cdot \overline{b}),$$

$$[\nabla \times \nabla \times \overline{a}] = \nabla(\nabla \cdot \overline{a}) - \nabla^2\overline{a},$$

$$(\nabla s \cdot \nabla s) = \left(\frac{\partial s}{\partial x}\right)^2 + \left(\frac{\partial s}{\partial y}\right)^2 + \left(\frac{\partial s}{\partial z}\right)^2,$$

$$(\nabla \cdot \nabla s) = \operatorname{div}(\operatorname{grad} s) = \nabla^2 s = \frac{\partial^2 s}{\partial x^2} + \frac{\partial^2 s}{\partial y^2} + \frac{\partial^2 s}{\partial z^2},$$

$$[\nabla \cdot \nabla\overline{v}] = [\nabla^2\overline{v}] = \frac{\partial^2\overline{v}}{\partial x^2} + \frac{\partial^2\overline{v}}{\partial y^2} + \frac{\partial^2\overline{v}}{\partial z^2} = \overline{i}\nabla^2 v_x + \overline{j}\nabla^2 v_y + \overline{k}\nabla^2 v_z,$$

$$[\nabla(rs)] = r\nabla s + s\nabla r,$$

$$(\nabla \cdot s[\overline{v}]) = \nabla s \cdot \overline{v} + s(\nabla \cdot \overline{v}),$$

$$\left(\nabla \cdot s\left\{\overline{\overline{\delta}}\right\}\right) = [\nabla s],$$

$$[\nabla \cdot \overline{vw}] = [v \cdot \nabla\overline{w}] + \overline{w}[\nabla \cdot \overline{v}],$$

$$(\nabla \cdot (\overline{a} \times \overline{b})) = \overline{b} \cdot [\nabla \times \overline{a}] - \overline{a} \cdot [\nabla \times \overline{b}],$$

$$\begin{aligned}[\nabla \times (\overline{a} \times \overline{b})] &= \overline{b} \cdot \nabla\overline{a} - \overline{a} \cdot \nabla\overline{b} + \overline{a}(\nabla \cdot \overline{b}) - \overline{b}(\nabla \cdot \overline{a}) \\ &= (\overline{b} \cdot \nabla)\overline{a} - (\overline{a} \cdot \nabla)\overline{b} + \overline{a}(\nabla \cdot \overline{b}) - \overline{b}(\nabla \cdot \overline{a}),\end{aligned}$$

$$\left(s\left\{\overline{\overline{\delta}}\right\} : \nabla\overline{v}\right) = s(\nabla \cdot \overline{v}),$$

$$[\nabla \cdot \overline{vv}] = [\overline{v} \cdot \nabla\overline{v}] + \overline{v}[\nabla \cdot \overline{v}] = (\overline{v} \cdot \nabla)\overline{v} + \overline{v}[\nabla \cdot \overline{v}],$$

$$(\overline{\overline{\tau}} : \nabla\overline{v}) = (\nabla \cdot [\overline{\overline{\tau}} \cdot \overline{v}]) - (\overline{v} \cdot [\nabla \cdot \overline{\overline{\tau}}]). \quad \text{(If } \overline{\overline{\tau}} \text{ is symmetrical)}$$

(iii) *Substantial derivative* is defined by $\frac{D}{Dt} = \frac{\partial}{\partial t} + \overline{v} \cdot \nabla$.

$$\frac{Ds}{Dt} = \frac{\partial s}{\partial t} + \overline{v} \cdot \nabla s = \frac{\partial s}{\partial t} + \sum_i v_i \frac{\partial s}{\partial x_i},$$

$$\frac{D\overline{v}}{Dt} = \frac{\partial \overline{v}}{\partial t} + (\overline{v} \cdot \nabla)\overline{v} = \sum_i \overline{\delta}_i \left(\frac{\partial v_i}{\partial t} + (\overline{v} \cdot \nabla) v_i \right).$$

A Few Remarks on the Classification of Differential Equations

Partial Differential Equations of First Order

Let us consider the partial differential equation of first order:

$$a(x, y) \frac{\partial u}{\partial x} + b(x, y) \frac{\partial u}{\partial y} = c(x, y)u + d(x, y), \tag{1}$$

where $u = u(x, y)$ is a differentiable function in the space (*x*,*y*). If the coefficients *a*, *b*, *c*, and *d* are constant or functions of *x* and *y* alone, Eq. (1) is linear; if the coefficients are functions of *x*, *y*, and *u* (but not of the derivatives $\partial u/\partial x$ and $\partial u/\partial y$), the equation is quasi-linear; if the coefficients are functions of the derivatives $\partial u/\partial x$ and $\partial u/\partial y$ or of derivatives of higher order, the equation is nonlinear.

Systems of Partial Differential Equations of First Order

Let us consider the following two-equation system linking two functions $u_1(x, y)$ and $u_2(x, y)$ of the independent variables *x* and *y*:

$$a_{11} \frac{\partial u_1}{\partial x} + a_{12} \frac{\partial u_1}{\partial y} + a_{13} \frac{\partial u_2}{\partial x} + a_{14} \frac{\partial u_2}{\partial y} = f_1, \tag{2a}$$

$$a_{21} \frac{\partial u_1}{\partial x} + a_{22} \frac{\partial u_1}{\partial y} + a_{23} \frac{\partial u_2}{\partial x} + a_{24} \frac{\partial u_2}{\partial y} = f_2. \tag{2b}$$

If the coefficients a and the right-hand sides f are constant or functions of the independent variables *x* and *y* only, the system is linear; if *a* and *f* depend also on the unknowns u_1, u_2 (but not on their derivatives), the system is quasi-linear; if a, f_1, and f_2 depend also on the derivatives $\partial u_1/\partial x$, $\partial u_1/\partial y$, $\partial u_2/\partial x$, and $\partial u_2/\partial y$, the system is nonlinear.

Let us define the discriminant:

$$\Delta = (a_{11}a_{24} - a_{21}a_{14} + a_{12}a_{23} - a_{22}a_{13})^2 - 4(a_{11}a_{23} - a_{21}a_{13})(a_{12}a_{24} - a_{22}a_{14}). \tag{3}$$

The system (3) is classified as elliptic in a point *P*(*x*,*y*) of the definition domain if $\Delta < 0$ in that point, parabolic if $\Delta = 0$, and hyperbolic if $\Delta > 0$. If the same holds in every point of the definition domain, the system is said to be simply of that type.

In most physical or engineering applications, it is found that close boundary problems are generally of elliptic type, whereas open boundary problems are generally of parabolic or hyperbolic types.

Partial Differential Equations of Second Order

Considering a two-dimensional space (*x*, *y*) and a differentiable function *u*(*x*,*y*) defined in this space, the most general partial differential equation of second order for this function can be written as:

$$a\frac{\partial^2 u}{\partial x^2} + b\frac{\partial^2 u}{\partial x \partial y} + a\frac{\partial^2 u}{\partial y^2} + d\frac{\partial u}{\partial x} + e\frac{\partial u}{\partial y} + fu + g = 0. \tag{4}$$

If the coefficients *a*, *b*, *c*. . . . . . are constants or functions of *x* and *y* only, Eq. (4) is linear; if the coefficients are functions of *x*, *y*, *u*, $\frac{\partial u}{\partial x}$, $\frac{\partial u}{\partial y}$ (but not of derivatives of higher order), the equation is quasi-linear; otherwise, when the coefficients are functions of second- or higher-order derivatives, the equation is nonlinear.

Equation (4) is classified as hyperbolic if the discriminant $\Delta = b^2 - 4ac > 0$ is positive, as parabolic if $\Delta = 0$, and as elliptic if $\Delta < 0$. Notice that only the coefficients of the second-order derivatives (*a*, *b*, *c*) determine the type of the equation.

Partial Differential Equations of Higher Order (n > 2)

Restricting ourselves to the two-dimensional case in a definition domain *D*(*x*,*y*), let:

$$F\left(x, y, u, \frac{\partial u}{\partial x}, \frac{\partial u}{\partial y}, \frac{\partial^2 u}{\partial x^2}, \frac{\partial^2 u}{\partial x \partial y}, \frac{\partial^2 u}{\partial y^2}, \ldots \frac{\partial^n u}{\partial x^n}, \frac{\partial^n u}{\partial x^{n-1} \partial y}, \ldots \frac{\partial^n u}{\partial y^n}\right) \tag{5}$$

be a function of the independent variables x, y, of the unknown function $u(x,y)$, and of its partial derivatives till order n. It is also assumed that the unknown function is continuous, derivable, with continuous derivatives.

If the coefficients in Eq. (5) are constant or functions of x and y alone, the differential equation is linear; if the coefficients are functions of x, y, $u(x,y)$ and of the derivatives of $u(x,y)$ till order $n–1$ (but not of order n), the equation is quasi-linear; if the coefficients are function also of the derivatives of maximum order n, the equation is nonlinear.

For the classification of systems of differential equations of every order, we remand to the classical reference (Courant & Hilbert, 1968) (Page. 141 ff.).

Nomenclature

A	Area (m^2)
	Differential operator
A_f	Cross-flow area (m^2)
a_{ij}	Anisotropy Reynolds stress tensor
a_β^g	Coefficients of g equation (kg/s)
a_β^{ij}	Coefficients of Reynolds stress equation (kg/s)
a_β^Φ	Coefficients of scalar fluxes equation (kg/s)
a_β^ε	Coefficients of ε equation (kg/s)
$a_\beta^{\Phi_{ij}}$	Coefficients of Reynolds stress equations (kg/s)
b_0^g	Right-hand side of g equation (kg k^2/s)
b_0^Φ	Right-hand side of scalar fluxes equations (kg mK/s^2)
b_0^ε	Right-hand side of ε equation (kg m^2/s^4)
$b_0^{\Phi_{ij}}$	Right-hand side of Reynolds stress equation (kg m^2/s^3)
"c"	Constants used in the turbulence models summarized in Table 8.2
$c = (\partial p/\partial \rho)^{1/2}$	Sound velocity (m/s)
c_p	Specific heat at constant pressure (J/kg K)
c_v	Specific heat at constant volume (J/kg K)
c_a	Geometry coefficient
C	Courant number
	Mean displacement of a particle (m)
D	Diffusion coefficient (m^3/s)
	Domain
D_h	Hydraulic diameter (m)
$D(k)$	Dissipation spectrum (m^3/s^3)
e	Error
$E(k)$	Energy spectrum (m^3/s^2)
f	Friction coefficient

M. Bottoni, *Physical Modeling and Computational Techniques for Thermal and Fluid-dynamics*, Mechanical Engineering Series,
https://doi.org/10.1007/978-3-030-79717-1

F	Volumetric flow (m^3/s) Convection coefficient (kg/s)
g	Gravity acceleration (m/s^2)
$g = \frac{1}{2}\langle \phi^2 \rangle$	One-half of the variance of temperature fluctuations (K^2)
G	Mass flow (kg/s) Amplification factor Production of turbulent kinetic energy due to buoyancy (m^2/s^3)
$G_{i\phi}$	Buoyancy production of turbulent kinetic energy in scalar flux equations (mK/s^2)
G_K	Buoyancy production of turbulent kinetic energy in K-equation (m^2/s^3)
G_{SI}	"Momentum slip" (kg/m^3)
h	Specific enthalpy (J/kg) Length of discrete interval (m)
h_{fg}	Vaporization enthalpy (J/kg)
H	Iteration matrix Distance between horizontal plates (m)
H_{SI}	"Enthalpy slip" (J/m^3)
H_v	Vapor specific enthalpy (J/kg)
i	Imaginary unit
I	Identity tensor
j	Imaginary unit
J	Imaginary unit
J^E	Energy flux vector (W/m^2)
$J_g = J_g^c + J_g^d$	Total (convective plus diffusive) flux in g equation (mK^2/s)
J_g^c	Convective flux in g equation (mK^2/s)
J_g^d	Diffusive flux in g equation (mK^2/s)
$\overline{J}_\phi = J_\phi^c + J_\phi^d$	Total (convective plus diffusive) flux in scalar fluxes equations (m^2K/s^2)
$J_\phi = \rho \overline{J}_\phi$	Total flux times density in scalar flux equations (Kkg/ms^2)
J_ϕ^c	Convective flux in scalar fluxes equations (m^2K/s^2)
J_ϕ^d	Diffusive flux in scalar fluxes equations (m^2K/s^2)
$J_{\varepsilon j}^c = \rho u_j \varepsilon$	Convective flux in ε equation (kg/s^4)
$J_{\varepsilon j}^{d\ell} = -\mu_\ell \frac{\partial \varepsilon}{\partial x_j}$	Laminar diffusive flux in ε equation (kg/s^4)
$J_{\varepsilon j}^{dt} = -\Gamma_{\varepsilon jm} \frac{\partial \varepsilon}{\partial x_m}$	Turbulent diffusive flux in ε equation (kg/s^4)
$k = 2\pi/\lambda$	Wavenumber (m^{-1})
K	Turbulent kinetic energy (m^2/s^2)
K_m	Turbulent diffusion coefficient for momentum (m^2/s)
K_n	Knudsen number
K^+	Dimensionless kinetic energy of turbulence
l	General coordinate direction (m)
ℓ	Macroscopic length scale of turbulence (m)

ℓ^*	Characteristic length (m)
L	General coordinate direction (m)
	Area coordinate
	Length scale (m)
	Differential operator
m	Mass flux by phase change (kg/m^{2}s)
Ma	Mach number
n_0	Loschmidt number (molecules/m^3)
N	Shape function
p	Pressure (N/m^2)
	Probability
$P_g = \frac{1}{2} P_\phi$	Mean field production in g equation (K^2/s)
P_{ij}	Stress Production due to mean flow (m^2/s^3)
P_K	Mean shear production in K and ε equations (m^2/s^3)
$P_{i\phi}$	Mean field production in scalar fluxes equations (mK/s^2)
p	Pressure (N/m^2)
Pr	Prandtl number
$\overline{q} = -\Gamma_h \nabla h_m$	Heat flux (W/m^2)
q	Energy flux (W/m^2)
	Probability
Q	Volumetric specific power (W/m^3)
r	Radial coordinate (m)
	Optimization parameter
	Residual
r_d	Droplet radius (m)
R	Body force/friction force (kg/m^2 s^2)
	Residual
Re	Reynolds number
$R_f = -G_K/P_K$	Richardson number
s	Azimuthal coordinate (m)
s_{ij}	Strain rate tensor (s^{-1})
S	Surface (m^2)
S_g	Source term in g equation (K^2/s)
S_{ij}^{Φ}	Source term in Reynolds stress equations (m^2/s^3)
S_ε	Source term in ε equation (kg/ms^4)
$S_{\phi i}$	Source term in scalar fluxes equations (mK/s^2)
t	Time (s)
t^*	Characteristic time (s)
t^+	Dimensionless time
T	Temperature (K)
	Truncation error
$T(k)$	Spectrum of energy flux (m^2/s^3)
$\overline{u}$	Velocity vector (m/s)
u	General velocity component (m/s)

	Radial velocity component (m/s)
	Unknown function
u_i	Fluctuation of velocity component (m/s)
u^*	Friction velocity (m/s)
u^+	Dimensionless velocity
$u_i u_j$	Reynolds stress (m^2/s^2)
$u_i \phi = \Phi$	Scalar flux (mK/s)
U	Perimeter (m)
	Main flow velocity (m/s)
U_i	Component of mean stream velocity (m/s)
U_{SI}	Slip velocity (m/s)
v	Azimuthal component of coolant velocity (m/s)
	Specific volume (m^3/kg)
v_K	Microscopic Kolmogorov scale of velocity (m/s)
V	Volume (m^3)
	Velocity (m/s)
V_f	Fluid volume in a control cell (m^3)
V_{SI}	Slip velocity (m/s)
w	Axial component of coolant velocity (m/s)
w^*	Friction velocity (m/s)
W	Mass flow (kg/s)
	Weight function
W^+	Dimensionless mass flow
W_{SI}	Slip velocity (m/s)
x	Thermodynamic quality
	Coordinate (m)
x_i	Coordinate (m)
x^+	Dimensionless distance
y_ℓ	Thickness of laminar sublayer (m)
z	Axial coordinate (m)

Greek Symbols

α	Thermal diffusivity (m^2/s) Volume fraction Dummy index
$\alpha = \partial\rho/\partial p$	Reciprocal of the sound speed (s/m)
β	Dummy index Angle between r-axis and azimuthal boundary surface of control volume
$\beta = -\frac{1}{\rho}\left(\frac{\partial\rho}{\partial T}\right)_p$	Volumetric expansion coefficient at constant pressure (K^{-1})
γ	Surface permeability
Γ_g	Total (laminar plus turbulent) diffusivity in g equation (m^2/s)
$\Gamma_{g\ell} = \frac{\lambda}{\rho c_p}$	Laminar diffusivity in g equation (m^2/s)
$\Gamma_{gt} = c'_{\phi}\frac{K^2}{\varepsilon}$	Turbulent diffusivity in g equation (m^2/s)
Γ_h	Thermal diffusivity (kg/ms)
$\Gamma_{\Phi_{ij}} = \nu_\ell + c_s\frac{K^2}{\varepsilon}_t$	Diffusion coefficient in Reynolds stress equations (m^2/s)
$\Gamma^{\nu}_{\phi} = \nu_\ell + c_{s\phi}\frac{K^2}{\varepsilon}$	Total (laminar plus turbulent) kinematic diffusivity in scalar fluxes equations (m^2/s)
$\Gamma_{\phi} = \rho\Gamma^{\nu}_{\phi} = \mu_\ell + c_{s\phi}\rho\frac{K^2}{\varepsilon}$	Total (laminar plus turbulent) dynamic diffusivity in scalar fluxes equations (kg/ms)
$\Gamma_{\varepsilon jm} = c_\varepsilon\rho\frac{K}{\varepsilon}\langle u_j u_m\rangle$	Turbulent diffusion coefficient in ε equation (kg/ms)
δ	Liquid film thickness (m)
δ_{ij}	Kronecker delta
δ_m	Mean spacing between molecules (m)
Δ	Difference operator Area of a triangle (m^2)

M. Bottoni, *Physical Modeling and Computational Techniques for Thermal and Fluid-dynamics*, Mechanical Engineering Series,
https://doi.org/10.1007/978-3-030-79717-1

∇	Nabla operator
ε	Volume porosity
	Dissipation of turbulent kinetic energy (m^2/s^3)
ε^+	Dimensionless dissipation of turbulent kinetic energy
$\varepsilon_g = \frac{1}{2}\varepsilon_\phi$	Dissipation in g equation (K^2/s)
ζ	Axial surface permeability
η	Local coordinate (m)
	Kolmogorov microscopic length scale of turbulence (m)
	Eigenvalue
θ	Time discretization parameter
λ	Thermal conductivity (W/m K)
	Eigenvalue
λ_m	Mean free path (m)
μ_{eff}	Effective dynamic viscosity (kg/ms)
μ	Dynamic viscosity (kg/ms)
	Eigenvalue
ν	Kinematic viscosity (m^2/s)
	Diffusion coefficient (m^2/s)
ν_K	Kolmogorov scale of velocity (m/s)
ξ	Azimuthal surface permeability
	Local coordinate (m)
π_{ij}	Pressure strain tensor (m^2/s^3)
$\pi_{i\phi}$	Pressure-scalar gradient correlation in scalar flux equations (mK/s^2)
ρ	Density (kg/m^3)
σ	Heat transfer coefficient (W/m^2 K)
τ	Shear stress (kg/m s^2)
	Diffusion time (s)
	Time interval between collision of particles (s)
	Microscopic Kolmogorov time scale (s)
τ_{ij}	Reynolds stress tensor (kg/ms^2)
ϕ	Temperature fluctuation (K)
$\overline{\phi}^2$	Variance of temperature fluctuations (K^2)
Φ	Shape function
$\Phi_{ij} = \langle u_i u_j \rangle$	Components of Reynolds stress tensor (m^2/s^2)
χ	von Karman constant
ψ_i	is an interfacial (liquid-vapor) friction coefficient
Ψ	Radial surface permeability
	Weight function
	Stream function (m^2/s)
ω	Vorticity (s^{-1})
	Relaxation parameter
Ω	Definition domain

Indices

α	General coordinate direction Dummy index
β	Dummy index
c	Convective
d	Diffusive
e	Element Eddy
f	Fluid Liquid film
i, j, k	Free or dummy indices for coordinate directions
ℓ	Laminar
l	Laminar Liquid
m	Mixture Momentum
n	Superscript for time discretization
o	Central node
NE	Number of elements
r	Iteration index
s	Saturation
S	Structure
Sl	Slip
t	Turbulent
v	Vapor
w	Wall
o	Initial value Central node Wall

M. Bottoni, *Physical Modeling and Computational Techniques for Thermal and Fluid-dynamics*, Mechanical Engineering Series,
https://doi.org/10.1007/978-3-030-79717-1

Special Symbols

$[a,b]$	$\max(a,b)$
∇	Gradient
$\nabla\cdot$	Divergence
$\cdot$	Scalar product
$:$	Double dot product
$\langle\rangle$	Average value
$\equiv$	Proportional
$+$	Dimensionless

M. Bottoni, *Physical Modeling and Computational Techniques for Thermal and Fluid-dynamics*, Mechanical Engineering Series,
https://doi.org/10.1007/978-3-030-79717-1

Historical Notes

Adams, John Couch (Laneast, UK, 1819–1892 Cambridge, UK)—English astronomer and mathematician. Nine months before the discovery of the planet Neptune by Le Verrier in 1846, based on perturbations of the orbit of Uranus, he had predicted the position of a new planet. His prediction, sent to the Royal Greenwich Observatory, was however ignored. His name is now associated to that of Le Verrier as co-discoverer of Neptune.

Amsden, Anthony A. Amsden was a member of the T3 theoretical group of fluid-dynamicists, headed by Harlow at the LANL (Los Alamos National Laboratory). He is known for his contribution to the development of the KIVA code which simulated fluid dynamics and combustion processes in automotive engines.

Bashforth, Francis (Thurnscoe, UK, 1819–1912 Woodhall, UK)—A British mathematician, studied at the University of Cambridge, and later became professor of mathematics at the Royal Military Academy of Woolwich. He studied ballistics, micro-physical phenomena in liquid drops, and surface tension. The Adams-Bashforth numerical integration method was also applied to study drop formation. Bashforth is also known for having first solved the Laplace-Young equation governing the interfacial pressure drop in a bubble.

Cauchy, Augustin-Louis (Baron) (Paris 1789–1857 Sceaux)—French mathematician, became professor at the École Polytechnique, Collège de France, and at the Sorbonne University. He is the founder of the theory of functions of complex variables and made valuable contributions in the theory of infinite series, differential equations, and elasticity theory. For political reasons, Cauchy was exiled for several years during which he lived in Switzerland and then in Turin and Prague. He published several books and almost 800 scientific reports or newspaper articles. His geniality became soon apparent from the simplicity of his solutions to a variety of problems.

Cholesky, André-Louis (Montguyon, France, 1875–1918 Bagneux)—French mathematician of Ukrainian origin. Studied at the École Polytechnique under Camille Jordan and Henri Becquerel and later at the École d'Application de

M. Bottoni, *Physical Modeling and Computational Techniques for Thermal and Fluid-dynamics*, Mechanical Engineering Series,
https://doi.org/10.1007/978-3-030-79717-1

l'Artillerie et du Génie. He is known especially for the matrix decomposition which has his name. Worked in the geographical service of the French army as officer of artillery. Cholesky died after being wounded on a military field during the World War I.

Clairaut, Alexis Claude (Paris 1713–1765 Paris)—French astronomer and mathematician. At age 18, he was admitted as a member of the French Scientific Academy. In 1759, based on perturbations of the orbit of the Halley comet, he advanced the hypothesis on the existence of a planet beyond Saturn, which turned out to be Uranus, discovered in 1781 by Herschel. In the same year, he published the cosine series (the sine series was obtained between 1762 and 1765 by Lagrange). He also published studies on the flattening of the Earth.

Crank, John (Hindley Lancashire, England, 1916–2006)—English mathematical physicist, studied at the Manchester University where he was awarded a PhD in 1953. He is best known for his work on numerical solution of partial differential equations and in particular for the solution of the heat diffusion equations, which resulted in the Crank-Nicolson method. Crank worked on ballistics during World War II. In 1957, he was appointed as the first Head of Department of Mathematics at Brunel College in Acton, where he remained up to his retirement.

Domanus, Henry (died in Downers Grove, Illinois, 2008)—Dr. Domanus was a US engineer who worked most of his time at the Argonne National Laboratory in the Energy Technology Division. For more than three decades, Dr. Domanus was the main developer of the series of COMMIX codes and started the publication in 1983 of the COMMIX-1A version. His colleague Dr. Lyczkowski wrote about him: *<Henry "Hank" Domanus was an unrecognized genius when he worked on the COMMIX series of codes at Argonne National Laboratory.>*

Doolittle, James Harold (Alameda, California, USA, 1986–1993, Pebble Beach, California, USA)—American aviator and Army Air Corps General. Studied at MIT (Massachusetts Institute of Technology) where he received a doctorate in Engineering. Participated actively in World War II in Japan and Germany.

Ekman, Vagn Walfrid (Stockholm, 1874–1954, Godstadt)—Swedish oceanographer. He studied at the Uppsala University where he followed fluid-dynamic courses under Vilhelm Bjerknes. He is known for the so-called Ekman layer in a fluid: the layer in which the flux results from the balance of pressure gradient, Coriolis force, and resistance due to turbulence. The Ekman theory was developed after observations of ice drifts made by Fridtjof Nansen aboard the ship Fram, during arctic expeditions.

Euler, Leonhard (Basel, 1707–1783, Saint Petersburg)—Swiss mathematician, born in Basel, where he studied and followed courses given by Johann Bernoulli. He was called to Petersburg Academy as Professor of Physics and later on by Frederick II and moved to Berlin where he was appointed as member of the Scientific Academy. He worked as mathematician, physicist, and astronomer giving many contributions to pure and applied mathematics, especially in the domain of differential equations, of variational calculus, and on the theory of the three-body problem.

Fourier, Jean Baptiste Joseph (Auxerre, 1768–1830, Paris)—French physicist and mathematician, he studied at the École Normale in Paris and became Professor at the École Polytechnique. He followed the Bonaparte expedition in Egypt and then became member and secretary of the French Scientific Academy. As physicist, Fourier gave strong contributions to the heat diffusion theory and, as mathematician, to the application of trigonometric functions to the harmonic analysis of time or space-dependent systems. After him is now named the Fourier series. The Fourier crater on the Moon has been named after him.

Frankel, Theodore (USA, 1929–2017)—American mathematician; he was educated at the University of California, at Berkeley, and became then professor at the University of California, at San Diego, and member of the Institute for Advanced Studies in Princeton, New Jersey. His major contribution to mathematics and physics was in the field of differential geometry and relativity theory.

Gauß, Carl Friedrich (Braunschweig, 1777–1855, Göttingen)—German mathematician and astronomer; he became director of the Astronomical Observatory in Göttingen and member of the Göttingen Science Academy. His major contributions to mathematics involved number theory, advanced algebra, the theory of elliptic functions, potential theory and non-Euclidian geometry in N-dimensional spaces. His contributions to physics involved geodesy, calculations of the orbits of celestial bodies, the theory of electromagnetic fields, and magnetism. Gauß cooperated with W. Weber in advanced research on Earth magnetism. The Gauß crater on the Moon has been named after him.

Hadamard Jacques (Versailles 1865–1963, Paris)—French mathematician, studied at the École Normale Supérieure of Paris. He is well-known for having introduced the concept of "well-posedness" of problems involving differential equations. Well-known are also the Hadamard inequality and Hadamard matrix, upon which the Hadard transform is based. The asteroid 10382 has been named after him.

Harlow, Francis Harvey (1928–2016)—American theoretical physicist, known for his work on numerical fluid dynamics, considered one of the founders of CFD (computational fluid dynamics). At Los Alamos National Laboratory (LANL), where he worked most of his time, he organized the theoretical T3 group. He developed several computational techniques, including PIC (particle-in-cell), MAC (Marker and Cell), and IMF (implicit multifield). Harlow, a fellow of the American Physical Society since 2003, was also a painter and sculptor of Native American Pueblo pottery.

Heun, Karl (Wiesbaden, 1859–1929, Karlsruhe)—German mathematician, studied in Göttingen and later in London. While in England, he taught in Uppingham. He then returned to Germany and completed his studies at the Munich University where he became a teacher, before moving to the Technical University of Karlsruhe. After him is named the Heun's equation, a second-order ordinary differential equation.

Hirt, C.W. (Living, USA)—American numerical mathematician, former member of the theoretical group at the Los Alamos Scientific Laboratory, NM, USA. Hirt is the founder of Flow Science Inc., located in Santa Fe, NM, and a developer of software for Computational Fluid Dynamic (CFD). Hirt has also pioneered the volume of fluid (VOF) method for tracking free surfaces or fluid-fluid interfaces.

Jacobi, Karl Gustav Jacob (Potsdam, 1804–1851, Berlin)—German mathematician and professor in Königsberg and later in Berlin. He contributed to the development of the theory of elliptical functions and of differential equations. His research on the elliptical functions was published in the basic work *Fundamenta nova theoriae functionum ellipticarum* (Königsberg, 1829) and in successive articles on the *Journal für die reine und angewandte Mathematik*. Jacobi was one of the first scientists to found the theory of determinants. The Jacobian determinant is named after him.

Kolmogorov, Andrey Nikolaevich (Tambov, 1903–1987, Moscow)—Russian mathematician, studied at Moscow University. He is considered one of the most influent mathematicians of the twentieth century. He made fundamental contributions in several domains: theory of probability, topology, classical mechanics, and theory of turbulence. Before World War II, he was professor in Berlin and then in Moscow. After the war, his research was dedicated mostly to the information theory.

Kutta, Martin Wilhelm (Pitschen, 1867–1944, Fürstenfeldbruck)—German mathematician, born in Pitschen, now in Poland, and studied at the University of Breslavia and then in Munich. Kutta became professor at the Technical School of Achen and then at the Stuttgart University. With his colleague Karl Runge, he developed the Runge-Kutta method for integrating numerically ordinary differential equations. He also made contributions in aerodynamics, formulating the Kutta-Zhukovsky theorem.

Lagrange, Joseph Louis (Turin, 1736–Paris, 1812)—Professor of mathematics in Turin; in 1766, he was invited by Frederick the Great to join the Scientific Academy in Berlin. After the death of Frederick the Great, he moved to Paris as professor at the Ecole Normale. Among his greatest contribution in the domain of applied mathematics are the variational calculus and the so-called Lagrangian equations of a system of points moving under geometrical constraints.

Le Verrier, Urbain-Jean-Joseph (Saint-Lo, France, 1811–1877, Paris)—French astronomer, member of the Academy of Science in Paris, and Director of the Astronomical Observatory in Paris. His most celebrated achievement is the discovery in 1846, together with John Couch Adams, of the planet Neptune, based on perturbations of the orbit of Uranus. Le Verrier became also known for his ephemerides of the planets. The Le Verrier crater on Mars, the Le Verrier crater on the Moon, one ring of Neptune, and asteroid 1997 have been named after him.

Lyczkowski, Robert (born 1941)—Dr. Robert W. Lyczkowski received a B.S. Ch.E. from the Cleveland State University in 1964 and an M.S. in Gas Engineering in 1966 and a Ph.D. in Gas Technology in 1970, both from the Illinois Institute of Technology. He made contributions in the fields of electro- and thermodynamics, nuclear reactor safety, coal and gas technology, and multiphase computational fluid dynamics (CFD). In 2008, Dr. Lyczkowski received the Ernest W. Thiele Award from the Chicago Section of the AIChE for his "Innovative application of multiphase theory and computational fluid dynamics in the areas of nuclear, fossil and biochemical engineering."

Milne, Edward Arthur (Kingston upon Hull, UK, 1896–1950, Dublin)—English astrophysicist and professor at Oxford University. He gave strong contributions to the theory of stellar composition and evolution and to cosmology and relativity theory. In 1932, in cooperation with Gerald James Whitrow, he developed a theory of relativity alternative to that proposed by Einstein, called Milne model.

Moulton, Forest Ray (Le Roy, Michigan, USA, 1872–1952 Wilmette, Illinois)—American astronomer and worked at the Chicago University. Together with the geologist T. C. Chamberlin, he advanced the theory of planetesimals according to which a star approaching our sun would have triggered the ejection of a mass which would then condense in the planets and asteroids. He was Secretary of the AAAS (American Association for the Advancement of Science). A crater of the Moon and the Adams-Moulton method for solving differential equations are named after him.

Nicolson, Phyllis (Macclesfield, UK, 1917–1968, Sheffield)—British mathematician most known for the Crank-Nicolson method of solving numerically the heat diffusion equation. Born Phyllis Lockett, she studied in Cambridge and married physicist Malcolm Nicolson in 1942. She moved to Leeds when her husband was appointed to a lectureship in Physics at Leeds University. When her husband died accidentally in 1951, Phyllis Nicolson was appointed to take over his lectureship.

Peaceman, Donald W. (Miami, 1926–2017)—Dr. Peaceman was a founder of modern numerical reservoir simulation and made many important contributions to solution of partial differential equations. He is especially remembered for his contribution to variants of the Alternating Direction Implicit (ADI) method. He worked at Exxon Production Research Corporation in Houston. Starting in 1983, Peaceman was adjunct professor of computational and applied mathematics at Rice University. In 2009, CAAM established the Peaceman Lectures on Numerical Mathematics to honor Dr. Peaceman's many contributions.

Peano, Giuseppe (Spinetta, Italy, 1858–1932, Turin)—Italian mathematician and glottologist, enrolled at the University of Turin where he later became professor. Meanwhile, he had taught at the Royal Military Academy. Peano was among the founders of the set theory and of mathematical logic, where he introduced the symbols for "union" and "intersection," still in use today. Peano is also known for editing in 1908 the *Formulario mathematico*, a collection of all formulas and theorems known at the time.

Picard, Emile Charles (Paris, 1856–1941, Paris)—French mathematician, studied at the Ecole Normale Supérieure (ENS), and became member of the French Scientific Academy. He was a professor at the University of Paris and Toulouse and later on professor of differential calculus at Sorbonne. He worked in the fields of function theory, differential equations, and theoretical physics. The asteroid 29,613 Charlespicard has been named after him.

Poisson, Siméon Denis (Pithiviers, France, 1781–1840, Paris)—French mathematician and physicist and studied at the Ecole Polytechnique of Paris where, with support from Laplace, he became professor succeeding Fourier. Later, he taught at the Sorbonne and became member of the French Scientific Academy. Poisson gave contributions to the potential theory, in which frame the Poisson equation was formulated, and to optics and statistics, where the Poison distribution is well-known. In the elasticity theory, the Poisson modulus has been named after him.

Prandtl, Ludwig (Freising, Germany, 1875–1953, Göttingen)—German engineer and physicist and studied at the Ludwig Maximilian University in Munich. He was appointed professor at the Technical High School in Hannover and then at the University of Göttingen, where Theodore von Kàrmàn was his PhD student. In Göttingen, he was appointed Director of the Aerodynamischen Versuchsanstalt (AVA) (Aerodynamic Test Facility). Prandtl became one of the founders of modern hydrodynamics and aerodynamics. His main contributions involve the boundary layer theory and the turbulence theory. The Prandtl number and one crater on the Moon have been named after him.

Reynolds, Osborne (Belfast, 1842–1912, Watchet, UK)—English physicist, born in Nord Ireland, and attended the Queen's College in Cambridge where he graduated in mathematics. He was later appointed at the Owen College in Manchester (now University of Manchester) where he became the first professor in UK history to hold the title of "Professor of Engineering." Reynolds became member of the Royal Society and was awarded the Royal Medal. His research is particularly known for the Reynolds number which characterizes flow equivalence of different fluids and the transition from laminar to turbulent regimes.

Riccati, Jacopo Francesco (Venice, 1676–1754, Treviso)—Italian mathematician and physicist and worked on problems of hydrodynamics and Newtonian mechanics. Studied the differential equation which was given his name in particular cases. His research was carried on by members of the Bernoulli family. One asteroid (14074 Riccati) has been named after him.

Richardson, Lewis Fry (Newcastle upon Tyne, UK, 1881–1953, Kilmun, UK)—English mathematician and physicist, educated at the Durham University and later at the King's College in Cambridge under J. J. Thompson. He was elected member of the Royal Society, and thanks to his research on the causes of wars and conditions of peace, he received a PhD in mathematical psychology from London University. Richardson pioneered modern numerical mathematical methods of weather forecasting. His research also involved fractals.

Runge, Carl David Tolmé (Bremen, 1856–1927, Göttingen)—German mathematician and physicist; after years spent in Havana, where his father was a Danish consul, he studied in Berlin under Karl Weierstrass and became professor in Hannover and Göttingen. Worked in the field of numerical mathematics, cooperating with the colleague Martin Wilhelm Kutta for the solution of differential equations. His research also involved spectroscopy, geodesy, and astrophysics. Together with Heinrich Kayser applied spectroscopy to astrophysical research. The crater Runge on the Moon has been named after him.

Taylor, Brook (Edmonton, UK, 1685–1731, London)—English mathematician, studied law at Cambridge University and mathematics under John Machin and John Keill. He was elected member of the Royal Society and later on appointed as Secretary. He pioneered the new field of calculus by finite differences and is mostly known for the Taylor series. Taylor also studied the principles of perspective, publishing his results in a book of difficule interpretation which was later re-published and explained by Josua Kirby and Daniel Foumier.

Thomas, Llewellyn Hilleth (London 1903–1992, Raleigh, North Carolina, USA)— Thomas was a British physicist who emigrated to the USA. Graduating from Cambridge University, he joined the Bohr's Institute in Copenhagen (1925–1926), where he proposed the Thomas precession to explain relativistic interactions between electron and atomic nucleus and the Thomas-Fermi model for electron-ion interactions. In numerical mathematics, he is known for the Thomas algorithm to solve exactly systems with tridiagonal matrix of coefficients. From 1929 to 1943, Thomas was professor of physics at the Ohio State University and from 1946 was member of the Watson Scientific Computing Laboratory at Columbia University.

von Kármán, Theodore (Budapest, 1881–1963, Aachen)—Engineer, physicist, and mathematician, born in Budapest, Hungary and later became a US citizen. He studied at the Technical University of Budapest and then in Göttingen under the leadership of Ludwig Prandtl. He became director of the Aeronautical Institute in Aachen before emigrating in 1930 to the USA where he directed the Guggenheim Aeronautical Laboratory at the California Institute of Technology (GALCIT). During World War II, he founded the Jet Propulsion Laboratory (JPL) which became a research center financed by NASA. His contributions in the fields of fluid dynamics, aeronautics, and space technology are major scientific achievements. The von Karman crater on Mars and the von Karman crater on the Moon were named after him.

von Neumann, John (Budapest, 1903–1957, Princeton)—Physicist and mathematician, born in Budapest, Hungary and became later a US citizen. While frequenting a gymnasium in Budapest, where Eugen Wigner was also a student, he was recognized as a prodigious boy. He studied mathematics in Budapest and chemical engineering at the Zurich Polytechnique. He moved then to Göttingen where he continued his studies under David Hilbert's supervision. In 1930, he moved to Princeton, USA, where he became professor of mathematics at the Institute for Advanced Studies, where also Albert Einstein taught. He is considered one of the greatest mathematicians of modern history. His research involved functional analysis, topology, informatics, fluid dynamics, and quantum physics.

von Seidel, Philipp Ludwig (Zweibrücken, Germany, 1821–1896, Munich)—German mathematician, studied at the University of Berlin under Dirichlet and Encke. He then moved to Königsberg, where he studied under Bessel, Jacobi, and Franz Neumann. Seidel is mostly known for the Gasuss-Seidel method of solving iteratively large systems of algebraic equations. He also studied aberration decomposing the first-order monochromatic aberration in the five components now known as Seidel aberrations. The crater Seidel on the Moon has been named after him.

References

Basque, B., Grand, D., Delapierre, L., Bottoni, M., Dorr, B., Homann, C., & Struwe, D. (1982). Thermohydraulic analysis of LMFBR subassemblies with the BACCHUS programme. In *Proceedings ANS/ENS international topical meeting on LMFBR Safety and Related Design and Operational Aspects*, Lyon, France, July 19–23, 1982.

Beam, R. M., & Warming, R. F. (1976). An implicit finite-difference algorithm for hyperbolic systems in conservation-law form. *Journal of Computational Physics, 22*, 87–110.

Bell C. R., Bleiweis P. B., Boudreau S. E., Parker F. R., & Smith L. L. (1977). *SIMMER-I: An S_n, implicit, multifield, multicomponent, Eulerian, recriticality code for LMFBR disruptive core analysis, LA_NUREG.6467.MS*. Los Alamos National Laboratory.

Binder, K., & Heermann, D. W. (1988). *Monte Carlo simulation in statistical physics*. Springer.

Bird, G. A. (1994). *Molecular gas dynamics and the direct simulation of gas flows*. Clarendon Press.

Bird, R. B., Stewart, W. E., & Lightfoot, E. N. (1960). *Transport phenomena*. Wiley.

Bottoni, M. (1982). Evaluation of the residual method of orthogonal collocations and solution of one-dimensional equations of sodium vapour flow. *Nuclear Science and Engineering, 82*, 1–18.

Bottoni, M., & Dorr, B. (1991). An advanced vectorizable variant of the ADI-method for two-phase flow calculations. In *Simulation multiconference, sponsored by the Society for Computer Simulation (SCS) International*, New Orleans, LA, April 1–5, 1991.

Bottoni, M., & Ninokata, H. (1998). An algorithm for attenuation of turbulence in particulate flow linked to the fluid-dynamic code COMMIX-M. *Journal of Nuclear Science and Technology, 35* (2), 101–112.

Bottoni, M., & Sengpiel, W. (1992). *Review of mathematical and physical basis of two-phase flow modeling*. KfK 4759.

Bottoni, M., & Struwe, D. (1982, December). *BLOW-3A, a theoretical model to describe transient two-phase flow conditions in LMFBR coolant channels*. KfK 3317.

Bottoni, M., & Willerding, G. (1987). Advanced solution algorithms for transient multidimensional thermohydraulics flow problems in complex geometries with the programme COMMIX-2/KfK. *Nuclear Engineering and Design, 100*, 351–365.

Bottoni, M., et al. (1985, September). *COMMIX-2: A three-dimensional transient computer program for thermal-hydraulic analysis of two-phase flow, Report NUREG/CR-4371, ANL-85-47*.

Bottoni, M., Dorr, B., Homann, C., & Struwe, D. (1987). State of development of the computer programme BACCHUS-3D/TP for the description of transient two-phase flow conditions in LMFBR fuel pin bundles. *Nuclear Engineering and Design, 100*, 321–349.

M. Bottoni, *Physical Modeling and Computational Techniques for Thermal and Fluid-dynamics*, Mechanical Engineering Series,
https://doi.org/10.1007/978-3-030-79717-1

Bottoni, M., Dorr, B., & Homann, C. (1992, April). *The three-dimensional transient two-phase flow computer programme BACCHUS-3D/TP*. KfK 4760.

Bottoni, M., Chang, F. C., Chien, T. H., Ding, J., & Sha, W. T. (1993). *Liquid film tracking model for the passive containment cooling system of the AP-600 reactor*. NUREG/CR-5985, ANL-92-22, R7.

Bottoni, M., Chang, F. C., & Ding, J. (1994, May). *First assessment of computation of turbulent bubbly flow and particulate flow with the COMMIX-M program*. ANL-95/6.

Bridgman, P. W. (1961). *The thermodynamics of electrical phenomena in metals and a condensed collection of thermodynamics formulas*. Dover Publications Inc.

Champagne, F. H., Harris, V. G., & Corrsin, S. (1970). Experiments on nearly homogeneous turbulent shear flow. *Journal of Fluid Mechanics, 41*(Part 1), 81–139.

Chandrasekhar, S. (1961). *Hydrodynamic and hydromagnetic stability*. Dover Publications, Inc.

Chang, F. C., & Bottoni, M. (1995, August). *Implementation and validation of a Reynolds stress model in the COMMIX-1C/RSM and CAPS-3D/RSM codes*. ANL-96/8.

Chang, F. C., Bottoni, M., & Sha, W. T. (1992, January). *Development and validation of the k-epsilon two-equation turbulence model and the six-equation anisotropic turbulence model in the COMMIX code (COMMIX-1C/ATM/ATHRP-44*.

Chapman, M. (1981). FRAM-Nonlinear damping algorithms for the continuity equation. *Journal of Computational Physics, 44*, 84.

Chen, C. J. (1990). *Prediction of turbulent flows*. Lecture Notes of the University of Iowa, Iowa City.

Courant, R., & Hilbert, D. (1968). *Methoden der mathematischen Physik II*. Springer.

Cugiani, M. (1967). *Metodi dell' Analisi Numerica*. UTET.

Demidovic, B. P. (1989). *Esercizi e problemi di analisi matematica (translation from Russian)*. Editori Riuniti.

Domanus, H. M., et al. (1990, November). *COMMIX-1C: A three-dimensional transient single-phase computer program for thermal-hydraulic analysis of single-components and multicomponent engineering systems*. Volume I: Equations and numerics: Volume II: User's guide manual, NUREG/CR-5649, Argonne National Laboratory Report ANL-90/33.

Douglas, J. (1962). Alternating direction methods for three space variables. *Numerische Mathematik, 4*, 41–63.

Dunn, F. E., Fischer, G. J., Heames, T. J., Pizzica, P. A., McNeal, M. A., Bohl, W. R., & Prastein, S. M. (1974, October). *The SAS2A LMFBR accident-analysis computer code, ANL-8138*.

Fehlberg, E. (1968, October). *Classical fifth-, sixth-, seventh- and eigth-order Runge-Kutta formulas with stepwise control, NASA-TR-R-287*.

Ferziger, J. H. (1983). Higher-level simulations of turbulent flows. In J. A. Essers (Ed.), *Computational methods for turbulent, transonic, and viscous flows*. Hemisphere Publishing Co.

Forsythe, G. E., & Wasow, W. R. (1960). *Finite-difference methods for partial differential equations*. Wiley.

Fox, L. (1964). *An introduction to numerical linear algebra*. Clarendon Press.

Grigorieff, R. D. (1972). *Numerik Gewönlicher Differentialgleichungen, Band 1 Einschrittverfahren*. Teubner Studienbücher.

Hadamard, J. (1923). *Lectures on Cauchy's problem*. Yale University Press.

Harlow, F. H., & Amsden, A. A. (1971). A numeric fluid dynamics calculation method for all flow speeds. *Journal of Computational Physics, 8*(2), 197–213.

Harlow, F. H., & Amsden, A. A. (1975). Numerical calculation of multiphase fluid flow. *Journal of Computational Physics Volume, 17*, 19–52.

Hofer, E. (1976, October). A partially implicit method for large stiff systems of ODEs with only a few equations introducing small time-constants. *SIAM Journal Numerical Analysis, 13*(5), 645–663.

Houstis, E. N., Lynch, R. E., Rice, J. R., & Papatheodorou, T. S. (1978). Evaluation of numerical methods for elliptic partial differential equations. *Journal of Computational Physics, 27*, 323–350.

Ishii, M. (1977, October). *One-dimensional drift-flux model and constitutive equations for relative motion between phases in various two-phase flow regimes, ANL-77-47.*

Krylov, V. I. (1962). *Approximate calculation of integrals*. The Macmillan Company.

Lax, P. D. (1954). Weak solutions of nonlinear hyperbolic equations and their numerical computation. *Communications on Pure and Applied Mathematics, II*, 159–193.

Leonard, B. P. (1979). A stable and accurate convective Modeling procedure based on quadratic interpolation. *Computer Methods in Applied Mechanics and Engineering, 19*, 59–98.

Lesieur, M. (1990). *Turbulence in fluids*. Kluwer Academic Publishers.

Lyczkowski, R. W. (2018). *The history of multiphase science and computational fluid dynamics – a personal memoir*. Springer International Publishing.

Lyczkowski, R. W., & Solbrig, C. W. (1980). Calculation of the governing equations for a seriated unequal velocity, equal temperature two-phase continuum. *AIChE Journal, 26*(1), 89–98.

Maekawa, Muramatsu, T., Matsumoto, M., Ninokata, H., & Bottoni, M. (1987). *Cooperation on the development of AQUA and COMMIX-KfK. Information exchange and numerical models discussions*. Power Reactor and Nuclear Fuel Development Corporation (PNC), O-Arai Engineering Center, O-Arai, Japan, Report PNC SN9410 87-167.

Marth, W. (1992, März). *Der schnelle Brüter SNR-300 im auf und ab seiner Geschichte*. KfK 4666.

McComb, W. B. (1990). *The physics of fluid turbulence*. Clarendon Press.

Messina, A., & Londrillo, P. (1988). *A N-body code for elliptical galaxies, super-computing in astrophysics*. Astronet Special Publications 88/1.

Norrie, D. H., & de Vries, G. (1973). *The finite element method*. Academic Press.

Noye, J. (Ed.). (1978). *Numerical simulation of fluid motion*. North-Holland Publishing Company.

Orszag, S. A. (1977). In R. Balian & J. L. Peube (Eds.), *Lectures on the statistical theory of turbulence, fluid dynamics* (pp. 235–374). Gordon and Breach Science Publishers.

Peaceman, D. W., & Rachford, H. H. (1955). The numerical solution of parabolic and elliptic differential equations. *Journal of the Society for Industrial and Applied Mathematics, 3*, 28–41.

Press, W. H., Flannery, B. P., Teukolsky, S. A., & Vetterling, W. T. (1986). *Numerical recipes*. Cambridge University Press.

Richtmyer, R. D., & Morton, K. W. (1967). *Difference methods for initial-value problems*. Interscience.

Rodi, W. (1980). *Turbulence models and their application in hydraulics*. IAHR.

Romstedt, P., & Werner, W. (1977). Efficient high-order method for the solution of fluid dynamics equations. *Nuclear Science and Engineering, 64*, 208–218.

Smith, G. D. (1969). *Numerical solution of partial differential equations*. Oxford University Press.

Solbrig, C. W. (1975, May). *Governing equations for a seriated continuum: An unequal velocity model for two-phase flow*. Aerojet Nuclear Company, ANCR-1193, UC-78b.

Strang, G., & Fix, G. J. (1973). *An analysis of the finite element method*. Prentice-Hall.

Stuhmiller, J. H. (1976, July). *A review of the rational approach to two-phase flow modeling*. EPRI NP-197.

Stuhmiller, J. H. (1977). The influence of the interfacial pressure forces on the character of two-phase flow model equations. *International Journal of Multiphase Flow, 3*, 551–560.

Tennekes, H., & Lumley, J. L. (1972). *A first course in turbulence*. The MIT Press.

Thomas, L. H. (1949). *Elliptic problems in linear differential equations over a network: Watson Scientific Computing Laboratory Report*. Columbia University.

Trapp, J. A. (1986). The mean flow character of two-phase flow equations. *International Journal Multiphase Flow, 12*(2), 263–276.

Travis, J. R., Harlow, F. H., & Amsden, A. A. (1976). Numerical calculations of two-phase flows. *Nuclear Science and Engineering, 61*, 1–10.

Tricomi, F. G. (1961). *Equazioni differenziali*. Boringhieri.

Wade, D. C., & Chang, Y. I. (1988). The integral fast reactor concept: Physics of operation and safety. *Nuclear Science and Engineering, 100*, 507–524.
Werner, W. (1979). *High order methods in fluid dynamics*. In *American Nuclear Society proceedings of the topical meeting on Computational Methods in Nuclear Engineering*, Williamsburg, Virginia, April 23–25, 1979, pp. 1–14.
Werner, W. (1981). Weighted residuals methods for the solution of fluid dynamics problems. In *Proc. Top. Mtg. Advances in mathematical methods for the solution of nuclear engineering problems*, München, April 27–29, 1981, Vol. I. American Nuclear Society.
Wilcox, D. C. (1998). *Turbulence modeling for CFD*. DCW Industries.
Wirtz, P. (1973, Oktober). *Ein Beitrag zur theoretischen Beschreibung des Siedens unter Störfallbedingungen in natriumgekühlten schnellen Reaktoren*. KFK 1858.
Young, D. M. (1971). *Iterative solution of large linear systems*. Academic Press.
Zauderer, E. (1983). *Partial differential equations of applied mathematics*. Wiley.
Zienkiewicz, O. C. (1971). *The finite element method in engineering science*. McGraw Hill.
Zurmühl, R., & Falk, S. (1984). *Matrizen und Ihre Anwendungen*. Springer.
Zwillinger, D. (Ed.). (1996). *Standard mathematical tables and formulae*. CRC Press.
Zwirner, G. (1960). *Istituzioni di Matematiche, Parte Seconda*. CEDAM.

Literature (Relevant for the Topics Treated but Not Quoted in the Text)

Binder, K., & Heermann, D. W. (1988). *Monte Carlo simulation in statistical physics*. Springer.
Bird, G. A. (1994). *Molecular gas dynamics and the direct simulation of gas flows*. Clarendon Press.
Chandrasekhar, S. (1961). *Hydrodynamic and hydromagnetic stability*. Dover Publications.
Chen, C. J., & Jaw, S.-Y. (1998). *Fundamentals of turbulence modeling*. Taylor & Francis.
Collatz, L. (1966). *The numerical treatment of differential equations*. Springer.
Collatz, L. (1968). *Funktionalanalysis und Numerische Mathematik*. Springer.
Courant, R., & Hilbert, D. (1968). *Methoden der mathematischen Physik I*. Springer.
Courant, R., & Hilbert, D. (1968). *Methoden der mathematischen Physik II*. Springer.
Lee Wen Ho. (2013) *Computational Methods for Two-Phase Flow and Particle Transport*. World Scientific Publishing Co. Ptc. Ltd, Singapore.
Launder, B. E., & Spalding, D. B. (1872). *Mathematical models of turbulence*. Academic Press.
Plesset, M. S., Zuber, N., & Catton, I. (Eds.). (1983). *Transient two-phase flow*. Hemisphere Publishing Company.
Pope, S. B. (2000). *Turbulent flows*. Cambridge University Press.
Potter, D. (1973). *Computational physics*. Wiley.
Roache P. J. (1998). *Fundamentals of Computational Fluid Dynamics*. Hermosa, Albuquerque, NM.
Roache, P. J. (1982). *Computational fluid dynamics*. Hermosa Publishers.

Index

M. Bottoni, *Physical Modeling and Computational Techniques for Thermal and Fluid-dynamics*, Mechanical Engineering Series,
https://doi.org/10.1007/978-3-030-79717-1